Adrian E. Scheidegger

Theoretical Geomorphology

Third, Completely Revised Edition

With 117 Figures

Springer-Verlag
Berlin Heidelberg New York
London Paris Tokyo
Hong Kong Barcelona

Professor Dr. ADRIAN E. SCHEIDEGGER
Institut für Theoretische Geodäsie
und Geophysik
Abteilung Geophysik
Technische Universität Wien
Gußhausstraße 27–29/128
1040 Wien, Austria

ISBN-13:978-3-642-75661-0   e-ISBN-13:978-3-642-75659-7
DOI: 10.1007/978-3-642-75659-7

Library of Congress Cataloging-in-Publication Data. Scheidegger, Adrian E., 1925– Theoretical geomorphology / Adrian E. Scheidegger. – 3rd completely rev. ed. p. cm. Includes bibliographical references and index. ISBN-13:978-3-642-75661-0   1. Geomorphology. I. Title. GB401.5.S34. 1990 551.4′1 – dc20

© Springer-Verlag Berlin Heidelberg 1961, 1970, 1991
Softcover reprint of the hardcover 3rd edition 1991

32/3145(3011)-543210 – Printed on acid-free paper

To the Memory of
My Parents

Preface

The surface features of the Earth are commonly split into two categories, the first of which comprises those features that are due to processes occurring inside the solid Earth (endogenic features) and the second those that are due to processes occurring outside the solid Earth (exogenic features). Specifically, the endogenic features are treated in the science of geodynamics, the exogenic features in the science of *geomorphology*.

I have treated the theoretical aspects of the endogenic features in my *Principles of Geodynamics*, and it is my aim to supplement my earlier book with a discussion of the theory of the exogenic features, the taxonomy of the latter having been discussed in my *Systematic Geomorphology*. It is my hope that the three books will together present a reasonably coherent, if necessarily incomplete, account of theoretical geology.

Contrary to endogenic phenomena, exogenic processes can often be directly observed as they occur: the action of a river, the development of a slope, and the evolution of a shore platform are all sufficiently rapid so that they can be seen as they take place. This has the result that in geomorphology one is generally on much less speculative ground regarding the mechanics of the processes at work than one is in geodynamics.

The book follows a pattern which is, mutatis mutandis, analogous to that of *Principles of Geodynamics*. First, a brief description is given of the physiographic facts of geomorphology, after which some of the basic physics is reviewed which is necessary for the understanding of the subsequent exposition. Then, the body of the book presents in sequence the pertinent subjects which are (1) the mechanics of slope formation, (2) the theory of river action, (3) the systems theory of landscape evolution, (4) the theory of aquatic effects, (5) the theory of niveal, glacial and periglacial forms, and (6) the theory of aeolian and desert features.

The present edition is the third of this book. A comparison with the earlier editions will show that many sections have been extensively rewritten. In the intervening 20 years, general system theory and such new mathematical fields as fractals have brought an entirely new approach to geomorphology.

Some of the sections in the present book are based upon articles of my own which appeared in the *Bulletin of the Geological Society* of America, in the *Journal of the Alberta Society of Petroleum Geo-logists*, in *Geofisica Pura e Applicata*, in the *Zeitschrift für Geo-morphologie*, in the *Bulletin of the International Society of Scientific Hydrology* and in *Water Resources Research*. I am grateful to the Edi-tors of these Journals for the permission to draw freely from my artic-les published therein. Ms. Sonja Mayer in Vienna has competently collated and ordered the Bibliographies for which my thanks are due. I am also indebted to the Springer-Verlag, which has again been most cooperative in effecting a speedy publication of the manuscript.

Vienna, Austria, October 1990   A. E. SCHEIDEGGER

Contents

1 Physical Geomorphology

1.1 Introduction

Geomorphology, in its widest sense, is that branch of the geosciences which concerns itself with the development of the surface features of the Earth. In a more restrictive sense, geomorphology is the science of those surface features whose shape is primarily determined by the action of *exogenic* processes, i.e., of processes which originate *outside* the solid Earth. It is with this latter concept of geomorphology that we shall concern ourselves. In contrast, the *endogenic* processes, i.e., the processes originating *inside* the solid Earth, are the subject of *geodynamics*, and have been treated by the writer in a separate volume (Scheidegger 1982a).

The landscapes depend in a large measure on the prevailing climate, which thus provides for a further taxonomic division of landscapes, into humid, glacial, and arid ones. In addition, the endogenic background (e.g., tectonism, volcanism) can also influence the type of landscape that results. A taxonomy of landscapes has been provided on this basis by the writer (Scheidegger 1987c) in a volume on systematic geomorphology.

In this first chapter of the book we shall give a review of those features of landscapes the knowledge of which is necessary for an understanding of the ensuing theoretical treatment.

1.2 Development of Slopes

1.2.1 General Remarks

Slopes are the most important constituent elements of a landscape. A landscape without any slopes would be a flat, level plain which, presumably, would hold very little interest for geomorphologists.

Thus, all the features that are characteristics of the geography of our globe: mountains, river banks, coasts, are characterized by slopes. Some of these slopes may have been formed by endogenic processes, such as by the thrusting up of a mountain range or by the opening up of a rift valley. However, the "primary" slopes, if one wishes to call them thus, will soon be acted upon by external ("exogenic") agents such as wind, water, ice, and residual stresses, so that their shape will change. If it can be understood how slopes change under the influence of exogenic processes, then it is obviously possible to explain physical geography.

1.2.2 Quantitative Description of Slopes

As always when quantitative methods of analysis are contemplated, it is necessary
to describe the field data in quantitative terms as well. Slopes are fundamentally
inclined surfaces, and as such one might think that all that one has to do is to give
the equation describing the surface in question in, say, Cartesian co-ordinates.
Needless to say, this is a very difficult undertaking: the surfaces in question are
extremely irregular; this becomes obvious if one thinks of the individual pebbles
etc. that might comprise them.

Thus, one has to take recourse (explicitly or implicitly) to some statistical
averages. For this purpose, the true hillside is approximated by a smooth surface
passing through such points as have been measured. Of this approximate surface
it is then possible to determine characteristic parameters.

Morever, one is generally not concerned with the entire hillside, but only with its
profile. In order to obtain a profile, a horizontal straight base line is chosen more or
less parallel to the general declivity of the area in question (there is a certain
amount of arbitrariness in choosing this line). Common parameters that are listed
are then the elevation H of the slope profile above the base line, the distance L
along the base line from an arbitrarily chosen initial point thereon, and the (local or
average) gradient given by

$$G = \Delta H / \Delta L. \tag{1.2.1}$$

Furthermore, the profile is generally smoothed even more; the smoothing that
has been done is indicated by introducing a roughness parameter; this is the length
ratio between the line corresponding to the actual measurements and the average
slope line corresponding to the smoothed model (Klein 1981).

An additional characterization of slope profiles can be obtained by making
statistical analyses on slope elements. Thus Speight (1971) has shown that the
tangents of the slope angles are lognormally distributed. Another possibility is to
set up a series of classes of slope types (Tschierske 1978; Parsons 1979) and to
classify the slopes in an area accordingly. It is then possible to establish
correlations of characteristic slope parameters with lithology (Tschierske 1978;
Christofoletti and Tavares 1976; Tandon 1974a; Kumar 1981a; Twidale and
Campbell 1986), and to find their rates of change with time (Sterr 1985) which can
be used as a basis for setting up dynamic slope models (cf. Chap. 3).

If one wants to proceed to a true three-dimensional analysis of slope surfaces,
recourse has to be taken to computers. Two approaches exist (Tipper 1977):
multivariate Fourier analysis and modeling by complex surfaces that can be easily
manipulated. In either case, sets of parameters are obtained for characterizing the
slope surfaces. The three-dimensional modeling procedures, naturally, can also be
used for a characterization of whole landscapes, as will be discussed in Sect. 1.5.

1.2.3 Agents in Slope Formation

As noted, slopes change under the influence of exogenic agents. The net effect is
called denudation. Depending on climate and season, it may occur rather rapidly
(Schumm 1964; Gerber 1961). A compilation of the various agents that might cause

the shapes of slopes to change has been given, for instance by Penck (1924). Accordingly, the processes that are effective in slope formation can be classified as follows: (1) reduction of rocks, (2) spontaneous mass movement, (3) corrasion, (4) erosion, (5) transport of mass and (6) accumulation. The terminology used by Penck is somewhat different from that in other writings on geomorphology, but the processes considered are usually of the same general nature as those listed above.

Looking at the various processes in somewhat greater detail, we note that the *reduction* (cf. also Sect. 3.2) of rocks represents their disintegration into small pieces. It takes place by weathering due to their exposure to wind and water. It may be mechanical or chemical.

The reduction of rocks alone does not produce any changes in the existing slopes. In order to produce such changes, it is necessary to have processes that can effect a transfer of mass. All such mass transfers are due to the action of gravity in some form. First of all, one has *spontaneous mass movement*. With no interference from any carrying medium, the debris produced by weathering may start to slide downhill. On steep rock walls, any loosened particles will immediately drop to the bottom and form a pile of debris.

The material moving over a slope by the above-mentioned process helps further in wearing down the slope. This wearing-down process has been termed *corrasion*. It occurs without any further intermediary. In contrast to corrosion one has *erosion*. This process is caused by the intermediary of some moving medium such as wind, water, or ice. It also causes the wearing down of the slope. The combined effect of the above agents is termed *denudation*.

In order to achieve further slope development, the material that has been loosened and that may have slid into the lower parts of the area under consideration must somehow be removed. This occurs by the various processes of *transport of mass*. In such transport processes, the appearance of a carrying medium is of prime importance.

The end stage of transport of mass is *accumulation*. The transporting agents (water, wind, ice) may dump material in some areas which by its very presence forms a slope. This occurs not only in alluvial plains and in sheet floods, but also in any place where the material is transported by external agents. Thus, near mountains, material might, for instance, be deposited in the form of alluvial fans which occur at the edge of hills and mountains. Their slopes are from 1 to 10 degrees; the finest deposits are always found at the periphery. The opposite arrangement with regard to the grading of deposits is found in alluvial cones and talus accumulations. Here the slopes are from 10 to 50 degrees. Boulders are found at the base, sand and gravel at the top. These features are caused by small intermittent streams. In fact, the mountain-talus environment is one of the classic cases demonstrating erosion-deposition morphology (Bull 1977; Gardner 1971; Whitehouse and McSaveney 1983).

1.2.4 Patterns of Slope Development

The discussion given earlier sets the agents which act upon the shape of slopes entirely apart from those processes that caused uplifted areas on the Earth's crust in the first place. It would therefore appear that one could treat the development of a landscape in terms of a *cycle* in which uplift and planation alternate. In fact, this is the old classic view amongst geomorphologists and will be treated in detail in Sect. 1.5.1.

However, a different point of view had already been taken by Penck (1924), according to whom there is little reason to believe that uplift and planation take place alternatively: rather, uplift and planation are concurrent phenomena, a view which is now commonly accepted by geomorphologists ("principle of antagonism"; cf. Scheidegger 1979). In addition, Penck distinguished between "waxing development", in which uplift is faster than the denudation (leading to convex slope profiles); "stationary development", in which uplift and denudation proceed at an equal rate (leading to straight slopes and parallel slope recession); and "waning development", in which the denudation rate exceeds the rate of uplift (leading to concave slopes).

In a vein similar to Penck (1924), White (1966) has noted that equilibrium slope profiles have generally a tripartite classification: an upper convex element, a middle straight element, and a lower concave element.

The tripartite classification of slope sections is quite general in nature, and is the expression of a "catena (chain) principle" (Scheidegger 1986) in landscape development. The latter principle states that *all* landscapes consist of sequences or chains (catenas) of flat-steep-flat elements. The reason for this being so will be discussed in the sections on landscape *systems* later on in this book.

1.2.5 Morphology of Mass Movements

The discussion above refers to the instantaneous form of slopes and slope profiles. It is necessary to discuss also the phenomenology of *changes* of slope forms.

Such changes are effected by mass movements. A complete review of the subject of mass movements has been given by the writer in connection with the study of the physics of natural catastrophes (Scheidegger 1975; updates Scheidegger 1984, 1987b). Accordingly, a phenomenological classification of mass movements on slopes has to be made according to their speed and according to the material in which they occur. In addition, they can be deep or surficial.

Thus, slow (up to m/day) deep (> 50 m) movements in hard rock appear as mountain fractures and valley closures. Surficial movements of this type occur on scree slopes. Fast movements (m/s) in hard rock are seen as rock avalanches (shallow) and as land slides (deep).

Plastic materials comport themselves differently from hard rock: they are *always* susceptible to creep; in this, however, phases of more rapid movement alternate with phases of quiescence. Mass movements in soft materials are usually quite shallow; in deep-seated mass movements there is not much difference between those in soft and in hard rock, since the rheology even of hard materials leads to creep phenomena under high pressure.

The mass movements described above occur in the moving mass itself. However, it is possible that masses are moved by *external* transporting agents. Thus, the slow transport of material on the edge and tongue of glaciers is well known: this leads to the positioning of moraines after the ice has melted. Faster mass movements are caused by torrents: this leads to mud and debris flows that can cause much damage.

It is the mass movement that cause the changes of the form of slopes. Studies of long-term change of slope forms have been reviewed by Dunkerley (1980). In general, a flattening of slope profiles has been found to occur with ongoing time. This corresponds to some of the principal tenets of system theory.

1.3 Curved Lines in Geomorphology

1.3.1 General Remarks

Many features in geomorphology are represented, on a map at least, by lines.

The most obvious of such "linear" features are rivers, although, in fact, a river has actually a finite width and is therefore, in nature, not the realization of a mathematical line.

A true linear feature is the course of each bank of a given river (for, say, mean water level). The course of the water's edge represents, in fact, a mathematical line. A similar true linear feature is a coastline. Here the natural feature is again a mathematical line: the course of the water's edge at mean sea level. This line represents, in effect, an isohypse; all other isohypses also represent geomorphic "lines". Another linear feature is the course of a mountain crest, or of any other watershed. This is again a mathematical line.

All the above geomorphic lines have one obvious characteristic: they are "wiggly". While the general trend of these lines is very often quite straight, the details, if one cares to look closely, are extremely complex and difficult to ascertain. It is the purpose of this chapter to touch upon some problems, old and new, which are caused by the "wiggliness" of geomorphic lines and by the task of quantitatively describing such lines.

1.3.2 The Length of Wiggly Lines

The problem of measuring the length of a line is an old one. In general, not much thought is given to this problem by geomorphologists, inasmuch as it is simply assumed that the length be measured by some type of integrating device.

However, the fact that natural geomorphological lines are wiggly, introduces certain complications. It is clear that more and more "wiggles" tend to disappear, the smaller the scale of the map is. Thus, the "length" of a geomorphic line does not have a meaning per se, inasmuch as it depends very much on the scale of map used: the better map, the greater the length of a given natural feature.

Richardson (1961) has studied this problem, based on the usual idea of Riemann integration: the curved line is approximated by polygons of smaller and smaller edge length l (in practice, a divider is "walked" along the line on a map), the last edge being taken as a fraction of l; then the length is approximated by the sum $\sum l$ of the lengths of all the polygon edges. From his numerical values, Richardson deduced empirically a law:

$$\sum l = l^{-\alpha}. \tag{1.3.1}$$

According to this law, the "length of a geomorphic feature is given by stating three parameters: the edge length l of the polygon, the length $\sum l$ measured using this edge length and the "Richardson parameter", α.

A different approach to the problem of length from that described above has been taken by Steinhaus (1954), who bases his argument on ideas of set theory and Lebesgue integration, and goes back to the possibility of defining the "length" of a point set by a double integral (Cauchy 1832) as the mean of the lengths of all projections of the set. Crofton (1868) described this idea by using an infinite sum which permitted a probabilistic interpretation. This leads to a practical way of measuring lengths of a geomorphic curve.

A further method for the determination of the length of a wiggly line has been suggested by Hakanson (1978, 1981), who proposed to count the intersections of the latter with the lines of a regular square grid. This works only for closed lines, but in the case of open lines a fictitious closure can be effected. Statistically, the ratio of the number of intersection points to the number of squares contained in the closed area can be used for obtaining a length value of the wiggly line. The method, however, is scale-dependent (scale of the map as well as scale of the squares); a normalization to a standard implies the implicit acceptance of internal laws of self-similarity.

The same problems as with the measurement of length would occur with the measurement of areas of geomorphic features, if the true surface area of the ground were desired. However, one defines as "area" generally the simple vertical projection onto a horizontal surface (geoid) whose area is always clearly defined.

1.3.3 Spectrum of a Wiggly Line

The length of a line is only one of its characteristic features. It is evidently desirable to use additional characteristics to describe it.

The most obvious one that comes to one's mind is the *power spectrum* of the line (Julian 1967).

In general, in order to define a power spectrum, one starts with a function of one variable

$$x = f(t), \tag{1.3.2}$$

where x is a univalued function of t. In practice, if x is measured for a sequence of t_i, one will only have a discrete "sequential" set of measurements x_i, with $0 \leqslant i \leqslant N$, say.

The first task is thus to define the geomorphic curve in terms of a sequential set x_i, t_i. Any plane curve, on course, can be given as a relationship between two variables ξ, η, say (e.g., the Cartesian coordinates of the arbitrary point on the curve) such that

$$F(\xi, \eta) = 0. \tag{1.3.3}$$

However, this representation will not generally be univalued for ξ as a function of η nor η as a function of ξ, so that it is unsuitable for calculating spectra.

One will therefore have to choose a parameter other than ξ or η for use as independent parameter t for the spectrum analysis. As such, the arc length along the curve, measured from a fixed point 0, comes to mind. However, here the difficulties enter which were encountered in defining "length" along geomorphic curves, discussed in the last section.

Thus, we choose a polygon length l, as described in the last section, and define measurement points along the curve by giving the distance $t_n = nl$ along the polygon, assuming that only measurement points with n = integer are admissible. Then, n will be our independent variable.

At each point thus defined, we define as dependent variable the angle φ_n which a straight line drawn through the points nl and $(n-1)\,l$ forms with a given datum line. The measured pairs of variables are thus (n, φ_n), for $n = 1, 2, \ldots, N$. On the sequence of values $\varphi_1 \cdots \varphi_n \cdots \varphi_N$, the customary types of spectral analyses can then be made (Ghosh and Scheidegger 1971).

Thus, we first define the correlation function $C_\varphi(r)$

$$C_\varphi(r) = \left\{ \sum_{S=1}^{N-r} \varphi_n \varphi_{n+r} \right\} - (N-r)\bar{\varphi}^2. \tag{1.3.4}$$

The spectral function $X(k)$ is then

$$X(k) = \frac{1}{N} \left[C_\varphi(0) + \sum_{r=1}^{m-1} C_\varphi(r) \left(1 + \cos\frac{\pi r}{m} \right) \cos\frac{\pi k r}{m} \right], \tag{1.3.5}$$

where m is the number of frequency bands of interest. Naturally, the above quantities depend very much on the polygon length l. If they are found not to depend on l, at least for a certain range of l, then the curve is self-similar (in statistical sense) over that range.

1.3.4 Fractals

1.3.4.1 General Remarks

Mandelbrot (1967) has shown that the strange regularity in natural wiggly lines expressed by Richardson's law (cf. Sect. 1.3.2) could be the outcome of a general self-similarity property of the geomorphic lines in question. In this case, the curves would have to be treated as nonrectifiable and as represented by sets with fractal (i.e., noninteger) dimensions.

1.3.4.2 Definition of Fractals

Such fractal dimensions can be introduced for sets Ω which are embedded in a space that contains a metric, so that a "length" L is in some way defined. In ordinary Euclidian space the variation of the number of points $n(L)$ of the set Ω contained in a region R of a "size" characterized by the length L (region of "size" L) increases as

$$n \sim L^{D},\tag{1.3.6}$$

where D is the ordinary Euclidian dimension.

In fractal sets, the above relation is retained, but D need no longer be integer. Evidently, if D is less that the Euclidian dimension E, the region R available increases as L^{E} which is faster than the number of points of the set Ω contained in R which increases like L^{D}. Hence, larger and larger "holes" appear and the set is concentrated on a decreasing fraction of the total region (this visualization of fractal sets was suggested by Lovejoy et al. 1986). If D is independent of L, the set is self-similar (or at least self-affine) for all sizes of regions R, and D has a well-defined value.

Conversely, if n is again the number of points of a set Ω covered by a region R of "size" L, then, evidently, the number N of regions R of "size" L needed to cover *all* points of the set Ω is proportional to $1/n$. Thus

$$N \sim L^{-D}\tag{1.3.7}$$

In this, the set Ω must have a topological dimension T (integer) which is less or equal to the Euclidian (or otherwise metrically defined) dimension E (integer) of the space in which it is embedded.

The above relations are quite general and apply to all sorts of "fractal" sets. In fact, Mandelbrot (1982) defined a set as fractal if

$$E > D > T.\tag{1.3.8}$$

1.3.4.3 Applications to the Geosciences

Applying the above equations to a geomorphic wiggly line, which is now considered as the set Ω, and taking L as the polygon length in the measuring process, we note that the total apparent length S of the line is

$$S = L \cdot N.\tag{1.3.9}$$

Thus

$$S(L) \sim L \cdot L^{-D} = L^{1-D}.\tag{1.3.10}$$

Therefore, the fractal dimension of a line with Richardson parameter α is

$$D = 1 + \alpha.\tag{1.3.11}$$

The empirical observation of Richardson (1961) has therefore been reduced by Mandelbrot (1967) to the postulate that a wiggly line is represented by a self-similar fractal set over the range over which Richardson's law holds. This is a range of a factor from 10 to 97. To postulate the validity of Richardson's law, and therewith of self-similarity, over *all* size ranges of L seems absurd, as it leads to "ultraviolet" and "infrared" catastrophes (divergences at very small and very large values of L). The self-similarity must therefore be confined to a reasonable interval or the fractal dimension may itself be scale-dependent.

As an aside, it may be noted that fractal analysis has been used in various instances in mathematics. Mandelbrot (1982) gives some general examples. Specific applications to the Earth sciences have been published by Burrough (1981) and Clarke (1986) in connection with landscape characterization, by Turcotte (1986) in connection with rock fragmentation, and by Brown (1987) in connection with the analysis of fracture surface roughnesses.

1.4 Fluvial Geomorphology

1.4.1 General Remarks

Rivers are very powerful agents in shaping our globe's surface. They act in various ways: by removing material from its confines, by transporting it and by depositing it in distant areas.

River courses cannot be considered as constant even in relatively short time intervals. Thus, Johnson (1974) has given evidence of the abandonment of river courses in comparatively short periods in certain cases.

The removal of material by flowing water from the confining channel can occur in two ways: either the channel is being scoured out and thereby deepened, or the removal occurs on the side. The latter case is referred to as lateral river action, the former is normally considered jointly with transportation phenomena, and the two referred to as river bed processes.

As an aside, it may be interesting to note that morphological features obviously of fluvial-type origin have been observed not only on Earth, but on Mars as well (cf. e.g., Baker 1978, 1979).

Fluvial geomorphology has been the subject of comprehensive monographs, such as those of Schumm (1977), Petts and Foster (1985), and Morisawa (1985). Some of the pertinent phenomenological features of river systems will be considered in their turn below.

1.4.2 River Bed Processes

Turning first to river bed processes, we note that this includes every kind of interaction of a river with its bed, such as the entrainment of particles of which the river bed is composed, the formation of bottom ripples, the silting up and scouring out of a channel, the contrition of bed particles, the gradation of pebbles, and so on. A scheme for the classification of river bed processes has been provided by Kellerhals et al. (1976).

The requirements for measurements on rivers have been set forth on several occasions (see, e.g., Tricart 1961; Schumm 1963a; Leopold and Skibitzke 1967; Morisawa 1968). Accordingly, many field measurements have been made of various typical river bed processes. However, these were usually made in connection with special mechanical investigations and it is therefore difficult to give a meaningful summary in connection with a general discussion of physiography. These investigations will be referred to when the appropriate mechanical theories will be discussed.

The phenomenologically most important feature of river bed processes is that they do not produce a uniform vertical action (scouring or deposition) on the river bed, but an intermittent one, which is not even stationary. Thus, a river channel appears as a sequence of riffles (high-velocity, low-depth sections) and pools (low-velocity, large-depth sections) whose position changes with time (Dolling 1968; Culbertson and Scott 1970). Morphologically, it turns out that riffles are topographic high areas with an accumulation of relatively coarse material, whereas pools are topographic low areas which contain mainly fine material on the surface of the channel bed.

A particularly detailed study of the morphology of riffle-pool sequences was made by Richards (1976). Thus, attempts at identifying riffle and pool sections in a river have been based on velocity (v) versus depth (h) ratios, such as v/h or v^2/h, but these are not dimensionless. More successful has been the characterization by the topographic deviation h from some mean level of the river bed defined by a fitted trend line. Then, this deviation h becomes a stochastic variable of the distance s along the thalweg of the river (choosing an arbitrary but fixed zero point), with which harmonic, spectral and autoregressive analyses can be made (cf. the discussion on wiggly lines in this book; Sect. 1.3.3).

Similar studies have been made by Alexander (1980; also Rendell and Alexander 1979).

River bed processes eventually influence and create the hydraulic geometry of rivers. In quite general terms, the latter is described by a set of equations (Leopold and Maddock 1953)

$$v = aQ^m$$
$$h = kQ^f$$
$$n = cQ^b$$
$$S = qQ^y,$$

where v is the mean velocity of the flow, h the depth of the water, w the width of the channel, and Q the discharge. The quantities a, k, c, q are proportionality constants and m, f, b, y are the so-called hydraulic geometry exponents. These latter can be determined from observations for particular rivers and can be plotted best (referring only to m, f, b) in a ternary diagram (Park 1977; Rhodes 1987). Critical values of b, f, and m can be delineated at which the river will present a response to changing external conditions.

Examples of establishing the hydraulic exponents have been published with regards to many rivers, e.g., in Arkansas (Lee and Henson 1977a), Ohio (Lee and Henson 1977b), Mississippi (Lee 1977a), Missouri (Lee 1977b), the Ardennes (Petit 1987), and Japan (Sawada et al. 1983).

1.4.3 Total Material Transport

The sediment load transported by a river consists of solid material and of material in solution. The solid material, in turn, may be transported (1) as bed load, (2) as material in suspension and (3) as material in flotation (Schmidt 1981).

Regarding the *solid* material transported by a river, it is noted that the bed load is generally (see review by Hadley and Walling 1984) measured by the installation of traps (Brown et al. 1970) at various points along it. These traps have to be emptied from time to time; the rate of accumulation of material in them is indicative of the bed load being transported. Recently, a new tracer technique using permanent magnets (Ergenzinger and Conrady 1982) has been developed: magnets are inserted into the bed load and move downstream with it; their motion can be monitored by appropriate magnetometers. Next, the suspended load in a river is measured by the installation of a series of splitters (Barnes and Johnson 1956) collecting a representative portion of the flow, again at various points along the river course. The proportion of the material in flotation is generally so negligible that it can be disregarded altogether (Schmidt 1981).

The content of material in *solution* is measured by monitoring the concentration of the respective ions in the river water. Generally, the rate of solutional transport is up to four times greater than that of the suspended load (Meybeck 1976; Schmidt 1981), except during single peak floods.

When the mass flux at two points in a river has been ascertained, an observed increase must be due to the denudation of the area between those two points. In this fashion, the denudation of the area in question can be obtained by dividing the volume of the moving mass (volume per unit time) by the area drained. The denudation rate is then represented by a velocity, usually mm/ 1000 years.

In making extrapolations of denudation rates to *type areas*, care must be taken that the sampling is unbiassed: errors can be due to the fact that most hydrologic stations are in anthropologically affected areas (e.g., near cities) so that inferences obtained therefrom should not be extrapolated to natural areas (Meade 1969). The denuadation rates depend on relief (mountains, lowlands, etc.) and climate (periglacial, humid, dry, warm, etc.). The literature on such rates is extremely large. Holeman (1968) has summarized the earlier work, Ohmori (1983a) the more recent. We present here (Table 1) the range of denudation rates gleaned from Ohmori's study on dependence of relief and climate.

Since, on occasion, the denudation rates reach one to several mm/a, equal to km/Ma, it is evident that corresponding uplift rates must be present if the mountains of the world are not to disappear in a geologically very short time. Thus, a geomorphodynamic mass balance must exist in all areas (Vondran 1979): tectonic activity must balance geomorphic denudation, and the whole landscape presents a case of dynamic equilibrium. We have discussed above the mass transport by rivers as the principal means of denudation; however, if a global balance is attempted, it should be noted that it is not the only possible transport mechanism: erosion can be caused by the cumulative action of landsliding (Aniya 1985) as well as by aeolian and glacial effects (see below in this book). An estimate

Table 1. Range of total denudation rates. Data after Ohmori (1983a) in mm/1000 years

Climate	Plains	Mountains
Cold dry (periglacial)	2–30	52–420
Cool dry (midlatitute forest)	4–119	18–5600
Warm humid (maritime forest)	1–216	14–7948
Warm dry (steppe)	0.05–1160	11–605
Hot dry (savannah)	3–32	—
Hot humid (tropical forest)	9–560	92

of the total material lost by erosion from the entirety of the nonsubmerged areas of the world has been made by Fournier (1960), who arrived at a value of 400 mm/1000 a.

The tectonic (endogenic) activity is generally taken as constructive (uplift of mountain ranges), but it can also be, by itself, destructive, as when rift valleys are formed or basins subside owing to its action. In that case, the exogenic processes are the constructive ones by accumulating deposits in the opening rifts (Tiercelin and Faure 1978) and in subsiding basins (Schwab 1976). The normal case, however, is the buildup of relief by endogenic processes and its destruction by exogenic denudation. In this connection, it has been found that the tectonic uplift rates just about balance the exogenic denudation rates (cf. Scheidegger 1982a, p. 43), at least over short time intervals. Over geological times, the rates may change.

1.4.4 Sideways Erosion

As noted earlier, rivers not only have a tendency to deepen their channels (under certain circumstances), but also to scour *sideways*.

It has been observed that the course of a river is almost never straight. Close to the source, there are V-shaped gorges with a sinuous course. Similarly curved courses occur in plains where rivers have a definite tendency to meander (i.e., to form loops) or to form braids (see, e.g., Schumm 1985). Innumerable individual studies have been made bearing out this fact, not only on Earth, but also on Mars (Weihaupt 1974).

The initiation of meander formation is caused by the instability of the riffle-and-pool sequences (Sect. 1.4.2) which induces the river to detour laterally (Tinkler 1970; Ori 1974), leading to actual bank erosion and undercutting (Brunsden and Kesel 1973; Hooke 1979). Thus, meanders have a tendency to grow and to wander (Leopold 1973; Brice 1974; Adams 1980a; Hooke 1984), presumably until they become so large that no further sideways erosion (scouring) can take place. On occasion, this causes meanders to touch and to short-circuit each other; the "dead" loops then form lakes which are called *oxbow lakes*.

If one wishes to make quantitative studies of meanders, one is first of all faced with the problem of having to describe a geomorphic *line*. The pertinent questions in this connection have been discussed in Section 1.3 of this monograph. Accordingly, the river is approximated by a smooth curve, called "thalweg", which is differentiable and rectifiable.

Once this is done, geometrical studies of meander loops can be made. The latter are generally assumed as symmetric (cf. Fig. 1), but this is probably an over-simplification (Carson and Lapointe 1983). A good review of the pertinent studies has been given by Callander (1978).

In characterizing a meander, it is common to use its length L, its amplitude A, its thalweg T, its sinuosity $P = T/L$ (Mueller 1968; Verma and Tandon 1971) and its radius of curvature R.

Then, the aim is to correlate these geometrical quantities with the channel width w, with the (average) depth d (or the width-to-depth ratio $F = w/d$) with the bankful discharge Q and with the bed slope S. Such correlations have been proposed by Inglis (1949), Zeller (1967), Schumm (1963b) Chang and Toebes

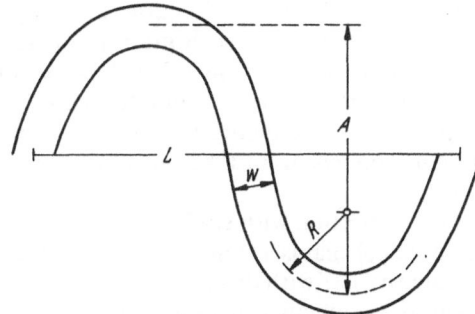

Fig. 1. Geometrical characteristics of a meander

(1970), Gregov (1976) and Parker et al. (1983). The general gist of such correlations is that the meander length ranges from seven to ten times the width of the stream. The amplitude correlates only poorly with meander length; the ratio R/w tends to lie near 2 or 3 (Leopold and Wolman 1960).

1.4.5 Morphometry of Particles

Since the river action is essentially concerned with the direct entrainment and deposition of particles, it may be well at this point to mention a few concepts which are of importance in the description of particles.

One generally likes to speak of the "size" of a particle. However, particles have generally an irregular shape and hence "size" does not have an obvious meaning. Nevertheless, the "largest diameter" d has always a geometrically clearly defined meaning: it is the largest distance between points on the surface of the particle. When speaking of "diameter", we always mean this quantity. This diameter has the dimension of a length, and thus may be given in cm or mm. However, one finds commonly in the literature a different type of unit, called the φ (phi) unit. The number of φ units is related to the diameter d in mm as follows:

$$d(mm) = (\tfrac{1}{2})^{\varphi}. \tag{1.4.1}$$

Diameters of particles other than the largest are much more difficult to define. One could introduce a "small" diameter as the diameter of the largest totally inscribed sphere, and an "intermediate" diameter as the diameter of the smallest circumscribed circular cylinder.

Thus, the characterization of grain shape poses a special problem. Ehrlich and Weinberg (1970) proposed the use of a Fourier series expansion about the center of mass utilizing coordinates of the peripheral points, but this is a most involved procedure. It is easier to introduce such concepts as "roundness" or "angularity" of the grains (cf. Krumbein and Pettijohn 1938; Hawksley 1951; Konzewitsch 1961).

The particle sizes are generally determined by sieve analyses. Theoretically, grains should pass through a given size mesh according to their smallest diameter, but in practice the grains rarely hit the sieve with their long axes normal thereto, so that some length in between the smallest and the largest diameter will be the determining factor in the passage through the sieve. In a mixture of grains, the weight fractions corresponding to various "sizes" are then given, leading to "populations". Statistical characteristics (means, modes, standard deviations, etc.) can be calculated for such grain populations (Charlier 1968; Roberts and Mark 1970). A further problem arises if the orientation of the grains in the mixture is of interest: this is generally not the case in *moving* river loads, but only in *fixed* deposits (tills, sediments, grains in igneous rocks). In the latter case, the statistical orientation tensor for the long axes has to be ascertained (Scheidegger 1965; Mark 1973).

In connection with studies of rivers, it has been noted that the (mean) grain size (and shape) changes along the thalweg. Generally, the sizes become smaller and the angularity less with distance from the source (Knighton 1982). This can be used to identify dominant transport conditions (cf. Sect. 4.6). In addition, the petrology and chemistry of the load may change along a river (Potter 1978; Li et al. 1984), indicating differential movements for the individual fractions.

1.4.6 Morphology of River Nets

1.4.6.1 General Remarks

Rivers do not usually occur as single features in an area, but in the form of nets. In order to proceed to a rational explanation of the development of such river nets, it is necessary to describe their geometrical features in numerical terms.

In principle, one can state that the geometrical properties of a figure (such as of a river net) can be topological or metric in nature; in the latter instance the metric may refer to lengths or angles (orientations). Marin and Riguidel (1978) have listed some general principles for the morphological characterization of various types of forms, Coffman et al. (1971) have listed the parameters that might be useful specifically for characterizing river nets and drainage basins. Recent reviews of drainage basin morphometry have been supplied by Gardiner (1981; cf. also Gardiner and Park 1978). In the discussion following we shall present the pertinent aspects of the subject in question.

1.4.6.2 Topological Aspects

Starting with the topological aspects of river net characterization, we note that one has to introduce a series of definitions. The *stream net* is defined as the interrelated drainage pattern formed by a set of streams in a certain area (omitting the possible downstream splitting of channels). A *junction* is the point where two channels meet. The *source* of a river is the beginning of the blue line on a map (in nature there may be some ambiguity in defining this term), a *link* is any unbroken stretch of river between two junctions (interior link) or between a source and the first junction (exterior link).

The river net has a hierarchical structure (Mayer 1970; Warntz 1975). Thus, it is necessary to label the position of a link within the hierarchy. This has been done in various ways.

The most common procedure is that due to Strahler (1957): We assign the order 1 to all the exterior links in the net. When two first-order links meet, they form a second-order link; when two second-order links meet, they form a third-order link, etc. However, when a N-th-order link meets a link of order P < N, the resulting link will have order N, i.e., the lower-order link gets "swallowed up". The formation of Strahler orders of links can be written as an algebraic operation. Let us denote the meeting of links of orders N, P by an asterisk; then we have

$$N * P = N + 1 \qquad \text{if } N = P, \tag{1.4.2}$$

$$N * P = \sup(N, P) \quad \text{if } N \neq P. \tag{1.4.3}$$

A stretch of river over which the Strahler order does not change is called a Strahler segment. The Strahler ordering of a hypothetical stream system is shown in Fig. 2B.

A slight modification of the above scheme was the (earlier!) notion of Horton (1945) order. In this scheme it is considered that the "main stream" in a river net should be denoted by the same order number all the way from its mouth to its headwaters. Thus, at every junction where the order changes, one of the lower-order streams (usually either the longest or the most direct upstream continuation of the main stream) is renumbered to the higher order. The resulting stream-ordering system is exemplified in Fig. 2A.

The stream-ordering systems above have the disadvantage that the distributive law does not hold: if, say, two third-order segments combine to form a fourth-order segment before joining another fourth-order segment, the result of the last junction is a fifth-order link. On the other hand, if the two third-order segments are simply individual tributaries to a fourth-order segment, the result is a fourth-order link. However, it could be argued that the hydraulic properties of a link should only depend on the number of lower-order tributaries, not on the sequence in which they join. Thus, a *consistent* ordering system of stream nets has been postulated by Scheidegger (1965a) in which the distributive law holds: Counting the number of source links that (eventually) contribute to an interior link, one obtains the latter's "magnitude" M (Shreve 1967). The "consistent order" N_c of Scheidegger (1965a) is then

$$N_c = \log_2 I, \tag{1.4.4}$$

with I = 2M; I is called "associated integer" of the link. Evidently, the distributive law holds for the thus defined stream order. The consistent ordering system is shown in Fig. 2C.

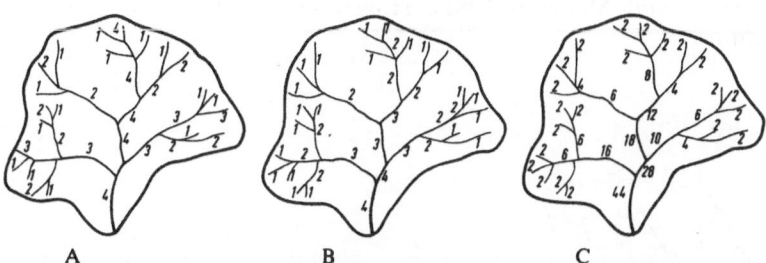

Fig. 2 A–C. Stream-ordering systems of Horton (A), Strahler (B) and Scheidegger (C) as applied to a hypothetical stream net. In C the italic numbers are the associate integers. (After Scheidegger 1965a)

The merits of the various types of ordering schemes have been discussed by Dunkerley (1977) and by Onesti and Miller (1978); it is clear that those schemes where some of the first-order links are "lost" give less information on the structure of the net than the "consistent" ones; but this is paid for by additional clumsiness. The various stream orders can also be taken as indicative of the sample source uncertainty as regards the provenance of the water (Sharp 1970).

As a further remark on "orders", it may be noted that the latter can also be assigned to the nodes (including the sources) of a stream net, not only to the links: the order of a node or junction is equal to that of the highest-order link issuing from it (Jarvis 1972).

The order parameter gives information on individual links in a network, but not on the topological structure of the net. This structure correponds to that of a special type of topological "graph": of a so-called bifurcating arborescence. Such graphs possess a root (outflow stem of the river net) and contain only nodes represented by bifurcations and free ends. In addition, the structure is such that there are no closed circuits within the graph. Such graphs can be represented by a so-called Lukasiewicz word (Berge 1958): the river net is "read" from its outflow end anticlockwise, setting 1 for each junction and 0 for each free end. Thus a net of n free ends is uniquely represented by a "word" consisting of n zeros and n − 1 ones. Conversely any word of n zeros and n − 1 ones represents a rooted, bifurcating arborescence; in order to obtain the direct correspondence with the Lukasiewicz representation, a cyclic permutation in the word may be required (Fig. 3). The Lukasiewicz word characterizes an *individual* network; this was extended by Jarvis (1972) to network *types* by the introduction of a topological index E. The latter is defined as

$$E = \sum_{\text{int}} MH \Big/ \sum_{\text{ext}} MH, \qquad (1.4.5)$$

where M is the magnitude of a point ($= \frac{1}{2}$ I, see definition above) and H the "link distance" to the outflow of the net, i.e., the number of links that lead from the point under consideration to the outflow. The sum in the numerator has to be taken over all interior points (nodes) of the net, the sum in the denominator over all exterior points (i.e., free ends or sources). The topological index is highest, given a fixed number of free ends, for those configurations that have the least lateral spread.

As a final remark, it may be stated that quite generally, the number of first-order links depends on the observer and on the map scale (Kheoruenromne et al. 1976; Suzuki and Shimano 1981). However, this does not affect the topological ratios.

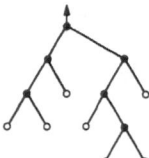

Fig. 3. Graph corresponding in Lukasiewicz representation to the word (1 1 1 0 0 0 1 1 0 1 0 0 0)

1.4.6.3 Symmetry Aspects

In addition to the topological aspects, the symmetry aspects in a river net may have a characteristic significance.

Thus, the usual topological analysis, expressed in the various ordering schemes, assumes tacitly that there is no difference between "left" and "right" tributaries. In general, this assumption is justified; but there are obviously exceptions; the most notable such case occurs in piedmont streams where most tributaries enter the main stream from one side, viz. from the side where the mountains are. James and Krumbein (1969) have introduced the concept of "cis" and "trans" link, according to whether the tributaries at the upstream and downstream ends enter it from the same or from opposite sides.

A classification of drainage nets based on symmetry properties (in part) has been proposed by Argialas et al. (1988), who distinguished between eight different fundamental drainage patterns. Some of these, however, are topologically identical. In fact, only two basic types of patterns distinguishable from symmetry properties remain: symmetrical and one-sided.

1.4.6.4 Metric Aspects

Next, we turn to the metric aspects of river nets. First of all, the *length l* of river segments or links is of interest. The problem of defining the length of such an element has been touched upon in Sect. 1.3. As noted there, it is necessary to approximate the natural stream course by a smooth curve which is differentiable and rectifiable. Then no further problems in stating stream lengths are encountered.

In a similar vein, the (drainage) *area*, a, belonging to any (or all) links or segments can be introduced. It is particularly useful to distinguish between the average length l_e and area a_e of the exterior links (free ends) and those, l_i and a_i of the interior links. The metric structure regarding links and areas of a drainage basin can then be represented by a series of dimensionless parameters (Smart 1972), viz.:

$$\lambda = l_e/l_i \tag{1.4.6}$$

$$\alpha = a_e/a_i \tag{1.4.7}$$

$$K_e = l_e^2/a_e \tag{1.4.8}$$

$$K_i = l_i^2/a_i. \tag{1.4.9}$$

In addition, the following dimensional parameter

$$D = \sum l/a$$

has been introduced; this is the sum total of all channel lengths divided by the total area drained. It has been called "drainage density". It can be easily and rapidly measured on maps (McCoy 1971; Christofoletti 1983). It ranges from as low as $1.5\,km^{-1}$ in massive sandstone to values of from 300 to $600\,km^{-1}$ and more for badlands.

A quantity related to the drainage density is the channel frequency F: if one counts all the stream segments up to a given order which are present in a drainage basin and divides this by the area drained by the streams up to that order, the quotient is called channel frequency.

Other quantities representing some metric aspects of drainage networks have been proposed by Gregory (1977), who introduced the notion of "network volume" by multiplying the total length of the channels in the network by their averaged cross-section and the notion of "network power" (Gregory 1979) by dividing the network volume by the height difference (the relief) obtaining in the basin.

It is further possible to give indices characterizing the *shape* of drainage basins (Jarvis 1976). The standard shape indices are all variations of ratios involving some combination of the longest axis length, the area, and the perimeter length of the drainage basin. Other indices are based on comparisons with shapes of models such as lemniscate loops (Anderson 1973), but these are somewhat arbitrary as they depend on the models chosen.

Finally, metric investigations in river nets can be made with regard to the orientation structure of the links and the channel confluences. Since the orientation structure of a river net seems to be directly related to the neotectonic stress field, the subject has been discussed by the writer at length in his treatise on geodynamics (Scheidegger 1982a). The morphology of confluences has been reviewed by Best (1986); as this refers to river dynamics, it will be treated in that context (Sect. 4.7.6.4).

1.5 Morphology of Landscape Systems

1.5.1 General Remarks

1.5.1.1 Introduction

A landscape consists of a great number of individual geomorphological forms such as slopes, river segments, valleys, peaks etc. In this instance, a landscape may be considered as a "system" consisting of many "elements". It is evidently impossible to build up a classification of landscapes on the basis of the morphology of their elements, since the latter can combine to form systems of a bewildering complexity. Attempts at such a classification must therefore be based on features of a larger scale.

1.5.1.2 Cycle Theory

An old attempt at a description of landscapes systems has been based on a hypothetical descriptive natural history of their genesis. Thus, Davis postulated at the end of the last century (for a summary, see Davis 1924) that each landscape represents a certain stage of an evolutionary cycle: It is clear that degradation by water, ice, and wind ("exogenic" agents) represents powerful destructive processes in a landscape. Since there has to be something present before it can be destroyed, a buildup must have occurred prior to the destruction: this is assumed to have been caused by tectonic uplift ("endogenic" processes). Thus, Davis (1924) arrived at his cycle theory: a geomorphological cycle has its beginning soon after an endogenic geodynamic process has completed creating an uplifted area, such as a mountain range. Weathering, erosion, and detrition begin to act on the uplifted area and gradually proceed to reduce it to a base level. This completes the cycle. A new cycle starts when a new endogenic diastrophism occurs.

Davis recognizes three distinct stages in the geomorphological cycle, which may be termed *youth*, *maturity*, and *old age*. In a humid climate, these are as follows:

In *youth*, one has some trunk streams but not many large tributaries. The valleys are strongly V-shaped, their depth depends on their height above sea level. Lithological variations cause waterfalls and rapids for which there has not been sufficient time to disappear.

In *maturity*, the drainage system becomes more integrated. Any waterfalls and rapids evident in youth have disappeared and most of the rivers are in a dynamic equilibrium condition. The extent of the relief represents the maximum that is possible.

In *old age*, valleys become very broad, most of the relief has disappeared due to continental planation. The level of the drainage basin approaches the base level of erosion. The final stage of the cycle is reached when *all* relief has been reduced to the base level, leading to a gently undulating plain which Davis called a peneplain (Leopold and Bull 1979).

1.5.1.3 Climatic Effects

In the development of a landscape, climate obviously plays a major role. Thus, Buedel (1977, 1982) introduced the idea of "generations of reliefs", each of which is characteristic of certain climatic conditions. Similarly, Jessen (1943), Wirthmann (1977), Ohmori (1979, 1983b), and Bremer (1985) investigated the dependence of slope and mountain development on climate, and showed that the former is very much affected by the latter. We will thus also have to analyze the effect of climate on geomorphology (see Sect. 5.3.4.2).

The most commonly encountered climate in inhabited land areas is a humid one, which consequently has been called "normal". This is the case that is generally tacitly assumed to underlies the discussions in this book.

However, there are instances where the climatic conditions are or have been quite different from "humid".

Thus the climate can be cold, leading to large amounts of ice on snow on the ground, which become then the principal exogenic geomorphic agents; they produce characteristic geomorphological forms and landscapes that are called glacial or peri-glacial. In this instance, it should be noted that the Pleistocene was punctuated by several periods during which large continental glaciations existed ("ice ages"). Thus, remnants of (peri-) glacial forms are frequently found in areas that are subject to a "normal" humid climate at the present time.

Another extreme is the arid climate. The latter is characterized by very frequent, but when they occur, terrific rainfalls. Wind action becomes a major geomorphic factor. The geomorphological effect of arid climatic conditions is that of producing desert forms and desert landscapes.

1.5.1.4 Criticisms of the Cycle Theory

Unfortunately, Davis' cycle theory is evidently based on a misinterpretation. Criticisms of the cycle theory appeared immediately after its invention (Tarr 1898; Shaler 1899). One outstanding critic of the cycle theory was Penck (1924), who viewed the development of a whole drainage basin as the results of the development of each individual slope it contains (cf. Sect. 1.2.4). This idea can also be expressed by the assumption of the existence of "cycles within cycles" (Klein 1985).

Another form of criticism of the cycle theory has been advanced by the adherents of the "equilibrium theory" (Hack 1960), who regard the present appearance of a landscape as the outcome of a dynamic equilibrium of the forces in action. Accordingly, a landscape preserves its character if the forces stay the same; some slopes will waste away whilst others are being created.

Similarly, Crickmay (1960) maintains that the activity of the exogenic agents is unequal. According to Crickmay, a slope bank can recede only if there is a river (or surf) cutting away laterally at its bottom. The lateral action of rivers is connected with their tendency to meander. Slope wastage without lateral action of a river (or surf) simply produces a slow decline in slope angle without a recession at the foot. Thus, according to Crickmay, there is again no distinct meaning to a geomorphic cycle. Endogenic movements may lift parts of the Earth's crust at a slow or rapid rate while denudation is also taking place. This is the *anagenic* stage of the development. Once the endogenic movements cease, denudation will continue and one has the *catagenic* stage of landscape development. Since denudation (according to Crickmay) is mostly achieved by rivers cutting away at the bottom of slopes while they meander in the valleys, some parts of a landscape may never be touched by slope recession (*stagnation* of development) and their "youthful" (in the Davisian sense) forms may locally subsist to a very late stage. Therefore one can no longer speak of the stages of youth, maturity, and old age.

1.5.2 Principle of Antagonism

1.5.2.1 General Argument

Thus, landscape development is not a temporally sequential affair with degradation following buildup, but rather degradation and buildup occur concurrently: a landscape represents a dynamic equilibrium between the antagonistic action of endogenic and exogenic processes (Principle of Antagonism; Scheidegger 1979). As was seen in Sect. 1.4.3, the uplift and degradation velocities (v_u and v_d) roughly balance each other. Thus, the apparent "age-stage" of a landscape is determined solely by the intensity of the antagonism expressed by $v_u = v_d$. One has the character of "youth" if the velocities are of the order of > 1 mm/a ("high-activity landscape"), "old age" if they are of the order of < 0.1 mm/a ("low-activity landscape") and "maturity" if they are in between ("medium-activity landscape"). A quantitative comparison between the Davisian scheme of landscape development and that corresponding to the principle of antagonism has also been suggested by Ohmori (1985a).

The fact that a landscape represents normally a state of (quasi-) equilibrium implies that the change of any one characteristic parameter induces the others to change in their turn in the direction of the new equilibrium state. Thus, a "process" affecting one landscape parameter elicits a "response" in the others ("process-response theory").

Deviations from a stationary "dynamic equilibrium" can occur in various ways. First of all, if the uplift velocity v_u is not equal to the denudation velocity v_d, one has non-stationary conditions. One can then define a stationarity index S as follows (Scheidegger 1987a)

$$S = v_u/v_d. \tag{1.5.1}$$

If the uplift is faster than the denudation, the stage is shifted to "greater" age. Thus, Adams (1980b) has shown that the peaks in the New Zealand Alps are flat-topped when $v_u > v_d$ (age form!), but spiky (youth form!) when $v_u < v_d$.

Furthermore, a nonuniformity of dynamic equilibirum is also the result if the antagonistic processes act in a temporally nonuniform fashion. Thus, Finkl (1982; also Fairbridge and Finkl 1979) noted that long stable "cratonic" periods (of the order of 100 mio. years) are punctuated by (comparatively) short (1 million years or less) unstable "rhexistatic" phases. The latter can assume catastrophic proportions and lower erstwhile peneplains by several meters in a few million years.

1.5.2.2 Endogenic Processes

The endogenic processes have their origin in plate tectonic phenomena. As is well known, one assumes that the lithosphere (consisting of about the uppermost 150 km of the solid Earth) is divided into a relatively small number of plates which undergo displacements with regard to each other. These displacements cause tension zones, over- and underthrust zones, and shear zones.

The operation of endogenic processes is rather obvious in the morphology of mountain chains, rift valleys, volcanic chains, etc., which are clearly of endogenic origin. Similarly, great earthquakes (which are evidently of endogenic origin) can have geomorphological effects: valley offsets, fissure formation, and the triggering of landslides (Ai et al. 1987; Carton et al. 1987). However, even manifestly exogenically affected features may have been subject to an endogenic or tectonic control. Since it is the movements of the lithospheric plates that exercise such control, there is a certain systematism in endogenically controlled features that reaches over distances of the order of the scale of the plates: in other words, the statistical pattern of the morphology of endogenically pre-designed features is *systematic* (or correlated) over distances that are comparable to the diameters of the plates.

Scheidegger and Ai (1986) have made a study of the relation of tectonic processes to geomorphological design. In it, it was shown that the orientation patterns of fluvial and glacial valleys are quite generally affected by neotectonic processes. Such patterns had indeed been used by Scheidegger (1982a) for the determination of the principal directions of the neotectonic stress field: the orientation of the valleys corresponds, in most instances, to the shear directions of the neotectonic stress field.

Furthermore, many landforms are direct reflections of the prevailing geodynamic conditions: these include mountain peaks, glacial cirques, and karst caves. In this instance, one has to distinguish between tectonic landforms proper which are caused by the plate tectonic stress field, and landforms which are caused by self-induced gravity stresses (Gerber and Scheidegger 1975, 1977). The term "structural landforms" has been used in part for endogenically caused features (Twidale 1971), but it also includes other forms which are due to the exploitation of weaknesses on the Earth's surface by exogenic agents. Recent reviews of the subject are due to Doornkamp (1986) and to Summerfield (1987). In essence, these subject matters, including the study of the results of volcanism and volcanic landscapes, are part of "geodynamics" (Scheidegger 1982a).

1.5.2.3 Exogenic Processes

By definition, the exogenic processes are caused by agents originating outside the solid Earth: water, ice, wind. Since exogenic processes operate on an inherently local scale and are, in the case of wind and water, connected with the turbulence (best described by a stochastic theory) in these media, they present the aspect of statistical *randomness* (Scheidegger 1979). Indeed, the characterization of landscape processes by their stochastic nature (random for exogenic nonrandom or systematic for endogenic processes) has been used successfully for the identification of the origin of geomorphic features.

Examples for random exogenic features are found particularly in the gully systems of badlands (cf. Bryan and Yair 1982), in the development of river meanders and in many others. However, the test for randomness or nonrandomness has to rely on general system theory. Therefore, specific examples will be found in the latter context in the appropriate chapters in this book.

1.5.3 Quantitative Landscape Description

1.5.3.1 Mathematical Representation of Forms

Geomorphology has progressed from an erstwhile descriptive science to a quantitative one. In this instance, it became necessary to replace qualitative terms (such as "round hills", "steep crags", etc.) by quantitative values. The general problems encountered in this endeavour have been outlined and discussed, for instance in books by Hoormann (1971) and by Doornkamp and King (1971).

In the quest for a quantitative landscape description, the first step is usually to provide a mathematical representation of (the surface of) the geomorphological "form" under consideration. Altitude, distance, and area are rather easily digitized and measured fundamental geomorphological attributes. From these, higher level attributes such as gradient, curvature, and relief can be derived (Ohmori 1985b). More sophisticated developments have used standard algebraic surface equations for the representation of geomorphic forms (Tipper 1976, 1977). In this fashion, slope maps (Engelen and Huybrechts 1981), trend maps for various attributes (Zuchiewicz 1981) etc. can be constructed. Data banks can store the values for several interdependent attributes as a matrix for every point (Zevenbergen and Thorne 1987).

However, it is generally not meaningful to have a numerical representation of the various landscape attributes mentioned above, but rather to know the values of certain parameters that characterize the landscape in some fashion. These parameters will generally represent some sort of statistical averages of such attributes as latitude, distance, and areas from which gradient, curvature, and relief can also be derived.

Quite general studies of the possibility of describing geomorphic form types by characteristic parameters have been made, e.g., by Anderson (1973) and, particularly, by Speight (1968, 1976, 1977), who also showed how these could be extracted from air photographs. Most of these studies were made in connection with the planning of (agricultural etc.) land use, since the characteristic parameters are thought to be connected with the latter. The parameters which are important in our study of *natural* landscape evolution will be dealt with in the appropriate sections of this book.

1.5.3.2 Hierarchies

Many landscape features show a hierarchical structure. We have already seen one such example during our discussion of drainage networks (Sect. 1.4.6): the ordering scheme of river segments is only possible because there is a corresponding type of hierarchical order in nature.

Since river nets are hierarchically ordered, it is easy to take the step to the corresponding drainage areas: an area of a certain order is simply the area drained by the stream segments up to that order of the river network under consideration.

The next step is evidently a search for more general hierarchical structures (other than drainage nets) in nature. Thus, Araya-Vergara (1977) proposed an analogy of geomorphological structure with plant or animal taxonomic structure, and has postulated classes ranging from kingdoms through phylum, family etc. down to species and individual. This type of procedure was formalized by Krcho (1986) on the basis of field theory, starting from a mathematically formalized expression of morphotopes as the smallest relatively homogeneous spatial units of the georelief. From the smallest units, the higher hierarchical levels are built up.

The idea that hierarchical structure is a fundamental property of many natural systems has been best formalized by Haigh (1987), who based his analysis on the key concept of the "holon" of Koestler (1978). A holon is any stable subwhole in a hierarchy which is subordinated to the larger entity of which it is a part, but also integrates the operations of the lower level subwholes. Many examples of such hierarchical structures can evidently be found in geomorphology (cf. river basins); whether they are fundamental to *all* geomorphic features, however, is an open question.

1.5.3.3 Hypsometry

In a large measure, the "character" of a landscape is determined by the intensity of its relief. The latter, in turn, is determined by the activity level of the antagonistically acting endogenic and exogenic processes.

The landscape "character" can be directly quantified by a study of the relief; this has been done by Strahler (1957) by the introduction of the *hypsometric curve*. The latter is obtained by calculating the fraction of the area under consideration that lies above a certain height level. In order to obtain dimensionless relations, it is customary to consider *relative* levels: the heights and areas are divided by the total height difference between the highest and lowest point in the area and by the total area under consideration, respectively. Strahler (1957) had shown that the hypsometric curves as defined above have characteristic shapes for so-called "youthful", "mature", and "old" landscapes: convex for youthful ones, straight (more or less) for mature ones, and concave for old age landscapes (see Fig. 4). This qualitative statement was further quantified by Scheidegger (1987b) by the introduction of a hypsometric index: this is the quotient of the area under the hypsometric curve and the area of the isosceles triangle obtained by drawing a straight line from the point (0.1) to the point (1.0) in the hypsometric graph. This hypsometric index can theoretically vary between 2 and 0. Together with the uplift/denudation rate it characterizes the landscape "stage" (cf. Sect. 1.5.2.1).

Incidentally, Wood and Snell (1960) introduced an "elevation-relief" ratio E by the formula

$$E = \frac{\text{Mean elevation} - \text{Minimum elevation}}{\text{Maximum elevation} - \text{Minimum elevation}} \tag{1.5.2}$$

It can be shown that this value E, is identical to the value of the area under the hypsometric curve for the same region (Pike and Wilson 1971), so that it does not represent a new parameter.

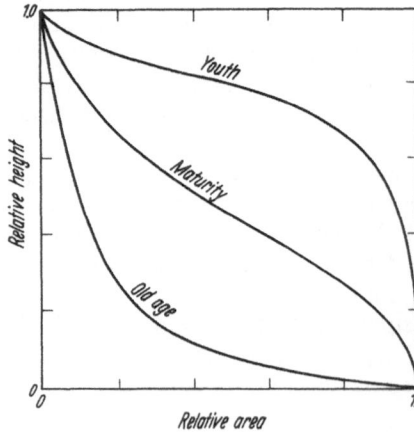

Fig. 4. Characteristic integrals of the hypsometric curve of drainage basins. (After Strahler 1957)

The hypsometric parameters introduced above are dimensionless, and thus do not give a direct indication of the absolute relief dimensions. To fix the latter, Ohmori (1978; also Ohmori and Hirano 1984) introduced a set of dimensional parameters such as mean altitude, maximum relief difference in an area, the dispersion of altitudes around the mean (average root-mean-square deviation of altitudes) and the mean gradient. For such parameters, empirical relationships were derived for type areas from map analyses (Ohmori and Hirano 1984; Ahnert 1984).

1.6 Aquatic Morphology

1.6.1 General Remarks

Water is a powerful exogenic agent. In Sect. 1.4 we have already discussed some aspects of the action of water, viz. the patterns that *rivers* form in a landscape. However, in that context the water nature of rivers was not of much importance; rather, it was assumed that rivers are geomorphological lines along which mass transport takes place. Now, we shall have to deal with water as a three-dimensional agent.

Indeed, most of the surface of the Earth is covered by water. Due to solar heating, water evaporates from the surface (in addition, some evaporation may occur as transpiration through plants), condenses again at higher altitudes, is transported over considerable distances, and is precipitated again to the surface of the Earth as rain or snow. If the point of precipitation is not in a water-covered area, the water proceeds downward toward such areas (through surface runoff) in rivers or (after seepage into the ground) in groundwater streams. In this fashion, water is the subject of a "hydrological cycle" which is driven by the energy of the Sun.

The "three-dimensional" action of water is seen, first of all, directly in the land areas. The linear effects of this surface runoff have already been dealt with in the section on river morphology (1.4); now we shall have to review some three-dimensional effects in connection with larger bodies of water (i.e., lakes) and with solution and precipitation effects on land.

Next, the morphology of the ocean bottom is of importance. Whereas the physiography of the land areas has been the object of the study of geologists for hundreds of years, the investigation of the water-covered areas had to await the advent of refined measuring techniques, such as echo-sounding, deep sea coring, underwater photography, and others.

During the course of modern investigations, it turned out that the deep sea is not just a bottomless abyss, as was originally thought, but that there is a rather varied topography. Many of the observed features are due to endogenic processes and will not be discussed here, but others are properly dealt with in a study of geomorphology.

Apart from the development of truly submarine phenomena, the dynamics of the sea has also had a pronounced effect upon some land areas, as seen in the evolution of coasts and river mouths.

1.6.2 Aquatic Land Morphology

1.6.2.1 Introduction

Beginning with the geomorphological effects of water on land areas, we note first of all the possibility of existence of large surface bodies of water: these are the lakes (Sect. 1.6.2.2).

Furthermore, as noted, one encounters solution and precipitation effects which are the result of a changing solution capacity of water in the hydrological cycle. This may occur on the surface or below the ground.

Solution effects can lead to the formation of karst phenomena which include flutes, sinkholes, and caves. Generally, these occur in limestone terrains, but gypsum and even siliceous materials can show karst-type phenomena (Sect. 1.6.2.3).

Precipitation effects are the opposite of solution effects: material is precipitated out of solution and deposited on or below the ground. Typical examples are the formation of spring tufa terracettes, the buildup of mounds of geyserite, and the growth of stalagmites and stalactites in caves (Sect. 1.6.2.4).

1.6.2.2 Limnology

The science studying the lakes is called limnology. The morphology of the lakes is to a large extent conditioned by the mechanism of their formation. Generally, the latter is due to a geotectonic predesign followed by the action of some exogenic process. In fact, Smith (1968) distinguishes between the following causes for the origin of lakes:

1. Tectonic movements, which include gentle crustal movements as well as folding and faulting. The resulting features are parts of "structural landscapes" (Sect. 1.5.2.2).
2. Volcanic activity, which leads to lakes in craters and calderas. The resulting features are parts of "volcanic landscapes" (Sect. 1.5.2.2).
3. Landslides, which may dam up a river valley so that a lake is formed behind the dam. The landslide itself may be of endogenic-seismic or exogenic-meteorological origin (cf. Sect. 3.3.3).
4. Glacial activity: the basins of many lakes (particularly in the Alpine foreland) have been hollowed out by ice age glaciers; an additional damming-up effect was achieved by the dumping of moraines by the same glaciers during their retreat. The corresponding types of landscapes are therefore basically "glacial landscapes" and will be discussed in the respective chapters of this book.
5. Solution effects: the dissolution of specific minerals can also lead to lake formation, such as in sinkholes etc. This subject will be discussed in the next section (1.6.6.3) in connection with other solution effects of water.
6. Wind action: water can accumulate in blow holes, leading to the formation of lakes therein. Such features are parts of "aeolian landscapes" (Sect. 1.8).
7. Meteorite impact, which causes a crater to be formed which is then filled with water. The cratering process is generally discussed in connection with geodynamics (Scheidegger 1982a).

8. Groundwater lakes. In some cases, a lake is simply the continuation of the groundwater table across a depression (Lake Balaton in Hungary, Lake Neusiedl in Austria).

The morphology of the shore of lakes is very similar to that of oceanic coasts and is created by similar processes: basically the interaction of waves and currents in the water with the land. The fundamental morphology of coastal features will be discussed in connection with the description of the ocean-land system (Sect. 1.6.3). What is different in the case of lakes is the *scale*: being relatively small bodies of water, lakes do not contain such large-scale movements as ocean currents and trans-Pacific surf. They do, however, show such specific features as eigen-oscillations (seiches) and rapid level changes owing to temporary imbalances between inflow and outflow.

1.6.2.3 Aqueous Solution Effects on Land

Solution effects by water occur on land in such areas where the solid material forming the upper layers of the Earth is water-soluble. The solubility is often not a direct one, but occurs by means of an intermediary: thus, limestone per se is not very soluble in pure water, but if the latter contains carbon dioxide, the limestone is soluble. The same holds true for dolomite. Indeed, limestone and dolomite terrains are those that are most prone to show geomorphic effects of the solution activity of water. However, solution effects do not only occur in carbonate rocks, but also in gypsum (here SO_2 is the solution intermediary), salt (Martinez et al. 1981), and even in rocks containing siliceous materials, such as granite (Watson 1985) or sandstone (Young 1986).

The geomorphic results of solution effects of water have collectively been called karst (in carbonates) or pseudokarst (in other soluble materials) features. There is an extensive literature on "karst geomorphology"; books on this subject have been published, e.g., by Sweeting (1973), Jakucs (1977), Jennings (1985), Trudgill (1985), and White (1988).

Thus, solution effects express themselves on the surface as well as underground.

Regarding surface effects, classifications of the resulting forms have been published, e.g., by Longman and Brownlee (1980) and by Fabre and Nicod (1982). Accordingly, one meets with karren structures (flutes or rills, up to some tens of centimeters wide and deep, up to some meters long) on naked limestone rock walls (Dunkerley 1979), large-scale karst towers in limestone (Jennings 1985) and sandstone (Young 1986), solution cauldrons ("opferkessel") in silicate rocks (Hedges 1969; Schipull 1978) and sinkholes ("dolines") (e.g., Florida: Foose 1981; Barbados: Day 1983), in salt (Martinez et al. 1981) and gypsum terrains. The morphology of sinkholes was discussed at length in a conference in Florida (Beck 1984), their spatial dispersion by Vincent (1987), who compared the average distance of each sink to its nearest neighbor in a specific study area (New Guinea) to the expected distance assuming various types of random distributions. Hereby he compared a result already found by Williams (1972) indicating that the distribution is *not* random: this, according to the statements in Section 1.5.2.2, indicates an endogenic-tectonic predesign of the sink locations (presumably by a preexisting system of fractures). A further difference in distribution patterns could be found according to whether the sinkholes were the result of cave-ins or of solution.

The most important solution effects on land occur below the ground, in form of the genesis of caves. As a matter of fact, a whole science, called speleology, has been developed around the study of caves. General textbooks on speleology have been written, e.g., by Trimmel (1968) and by Ford and Cullingford (1972).

Qualitatively speaking, caves are generally long, winding passages, but they may also be in the form of big rooms.

Quantitative morphometric studies of caves are more difficult to make than of rivers, because cave systems are inherently three-dimensional. However, it is often sufficient to project a cave system onto a horizontal plane. In this fashion, Deike and White (1969) have compared cave ducts to surface meanders of rivers and have calculated prevailing wave lengths etc. This approach has been carried further by Brown (1973) to the actual application of spectral analysis. The topological structure of caves has been studied by Howard (1971) and again by Brown (1973). Cave systems are not bifurcating arborescences, like river-drainage networks, so that the internal connectivity becomes of importance. Appropriate parameters of connectivity have to be introduced.

Finally, karst forms have a relation to the prevailing climate, but in a subtle way that is not yet well understood (Sweeting 1980). In this connection, it is to be noted that the climate inside a karst cave may be quite different from that on the outside, as is evidenced by the numerous "ice caves" in middle latitudes. The history of ice development in such a cave (Coulthard Cave, Alberta) was studied by following the oxygen isotope ratio through the Pleistocene (Marshall and Brown 1974). Thus, it was found that temperatures remain negative throughout the year (in spite of a higher mean temperature outside), resulting in actual ice accumulation during warm, humid periods, and ablation (by sublimation) during cold periods. Only when the outside temperature was quite high at the height of interglacial warming (cf. Sect. 2.5) did all the ice melt.

1.6.2.4 Aqueous Deposition Effects on Land

Like solution effects, deposition effects occur on as well as below the ground: material is precipitated out of solution due to a decrease of solubility. The mechanism is the reverse of that in solution: since the solubility is tied to CO_2 or SO_2 content in the water, a loss of these gases will decrease the solubility. Other possibilities are a temperature decrease of the water which affects the solubility of many minerals, or an increase of solute concentration due to evaporation of the solvent water, to the point where material precipitates out.

On the ground, solute precipitation leads to the formation of various sorts of "mounds". Best known are spring tufa terracettes: sequences of steps (wave lengths of the order of 1 m) consisting of porous limestone precipitates (formed by the decrease of solubility owing to the escape of CO_2 from the flowing water) in the source region of a brook.

Underground, precipitation effects are well known from the formation of speleothems (conical forms) in caves: stalactites which hang from the ceiling, stalagmites which stand on the floor. The mechanism is similar to that of precipitation effects on the ground: the loss of the solution intermediaries CO_2 or SO_2 or the simple evaporation of the solvent water, all of which induce precipitation. Thus, such features occur in limestone and gypsum caves. Maximum growth rates of 0.01 mm/a are possible for stalagmites (Dreybrodt 1982). Speleothems are normally vertical, but asymmetries may also occur (Bender 1969; Reinboth and Göbel 1975).

Finally, precipitation effects can cause concretions and nodules within porous sediments.

1.6.3 Shorelines and Coasts

1.6.3.1 General Remarks

We turn now to a discussion of the physical geomorphology of coasts and shorelines. This has become a much-stressed subject as of late; books on it have been written recently by Bird (1969), Clayton (1979), Pethick (1984) and Trenhaille (1987).

First, we have to clarify the term *coast* somewhat, since the line between the water and the land (owing to tides, storms etc.) is not entirely definite. Thus, the *coast proper* is that part near the water's edge which *never* gets wet. The part which is affected by the action of the water (generally this may be above or below the water line) is properly called *shore*. A normal shore profile, in which the various features involved are explained, is shown in Fig. 5. The dynamic interaction between the water and the land takes place on the shore.

When talking about "coasts" or "shores,.it should be noted that over longer time spans, not even the *average* position of the water line is constant: the mean sea level has been changing in relation to land by substantial amounts even during a few thousand years. Such movements of the water line have been called eustatic movements; they will be discussed more thoroughly in Sect. 2.5.2.5 of this book.

The mapping of shore features has been done by traditional mapping procedures. However, recently remote sensing has also been used for this purpose. A comparison of shoreline mapping techniques has been made by Leatherman (1983).

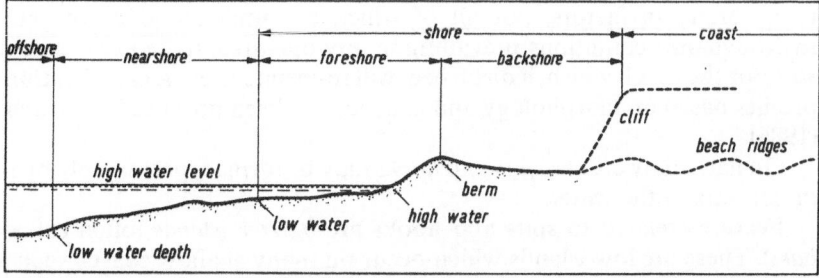

Fig. 5. Schematic diagram showing features usually associated with a normal shore profile. (After Alexander 1966)

From the prevalence of certain features on coasts, it is possible to attempt a classification of various types thereof (Inman and Nordstrom 1971). It turns out that there are several methods for doing this.

a) *Genetically*, one distinguishes between coasts of emergence and coasts of submergence (Johnson 1919). The meaning of these terms is self-evident. The phenomena of emergence and submergence of land are primarily due to endogenic causes, but fluctuations of sea level due to climatic changes may also be involved. Thus, Hovermann (1985) has attempted a classification of coasts on climatic conditions. More sophisticated schemes of genetic classification introduce also such terms as "advancing" and "retreating" coasts (Valentin 1952; Bloom 1965).

b) *Tectonically*, one distinguishes coasts of the Atlantic type, where the axes of the tectonic system near the coast intersect the latter at more or less rightangles, from the coasts of the Pacific type, where the tectonic axes are more or less parallel to the shore line (Richthofen 1901). This can be specified further by stating that Pacific-type coasts arise at (destructive) plate margins, Atlantic-type coasts (De Almeida 1975) at plate interiors.

c) *Morphologically*, a classification of shore lines can be made which is entirely independent of the genetic or tectonic interpretation of the surrounding area (Alexander 1962). Thus, the shore profile may be either cliffed or noncliffed, and in plan the coast may be regular or irregular. A characteristic difference in shore morphology also results from the type of material underlying it: friable sand or hard rock. Specific features also occur where a river empties itself into a large body of water.

In the following sections, we shall discuss some of the specific coastal features that may be encountered.

1.6.3.2 Friable Material Coasts

Turning first to coasts consisting of friable (soft) materials, we note that often in such cases a beach is formed which may consist of sand or shingle. It slopes gently towards the sea with somewhat of a step at the line where the waves commonly break. The sand in beaches appears to be graded, although no general empirical laws can be established regarding the grading. Beaches may change their shape due to a variety of factors, not all of which are understood as of yet. For the hydrodynamic conditions prevailing in any one area, there exists an *equilibrium state* for the beach which, if disturbed, will re-instate itself. A classification of beach profiles based on morphology and genetics has been proposed by Araya-Vergara (1986).

On flat, sandy coasts, *spits* and *hooks* may be formed, presumably by the action of currents in the water.

Features related to spits and hooks are *barrier islands* (often called *offshore bars*). These are low islands which occur on many shallow coasts, such as on the Gulf of Mexico. An extensive bibliography of the physiography of barrier islands (offshore bars) has been published by Shepard (1960). An extensive exposition of the classification of shallow-coast features, based on morphology, has been presented by Leatherman (1982). Offshore bars are often linked with the emergence of a shore line. Thus, there is an important tectonic (endogenic) element in the global distribution of barrier islands: they occur in greatest abundance where broad, low-relief coastal plains lie close to the inner shelf zone of the sea (Glaeser 1978). This is characteristic of coast lines that are far removed from plate margins.

Beaches are commonly found on coasts where the land *gently dips* toward the sea. On steep coasts, beaches may or may not form. The principal action of the waves in this case is one of undercutting the coastal slope. If the material is hard rock, *cliffs* are formed which gradually recede. This may occur in friable as well as in hard rock (cf. Sect. 1.6.3.3). In friable materials (e.g., clay), there is usually a beach at the foot of the cliff.

A classic case is that of a coastline consisting of a sequence of headlands and bay beaches. In fact, headland bay beaches are well studied geomorphic features (Yasso 1965); their geometry is best described by logarithmic spirals.

1.6.3.3 Hard Rock Coasts

In cliff recession on a hard rock coast, one notices that a shore platform consisting of solid rock develops at its foot rather than a beach. This is due to the specific mechanism of rock attrition by sea water (Bigelow 1982). The process of cliff development on a rocky coast is schematically shown in Fig. 6. On coasts of emergence, a series of former shore platforms may often be observed high above the present water level. Their morphometry is determined by the elevational distribution of wave action, particularly during storms (Trenhaile 1978).

Other actions of the sea are connected with the scooping out of bays (cf. Sect. 1.6.3.2) and, indirectly, by providing the right environment for corals to grow, resulting in the development of barrier reefs near a coast line.

The understanding of the evolution of reefs (cf. Newell 1972; Braithwaite 1982) is quite a complicated one, inasmuch as biological processes are involved which require special climatic (tropical) conditions. The actual reef tract facies and their geometries, thus, record interaction rates of reefgrowth, glacio-eustatic sea level fluctuations (cf. Sect. 2.5.2.5) and tectonism (Mesolella et al. 1970). The final relief of an uplifted reef terrace may resemble that of a karst plateau (Strecker et al. 1986).

Coastal Geomorphology

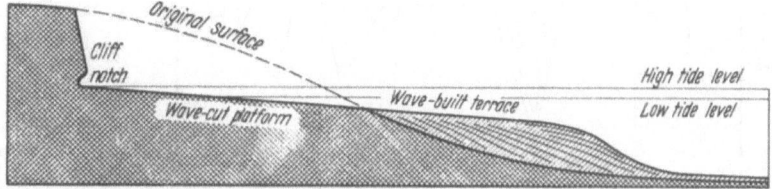

Fig. 6. Cliff recession and formation of a shore platform. (After Holmes 1944)

1.6.3.4 Morphology of River Mouths

In the vicinity of the places where rivers empty into the sea, the coasts are disturbed in a peculiar fashion. In such places, the steady trend of the shoreline is generally disrupted by either an *estuary* or a *delta* which represent the two principal types of river mouths. In an estuary, the sea forms a funnel into the land, and on a delta the land protrudes into the sea. A physiographic study of river mouths has been made, e.g., by Samoilov (1954).

Geomorphologists discern various regions on a river mouth. For the two basic types of the latter (i.e., deltas and estuaries), the various regions are shown in Fig. 7. The general river mouth region comprises the area between the first widening or division of the river channel and the underwater step some distance away from the mouth in the sea. Under certain circumstances, a river may give rise to a *land spit*.

Whether a delta or an estuary is formed depends on the presence or absence of lateral currents in the sea: if the currents in the sea are weak, the river will dump its load near its mouth and thus form a delta. Otherwise, the sediments introduced may be removed from the area and an esturary is the result (Wright and Coleman 1973).

Of the two types of river mouths, *deltas* have been studied much more than estuaries. The very term and concept of "delta" has been scrutinized and redefined by Moore and Asquith (1971) as the "subaerial and submerged contiguous sediment mass deposited in a body of water primarily by the action of a river". Modern deltas are mainly located in (1) intracratonic regions, (2) at rifted continental margins, (3) in marginal basins, and (4) at the boundaries of a craton with Mesozoic-Cenozoic mountain belts (Audley-Charles et al. 1977). It was observed long ago that two types of deltas are possible, which have been called *arcuate* deltas (example: Nile delta) and *bird's foot* deltas (example: Mississippi delta; cf. Ritchie and Brunsden 1973). The two types of deltas are illustrated in Fig. 8. The steeper the slopes of a delta are, or the smaller it is, the coarser are usually the materials of which it consists. Detailed studied of the deltaic environments of deposition have been reviewed by Coleman (1976). Standard profiles of modern prograding deltas have been measured by Matyas (1984), who

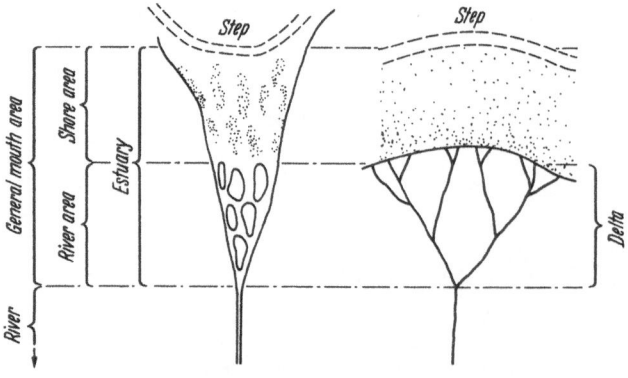

Fig. 7. Morphology of a river mouth. (After Samoilov 1956)

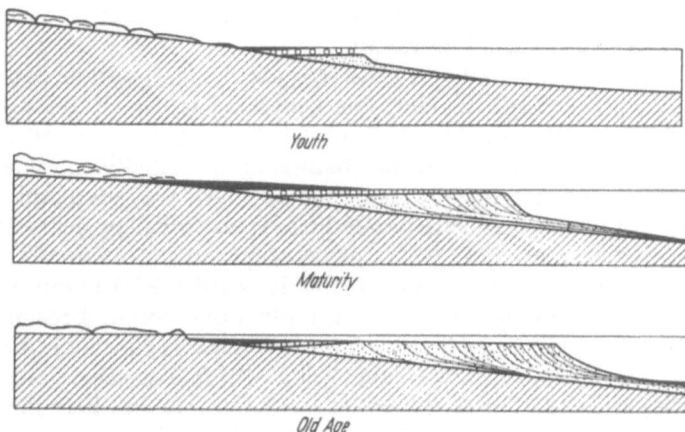

Youth

Maturity

Old Age

Fig. 8. Arcuate delta (*top*) and bird's foot delta (*bottom*)

found that the latter can be described by the equation

$$h = ax^b, \tag{1.6.1}$$

where h is the depth below the water at the distance x from the origin point of the delta. There is a linear relationship between b and log a; b is around 0.4–0.7.

Morphometric studies of *estuaries*, in contrast to those of deltas, are (as noted) much fewer in number. Fairbridge (1968b) has presented a classification of estuaries and has shown that there are essentially two types: funnel-shaped and barred estuaries. The morphogenesis of estuaries occurs through overdeepening of river channels. Primarily, this is initiated either by a rise of sea level or by subsidence of the land, but must subsequently be maintained by the depositional conditions at the river mouth, otherwise a delta is formed. The bed of a river estuary is similar to that of the river before entering the larger body of water: it contains wandering riffles and dunes (Castaing and Froidefond 1978: Gironde estuary). The sediment transport is essentially riverine.

1.6.3.5 Coast Line Changes

Above, we have treated coasts as static geomorphic features. However, it is evident that this is an oversimplification, inasmuch as coastal features, like all landscape features, are subject to change.

On the one hand, such changes are part and parcel of the normal landscape evolution: they occur slowly over a long period of time in consequence of the operation of the principle of antagonism. The endogenic processes effect a gradual change of the relative water level ("eustatic" changes: up or down), the exogenic processes are seen in the shore erosion.

On the other hand, coast line changes may occur in consequence of specific catastrophes: storm surges due to bad weather or due to a combination of bad weather and high tides, river floods (in deltas, estuaries) or tsunamis (earthquake-generated devastating waves).

Turning first to the "regular" coastal changes, we note that Ellenberg (1982) has set up a general classification, according to which different geomorphological processes take place on shallow, flat shores, on river mouths and on steep shores. On shallow shore lines, the changes are essentially continuous, at river mouths they occur rhythmically, and on steep coasts they occur episodically whereby the active phases may be separated from each other by years or decades.

Indeed, this proposed scheme seems to be essentially confirmed by many observations. Thus, Winkler and Howard (1977) found a regular gradational change on the Atlantic sea board plain of Georgia. Erosion rates on the North Shore of Lake Erie where the latter is formed by clay bluffs, were found to be cyclic, averaging about 2 m/a (Quigley et al. 1977). The same pattern was found on the clay cliffs near Scarborough on Lake Ontario: but here the slope retreat occurs sporadically (Bryan and Price 1980). Thus, in this case an "average" retreat rate is no longer meaningful, since the retreat occurs by sudden massive slumps where 15 m of shore can be "lost" in one event.

Evidently, the individual slides are triggered, if not caused, by occasional rainfall and storm conditions. These can act directly on the coastal land causing slope instability (cf. Chap. 3 of this book) or river flooding (Nichols 1977). More specifically, they can create storm surge waves in the adjacent sea or lake. Notorious floods occurred in this fashion near Hamburg in 1717 and 1825 (Schröder 1969, 1972) and in 1978 on the East Coast of England (Steers et al. 1979). Around the Pacific and in the Caribbean, hurricanes (typhoons) are known to cause storm surges that cause heavy damage (Tsuchiya and Yasuda 1980; Ueno 1981). The process is aggravated if the storm surge occurs concurrently with a high spring tide in the sea. In storm surges, the accumulating water masses affect also the tides of the solid Earth (Zschau et al. 1979). Solid Earth tide measurements can thus be used to predict the arrival of storm surges on a threatened coastal area (Zschau and Kümpel 1979). Finally, it should be mentioned that water surges, as "tsunamis", can also be the consequence of the occurrence of a marine earthquake (cf. e.g., Scheidegger 1975). All these occurrences cause catastrophic changes on coast lines. Thus, coast lines and coastal features are anything but static.

1.6.4 Subaqueous Geomorphology

1.6.4.1 General Remarks

The subaqueous areas of the Earth are those that are covered by bodies of water other than rivers. They include the bottoms of lakes, seas, and oceans. In this instance, lake bottoms show forms similar to the bottoms of the shallow parts of the seas and oceans. These areas can thus be considered together.

In comparison with the land surface of the Earth, the detailed morphology of the subaqueous regions is much less known. It is only recently that efficient exploration tools such as underwater cameras, coring techniques (e.g., Dietrich 1959; Menard 1959; Shipek 1960; Gibson 1960), and side-scan sonar (Prior et al.

1979, 1981) have become available for this purpose. The resulting facts have been summarized in monographs, e.g., by Shepard (1948), Kuenen (1950), Guilcher (1958), Defant (1961), and Weyl (1970).

The general large-scale morphology of the oceans has been known from soundings for a long time. Thus, adjacent to the coast, there is a shallow continential shelf in many instances, followed by a steep (continental) slope flattening out into an abyssal plain.

Accordingly, it will be convenient to consider the three characteristic areas (shelf, steep slope, deep sea regions) separately.

1.6.4.2 Shelves and Other Shallow Regions

Starting with the shallow subaqueous regions, we note that these show many features that have a great resemblance with subaerial features (Heezen 1959).

Near the mouths of rivers, the latter can cause large depositionary fans (Normark and Piper 1969) to be built up with a structure of meander channels (Mississippi fan: Kastens and Shor 1986) that appear almost as subaerial. Alternatively, submarine canyons can be formed (Shepard and Dill 1966; Laughton 1968; Rona 1970; Twichell 1982) that are also in appearance very similar to subaerial features, namely valleys that seem to be a continuation of the respective river courses. Consequently, there is a controversy regarding the formation of submarine canyons: whether they have indeed been formed subaerially during a period of low water level or whether they have been eroded in and by the water itself (Dietz 1958; Deluca 1968). However, the largest canyon known stretches over the entire Northwest Atlantic Ocean (Elmendorf and Heezen 1957); since it is impossible to conceive that the oceans at one time or another were empty of water to the extent necessary for such a feature to develop by subaerial erosion, subaqueous mechanisms of erosion and deposition must be assumed to exist.

The existence of such processes is further evidenced by the fact that shallow subaqueous regions show, in part, features similar to those found on river beds. Thus, underwater ripples, dunes, and ridges are common: in some instances, steady bottom currents may indeed be present on the shelf, but in others, the direction of the current may be fluctuating. Nevertheless, the outcome is a rippled appearance of the sea floor on which a spectral analysis can be performed (Clarke et al. 1983). Small scales are frequent (Lacombe 1960), but many characteristic dune features have wavelengths from 3 to 12 km (shelf off Cape Hatteras: Rona 1969a; Bermuda Rise: Embley et al. 1980). This is, of course, much more that what is observed in rivers, but the existence of subaqueous entrainment and deposition processes is thereby nevertheless indicated.

1.6.4.3 Steep Subaqueous Slope Regions

Steep subaqueous slopes are found primarily on the continental shelf edges. However, there are also other instances, like the front of a river delta deposit or the canyon walls on a shelf, in which such steep underwater slopes are found.

One of the main features of steep underwater slopes is that they, like subaerial slopes, are subject to wasting. Thus, mass movements do occur on them. In fact, there is an analogy between subaerial and subaqueous mass movement *mechanisms* based on the theory of effective pressure of Terzaghi (correspondence principle: cf. Scheidegger 1982b); thus, it may be presumed that the *phenotypes* of mass movements on land (cf. Sect. 1.2.5) have their counterparts on submarine slopes. However, observations of creep below water are difficult to make; events become obvious, though, if they are large and sudden, i.e., if a submarine *slide* occurs.

In fact, submarine slides are extremely common. Rona (1969b) made a survey of such slides along the North American Atlantic continental slope; Embley and Jacoby (1975) investigated the distribution of 14 major slides on Atlantic coastal margins. The best-studied submarine slide area is undoubtedly the front of the Mississippi Delta, on which Prior and Coleman (1978a,b, 1980) have written a series of papers. The general terminology and morphometry applicable to such events have also been suggested by Prior and Coleman (1979) which will be of importance in connection with the exposition of the *theory* of such features (Chap. 6.6.3).

The consequence of a subaqueous slide is that it may trigger a *turbidity current*. Such turbidity currents are far-reaching flows of turbid water, i.e., water mixed with sediment; they are thus currents of a dense fluid inside a less dense one. When they reach their end, the sediment falls out. In this fashion, they give rise to graded deposits. Typical grading curves are shown in Fig. 9.

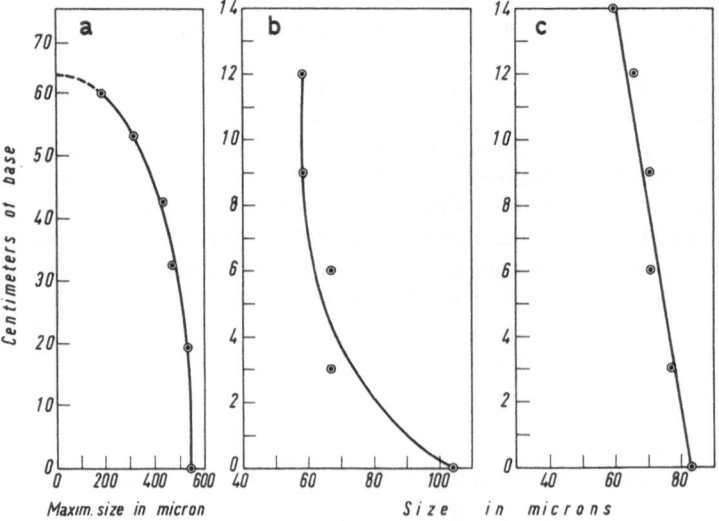

Fig. 9 a–c. Vertical grading curves in turbidites. **a** Martinsburg formation (Ordovician) at Hamburg, Berks Co., Pa., U.S.A., **b** Upper Broverian turbidite (Dangeard, et al. 1961 sample R2, Table 1). **c** same as **b**, but sample R1. (After Scheidegger and Potter 1965)

Turbidity currents have been observed directly in Lake Mead (Gould 1951) and indirectly (inferred from submarine cable breaks) on the Grand Banks of Newfoundland (Heezen et al. 1954) and in the West New Britain Trench (Krause et al. 1970). They have also been observed *directly* in the ocean (Heezen 1958). They occur, for instance, quite frequently off the mouths of great rivers, such as the Amazon (Damuth and Embley 1979), the Congo, the Rio Magdalena (Heezen 1956) and the Rhine in Lake Constance (Lambert 1982). These turbidity currents are rather episodic, spectacular events. However, currents of speeds of little more that 0.5 m/s are also quite common, primarily in submarine canyons (Shepard et al. 1977). Over a long period, these slow currents have a substantial sedimentological effect on underwater slopes and adjacent regions.

1.6.4.4 Deep Sea Regions

It turns out that the ocean bottom exhibits a rather varied topography. Excluding discussions of features that are presumably caused by endogenic processes such as mid-ocean ridges, we note that some of the large-scale features affected by exogenic agents are the sediment-covered *abyssal plains*. In their original state, they show a hilly relief of roughly 300 m undulation (particularly in the Pacific), but in many parts of the world they are covered with sediments. Arrhenius (1952) quotes sedimentation rates from 1 mm to several cm per 1000 years. In ponded areas, rates are much greater (Van Andel and Komar 1969). The average depth of submersion of abyssal plains is 5000 to 6000 m. The sediment cover may average 600 m in thickness over the deep areas of the ocean floor (Elmendorf and Heezen 1957); ripple marks are common (Lacombe 1960; Boggs 1974). On the mid-ocean ridges, sediment covers are much less, viz. of the order of 100–200 m (Ewing et al. 1964). The contrary to sediment deposition is erosion (McCave 1982; Heezen and Rawson 1977). In deposition, as well as in erosion, the mass transfer is caused by existing bottom currents (Boggs 1974) which carry suspended particulate loads in the form of turbid water (Eittreim et al. 1976; Biscaye and Eittreim 1977; Silverberg and Sundby 1979).

1.7 Glacial and Periglacial Morphology

1.7.1 Introduction

We turn our attention now to features that are due to the action of snow and ice. The complex of geomorphic features resulting from such action is called *glacial* or *periglacial* morphology and has been the subject of many investigations. Books on it have been written, e.g., by Eyles (1983), Semmel (1985), and Drewry (1986).

The action of snow and ice manifests itself in a variety of ways. Snow, if it accumulates, consolidates eventually into ice, which may form giant streams proceeding downhill in a manner similar to that of rivers. The giant moving masses of ice are called *glaciers*. We shall describe (in Sect. 1.7.2) some features of the morphology of snow and ice accumulations as this is important for the understanding of the geomorphological action of these materials.

The effect of glaciers upon the appearance of the Earth's surface occurs in two ways: by erosion and by deposition. The erosional activity of glaciers manifests itself in the scouring out of valleys and hollows in a special manner, the depositional activity in the dumping of moraines and similar features. Both types of glacier activity will be discussed in some detail (Sect. 1.7.3).

Next, we shall discuss some features that arise because of the meltwater effects under and near glaciers. These are the fluvioglacial effects (Sect. 1.7.4).

Finally, we shall discuss some peculiar features which also have been ascribed to the action of ice: niveal soliflunction, the upheaval of pingos, and the formation of pressure ridges. These features are peculiar to specific areas of the world as they require the presence of permafrost in the ground. They will be described in Sect. 1.7.5.

Before starting with a description of the various geomorphological features caused by ice and snow, it may be well to note that the abundance of these substances upon the Earth's surface has been much greater during certain earlier periods during the Earth's evolution than it is at present. Such periods are called *ice ages*. Thus, glacial features may be found in areas that are now far removed from glaciated regions.

There is a large literature on the effects and causes of these ice ages. They will be considered more fully in connection with the exposition of the theory of climatic fluctuations (Sect. 2.5) in this book.

1.7.2 The Snow and Ice Cover

1.7.2.1 General Remarks

Under cold weather conditions, precipitation falls from the sky onto the ground in form of the solid phase of water. This occurs generally as snow, which consists of tiny crystals of ice, usually arranged in a complex dendritic hexagonal pattern (Nakaya 1954). On occasion, the ice particles may be of a larger size; one then speaks of hail stones.

On the ground, the snow may melt immediately or else, if the ambient temperature is low enough, stay in form of a snow cover. The latter forms first a loose, porous layer, which, owing to the action of sunlight and percolating water (Schommer 1977), is slowly transformed into a more dense material, until, after recrystallization, it presents the aspect of solid ice.

1.7.2.2 Snow

The snow cover, particularly on slopes, forms a mantle that is subject to displacements. Of particular importance is the case where a part of the snow cover detaches itself from the substratum and moves downslopes as a rapid *avalanche*.

The dynamics of avalanches is a subject unto itself which is of importance in connection with natural hazards and their mitigation (Scheidegger 1975). As geomorphic agents, avalanches are important as carriers of assorted sizes of rock materials on debris slopes (Gardner 1970) and as initiators of erosion processes (Gardner 1983).

1.7.2.3 Ice

As geomorphological agent, ice is more important than snow.

Ice occurs as a sheet cover on the sea or on the ground, or it collects in hollows and valleys in which, owing to its creep properties, it moves very much like a slow river. In all these cases, one speaks of *glaciers*.

Turning to the morphological classification of glaciers, we note that one generally (Holmes 1944) discerns three classes of such features.

The *first* class contains ice sheets and ice caps covering large continental areas such as Greenland and Antarctica (Benson 1969; Denton et al. 1984). The ice creeps slowly towards the margin of these features. Ice sheets may be up to 3 km thick. At the water's edge, ice sheets push out to the sea where ice shelves are formed: icebergs calve off them. Specific processes occur on a glacier-dominated coast (Molnia 1985): beach features may be ice-formed (Dionne and Laverdiere 1972). During the ice ages, ice sheets existed in many now ice-free areas.

The *second* class contains mountain or valley glaciers. These have again been split into subgroups which represent glaciers of the first and of the second order, respectively. Glaciers of the first order represent the true valley glaciers which are veritable rivers of ice extending for large distances along a mountain valley. On their surface, longitudinal, transverse, and marginal crevasses may arise. The longitudinal crevasses are roughly parallel to the direction of flow, transverse crevasses cut across a valley glacier wherever a steepening of its course (in the extreme up to the formation of an ice fall; cf. Hambrey and Milnes 1977) occurs, and marginal crevasses may point upstream from its rim. Rhomboid parallelogram patterns are common in the upper part of a valley glacier (Waag 1979); at its end there is commonly a snout. Glaciers of the second order are those that are cut off from the main valley and thus hang from its side, unable to reach the outflow of the drainage area. These features include cirque glaciers.

Finally, the *third* class of glaciers contains piedmont glaciers which are formed by the coalescence when several valley glaciers reach a plain at the foot of a mountain range.

Over decades and centuries, glaciers advance and retreat. Thus, many studies have been made of the (thermal) energy and mass balance of individual glaciers. More general investigations of the thermal and the mass balance problem in glaciers have been published by Lliboutry (1974) and by Vischer (1988).

The normal flow velocities of glaciers are in the Alps 20–150, in the Himalayas 700–1300 and in Greenland 1000–3000 m per year (Neumayr and Suess 1920). In fact, these velocities are by no means constant, but fluctuate substantially (Iken 1974). Moreover, at times glaciers are subject to sudden "surges" (Clarke et al. 1986) in which (during days or weeks) the normal velocities are greatly increased, up to a factor 100. Thus, for instance, the Muldrow Glacier in Alaska suddenly surged with a velocity of 20 m/day (Post 1960). Surges have the consequence that the glacier tongue may become unstable so that ice avalanches are created (Alean 1984). Other morphological effects of surges are push structures, erosional forms, and drainage changes (Johnson 1972). Surges have not only been observed on valley glaciers, but also on continental ice caps (e.g., Barnes Icecap, Baffin Land: Holdsworth 1973).

1.7.3 Geomorphological Effects of Glacier Motion

1.7.3.1 General Remarks

Glaciers move and cause thereby a variety of geomorphological effects.

In the first place, these effects are due to erosion and scour on the bed (Sect. 1.7.3.2). In glaciated valleys, a characteristics longitudinal and transverse profile is produced. In catchment areas and hollows the erosion of glaciers produces glacial cirques. On the bed, the erosion produces roches moutonnées on a rocky and drumlins on a friable substratum.

Erosion, however, is not the only result of glacier movements: the latter can also induce an accumulation of material to occur (Sect. 1.7.3.3). Thus, subglacial till is formed and moraines and kames are built up at the edge of glaciers.

1.7.3.2 Erosional Effects of Moving Glaciers

The principal effect of glacial erosion in a valley is the development of a step-shaped longitudinal profile. The situation is described in most textbooks of geomorphology (see, e.g., Machatschek 1952) and has been analyzed in great detail by Gerber (1956). It is illustrated in Fig. 10. It appears that at least some of the steps were present before the existence of a glacier and that the glacier accentuated the structure (Trenhaile 1979).

A second effect due to the scouring of a glacier is the characteristic transverse profile of a valley that was carved out by the glaciers of the ice ages. Such valleys are generally U-shaped, best described by a parabola (Wheeler 1984) with broad shoulders (see Fig. 11). There is thus a general thought that glacial valleys are

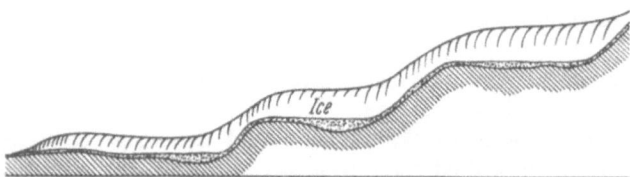

Fig. 10. Step-shaped longitudinal profile of a glaciated valley. (After Machatschek 1952)

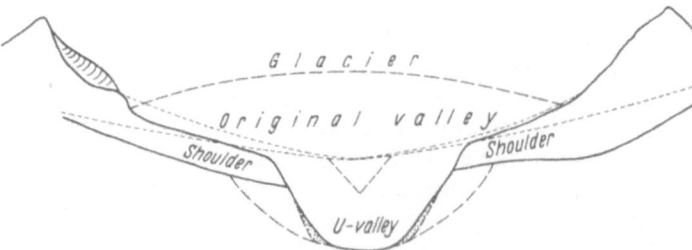

Fig. 11. Cross-section through a glaciated valley. (After Machatschek 1952)

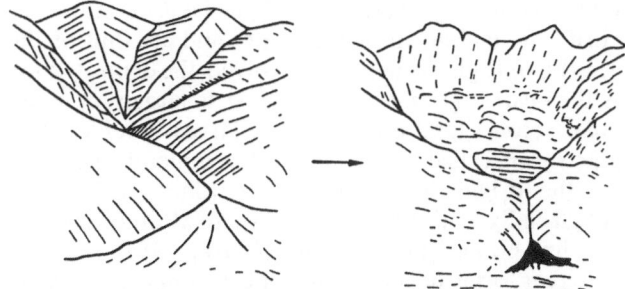

Fig. 12. Transformation of a drainage basin into a glacial cirque. (After Machatschek 1952)

U-shaped, fluvials valleys V-shaped; this inference, however, may well be wrong (cf. Sect. 7.3.4.5).

A further effect due to the scouring of glaciers is the transformation of an erosional basin into a *glacial cirque* (see Fig. 12). This type of transformation occurred many times in the Alps during the ice ages, mainly at the then existing snow lines (Trenhaile 1975b, 1977) and has been well documented. This transformation, however, may be partly stressed-induced (Gerber and Scheidegger 1969). The size of a cirque seems to depend on its position in the valley (Trenhaile 1976; Vilborg 1984), being the larger the farther from the valley head it is located.

On a smaller scale, scouring effects are seen on the bed of a glacier. If the latter consists of bare rock, striations and similar features are the most obvious indicators of scour (Laverdiere et al. 1985; Lassila 1986). However, due to instabilities in the ice flow, larger hummocks can also be formed which are called *roches moutonnées*. They are elliptical bodies with their long axes oriented in the ice-flow direction, the upstream face being less steep than the downstream side. The ratio of long to short axes varies from 2.0 in igneous rocks to 4.9 in sedimentary rocks. In part, roches moutonnées may have been predesigned by the neotectonic joints present in the glacier bed (Bär 1957).

In soft, friable materials, the erosive action of moving ice produces bed instabilities in the form of *drumlins*. These are features which have been described by a variety of authors; a bibliography of the pertinent recent literature on drumlins has been published by Menzies (1984). Accordingly, one may regard as a typical drumlin a hill of glacial bed material which approximates the form of an elongated ovoid (Alden 1905; Mills 1980). In plan, they are best represented by lemniscate loops (Chorley 1959; Komar 1984). The highest point coincides with the center of the greatest breadth. Chorley (1959) made basically visual observations; Komar (1984) extended these to more quantitative studies and found that about one-half of the drumlins did indeed fit lemniscate loops, but an equal number is more closely approximated by ellipses. Inside a drumlin there are inverse-graded units of material (Menzies 1986). Drumlins occur commonly in fields (Trenhaile 1975; Boots and Burns 1984).

1.7.3.3 Depositional Effects of Moving Glaciers

The principal depositional effect of moving glaciers is that they carry with them loose material which is deposited eventually. The material carried by and deposited from glaciers is called glacial *drift*. However, the stratified forms of this "drift" are laid down by meltwater and are thus of "fluvioglacial" origin (cf. Sect. 1.7.4); if only glacial action is involved, the deposited material is not stratified and it is then called glacial *till*. Till consists largely of mixtures of sand, silt, clay with coarse stones, and boulders. The largest boulders transported and deposited by glaciers are called *erratic blocks*. The mineral composition of till materials depends on the origin and corradability (Shilts 1984). The final deposition occurs subglacially as a three-component mixture of water, coarse-grained clast solids, and fine-grained matrix solids (Clarke 1987).

Other forms of subglacial deposits appear as crevasse-fill ridges (Sharp 1985a) and as till wedges (Amark 1986). In the former, ice crevasses are filled with debris which are left on the ground upon melting, in the latter, deposits are formed in the fractured bed of an active glacier.

The principal depositional effects due to moving ice masses, however, occur at the ice edge. Thus we note the existence of *moraines* which are symmetrical heaps of debris that are left by glaciers after their retreat. Most conspicuous are terminal moraines consisting of material that a glacier accumulates on its tongue and leaves in an arcuate position. Terminal moraines often cause the formation of a lake. Of interest are also longitudinal moraines which may have been left along the walls of a formerly glaciated valley, creating several parallel valleys. Moraines, in fact, are ice-push features which become gradually built up by the continued advection of debris by the moving glacier which then melts off in the edge zone (Kälin 1971; Rogerson and Eyles 1979; Boulton 1986). If the debris accumulations are asymmetric, one speaks of *kames*.

Finally, features which are also due to glacier motion are those described as ice-thrust features by Charlesworth (1957) Flint (1957), and Mackay (1959). These represent tilted, folded, and contorted sediments and bed rock with truncated layers, thrust faults, closely spaced shear planes, clastic dykes and ice-pushed caves (Schroeder et al. 1986). They also include ice-pushed pressure ridges which are conspicuous features in plains of Europe and northwestern Canada (Rutten 1960). They are structures of considerable dimensions. Their height may reach from 50–200 m above their surroundings and their linear extent may attain 100 km. They are an indication that a glacier may be able to produce considerable forces when it is moving over an area.

1.7.4 Glaciohydrological Effects

1.7.4.1 General Remarks

When glacial meltwater is involved in the shaping of landscape features, the latter have different characteristics from those that are formed directly by the moving ice. They are comprised under the term fluvioglacial effects.

There are usually two domains of meltwater action: either beneath the ice (subglacial features; Sect. 1.7.4.2) or outside the ice (periglacial features; Sect. 1.7.4.3). A more distant meltwater effect occurs in periglacial lakes and seas, leading to glaciolacustrine and glaciomarine phenomena (Sect. 1.7.4.4).

1.7.4.2 Subglacial Meltwater Effects

One of the best-known subglacial meltwater features are *eskers*: they are formed by sediment deposition in meltwater tunnels along the beds of ice masses. Such tunnel systems were quite extensive in many areas during the Pleistocene ice ages. After the ice retreated, the sediment-filled tunnels remained as winding gravel walls, as much as 30 or more meters high and stretching in trunk and tributary patterns for many tens of kilometers (Price 1973).

1.7.4.3 Periglacial Runoff Effects

Runoff effects of glacial meltwater occur first of all on the ice surface itself where sinuous channels are formed (Ferguson 1973). Because such features are quite ephemeral, their geomorphological significance is rather limited.

Of more interest in the context of our subject are the glacial outwash rivers at the edge of an ice mass: these are generally braided and of low sinuosity (Ashworth and Ferguson 1986). The braiding is due to the extreme variation of the discharges, so that the glacial runoff river is most of the time unable to transport more than a small fraction of the available outwash material, leading to the formation of bars and therewith braids.

The glacial outwash consists often of very fine material. Being water-deposited, it is stratified (Kolstrup and Jorgensen 1982), possibly cross-bedded.

Furthermore, the asymmetrical accumulations at ice edges which we had earlier referred to as *kames*, and which we have assumed in Sect. 1.7.3.3 to be ice-deposited, may in fact also have been affected and partly deposited by meltwater runoff from the glaciers.

1.7.4.4 Glaciolacustrine and Glaciomarine Phenomena

Finally, we are looking at some effects that arise when glacial meltwater enters a large body of water, i.e., a lake or the sea.

Into this category fall ice-contact deltas (Shilts 1984) and hollows formed by the presence of stagnant ("dead") ice in the lake bottom ("kettle holes"; cf. Larocque 1985).

The sediment input into the glaciolacustrine (or glaciomarine) environment is highly variable, temporally and spatially; this fact is explainable in terms of the thermal regime and behaviour of the source glaciers (Powell 1984; Sharp 1985b).

On an annual basis, the sedimentation of fine materials (clays) varies so as to produce *varves* (e.g., Ignatius 1958): these are sedimentary laminae which are often observed in postglacial basins, such as the Baltic Sea. In a vertical section, dark- and light-colored bands alternate and the contention is that these bands correspond to yearly cycles of deposition, like the rings on a tree.

It has been observed that, in varved clays, the yearly sedimentation rate decreases upwards. Sedimentation curves (annual rate *versus* height in years above base) can be constructed; typical curves have, for instance, been published by Ignatius (1958). On semilogarithmic paper (abscissa: logarithm of varve thickness;

ordinate: height or years above base of core) the latter are generally straight lines. Evidently, these observed features require a physical explanation.

1.7.5 Ground Freezing Effects

1.7.5.1 General Remarks

With regard to periglacial features, it remains to discuss the effects that are caused by ground freezing phenomena.

Soil, being a porous medium, shows a peculiar behavior when the water freezes in its pores. Since water expands on freezing, large pressures and stresses are built up in the ground which cause the latter to undergo displacements leading to solifluction, to crack patterns and to frost heave. Finally, rocks on a slope affected by frost action move downhill in form of a rock glacier.

We shall discuss these features in detail.

1.7.5.2 Permafrost and Solifluction

The first feature to be discussed in that of *niveal solifluction*. Under arctic conditions, it is observed that soil may be slowly creeping downhill on a slope. This is called solifluction. The presence of frost action is essential for this phenomenon to occur. It has been described in the literature on many occasions (cf. Holmes 1944). Movements of the soil are particularly noticeable in spring. Williams (1957), in an investigation of the solifluction phenomenon, measured displacements of 20 cm in one spring season on a slope of slope angle of 19°. The movements reached to a depth of 75 cm. Niveal solifluction should be distinguished from *aqueous solifluction*, which also refers to soil creep, but without the presence of frost action in water-logged soils. The two phenomena, although superficially similar, are genetically different.

The result of the action of solifluction produces some characteristic phenomenological features. Thus, frost action has been observed to produce involuted structures in sediments (French 1986; Kolstrup 1987b) and other rapid mass movement forms (Johnson 1984). Clasts are oriented nonuniformly in various parts of solifluction lobes (Nelson 1985). On a larger scale, the freeze-thaw activity at snow patch sites (Nyberg 1986) and above buried glacier ice (Haeberli and Epifani 1986) makes it possible to map the latter from air photographs.

On a very large scale, Buedel (1977) held a frozen layer ("ice rind") responsible for the large, excessive valley erosion observed in such places as Spitsbergen. This view was extended by Späth (1986) to the explanation of periglacial mass wasting generally, leading to *stepped slopes* in such once-glaciated areas as upper Bavaria.

1.7.5.3 Patterned Ground

In niveal regions, one can often observe strange polygon patterns on the surface. Their largest diameters are usually between 1 and 2 m. Such polygons are cracks in the ground; sometimes they manifest themselves simply by a consistent sorting of the surface gravel or by the arrangement of larger stones along lines. A description of such features as subniveal phenomena was given for instance, by Furrer (1965); a summary of the literature on them was made by Dylik and Maarleveld (1967).

As noted, the genesis of patterned ground features is generally sought in some process connected with freezing and thawing. Vitek and Tarquin (1984) leave the exact mode of genesis of the polygons open, but Hauner (1985) thinks that the swelling of the underlying material due to humidity in summer and the crystallization of ice granules in winter is of morphogenetic significance. Furthermore, Abt et al. (1971) assume that an uneven drainage in the underlying frozen soil causes the polygons. In permafrost regions, the mechanism usually invoked is ice-wedge growth in cracks (Worsley 1986a; Rapp et al. 1986; Kolstrup and Mejdahl 1986; Kolstrup 1986, 1987a).

However, polygonal-patterned ground features need not exclusively be of glacial origin. Rossbacher (1986) gives a table of the statistics of such features, some of which are definitely not due to glacial effects, but to thermal contraction. The existence of desiccation polygons in drying clay is well known. Elbersen (1983) presents some patterned ground features from Texas which he assumes to be caused by an unstable erosion pattern, and Kelletat (1985) shows surface patterns in Utah that developed by rainstorm erosion.

Thus, whilst many ground polygons, particularly in Arctic or high-altitude regions, are undoubtedly connected with frost action, conditions producing them may also be present in non-niveal climatic conditions.

1.7.5.4 Frost Heave Phenomena: Pingos and Palsas

Next, we shall discuss some types of geomorphological features which are presumably caused by frost heave action. A general review of the terms applying to such features has been given by Worsley (1986b).

The first of these features are *pingos*. Pingos, sometimes also called *hydrolaccoliths*, are ground-ice mounts which occur in areas where permafrost is prevalent. They have been described, e.g., by Sharp (1942), Pihlainen et al. (1956), Maarleveld (1965), Mackay (1963), and, in a very extensive study by Müller (1959). Pingos have the appearance of more or less regular cones up to 100 m in height with a crater-like depression in the center. Their body contains an ice lens. Tensional cracks on the side of the cones indicate that pingos were formed by some kind of diapirism. Evidence of fossil pingos (from the last ice age) has also been discovered (Wiegand 1965; Slotboom 1963; Marsh 1987). They have also been found on the bottom of the Beaufort Sea (Gendzwill 1983).

The second of the frost heave phenomena to be discussed here are *palsas*. These are peaty permafrost mounds, i.e., hummocks rising out of a bog (Seppälä 1972; Washburn 1983). They contain a core of ice or of frozen peat or silt (Seppälä 1986) which can survive the summer heat. There is, in fact, a gradual transition from palsa to pingo, depending on the fraction of organic to inorganic material contained in it.

1.7.5.5 Rock Glaciers

Finally, we consider the phenomenon of rock glaciers. A good review of the present state of the knowledge on these features has recently been given by Haeberli (1985).

Rock glaciers are masses of ice-rich, frozen talus that move downslope (King et al. 1987). Their average (over a year) creep velocities are between 0.05 and 3.57 m/a (Kerschner 1976), and are almost constant over decades, centuries, and even millennia. Substantial short-term variations of the surface speed do exist, however; they are correlated with the seasons of any one year. Because of their mobility, rock glaciers present a considerable engineering hazard (Giardino and Vick 1985).

In plan, rock glaciers may be (Wahrhaftig and Cox 1959) lobate (length about equal to width) or tongue-shaped (length much greater than width). In the Alps, their extensions are of the order of some hundred meters up to a kilometer or so. Their maximum thickness is up to 50 m (Giardino and Vick 1985).

1.8 Aeolian and Desert Morphology

1.8.1 Introduction

We consider finally some geomorphological features which are characteristic of the arid and semiarid regions of the world.

Inasmuch as *wind* is a powerful force under desert conditions, it is convenient to start our review with a general discussion of "aeolian" (= wind) effects: these occur mainly, but not exclusively, in arid regions.

Next, we turn our attention to some specific desert features. Whilst wind is a powerful agent in arid regions, it is by no means the only one. Infrequent but heavy rainfalls, high rates of evaporation and extreme temperatures combine to produce the typical morphology of a "desert land-scape" (see, e.g. Mabbutt 1977, for a general review).

Finally, we note that the transition between desert and humid conditions cannot be an abrupt one: it is represented by semiarid conditions which manifest themselves in semidesert features. These also merit a review.

1.8.2 Occurrence of Effects Due to Wind

1.8.2.1 General Remarks

The force of the wind can have great geomorphological significance. This occurs mostly in the dry areas of the world because wind forces have little room for attack on the soil if the surface of the Earth is covered by plants. Wind effects can also become significant if a plant cover has been prevented by some means other than lack of precipitation. This may have occurred because of human activity, because of wave action on an oceanic coast, or because of the severity of the climate in high mountain regions.

The action of the wind consists in the transportation of loose dust or sand particles. It is important to distinguish between "dust" and "sand"; the former is lifted by the turbulence of the air like suspended sediment in rivers, whereas the latter is dragged along like the bed load in a water stream. In the two cases, the morphological effects are entirely different.

Considering first the movement of sand, we note that the latter is particularly significant in *deserts*. Sand ripples, dunes, and corrasive features are caused in this fashion. We shall discuss the morphology of such features in some greater detail in Sect. 1.8.2.2.

We shall then proceed to a discussion of dust movement. Soil erosion and the deposition of loess are probably the best known effects thereof. This will be described in Sect. 1.8.2.3.

Wind-borne features are also created after a volcanic eruption: Material is ejected from the volcano and then travels through the air. We shall discuss this in Sect. 1.8.2.4.

1.8.2.2 Movement of Sand

Blowing wind can pick up and transport grains of sand over large distances, depositing them again at some other place. The physics of blown sand has been studied extensively by Bagnold (1941); the theoretical aspects of his work will be reviewed in Sect. 8.2.2 of this book. At this juncture, we discuss only the phenomenological effects of the movement of sand.

Thus, winds of varying strength and direction may cause *ripples* on a sand surface. These ripples are entirely ephemeral; they change their size and direction with the wind that happens to be blowing. They are distinguished from larger features not only by their impermanence, but also by the fact that the fine sand particles are always found in the troughs between the ripples, the coarse particles at the crests. This type of particle size grading is reverse to that observed in larger scale features. The ratio of height to wave length in ripples ranges from 1:10 to 1:70 (cf. Bagnold 1941), depending on the range of the grain size distribution of the sand; the ratio of height to wavelength of the ripples increases with the range of the grain size distribution of the sand. If ripples become large, they are called *sand ridges*.

The above-mentioned features answer to the criteria of impermanence and sorting of the fine grains into the troughs, coarse grains onto the crests. With the so-called large-scale features, the reverse is true. These large-scale features are represented by *dunes* and by modifications of dunes. There are essentially two types of dunes, which have been called *barchan dunes* and *seif dunes*. Barchan dunes are crescentic mounds of sand which may occur singly or in groups, the horns of the crescents facing leeward (referring to the prevailing wind direction). The part of the dune between the horns on the leeward side is steep and represents a *slip face*. A schematic drawing of a barchan dune is shown in Fig. 13. The height of barchans may vary. Barchan dunes are observed to progress ("wander") downwind. Their speeds are variable and may be related to their heights. Some measurements have been reported by Breadnell (1910) whose results (for five dunes in the area of the Karga Oasis in Egypt) are shown in Table 2.

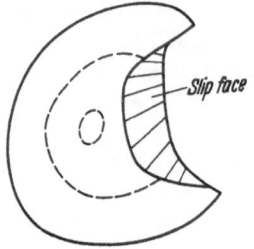

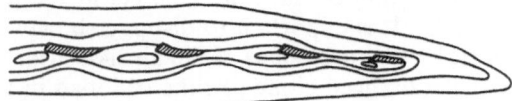

Fig. 14. A chain of seif dunes. Note the slip faces

Fig. 13. Barchan dune: view in plan

Table 2. Height (H), total advance (L) in one year and speed (C) of five barchan dunes in Egypt. (After Breadnell 1910)

H(cm)	L(cm)	C(cm/s)
2000	1090	3.46×10^{-5}
1700	1080	3.42×10^{-5}
1100	1620	5.14×10^{-5}
1050	1880	5.96×10^{-5}
400	1840	5.83×10^{-5}

As noted above, in addition to barchan dunes, one may observe seif dunes (Fig. 14). These are longitudinal features which do not show the crescentic shape of the barchans. It may be that a barchan dune can develop into a seif dune if it becomes asymetrical (cf. Sect. 8.2.3.5). Holmes (1944) states that seif dunes commonly occur in parallel ranges of immense length (see Fig. 14).

Modifications of the two basic types of dunes have been classified as aklé or star dunes (if the two horns of a barchan dune meet and coalesce; cf. Besler and Gerster 1985), as sand levees (or whale-backs: low connecting ridges between high dune chains; cf. Bagnold 1941) and as draas (huge stationary immobile longitudinal dunes; cf. Besler and Gerster 1985).

Incidentally, it may be noted that it is also possible to find "'fossil" dune fields, inasmuch as the climate has changed over the ages. Such fossil dune fields have been reported, e.g., from Nova Scotia (Hubert 1980) and South Africa (Thomas 1986).

Finally, one may note that blowing sand can act as a corrasive agent even in hard rocks in a desert. Experimentally, Sharp (1980) has found that abrasion rates of about 0.1 mm/a can be attained on bricks. Sweeting and Lancaster (1982) described wind-erosion forms in the Namib Desert. The most spectacular wind-erosion features, however, occur in more friable terrains, such as unconsolidated silts in Iran: ridges 200 m high above the scoured-out troughs. They have been called *yardangs* (Hedin 1903; Blackwelder 1934).

1.8.2.3 Dust Movement

The desert features discussed in the last section (1.8.2.2) are caused by the movement of *sand*. The finer particles, commonly called *dust*, also occur in deserts but they do not seem to be prominently involved in the formation of desert features.

However, the movement of dust has great significance in the plains areas of the world. It may become significant in two possible manifestations: the first representing removal of material and the second deposition.

The removal of dust from the surface of the earth is particularly evident where the phenomenon of *soil erosion* occurs. If the plant cover is destroyed by some means (such as human activity), the Earth is laid open to attack by the wind, and great quantities of material may be removed. Dust removal, of course, also occurs in deserts but it is not so conspicuous there.

The *deposition* of dust may lead to a very characteristic phenomenon: the deposition of loess. It matters little whence the dust was taken, from a desert or from the rock flour of glacial deposits (cf. Whalley et al. 1982) the result of massive deposition in a plains area is in each case a *blanket of loess*. The dust is trapped by the vegetation that is present; each year the blanket of loess grows a little higher as new dust is deposited. The loess blankets may stretch over immense areas. A recent review of the loess phenomenon has been given by Pye (1984).

1.8.2.4 Wind Transport of Volcanic Materials

During a volcanic eruption, large amounts of material may be ejected into the air. There are then two types of aeolian mechanisms by which this material may be transported and deposited (Murai 1961).

The first is that of the *ash flow* represented by nuées ardentes. Such nuées represent downward-rolling hot clouds; a particularly notorious example of a nuée ardente occurred in 1902 after an eruption of Mt. Pelée on the island of Martinique and destroyed the city of St. Pierre. A classical description of nuées ardentes has been given by Lacroix (1904), who visited Martinique after the eruption of Mt. Pelée. Nuées ardentes are also thought to give rise to welded tuffs (cf. Smith 1960; Ross and Smith 1961). Ash flows may travel a distance of up to 120 km from their source, which represents a considerable distance. It thus appears that nuées ardentes are extremely mobile and can travel over large distances. Velocities of nuées ardentes have been estimated to reach up to 130 m/s (during the Mt. Pelée eruption the distance of 8 km from the crater to the city was covered in somewhat less than a minute), their length 1 km and their height 4 km, where, however, perhaps only a very small fraction of the total height contains coarse material. A similar spectacular mobility of ash flows has been observed in Alaska (Miller and Smith 1977) and near Mount St. Helens (Walker and McBroome 1983), albeit in the latter case the temperature was lower so that the cloud was not incandescent. It has also been purported that ash flows can erode veritable U-shaped channels (Fisher 1977).

The second type of mechanism is the *ash fall*. Ash falls are the products of aeolian differentiation wherein particles are segregated by their fall velocity as they are transported downwind. Ash fall deposits tend to have an exponential decay of thickness downwind and a corresponding decrease in grain size. A review of the available observations may be found in papers by Fisher (1964) and by Eaton (1964).

1.8.3 Further Specific Desert Features

1.8.3.1 General Remarks

In addition to landscape features that are caused by blown sand and dust (and which are not only found in deserts), there exist some features that are specific to desert conditions. For the latter to occur, the scarcity of precipitation and the high evaporation rates are significant.

On the *surface* of the Earth, such conditions produce some typical small-scale phenomena: desert varnish and lateritization. Furthermore, a high evaporation rate is conducive to the formation of evaporites and desiccation cracks.

On a more massive scale, characteristic island mounts have been found in many deserts. In some fashion, these are also in part the result of the effect of the climatic conditions prevailing in arid regions.

1.8.3.2 Small-Scale Surface Phenomena in Deserts

In a desert, pebbles and rocks on the surface are generally covered by a thin film of a lacquer-like patina (*desert varnish*: Dorn and Oberlander 1982). The patination process is probably linked to dew and capillarity (Fairbridge 1968a): the soft halides are rapidly removed by aeolian abrasion, whilst the durable iron and manganese compounds remain and combine to form the lacquer (Engel and Sharp 1958).

Sometimes, the ground surface in deserts (and in other tropical regions) is covered by a crust thicker than is represented by desert varnish: iron is leached out below the water table and precipitated in the zone between high and low groundwater level, where it produces a hard crust (*laterite*; cf., e.g., Schroeder 1978). The iron comes from the mafic components of the underlying rocks.

Quite generally, if the evaporation is faster than the runoff and the water does not reach an outflow from the area, *evaporites* are precipitated. Such conditions prevail, for instance, in the Great Salt Lake region of Utah (Cohenour 1966). On the dried-up flats, the evaporite forms a pearly crust (cf. Scheidegger 1987c, p. 247 Fig. 9.21B); at the same time, desiccation cracks are formed.

1.8.3.3 Island Mounts

Finally, we have to discuss some features which occur frequently in dry areas with a hard-rock (igneous or arkose) base. The rocks, where they are exposed to the elements, are eroded in the form of island mounts (or "inselbergs", by the German name) with a characteristic morphology: they present themselves as elongated *whalebacks*, or as smooth rounded mounts (*bornhardts*), or sometimes as structures of a bizarre form (*tors*). Occasionally, there are several levels of island mounts superimposed upon each other.

1.8.4 Semidesert Features

1.8.4.1 General Remarks

It remains to discuss some landscape features that are transitional between arid and humid conditions.

First of all, the existence of intermittent streams (*wadis*) is of importance in such regions. Second, in poor soil areas, a specific type of landscape (*badlands*) evolves.

We shall discuss these features in turn below.

1.8.4.2 Wadis

Many semiarid regions contain dry valleys which have been carved out by rivers that rarely run. In fact, such intermittent rivers (*wadis*) seem to be a significant factor in shaping the (semi-) desert morphology.

The beds of wadis may remain dry for many years, but when rain does occur, the fluviatile action of the water is highly dynamic and significant (Reid and Frostick 1984). Thus, deep canyons and gorges are carved out, presenting generally the aspect of the surroundings of an ordinary humid-climate valley.

Since aeolian action is significant during the long dry spells, fluvial and aeolian deposits are interbedded in such ephemeral stream areas. Morphologically, it is quite difficult to discriminate between the two types of deposits, but Shepherd and Macke (1978) have shown that aeolian and fluviatile processes produce deposits with quite different grain size-frequency distribution curves. This fact can be used in morphological studies.

1.8.4.3 Badland Erosion

In semiarid, sandy, or clayey areas where there is not enough moisture and time available between cloudbursts to allow vegetation to grow profusely, the water washes gulleys and valleys into otherwise undisturbed, flat strata. Due to the lack of vegetation, the sides of the gulleys remain bare of plant growth although the more level parts show some cover with such low-lying plants as prairie grass and cactus. The whole area thus takes on a bleak appearance; the type of landscape it represents is therefore referred to as badlands.

Since erosion in badlands does not proceed at the same pace in all localities, characteristic and sometimes fantastic features result. At times, strata are encountered which present slightly more resistance to ablation and dissolution by water than others, so that "islands" are formed around which erosion takes place at a faster pace. The water now collects even more in the deeper places and the more resistive top of the developing feature acts as a protection. Thus, a series of features will eventually stand out in an area which all around has been eroded to a lower level. In general, the features thus created are pyramidal structures and are referred to as *mesas* or *buttes*.

In addition to the pyramidal structures just mentioned, one occasionally finds clusters of more unusual structures which have a strange, mushroom-shaped form.

Instead of being pyramidal, they have an overhanging "hat" so that they have the general appearance of giant mushrooms. Such structures are called *hoodoos*. Their heights run from 1.6 to 3.0 m, their waists from 0.6 to 1.5 m (in diameter); the overhang is 0.2–0.5 m (Scheidegger 1958).

The occurrence of hoodoos in badlands requires an explanation. It is obvious that they are erosional features, but so are mesas and buttes, which have a pyramidal structure. The reasons for the different appearance of *hoodoos* are not at all *a priori* evident.

References

Abt, P. et al.: Geogr. Helv. 26(3), 115 (1971).
Adams, J.: Geology 8, 442 (1980a)
Adams, J.: Bull. Geol. Soc. Am. Pt. II 91(1), 1 (1980b)
Ahnert, F.: Am. J. Sci. 284, 1035 (1984)
Ai, N.S., B. Lui and J. Zitou: Z. Geomorph. Suppl. 63, 169 (1987)
Alden, W.C.: U.S. Geol. Surv. Bull. 273, 18 (1905)
Alean, J.C.: Untersuchungen über Entstehungsbedingungen und Reichweiten von Eislawinen. Zürich: Vers. Anst. Wasserb. (1984)
Alexander, C.S.: Calif, Geogr. 3, 131 (1962)
Alexander, C.S.: Ann. Assoc. Am. Geogr. 56(1), 128 (1966)
Alexander, D.E.: Mathemat. Geol. 12(1), 79 (1980)
Amark, M.: Geolog. Fören. Stockholm Förh. 108(1), 13 (1986)
Anderson, M.G.: Water Resour. Res. 9(2), 378 (1973)
Aniya, M.: Z. Geomorph. 29(3), 301 (1985)
Araya-Vergara, J.F.: Not. Geomorf. 17(33), 19 (1977)
Araya-Vergara, J.F.: J. Coastal Res. 2(2), 159 (1986)
Argialas, D., J.G. Lyon, and O.W. Mintzer: Photogram. Eng. 54(4) 505 (1988)
Arrhenius, G.: Rept. Swedish Deep Sea Exp. 5, 1 (1952)
Asworth, P.J. and R.I. Ferguson: Geografiska Ann. 68A(4), 361 (1986)
Audley-Charles, M.G., J.R. Curray and G. Evans: Geology 5, 341 (1977)
Bagnold, R.A.: The physics of blown sand and desert dunes. London: Muthuen (1941)
Baker, V.R.: Proc. Lunar Planet. Sci. Conf. 9th, 3205 (1978)
Baker, V.R.: J. Geoph. Res. 84(B14), 7985 (1979)
Bally, A.: J. Alberta Soc. Petrol. Geol. 5, 89 (1957)
Bär, O.: Geogr. Helv. 12(1), 1 (1957)
Barnes, K.K. and H.P. Johnson: Agr. Eng. 37, 813 (1956)
Beck, B.F. (ed.): Sinkholes, their geology, engineering and environmental impact. Proc. Conf. Orlando, Florida, Rotterdam: Balkema (1984)
Bender, H.: Die Höhle 20(1), I (1969)
Benson, C.A.: Antarctic J. U.S. 4(5), 217 (1969)
Berge, C.: Théorie des graphes et ses applications. Paris: Dunod (1958)
Besler, H., and G. Gerster: Bild d. Wiss. 1985(5), 38 (1985)
Best, J.L.: Progr. Phys. Geogr. 10(2), 157 (1986)
Bigelow, G.E.: Z. Geomorph. 26(2), 225 (1982)
Bird, E.C.F.: Coasts, Cambridge Mass.: MIT Press (1969)
Biscaye, P.E. and S.L. Eittreim: Mar Geol. 23, 155 (1977)
Blackman, R.B. and J.W. Tukey: The measurement of power spectra. New York: Dover (1958)
Blackwelder, E.: Bull. Geol. Soc. Am. 45, 159 (1934)
Bloom, A.L.: Z. Geomorph. 9, 422 (1965)
Boggs, S.: Geology 2(5), 251 (1974)
Boots, B.N. and R.K. Burns: J. Glaciol. 30(106), 302 (1984)
Boulton, G.S.: Sedimentology 33(5), 677 (1986)
Bouma, A.M., and A. Brouwer (ed.): Turbidities, developments in sedimentology, Vol. 3, Amsterdam: Elsevier (1964)

Braithwaite, C.J.R.: Progr. Phys. Geogr. 6(4), 505 (1982)
Breadnell, H.J.L.: Geogr. J. 35, 379 (1910)
Bremer, H.: Z. Geomorph. Suppl. 54, 11 (1985)
Brice, J.C.: Bull. Geol. Soc. Am. 85, 581 (1974)
Brown, H.E., E.A. Hansen and N. Champagne: Water Resour. Res. 6, 818 (1970)
Brown, M.C.: Water Resour. Res. 9(3), 749 (1973)
Brown, S.R.: Geoph. Res. Lett. 14(11), 1095 (1987)
Brunsden, D. and R.H. Kesel: J. Geol. 81, 576 (1973)
Bryan, R.B. and A.G. Price: Z. Geomorph. Suppl. 34, 48 (1980)
Bryan, R.B. and A. Yair (ed.): Badland geomorphology and piping, Norwich: Geobooks (1982)
Buedel, J.: Klima-Geomorphologie, Berlin, Borntraeger (1977)
Buedel, J.: Climatic geomorphology, Princeton: univ. Press (1982)
Bull, B.: Progr. Phys. Geogr. 1(2), 227 (1977)
Burrough, P.A.: Nature 294, 240 (1981)
Burt, R.F. and D.E. Walling (ed.): Erosion and sediment yield, Norwich: Geohooks 1984
Callander, R.A.: Annu. Rev. Fluid Mech. 10, 129 (1978)
Carson, M.A. and M.F. Lapointe: J. Geol. 92, 41 (1983)
Carton, A., F. Dramis and M. Sorriso-Salvo: Z. Geomorph. Suppl. 63, 149 (1987)
Castaing, P. and J.M. Froidefond: Bull. Inst. Geol. Bass. Aquitaine 24, 131 (1978)
Cauchy, L.A.: Oevres Compl. Ser. 1, 2, 167 (1832)
Chang, T.P. and G.H. Toebes: Water Resour. Res. 6(2), 537 (1970)
Charlesworth, J.K.: The Quaternary Era, 2 vols. London: E. Arnold & Co. (1957)
Charlier, R.H.: Z. Geomorph. 12(4), 375 (1968)
Chorley, R.J.: J. Glaciol. 3, 339 (1959)
Christofoletti, A.: Bol. Geogr. Teor. Rio Claro 13(26), 27 (1983)
Christofoletti, A. and A.C. Tavares: Geografia 1(2), 67 (1976)
Clarke, G.K.C.: J. Geophys. Res. 92(B9), 9023 (1987)
Clarke, G.K.C. and 3 others: J. Geophys. Res. 91(B7), 7165 (1986)
Clarke, K.C.: Comput. Geosci. 12(5), 713 (1986)
Clarke, T.L., W.L. Stubblefield and D.J.P. Swift: J. Geol. 91(1), 93 (1983)
Clayton, K.: Coastal Geomorphology, London: Macmillan (1979)
Coffman, D.M., A.K. Turner and W.N. Melhofn: The W.A.T.E.R. system: computer programs for stream analysis. Lafayette: Purdue Univ. Water Resources Res. Ctr. (1971)
Cohenour, R.E.: Second Symp. Salt, Northern Ohio, Geol. Soc. 1, 201 (1966)
Coleman, J.M.: Deltas, processes of deposition and models for exploration. Minneapolis: Burgess Pub. Co. (1976)
Corbel, J.: Z. Geomorph. 3, 1 (1959)
Crickmay, C.H.: J. Geol. 68, 377 (1960)
Crofton, M.W.: Philos. Trans. R. Soc. 158, 181 (1968)
Culbertson, J.K. and C.H. Scott: U.S. Geol. Surv. Prof. Pap. 7008, B 237 (1970)
Damuth, J.E. and R.W. Embley: Sedimentology 26, 825 (1979)
Dangeard, L. et al.: Rev. Geogr. Phys. Geol. Dyn. Ser. 4, 251 (1961)
Davis, W.N.: Die erklärende Beschreibung der Landformen. 2nd edn. Leipzig: Teubner (1924)
Day, M.: Ann. Assoc. Am. Geogr. 73(2), 206 (1983)
De Almeida, F.F.M. (ed.): Simposio Internacional sobre las mergens contentais de tipo atlantico. Sao Paulo: Acad. Brasil. de Ciencias (1975)
Defant, A.: Physical Oceanography. 2 vols. London: Pergamon (1961)
Deike, G.H. and W.B. White: Am. J. Sci. 267(2), 230 (1969)
Deluca, F.P.: Compass 45(4), 236 (1968)
Denton, G.D. and 3 others: Geology 12(5), 263 (1984)
Dietrich, G.: Dtsch. Hydrogr. Z. Erg.-H. B, No. 3, 26 (1959)
Dietz, R.S.: New Sci. 4, 946 (1958)
Dionne, J.C. and C. Laverdiére: Can. J. Earth Sci. 9(8), 979 (1972)
Dolling, R.: Ontario Geogr. 2, 3 (1968)
Doornkamp, J.C.: J. Geol. Soc. Lond. 143(2), 335 (1986)
Doornkamp, J.C. and C.A.M. King: Numerical Analysis of geomorphology London: Arnold (1971)
Dorn, R.I. and T.M. Oberlander: Progr. Phys. Geogr. 6(3), 317 (1982)
Drewry, D.: Glacial geological processes. London: Arnold (1986)

Dreybrodt, W.: Earth and Planet. Sci. Let. 58(2), 293 (1982)
Dylik, J. and G.C. Maarleveld: Meded. Geol. Sticht 18, 7 (1967)
Dunkerley, D.L.: Geogr. Anal. 9(4), (1977)
Dunkerley, D.L.: Z. Geomorph. 23(3), 332 (1979)
Dunkerley, D.L.: Z. Geomorph. 24(1), 52 (1980)
Eaton, G.P.: J. Geol. 72, 1 (1964)
Ehrlich, R. and B. Weinberg: J. Sediment. Petrol. 40(1), 205 (1970)
Eittreim, S., E.M. Thorndike and L. Sullivan: Deep Sea Res. 23, 1115 (1976)
Elbersen, W.: ITC Journal 4, 322 (1983)
Ellenberg, L.: Z. Geomorph. 26(1), 103 (1982)
Elmendorf, C.H. and B.C. Heezen: Bull. Syst. Tech. J., 36, 1047 (1957)
Embley, R.W. and R.D. Jacobi: Mar. Geotech. 2, 205 (1975)
Embley, R.W. and 4 others: Bull. Geol. Soc. Am. Pt. I, 91, 731 (1980)
Engel, C.G. and R.P. Sharp: Bull. Geol. Soc. Am. 69, 478 (1958)
Engelen, G. and W. Huybrechts: Catena 8, 239 (1981)
Ergenzinger, P. and J. Conrady: Catena 9, 77 (1982)
Ewing, M. et al.: Bull. Geol. Soc. Am. 75, 17 (1964)
Eyles, N.: Glacial Geol. Oxford: Pergamon (1983)
Fabre, G. and G. Nicod: Z. Geomorph. 26(2), 209 (1982)
Fairbridge, R.W.: In: Encyclopedia of Geomorphology, ed. R.W. Fairbridge, p. 279, New York: Reinhold (1968a)
Fairbridge, R.W.: In: Encyclopedia of Geomorphology, ed. R.W. Fairbridge, p. 325, New York: Reinhold (1968b)
Fairbridge, R.W., and Finkl, C.W.: J. Geol. 88, 69 (1979)
Ferguson, R.I.: Bull. Geol. Soc. Am. 84, 251 (1973)
Finkl, C.W.: Z. Geomorph. 26(2), 137 (1982)
Fisher, R.V.: J. Geophys. Res. 69, 341 (1964)
Fisher, R.V.: Bull. Geol. Soc. Am. 88, 1287 (1977)
Flint, R.F.: Glacial and Pleistocene geology. New York: J. Wiley & Sons (1957)
Foose, R.M.: Geotimes 26(8), 20 (1981)
Ford, T.D. and C.H.D. Cullingford: The science of speleology. New York: Academic press (1972)
Fournier, F.: Débit solide des course d'eau. Paper presented at the 12th General Assembly, Assoc. of Scientific Hydrology, U.G.G.I., Helsinki (1960)
French, H.M.: Geogr. Ann 86A(3), 167 (1986)
Furrer, G.J.: Die Höhenlage von subnivalen Bodenformen. Habilitationsschrift Phil. Fak. II, Univ. Zürich. Pfäffikon ZH: Kunz (1965)
Gardiner, V.: In: Geomorphological Techniques, ed. Goudie et al. p. 47, London: Allen & Unwin (1981)
Gardiner, V. and C.C. Park: Prog. Phys. Geogr. 2(1), 1 (1978)
Gardner, J.S.: Arctic and Alp. Res. 2(2), 135 (1970)
Gardner, J.S.: Z. Geomorph. 15(4), 390 (1971)
Gardner, J.S.: Arct. Alp. Res. 15(2), 271 (1983)
Gendzwill, D.J.: Musk-ox 32, 1 (1983)
Gerber, E.: Geogr. Helv. 11, 160 (1956)
Gerber, E.: Mitt. Aargauischen Naturforsch. Ges. 26, 86 (1961)
Gerber, E. and A.E. Scheidegger: Eclogae Geol. Helv. 62(2), 401 (1969)
Gerber, E. and A.E. Scheidegger: Riv. Ital. Geofis. 2(1), 47 (1975)
Gerber, E. and A.E. Scheidegger: Verh. Geol. Bundes-Anst. Wien 1977(2), 165 (1977)
Ghosh, A.K. and A.E. Scheidegger: J. Hydrol. 13, 101 (1971)
Giardino, J.R. and S.G. Vick: Bull. Assoc. Eng. Geol. 22(2), 201 (1985)
Gibson, W.M.: Bull. Geol. Soc. Am. 71, 1087 (1960)
Glaeser, J.D.: J. Geol. 86(3), 283 (1978)
Gould, H.R.: Soc. Econ. Paleontol. Mineral. Spec. Publ. 2, 34 (1951)
Green, R.S., D.P. Dubois and C.W. Tutwiler: J. San. Eng. Div., ASCE 92(SA), 55 (1966)
Gregory, K.J.: Bull. Geol. Soc. Am. 88(8), 1075 (1977)
Gregory, K.J.: Water Resour. Res. 15(4), 775 (1979)
Gregov, G.: Riv. Ital. Geofis. 3, 97 (1976)
Guilcher, A.: Coastal and submarine morphology (Transl. from French) London: Methuen & Co. (1958)

Hack, J.T.: Am. J. Sci. 258 A, 80 (1960)
Hadley, R.F. and D.E. Walling (eds): Erosion and sediment yield. Norwich: Geobooks (1984)
Haeberli, W.: Creep of mountain permafrost: internal structure and flow of Alpine rock glaciers. Zürich: Mitt. vers. Anst. Wasserb. & c. d. ETH Nr. 77 (1985)
Haeberli, W. and F. Epifani: Ann. Glaciol. 8, 78 (1986)
Haigh, M.J.: Catena Suppl. 10, 181 (1987)
Hakanson, L.: Math. Geol. 10(2), 141 (1978)
Hakanson, L.: Z. Geomorph. 25(4), 369 (1981)
Hambrey, M.J. and A.G. Milnes: Ecologae Geol. Helv. 70(3), 667 (1977)
Hauner, U.: Eiszeitalter u. Gegenw. 35, 205 (1985)
Hawksley, P.G.: Bull. Br. Coal. Util. Res. Ass. 15, 105 (1951)
Hedges, J.: Z. Geomorph. 13(1), 22 (1969)
Hedin, S.: Central Asia and Tibet, Vol. 1, see p. 350. London (1903)
Heezen, B.C.: Bol. Soc. Geogr. Colombia 51/52, 135 (1956)
Heezen, B.C.: Ecologae Geol. Helv. 51, 521 (1958)
Heezen, B.C.: Geoph. J. R. Astr. Soc. 2, 142 (1959)
Heezen, B.C. and M. Rawson: Mar. Geol. 23, 173 (1977)
Heezen, B.C., D.B. Ericson and M. Ewing: Deep Sea Res. 1, 193 (1954)
Holdsworth, G.: Can. J. Earth Sci. 10(10), 1565 (1973)
Holeman, J.N.: Water Resour. Res. 4, 737 (1968)
Holmes, A.: Principles of physical geology, London: T. Nelson & Sons (1944)
Hooke, J.M.: J. Hydrol. 42, 39 (1979)
Hooke, J.M.: Progr. Phys. Geogr. 8(4), 473 (1984)
Hoormann, K.: Morphometrie der Erdoberfläche, Kiel: Verl. Geogr. Inst. Univ. (1971)
Horton, R.E.: Bull. Geol. Soc. Am. 56, 275 (1945)
Hovermann, J.: Kiel. Geogr. Schr. 62, 145 (1985)
Howard, A.D.: Caves and Karst 13(1), 1 (1971)
Hubert, J.F.: Geology 8(11), 516 (1980)
Ignatius, H.: C.R. Soc. Geol. Finland 30, 135 (1958)
Iken, A.: Velocity fluctuations of an Arctic valley glacier. Ottawa: National Research Council Canada (1974)
Inglis, C.C.: The behaviour and control of rivers and canals. Poona: Res. Pub. Centr. Waterpower, Irrigation and Navigation Res. Stat. (1949)
Inman, D.L. and C.E. Nordstrom: J. Geol. 79, 1 (1971)
Jakucs, L.: Morphogenetics of karst regions, Letchworth (England): Adam Hilger (1977)
James, W.R. and W.C. Krumbein: J. Geol. 77, 544 (1969)
Jarvis, R.J.: Water Resour. Res. 8(5), 1265 (1972)
Jarvis, R.J.: Water Resour. Res. 12(6), 1151 (1976)
Jennings, J.N.: Karst Geomorphology, 2nd edn. London: Blackwell (1985)
Jessen, O.: Peterm. Geogr. Mitt. Erg. h. 241, 1 (1943)
Johnson, D.W.: Shore processes and shoreline development. New York: Wiley (1919)
Johnson, P.G.: J. Glaciol. 11 (62), 227 (1972)
Johnson, P.G.: Proc. Yorksh. Geol. Soc. 40(2), 223 (1974)
Johnson, P.G.: Z. Geomorph. 28(2), 235 (1984)
Julian, P.R.: Water Resour. Res. 3(3), 831 (1967)
Kälin, M.: The active push moraine of the Thompson Glacier, Ottawa: Nat. Res. Counc. Canada Proj. A-2662 Report (1971)
Kastens, K.A. and A.N. Shor: Mar. Geol. 71, 165 (1986)
Kellerhals, R., M. Church and D.K. Bray: J. Hydraul. Div. ASCE 102 (HY7), 813 (1976)
Kelletat, D.: Catena 12, 255 (1985)
Kerschner, H.: Das Daun- und das Egesen-Stadium in ausgewählten Tälern der Zentralalpen von Nordtirol und Graubünden. Ph.D. Diss. Univ. Innsbruck (1976)
Kheoruenromne, I., W.E. Sharp and L.R. Gradner: Water Resour. Res. 12(5), 919 (1976)
King, L. and 3 others: Z. Gletscherkunde Glazialgeol. 23(1), 77 (1987)
Klein, C.: Rev. Geol. Dyn. Geogr. Phys. 26(2), 95 (1985)
Klein, M.: Catena 8, 281 (1981)
Knighton, A.D.: Catena 9, 25 (1982)
Koestler, A.: Janus, a summing-Up., London: Hutchinson (1978)

Kolstrup, E.: Palaeogeogr., Palaeoclimatol., Palaeoecol. 56, 237 (1986)
Kolstrup, E.: Z. Geomorph. 31(4), 449 (1987a)
Kolstrup, E.: Dan. Geol. Forening. Arsskr. for 1986, 67 (1987b)
Kolstrup, E. and J.B. Jorgensen: Bull. Geol. Soc. Denmark 30, 71 (1982)
Kolstrup, E. and V. Mejdahl: Boreas 15, 311 (1986)
Komar, P.D.: J. Geol. 92, 133 (1984)
Konzewitsch, N.: La Forma de los Clastos. Buenos Aires: Servicio Hidrograf. Naval (1961)
Krause, D.C. and 3 others: Bull. Geol. Soc. Am. 81, 2153 (1970)
Krcho, J.: Geograf. Cas. Bratislava 38 (2/3), 210 (1986)
Krumbein, W.C. and F. Pettijohn: Manual of sedimentary petrography, London: Appleton-Century
    (1938)
Kuenen, P.H.: Marine geology, New York: J. Wiley and Sons (1950)
Kumar, A.: Z. Geomorph. 25(4), 391 (1981)
Lacombe, H.: Deep-Sea Res. 6, 211 (1960)
Lacroix, A.: La montagne Pelée et ses éruptions. Paris: Masson & Cie. (1904)
Lambert, A.: Wasserwirtsch. 72(4), 1 (1982)
Laroque, A.C.L.: Current Res. Geol. Surv. Canada 85 B(1), 431 (1985)
Lassila, M.: Z. Geomorph. 30(2), 129 (1986)
Laughton, A.S.: Deep Sea Res. 15, 2 (1968)
Laverdiere, C., P. Guimont and J.C. Dionne: Palaeogeogr., Palaeoclimatol., Palaeoecol. 51 (1–4), 365
    (1985)
Leatherman, S.P.: Barrier Island Handbook (2nd edn.) Charlotte, N.C.: Coastal Publications (1982)
Leatherman, S.P.: Shore and beach 1983(7), 28 (10983)
Lee, L.J.: Z. Geomorph. 21(1), 37 (1977a)
Lee, L.J.: J. Hydrol. 33, 123 (1977b)
Lee, L.J. and B.L. Henson: Rev. Geomorph. Dynam. 16, 29 (1977a)
Lee, L.J. and B.L. Henson: Water Resour. Res. 13(6), 1006 (1977b)
Leopold, L.B.: Bull. Geol. Soc. Am. 84, 1845 (1973)
Leopold, L.B. and W.B. Bull: Proc. Am. Philos. Soc. 123(2), 168 (1979)
Leopold, L.B. and T. Maddock: U.S. Geol. Surv. Prof. Pap. 252, 1 (1953)
Leopold, L.B. and H.E. Skibitzke: Geogr. Ann. 49A, 2 (1967)
Leopold, L.B. and M.G. Wolman: U.S. Geol. Surv. Prof. Pap. 282-B (1957)
Leopold, L.B. and M.G. Wolman: Bull. Geol. Soc. Am. 71, 769 (1960)
Li, Y.H., H. Teraoka, T.S. Yang and J.S. Chen: Geochim. Cosmochim. Acta 48, 1561 (1984)
Lliboutry, L.J.: Glaciol. 13(69), 371 (1974)
Longman, M.W. and D.N. Brownlee: Z. Geomorph. 24(3), 299 (1980)
Lovejoy, S., D. Schertzer and P. Ladoy: Nature 319, 43 (1986)
Maarleveld, G.C.: Meded. Geol. Stich., N.S. No. 17 (1965)
Mabbutt, J.A.: Desert landforms. Cambrige, Mass.: MIT Press (1977)
Machatschek, F.: Geomorphologie. 5th edn. Leipzig: Teubner (1952)
Mackay, J.R.: Geograph. Bull. 13, 5 (1959)
Mackay, J.R.: Proc. Permafrost Int. Conf., Lafayette 71 (1963)
Mandelbrot, B.B.: Science 156, 606 (1967)
Mandelbrot, B.B.: The fractal geometry of nature, San Francisco: Freeman (1982)
Marin, A.F. and M.J. Riguidel: Éléments de morphologie généralisée; Notes et Mem. Comp. Franç.
    Petroles Paris 14, 1 (1978)
Mark, D.M.: Bull. Geol. Soc. Am. 84, 1369 (1973)
Marsh, B.: Geology 15, 945 (1987)
Marshall, P. and M.C. Brown: Can. J. Earth Sci. 11(4), 510 (1974)
Martinez, J.D. et al.: Geotimes 26(3), 14 (1981)
Matyas, E.L.: Can. J. Earth Sci. 21, 1156 (1984)
Mayer, H.E.: Water Resour. Res. 6(1), 303 (1970)
Mc Cave, I.N.: Bull. Inst. Geol. Bassin d'Aquitaine 31–32, 47 (1982)
Mc Coy, R.M.: Bull. Geol. Soc. Am. 82, 757 (1971)
Meade, R.H.: Bull. Geol. Soc. Am. 80, 1265 (1969)
Melton, M.A.: J. Geol. 66, 35 (1958)
Menard, H.W.: Experientia (Basel) 15, No. 6, 205 (1959)
Menzies, J.: Drumlins: a bibliography. Norwich: Geobooks (1984)

Menzies, J.: Can. J. Earth Sci. 23 (6), 774 (1986)
Mesolella, K.J., H.A. Sealy and R.K. Matthews: Bull. Am. Assoc. Pet. Geol. 54 (10), 1899 (1970)
Meybeck, M.: Hydrol. Sci. Bull. 21 (2), 265 (1976)
Miller, T.P. and R.L. Smith: Geology 5 (3), 173 (1977)
Mills, H.H.: Bull. Geol. Soc. Am. Part I, 91, 637 (1980)
Milton, L.E. and C.D. Ollier: J. Hydrol. 3, 66 (1965)
Molnia, B.F.: Z. Geomorph. Suppl. 57, 141 (1985)
Moore, G.T. and D.O. Asquith: Bull. Geol. Soc. Am. 82, 2563 (1971)
Morandi, M.C., A. Del Grosso and B. Limonelli: Geol. Appl. Indrogeol. (Bari) 9 (1), 7 (1976)
Morisawa, M.: Streams, their dynamics and morphology. New York: Mc Graw-Hill (1968)
Morisawa, M.E.: Rivers, form and process. London: Longman (1985)
Mueller, F.: Medd. Grønl. 153 (3), 1 (1959)
Mueller, J.E.: Ann. Assoc. Am. Geogr. 58 (2), 371 (1968)
Murai, I.: Bull. Earthque. Res. Inst. 39, 133 (1961)
Nakaya, U.: Snow crystals: natural and artificial. Cambridge, Mass.: Harvard Univ. Press (1954)
Nelson, F.E.: Catena 12, 23 (1985)
Neumayr, M. and F.E. Suess: Erdgeschichte, 3rd. edn. Stuttgart: Bibliogr. Inst. (1920)
Newell, N.D.: Sci. Am. 226 (6), 54 (1972)
Nichols, M.M.: J. Sediment. Petrol. 47 (3), 1171 (1977)
Normark, W.R. and D.J.W. Piper: Bull. Geol. Soc. Am. 80 (9), 1859 (1969)
Nyberg, R.: Geogr. Ann. 68A (3), 207 (1986)
Ohmori, H.: Bull. Dept. Geogr. Univ. Tokyo 10, 31 (1978)
Ohmori, H.: Bull. Dept. Geogr. Univ. Tokyo 11, 77 (1979)
Ohmori, H.: Bull. Dept. Geogr. Univ. Tokyo 15, 77 (1983a)
Ohmori, H.: Z. Geomorph. Suppl. 46, 1 (1983b)
Ohmori, H.: Bull. Dept. Geogr. Univ. Tokyo 17, 19 (1985a)
Ohmori, H.: Trans. Jpn. Geomorph. Un. 6 (3), 225 (1985b)
Ohmori, H. and M. Hirano: Trans. Jpn. Geomorph. Un. 5 (4), 293 (1984)
Onesti, L.J. and T.K. Miller: Water Resour. Res. 14 (1), 144 (1978)
Ori, G.G.: Boll. Soc. Geol. Ital. 98, 35 (1974)
Park, C.C.: J. Hydrol. 33, 133 (1977)
Parker, G., P. Diplas and J. Akiyama: J. Hydrol. Eng. ASCE 109 (10), 1323 (1983)
Parsons, A.J.: Earth Surf. Proc. 4 (4), 395 (1979)
Penck, W.: Die geomorphologische Analyse. Stuttgart: J. Engelhorns Nachf. (1924)
Pethick, J.: An introduction to coastal geomorphology. London: Arnold (1984)
Petit, F.: Int. Geomorph. 1986 (1), 611 (1987)
Petts, G. and I. Foster: Rivers and Landscape, London: Arnold (1985)
Pihlainen, J.A., R.J.E. Brown and R.F. Legget: Bull. Geol. Soc. Am. 67, 1119 (1956)
Pike, R.J. and S.E. Wilson: Bull. Geol. Soc. Am. 82, 1079 (1971)
Post, A.: J. Geophys. Res. 65, 3703 (1960)
Potter, P.E.: J. Geol. 86, 423 (1978)
Powell, R.D.: Mar. Geol. 57, 1 (1984)
Price, R.J.: Glacial and fluvioglacial Landforms, Edingburgh: Oliver & Boyd (1973)
Prior, D.B. and J.H. Coleman: Geosci. & Man 19 (30 June), 41 (1978a)
Prior, D.B. and J.H. Coleman: Mar. Geotech. 3 (1), 37 (1978b)
Prior, D.B. and J.H. Coleman: Z. Geomorph. 23 (4), 415 (1979)
Prior, D.B. and J.H. Coleman: Mar. Geol. 36, 227 (1980)
Prior, D.B. and 3 others: Proc. 13th Int. Sympos. Remote Sensing, Ann Arbor, Mich. p. 195 (1979)
Prior, D.B., J.M. Coleman and H.H. Roberts: Offshore 1981, 4 (1981)
Pye, K.: Progr. Phys. Geogr. 8 (2), 176 (1984)
Quigley, R.M.: Can. Geotech. J. 14 (3), 310 (1977)
Ranalli, G. and A.E. Scheidegger: Bull. Int. Assoc. Sci. Hydrol. 13 (2), 142 (1968)
Rapp, A., R. Nyberg and L. Lindh: Geograf. Ann. 68A (3), 197 (1986)
Reid, I. and L. Frostick: Geogr. Mag. 56 (4), 178 (1984)
Reinboth, F. and F. Göbel: Die Höhle 26 (1), 123 (1975)
Rendell, H. and D. Alexander: Bull. Geol. Soc. Am. 90, 761 (1979)
Rhodes, D.D.: Geogr. Ann. 69A (1), 147 (1987)
Richards, K.S.: Earth Surface Proc. 1, 71 (1976)

Richardson, L.F.: Gen. Systems Yearb. 6, 139 (1961)
Richthofen, F.F.v.: Führer für Forschungsreisende, Hannover (1901)
Ritchie, W. and D. Brunsden: Geogr. Mag. 45(7), 511 (1973)
Roberts, M.C. and D.M. Mark: Can. J. Earth Sci. 7, 1179 (1970)
Rogerson, R.J. and N. Eyles: Berendon Glacier medial moraines. St. Johns Nfdl.: Mem. Univ. Dept.
    Geogr. 1979
Rona, P.A.: J. Dediment Petrol. 39(3), 1132 (1969a)
Rona, P.A.: Bull. Am. Assoc. Pet. Geol. 53(7), 1453 (1969b)
Rona, P.A.: J. Geol. 78(2), 141 (1970)
Ross, C.S. and R.L. Smith: U.S. Geol. Surv. Prof. Pap. 366, 1 (1961)
Rossbacher, L.: Geograf. Ann. 68A(1–2), 101 (1986)
Rutten, M.G.: Am. J. Sci. 258, 293 (1960)
Samoilov, I.V.: Ust' ya ryok. Moscow: Geografgiz (1954). German translation by F. Tutenberg under
    the title: Die Flußmündungen. Gotha: Herm. Haack Verlag (1956)
Sawada, T., K. Ashida and T. Takahashi: Z. Geomorph. Suppl. 46, 55 (1983)
Scheidegger, A.E.: Geofis. Pura e Appl. 41, 101 (1958)
Scheidegger, A.E.: U.S. Geol. Surv. Prof. Pap. 525-B, B 187 (1965a)
Scheidegger, A.E.: U.S. Geol. Surv. Prof. Pap. 525-C, C 164 (1965b)
Scheidegger, A.E.: Physical aspects of natural catastrophes. Amsterdam: Elsevier (1975)
Scheidegger, A.E.: Tectonophysics 55, 7–10 (1979)
Scheidegger, A.E.: Principles of geodynamics 3rd edn. Berlin Heidelberg New York: Springer (1982a)
Scheidegger, A.E.: In: Marine slides and other mass movements, ed. Saxov & Nieuwenhuis. New York:
    Plenum, p. 11 (1982b)
Scheidegger, A.E.: Earth Sci. Revs. 21, 225 (1984)
Scheidegger, A.E.: Z. Geomorph. 29(2), 223 (1985)
Scheidegger, A.E.: Z. Geomorph. 30, 257 (1986)
Scheidegger, A.E.: Catena Suppl. 10, 199 (1987a)
Scheidegger, A.E.: Wildbach- und Lawinenverb. 51 (105), 1 (1987b)
Scheidegger, A.E.: Systematic geomorphology, Vienna New York: Springer (1987c)
Scheidegger, A.E. and N.S. Ai: Tectonophysics 126, 285 (1986)
Scheidegger, A.E. and P.E. Potter: Sedimentology 5, 289 (1965)
Schipull, K.: Z. Geomorph. 22(4), 426 (1978)
Schmidt, K.H.: Z. Geomorph. 39, 59 (1981)
Schommer, P.: Z. Gletscherk. Glazialgeol. 12(2), 125 (1977)
Schroeder, D.: Bodenkunde in Stichworten, 3rd edn. Kiel: Hirt (1978)
Schroeder, J., M. Beaupre and M. Cloutier: Can. J. Earth Sci. 23, 1842 (1986)
Schroeder, W.: Acta Hydrophys. 14(1), 237 (1969)
Schroeder, W.: Acta Hydrophys. 17(1), 47 (1972)
Schumm, S.A.: U.S. Geol. Survey Circ., 477 (1963a)
Schumm, S.A.: Bull. Geol. Soc. Am. 74, 1089 (1963b)
Schumm, S.A.: Z. Geomorph. Suppl. 5, 215 (1964)
Schumm, S.A.: The Fluvial System. New York: Wiley (1977)
Schumm, S.A.: Annu. Rev. Earth Planet. Sci. 13, 5 (1985)
Schwab, F.L.: Geology 4, 723 (1976)
Semmel, A.: Periglazialmorphologie, Darmstadt: Wiss. Buchgem. (1985)
Seppälä, M.: Z. Geomorph. 16, 463 (1972)
Seppälä, M.: Geogr. Ann. 68A(3), 141 (1986)
Shaler, N.S.: Bull. Geol. Soc. Am. 10, 263 (1899)
Sharp, M.: Geogr. Ann. 67A(3–4), 213 (1985a)
Sharp, M.: Progr. Phys. Geogr. 9(2), 291 (1985b)
Sharp, R.P.: Geogr. Rev. 32, 417 (1942)
Sharp, R.P.: Bull. Geol. Soc. Am. Part I 91, 724 (1980)
Sharp, W.E.: Water Resour. Res. 6(3), 919 (1970)
Shepard, F.P.: Submarine Geol. New York: Harper (1948)
Shepard, F.P.: In: Recent sediments, Northwest Gulf of Mexico, 1951–1958, p. 197 ff. Tulsa: Amer.
    Assoc. Petroleum geol. Spec. Pub. (1960)
Shepard, F.P. and R.F. Dill: Submarine canyons and other sea valleys, Chicago: Rand Mc Nally (1966)
Shepard, F.P. and 3 others: Geology 5, 297 (1977)

Shepherd, R.G. and D.L. Macke: J. Res. U.S. Geol. Surv. 6(4), 499 (1978)
Shilts, W.W.: J. Geochem. Expl. 21, 95 (1984)
Shilts, W.W.: Curr. Res. Geol. Surv. Can. 84B(1), 217 (1984)
Shipek, C.J.: Bull. Geol. Soc. Am. 71, 1067 (1960)
Shreve, R.L.: J. Geol. 75, 178 (1967)
Silverberg, N. and B. Sundby: Can. J. Earth Sci. 16(4), 939 (1979)
Slotboom, R.T.: Comparative geomorphological and palynological investigation of the pingos (Viviers) in the Hautes-Fagnes (Belgium) and the Mardellen in the Gutland (Luxemburg). Amsterdam: Univ.-Diss. (1963)
Smart, J.S.: Water Resour. Res. 8, 1487 (1972)
Smith, A.J.: In: Encycl. Geom. ed. Fairbridge, p. 598, New York: Reinhold (1968)
Smith, R.L.: Bull. Geol. Soc. Am. 71, 795 (1960)
Späth, H.: Z. Geomorph. Suppl. 61, 3 (1986)
Speight, J.G.: J. Hydrol. 3(1), 1 (1965)
Speight, J.G.: In: Land evaluation, CSIRO Symp. 26–31 Aug. 1968, ed. G.A. Steward. Melbourne: Mac Millan, p. 239 (1968)
Speight, J.G.: Z. Geomorph. 15(3), 290 (1971)
Speight, J.G.: Z. Geomorph. Suppl. 25, 154 (1976)
Speight, J.G.: Photogr. 32, 161 (1977)
Steers, J.A. and 4 others: Geogr. J. 145(2), 192 (1979)
Steinhaus, H.: Coll. Math. 3(1), 1 (1954)
Sterr, H.: Z. Geomorph. 29(3), 315 (1985)
Strahler, A.N.: Trans. Am. Geoph. Un. 38, 913 (1957)
Strecker, M., A. Bloom and L. Gilpin: Z. Geomorph. 30(4), 387 (1986)
Summerfield, M.A.: Progr. Phys. Geogr. 22(3), 364 (1987)
Suzuki, Y. and Y. Shimano: Annu. Rep. Geosci. Univ. Tsukub a 7, 26 (1981)
Sweeting, M.M.: Karst landforms, New York: Columbia Univ. Press. (1973)
Sweeting, M.M.: Z. Geomorph. Suppl. 36, 203 (1980)
Sweeting, M.M. and N. Lancaster: Z. Geomorph. 26(2), 197 (1982)
Tandon, S.K.: Z. Geomorph. 15(3), 290 (1974a)
Tandon, S.K.: Z. Geomorph. 18(4), 460 (1974b)
Tarr, R.S.: Am. Geol. 21, 341 (1898)
Taylor, P.T., C.A. Wood and T.J. O'Hearn: Geology 8, 390 (1980)
Thomas, D.S.G.: Z. Geomorph. 30(2), 231 (1986)
Tiercelin, J.J. and H. Faure: In: Tectonics and geophysics of continental riffes, ed. Ramberg and Neumann, p. 41, Dordrecht: Reidel (1978)
Tinkler, K.J.: Bull. Geol. Soc. Am. 81, 547 (1970)
Tipper, J.C.: J. Geol. 84, 476 (1976)
Tipper, J.C.: J. Geol. 85(5), 591 (1977)
Trenhaile, A.S.: Ann. Assoc. Am. Geogr. 65(2), 297 (1975a)
Trenhaile, A.S.: Ann. Assoc. Am. Geogr. 65(4), 517 (1975b)
Trenhaile, A.S.: Ann. Assoc. Am. Geogr. 66(3), 451 (1976)
Trenhaile, A.S.: Z. Geomorph. 21(4), 445 (1977)
Trenhaile, A.S.: Ann. Assoc. Am. Geogr. 68(1), 95 (1978)
Trenhaile, A.S.: Z. Geomorph. 23(1), 27 (1979)
Trenhaile, A.S.: The geomorphology of rock coasts. Oxford: Univ. Press (1987)
Tricart, J.: Inform. Geol. 24(5), 210 (1961)
Trimmel, H.: Höhlenkunde, Braunschweig: Vieweg (1968)
Trudgill, S.: Limestone geomorphology, London: Longmans (1985)
Tschierske, N.: Rock Mech. 10, 113 (1978)
Tsuchiya, Y. and T. Yasuda: J. Nat. Disaster Sci. 2(2), 27 (1980)
Turcotte, D.L.: J. Geoph. Res. 91 (B2), 1921 (1986)
Twichell, D.C.: Geology 10(8), 408 (1982)
Twidale, C.R.: Structural landforms, Cambridge Mass.: MIT Press, (1971)
Twidale, C.R. and E.M. Campbell: Z. Geomorph. 30(1), 35 (1986)
Ueno, T.: J. Oceanogr. Soc. Jpn 37(2), 61 (1981)
Valentin, H.: Peterm. Mitt. Erg.-H. 246 (118p) (1952)
Van Andel, T.H. and P.D. Komar: Bull. Geol. Soc. Am. 80(7), 1163 (1969)

Verma, V.K. and S.K. Tandon: Z. Geomorph. Suppl. 12, 165 (1971)
Vilborg, L.: Geogr. Ann. 66A (1–2), 41 (1984)
Vincent, P.J.: Z. Geomorph. 31 (1), 65 (1987)
Vischer, D. (ed.): Schnee, Eis und Wasser alpiner Gletscher; Festschrift H. Röthlisberger. Zürich: Vers.
    Anst. Wasserb., (1988)
Vitek, J.D. and P. Tarquin: Z. Geomorph. 28 (4), 445 (1984)
Vondran, G.: Augsburger Geogr. H. 1, 1 (1979)
Waag, C.J.: J. Glaciol. 22 (87), 247 (1979)
Wahrhaftig, C. and A. Cox: Bull. Geol. Soc. Am. 70 (4), 383 (1959)
Walker, G.P. and L.A. McBroome: Geology 11, 571 (1983)
Warntz, W.: J. Hydrol. 25, 209 (1975)
Washburn, A.L.: Abh. Akad. Wiss. Göttingen, Math.-Physik. Kl. (3) 35, 34 (1983)
Watson, A.: Z. Geomorph. 29 (3), 285 (1985)
Weihaupt, J.G.: J. Geophys. Res. 79 (14), 2073 (1974)
Weyl, P.: Oceanography, an introduction to the marine environment. New York: Wiley (1970)
Whalley, W.B., J.R. Marshall and B.J. Smith: Nature 300, 433 (1982)
Wheeler, D.A.: Earth Surf. Landforms 9 (4), 391 (1984)
White, J.F.: Ohio J. Sci. 66, 592 (1966)
White, W.B.: Geomorphology and hydrology of karst terains. Oxford: Univ. Press (1988)
Whitehouse, J.E. and M.J. McSaveney: Arctic and Alp. Res. 15 (1), 53 (1983)
Wiegand, G.: Fossile Pingos in Mitteleuropa. Würzburg: Verl. Geogr. Inst. Univ. No. 16 (1965)
Williams, P.J.: Am. J. Sci. 255, 705 (1957)
Williams, P.W.: Bull. Geol. Soc. Am. 83, 761 (1972)
Winkler, C.D. and J.D. Howard: Geology 5, 123 (1977)
Wirthmann, A.: Z. Geomorph. Suppl. 28, 42 (1977)
Wood, W.F. and J.B. Snell: US Army Natick Lab., Tech. Rep. EP-124 (1960)
Worsley, P.: Nature 322, 683 (1986a)
Worsley, P.: Progr. Phys. Geogr. 10 (2), 265 (1986b)
Wright, L.D. and J.M. Coleman: Bull. Am. Assoc. Petrol. Geol. 57 (2), 370 (1973)
Yasso, W.E.: J. Geol. 73 (5), 702 (1965)
Young, R. W.: Z. Geomorph. 30 (2), 189 (1986)
Zeller, J.: Int. Assoc. Sci. Hydrol., Symposium on River Morphology, Gen. Ass. Bern. Trans. p. 174
    (1967)
Zenkovich, V.P.: Trudy Inst. Okeanologii Akad. Nauk SSSR 21, 3 (1957)
Zevenbergen, L.W. and C.R. Thorne: Earth Surf. Proc. Landf. 12 (1), 47 (1987)
Zschau, J. and 3 others: Die Küste 34 (71) (1979)
Zschau, J. and H.J. Kümpel: Geoph. Astroph. Fluid Dyn. 13, 245 (1979)
Zuchiewicz, W.: Ann. Soc. Geol. Pol. 51 (1/2), 99 (1981)

2 Physical Background

2.1 Introduction

2.1.1 General Remarks

The materials that are involved in the evolution of exogenic landscape features are soil, water, air, and ice. Of these materials, water and air can be treated as viscous fluids to a high degree of approximation; soil and ice, on the other hand, are "solids", which must be treated by the general methods of rheology.

The present chapter will be devoted to a brief review of the basic dynamics of these substances, as far as this is of importance regarding geomorphological effects.

2.1.2 Hydrodynamics of Viscous Fluids

The general hydrodynamics of viscous fluids is well known Since many treatises exist on the subject, there is no need to give many details here. We shall simply briefly review a few of the basic concepts which will be of importance later (cf. Lamb 1932; Pai 1957; Goldstein 1938).

Thus, we may note that for the complete description of the behavior of continuous matter (of which viscous fluids represent an example), one needs four types of relationships. The first is a kinematic condition, the second a continuity condition, the third an equation of motion and the fourth an equation of state. Since the above conditions lead to a system of differential equations, one will have to add suitable boundary and initial conditions to make a problem determined.

The set of above conditions can be combined to yield differential equations which are applicable under various conditions. A well-known equation of this type is the so-called Navier–Stokes equation which is

$$\mathbf{v}\operatorname{grad}\mathbf{v} + \frac{\partial \mathbf{v}}{\partial t} = \mathbf{F} - \frac{1}{\rho}\operatorname{grad}p - \frac{\eta}{\rho}\operatorname{curl}\operatorname{curl}\mathbf{v}. \tag{2.1.1}$$

Here, $\mathbf{v}$ is the local velocity vector of a point of the fluid, t is time, $\mathbf{F}$ is the volume force per unit mass, ρ is the density and η is the viscosity of the fluid. The Navier–Stokes equation applies to incompressible fluids.

It has been observed that the flow pattern of a fluid becomes transient at high flow velocities although the boundary conditions remain steady: eddies are formed which proceed into the fluid at intervals. The transient flow pattern is termed *turbulent*. It appears that, for any one flow system, a transition velocity exists at which the flow becomes turbulent; at lower velocities the flow remains steady (*laminar*). The best-known criterion for this transition velocity is based on the *Reynolds number*, Re, which is defined as follows (Barr 1980)

$$Re = \frac{\rho v d}{\eta} \tag{2.1.2}$$

where d denotes a characteristic diameter of the flow system. The criterion states that turbulence will occur if the Reynolds number reaches a critical value, commonly quoted as in the neighborhood of 2200. However, this criterion is strictly applicable to straight tubes only; in other types of flow systems the critical Reynolds number may be different.

For further details regarding the hydrodyanmics of viscous fluids, the reader is referred to standard textbooks.

2.1.3 Rheology of Solids

The theory of deformation of a continuous solid medium involves (a) a measure of inner displacements (strains), (b) a continuity condition, (c) Newtonian equations of motion, and (d) a rheological condition. The latter is a relation between the forces (and their derivatives) acting inside the medium ("stresses") and the measure of displacements (and its derivatives). It is the rheological condition which describes the "nature" of the substance and therefore has also been called "equation of state". The literature on rheology is large. In the Earth sciences, the basic rheological theory is just as important for geodynamics as it is for geomorphology. It has thus been reviewed at length by the writer (Scheidegger 1982) in his volume on the former subject, to which the reader is referred. We just summarize here some of the most important basic implications.

Thus, we recall that the solids involved in geoscience have a response behavior which falls into three ranges: the elastic range, the creep range, and the failure range.

The elastic range describes the behavior of most solids for small stress and strain changes. If the latter become larger or last a long time, the response is by some fluid type of behavior: displacements occur slowly and permanently. This has been called the creep range. Finally, if the stresses exceed a certain limit, the displacements become discontinuous: the material fails or fractures.

Individual cases will be discussed in this book in connection with studies of specific materials and agents involved in landscape evolution.

2.2 Dynamics of Flowing Water

2.2.1 Introduction

We turn our attention first to the dynamics of water. The latter substance is a geomorphological agent which occurs on the Earth's surface as well as underground.

Surface water is generally in a state of *turbulence*. It is therefore important to review the fundamentals of turbulence theory. Applications of the latter, incidentally, are found not only in connection with water, but also in connection with air. It is turbulence which picks up solid particles and holds them in suspension, transporting them over large distances.

Furthermore, water flow can be stratified, on the surface as sheet flow as well as in large body of water.

Finally, underground water occurs in the pore spaces of the rocks in question. In this context, it can greatly decrease the stability of a slope bank. While flowing, it causes consolidation and caverns through solution activity.

2.2.2 Theory of Turbulence

2.2.2.1 Principles of the Statistical Theory

Fluids that cause hydraulic actions which have an effect upon geodynamics are usually in a state of turbulence.

The fluids in question are generally ordinary, viscous (Newtonian) fluids. In principle, it should therefore be possible to solve any flow problems simply by solving the basic Navier–Stokes differential equations for the correct boundary conditions. Unfortunately, for the conditions prevailing in turbulent flow, this is impossible. Turbulent flow is characterized by irregular velocity fluctuations which are much too complicated to be followed in detail. It has therefore proven convenient to use the methods of statistical mechanics to deal with the problem (see, e.g., Batchelor 1953).

Accordingly, the velocity field $\mathbf{u}(\mathbf{x})$ (where **boldface** denotes vectors, $\mathbf{x}$ is the space coordinate) of a fluid in turbulent motion is considered as a field of a random variable. One can then define various types of averages. First of all, one has the average velocity $\bar{\mathbf{u}}(t)$, where the average has been taken with regard to the space coordinate $\mathbf{x}$. The fluctuation of the velocity is then given by

$$\mathbf{u}' = \mathbf{u} - \bar{\mathbf{u}}, \tag{2.2.1}$$

or, in components

$$u'_k = u_k - \bar{u}_k, \tag{2.2.2}$$

where $k = 1,2,3$. It is then customary to define the *correlation tensor* R_{ik} by

$$R_{ik}(\mathbf{r}) = \overline{u'_i(\mathbf{x})u'_k(\mathbf{x} + \mathbf{r})}. \tag{2.2.3}$$

Eq. (2.2.3) represents a tensor field.

The correlation tensor, and most significantly, its Fourier transform (called *spectrum* tensor) can be taken as fundamental kinematic variables in (homogeneous) turbulence problems.

2.2.2.2 Turbulent Stresses

Let us assume that turbulence has been established in a system of two-dimensional flow, such as in a channel or in a pipe. Let the coordinate parallel to which the mean flow is taking place be denoted by x, the coordinate orthogonal to this by y. Let the mean velocity in the x-direction be $\bar{u}_x(y)$, the fluctuation u'_x, and the fluctuation in the y-direction u'_y. The excess of momentum parallel to the mean flow is then $\rho u'_x$ per unit volume, and, consequently, the force per unit area (i.e., the shearing stress) is given by

$$\sigma = \overline{\rho u'_x u'_y}. \tag{2.2.4}$$

This is the general expression for the turbulent shearing stress.

In an ordinary, viscous fluid in laminar motion, the viscosity is defined in terms of the velocity gradient

$$\rho = \eta \frac{d\bar{u}_x}{dy}. \tag{2.2.5}$$

If, therefore, the turbulent shearing stress is expressed in terms of the average velocity gradient, then it can be said to be due to an "*eddy viscosity*". This "eddy viscosity" is a fictitious quantity; it can be used, however, to indicate the relationship between the turbulent stresses and the (average) velocity gradient.

The form [Eq. (2.2.4)] of the stress formula suggests that the drag **R** (a *force*) experienced by an object of linear dimension d immersed in turbulent flow is proportional to (for the x-component)

$$R_x \sim d^2 \rho \bar{u}_x^2. \tag{2.2.6}$$

The drag formula is commonly written as follows

$$R_x = \frac{\pi}{8} C_D d^2 \rho \bar{u}_x^2 \tag{2.2.7}$$

where C_D contains the correlation. The coefficient C_D is called drag coefficient. Prandtl (1926) approached the problem of turbulent momentum exchange from a different angle. He introduced a *mixing length l*, which he regarded as the mean distance which a small volume of fluid may travel normal to the main stream until it loses its identity by mixing. Assuming two-dimensional flow, with the mean velocity $\bar{u}$ being parallel to, say, the x-direction, but with a velocity gradient $d\bar{u}/dy$ being present in the y-direction, then the velocity of a mass of fluid arriving at a certain position will be proportional, in the first approximation, to $\bar{u} \pm l d\bar{u}/dy$, because l is precisely that (average) distance which it can travel without losing its identity. Hence the average velocity *fluctuation*, u', will be proportional to $l d\bar{u}/dy$. In isotropic turbulence, the lateral velocity fluctuation v' will, *ab hypothesi*, be proportional to the same quantity. Hence one obtains for the stress according to general principles [cf. Eq. (2.2.4)]:

$$\sigma = \overline{\rho u'v'} = \text{const. } \rho l^2 \left(\frac{d\bar{u}}{dy}\right)^2, \tag{2.2.8}$$

where any correlation factor between u' and v' may be incorporated into the (unknown) quantity l. Taking into consideration the fact that the stress σ must change its sign if the velocity $\bar{u}$ does, one can write (incorporating the *const.* also into *l*):

$$\sigma = \rho l^2 \left|\frac{d\bar{u}}{dy}\right|\frac{d\bar{u}}{dy} = \rho\varepsilon\frac{d\bar{u}}{dy} = \varepsilon'\frac{d\bar{u}}{dy}. \tag{2.2.9}$$

Here, one can call ε' the "exchange coefficient" for the momentum. The last expression is that which was suggested by Prandtl as describing the turbulent (shearing) stress. Comparison with Eq. (2.2.5), suggests that, in the present case, $\varepsilon' = \rho\varepsilon$ could also be called *eddy viscosity*, since it occupies the same position in the stress-velocity gradient relationship as does the viscosity in laminar flow.

If there are boundaries in a (viscous) fluid which is in turbulent motion, then it was noted long ago by Prandtl (1926) that the amount of turbulence present in any one region of the fluid must be affected by the proximity of such a boundary.

In fact, the boundary condition usually applied in the case of a viscous fluid is that the latter must stick to the walls. Hence, the velocity of the fluid near the boundary must be reduced by viscous drag. One usually defines as "boundary layer" that region near the boundary in which the velocity differs by 1% from the mean fluid velocity.

The motion in the boundary layer may be either turbulent or laminar, depending on the velocity of the fluid. If it be turbulent, then it is clear that there must be a sublayer very close to the wall in which the velocity is so small that the Reynolds number is smaller than the critical Reynolds number necessary for the maintenance of turbulence. This region is then called the laminar sublayer.

2.2.2.3 Homogeneous Turbulence

In a steady state, the turbulence has to be maintained at a constant level. Because energy is dissipated into heat during the turbulent motion, turbulent energy has to be introduced at a rate which equals that of the dissipation. In this fashion, a steady state can develop.

The mechanism is therefore clear in principle: turbulent energy is dissipated (into heat) at one end of the eddy spectrum (at high wave numbers) and introduced at the other end (low wave numbers) from some external source; it thus "flows" from one end of the spectrum to the other. According to Batchelor (1953), the energy relation in the linear case, for homogeneous, isotropic turbulence, originally due to Bass (1949) is given by

$$E = A\varepsilon^{2/3}k^{-5/3}, \tag{2.2.10}$$

where A is some constant, E is the total energy, k is the wave number (the Fourier transform of the space coordinate), and ε the rate of energy flux (dissipation).

In homogeneous turbulence, the kinematic structure of the eddies is given by the velocity correlation tensor [Eq. (2.2.3)]. Pressure fluctuations are associated with the velocity fluctuations (Mulhearn 1975), although they have not received much attention. Attempts have been made to model turbulent flows numerically (Fox and Lilly 1972), starting from the Navier–Stokes equations, assuming an initial Gaussian distribution of Fourier modes.

The displacement of a particular volume ("particle") of fluid in stationary homogeneous turbulence corresponds in the long run to that of Brownian motion, i.e., the mean square displacement $\overline{x^2}$ is proportional to the time t elapsed:

$$\overline{x^2} = \text{const.t.} \tag{2.2.11}$$

There are indications (Nordin et al. 1972), however, that the relation could also be different, viz.

$$\overline{x^2} = \text{const.t}^{2H}, \tag{2.2.12}$$

where H is a parameter between 0.5 and 1 (Hurst phenomenon; H is the Hurst parameter; after Hurst 1956, who discovered an analogous case in hydrological time series). Mandelbrot and Wallis (1969) showed that the Hurst phenomenon, like Richardson's law in wiggly lines, could be the outcome of a self-similarity model applying to the scales of the corresponding features.

When no more energy is introduced into a fluid volume in homogeneous, isotropic turbulent motion, the latter will start to decay. The decay laws for turbulence are fairly well understood (Hinze 1959). In homogeneous turbulence, if the velocity fluctuations are denoted by u', the decay law may be written as follows

$$\overline{u'^2} = \text{const.t}^{-m} \tag{2.2.13}$$

where

$$m = 1 \text{ for the "initial range",} \tag{2.2.14}$$

$$m = \tfrac{5}{2} \text{ for the "terminal" range.} \tag{2.2.15}$$

Numerical values for the instant when the "initial" time range ends and the "terminal" range begins, cannot easily be given. For turbulence created by obstacles of linear dimension M in a stream of mean velocity v, it was found that for the initial range

$$\frac{vt}{M} \leqslant 100 \tag{2.2.16}$$

and for the terminal range

$$\frac{vt}{M} \geqslant 500. \tag{2.2.17}$$

Writing $vt = x$, the above relations yield, together with some laws for the generation of turbulence, that the turbulent velocity fluctuations behind an obstacle are given by (Hinze 1959)

$$\left(\frac{v}{u'}\right)^2 = \alpha \frac{x}{\delta} - \beta, \tag{2.2.18}$$

where α and β are constants; β is a small number.

2.2.3 Stratified Flows

2.2.3.1 General Remarks

Stratified flows of water have a great geomorphic significance in connection with sheet floods on slopes and as density currents (transporting agents) in large bodies of water. The basic hydrodynamic theories concerning such flows will be reviewed in this section; specific applications will be presented in context with the corresponding landscape features.

2.2.3.2 Sheet Flows

A stratified flow regime occurs when a fluid is moving as a sheet down an inclined plane. In nature, this type of phenomenon occurs during the development of an areal "sheet flood".

The dynamics of the buildup of a sheet flow (due to the onset of a steady rainfall) was studied by Henderson and Wooding (1964), and has been reviewed at length by the writer in his book on natural catastrophes (Scheidegger 1975; see p. 195 therein). Accordingly, rainfall causes the buildup of a water plateau parallel to the underlying slope whose front (height h, horizontal distance x) has the form of

$$h = \text{const.} x^n \tag{2.2.19}$$

with $n = 2/3$.

A further study of the problem has since been made by Huppert (1982), who obtained an equation similar to (2.2.19) for the longitudinal flow profile, but with $n = 1/2$. Furthermore, he showed that it is not sufficient to consider only the longitudinal profile of the flow: the flow front begins spontaneously to develop instabilities in the form of a series of small amplitude waves across the slope which grow eventually into large tongues or lobes.

2.2.3.3 Superposed Streams of Different Densities

A further problem of fluid dynamics which has a bearing upon geomorphological effects is the question of stability of superposed streams of different density.

Let us suppose that one has two fluids of different densities, one beneath the other, moving parallel to, say, the x-direction. Let us assume that the interface is a horizontal plane and that it represents a discontinuity in the velocity by a finite amount. The stability of such a system was analyzed long ago by Helmholtz; his analysis was reproduced, for instance, by Lamb (1945, see p. 373 ff. therein).

It turns out (see Lamb 1945) that the common boundary is *always* unstable for sufficiently small wave lengths of the perturbations. If the relative velocity between the two fluids be V, and the densities ρ_1 and ρ_2, respectively, then those waves whose wave lengths are shorter than

$$\lambda < 2\pi \rho_1 \rho_2 V^2 / [g(\rho_1^2 - \rho_2^2)] \tag{2.2.20}$$

are unstable. Longer waves are propagated at a constant speed and constant amplitude. This result would indicate that, if there were no modifying influences, density currents would not be possible.

2.2.4 Underground Flow

2.2.4.1 General Remarks

Underground water exists in and travels through the pore space of soil and rocks. Pores are holes that are small compared with bulk volumes of the medium concerned. The "ground", is thus a porous medium.

A porous medium is characterized first of all by the fraction (or percentage) of voids it contains (*porosity*). However, in connection with hydrodynamics, it should be noted that the pores may be interconnected (in which case groundwater flow is possible) or noninterconnected (this occurs in clays, which are highly porous but do not permit flow). Thus, a second quantity, called *permeability*, is of importance in characterizing porous media.

The writer (Scheidegger 1974) has written a monograph on the physics of flow through porous media. Thus, we shall only briefly review here those facts of hydrostatics and hydrodynamics in porous media which are of importance with regard to geomorphology.

2.2.4.2 Hydrostatics in Porous Media

The statics of a fluid contained in the pore space of the ground corresponds to a large extent to that in an open container. Thus, water forms a surface in the ground (this is the groundwater level) as it does in a lake.

What is specific to porous media is the interaction of the fluid with the walls of the medium. First of all, there are capillary effects; these are, however, not of great importance on a large geomorphic scale. Second, there is the action of the pore-fluid pressure on the porous medium, and this may greatly influence the rheology and stability of the latter.

Thus, there is a famous hypothesis due to Terzaghi (1923), which states that the deformation of a porous medium is caused solely by the difference between the bulk (or overburden) pressure p_T in the medium and the pressure in the pore fluid p_F. This difference has been called effective pressure p_E

$$p_E = p_T - p_F. \tag{2.2.21}$$

If the overburden effect has to be considered as a stress tensor $(\sigma_T)_{ik}$ rather than as a pressure p_T, the Terzaghi relation becomes

$$(\sigma_E)_{ik} = (\sigma_T)_{ik} - p_F \delta_{ik}, \tag{2.2.22}$$

where δ_{ik} is the usual Kronecker symbol and $(\sigma_E)_{ik}$ is now called effective *stress*. A justification of the Terzaghi hypothesis has been presented by the writer (Scheidegger 1982, p. 131). Its validity has been confirmed by many experimental investigations. In particular, it has been checked repeatedly for geological materials, including crystalline rocks with "fracture porosity".

2.2.4.3 Hydrodynamics in Porous Media

The fundamental flow law for groundwater was established by Darcy (1856) in connection with a study of water filters. Consider a homogeneous filter bed of height h bounded by horizontal plane areas of equal size A. The bed is percolated by water; manometer tubes are attached at the upper (z_2) and lower (z_1) levels of the bed. Then the heights h_2 and h_1 to which the water rises in the manometer tubes (hydraulic heads) are connected with the total volume Q percolating in unit time through the areas A by the following relation (Darcy's law):

$$Q = - KA\frac{h_2 - h_1}{h},$$ (2.2.23)

where K is a constant. For small differences

$$z_2 - z_1 = dz,$$ (2.2.24)

Darcy's law becomes ($dh = h_2 - h_1$)

$$q \equiv \frac{Q}{A} = - K\,dh/dz.$$ (2.2.25)

The constant K (dimension: velocity) is called the *permeability* of the filter bed.

Darcy's law shows some peculiarities. It refers to the gradients of hydraulic heads (not of pressure) and it applies to a standard substance (water). It can be generalized to refer to pressure gradients and to variable densities and viscosities (cf. Scheidegger 1974), but in connection with groundwater and landscape problems (in contrast to problems arising in the petroleum and natural gas industries) the form given above is sufficient.

2.3 Geocryology

2.3.1 Introduction

Water, when it freezes, becomes ice. The solid phase "ice" may occur as a loose accumulation of crystals; it is then called "snow".

Alternatively, ice forms a polycrystalline solid which has some rather strange (compared with other types of solids) creep properties. The physics of poly-crystalline ice is therefore a subject in its own right.

Freezing water inside a porous medium reacts rather drastically with the latter. The solid phase "ice" is less dense than the liquid phase "water"; therefore a given volume of water (e.g., that contained in a certain pore system) expands upon freezing. Thus, large internal stresses are created in the ground when water freezes in its pore space.

2.3.2 Physics of Snow

2.3.2.1 General Remarks

Basically, snow is a loose, granular solid. As such, it behaves like a friable solid and shows creep and fracture behavior.

As a loose solid, it contains a large number of interstices or pores. Air and water can seep through these pores, effecting mass and heat transfer.

Water percolating through the pores of a snow mass can refreeze within the latter, thereby effecting a recrystalization of the snow until it becomes ice.

2.3.2.2 Mechanical Properties of Snow

As noted above, snow is a granular solid. As such it is subject to the three standard response behavior ranges (cf. Sect. 2.1.3) of a homogeneous medium: elastic, creep, and fracture ranges. The mechanical properties of snow in this context have been summarized, e.g., by Salm (1982).

The elastic range is geomorphologically insignificant, as it is almost nonexistent, i.e., there is (almost) no threshold until creep sets in. Nevertheless, stress waves from explosions do travel in snow packs as elastic-plastic waves (Brown 1980).

Geomorphologically of great importance is the creep behavior of snow. Generally, it is described by a viscous flow model (Haefeli 1967), although it is difficult to give an unequivocal viscosity for the snow since it varies widely. Furthermore, snow can also flow by an internal or interfacial friction mechanism (Lang and Dent 1982) which is important with regard to the dynamics of avalanches (cf. Scheidegger 1975). A nonlinear creep law was proposed by Ambach and Eisner (1985, 1986).

Finally, the failure range of snow can, first of all, be described by assuming a critical shearing strength (Ballard and McGraw 1966). The latter lies in the range of 0.06 bar but may, however, show features of strain softening and strain hardening depending on displacement rates (McClung 1977). Secondly, tensile strength limits of snow slabs are also of interest. In this context, snow behaves as a brittle material so that its mean tensile strength is a function of sample volume (Sommerfeld 1974) and strain rate (Narita 1980). General values can therefore not be given.

2.3.2.3 Seepage of Air and Water Through Snow

In dry snow layers, air and water vapor move by thermal convection (Palm and Tweitereid 1979). The mass flux is governed by a diffusivity equation; it carries with it a heat (energy) flux which is in addition to the heat flux due to conduction (the latter is also governed by a diffusivity equation). The net result of thermal convection in snow is that it lowers the insulating power of the layer.

More important is the water seepage through a snow pack (Colbeck 1975, 1976) as this can be initiation of a recrystallization process.

The percolation of water through snow corresponds to undersaturated flow in a porous medium (cf. Scheidegger 1974) and is thus not simple Darcy flow. In effect, the notion of relative permeability comes into play and the water saturation builds up a shock front (Scheidegger 1974, p. 257). This fact expresses itself in the hydrographs arising during snow percolation processes having discontinuities (Colbeck 1975).

The matter is even more complicated in layered snowpacks, inasmuch as the permeability becomes a tensorial quantity. The basic equations are well known from investigations in the petroleum industry (Scheidegger 1956).

2.3.2.4 Metamorphosis of Snow to Ice

The percolation and diffusion of water vapor through snow packs is a contributory factor in the transformation of snow to ice.

The growth of ice particles in snow pores is caused by vapor diffusion: the latter is calculated according to the methods outlined in the last section (Colbeck 1983). Alternatively, Lliboutry (1971) developed a metamorphism theory based on the gradual closing of the pores due to ice creep of the snow crystals. Lliboutry's (1971) formulas do not, however, seem to explain the conditions correctly, so that percolating and refreezing meltwater must have a profound influence (Shumskyi 1964).

2.3.3 Physics of Ice

2.3.3.1 General Remarks

As noted, ice is the solid phase of water. Whilst snow is a loose accumulation of single crystals of that substance, ice is a solid concretion of such crystals and, thus, is best treated as a polycrystalline substance.

Ice occurs in vast quantities in certain regions of the Earth. Its action is then that of an eroding and transporting agent which is similar to the action of water: for, as a polycrystalline aggregate, ice has certain rheological properties by which it is able to *flow*. In order to understand the transporting action of ice, it is necessary to give a brief review of its physical properties (Sect. 2.3.3.2) and of the various flow laws (Sect. 2.3.3.3) that have been suggested in the literature. These laws will form the basis for an understanding of the geomorphological significance of ice.

2.3.3.2 Some Physical Properties of Ice

Ice is a polycrystalline substance which exhibits a rather complicated mechanical behavior. The crystals belong to the hexagonal system which, in the aggregate, may be from less than 1 mm to over 1 m long. The physical properties of ice have been extensively studied, as witness for instance the reviews of Bernal (1958), Butkovich (1959), Paterson (1969) and Hutter (1983).

Upon being subjected to an external load, ice responds instantaneously by an *elastic* deformation; this means that in the elastic range, the stress depends on the deformation gradient only. Since the basic crystal structure in hexagonal, the elastic behavior of ice is described by five elastic constants. The latter depend somewhat on the temperature, but are generally of the order of 10^4 MPa (Hutter 1983).

However, more important is the *creep* behavior of ice. It is experimentally proven that ice does not have a creep threshold so that its creep behavior is best described by some pseudoviscous (i.e., nonlinear) law.

On occasion, instabilities may occur in the rheological behavior of ice. Thus, Clarke et al. (1977) discuss thermodynamic creep instabilities (runaway increase of internal temperature and deformation rate) and the development of superplasticity owing to grain growth (Duval andLliboutry 1985). These problems can no longer be treated by rheology alone, but involve thermodynamic relationships. They are of importance in connection with sudden glacier surges (cf. Scheidegger 1975).

For the ultimate *strength* of polycrystalline ice, rather divergent values have been obtained. The compressive strength of lake ice was found (Butkovich 1959) to range from 3.5 to 6.0 MPa (in the temperature range from -5 to $-15\,°C$); for the tensile strength, values between 1.4 and 1.7 MPa are generally accepted (Gold 1960). At high strain rates, the tensile strength may increase to 8 MPa (Hawkes and Mellor 1972).

2.3.3.3 Various Flow Laws

Experience shows that the flow behavior of ice cannot be characterized in any simple fashion. It is possible to define a pseudoviscosity (denoted by η). Therefore, one would write (Paterson 1969)

$$\tau = 2\eta\dot{\varepsilon}, \tag{2.3.1}$$

but with a viscosity η that does depend on stresses (shear stress τ). The following propositions have been made (Lliboutry (1970, 1987):

$1/\eta = 0.296\tau^{4.2}$ (Glen 1952)
$1/\eta = 0.25\tau^2$ (Haefeli 1967)
$1/\eta = 0.036 + 0.128\tau^{4.5}$ (Meier 1960)
$1/\eta = 0.036 + 0.080\tau^2 + 0.053\tau^4$,   (Lliboutry 1970)

where the numerical values correspond to stresses in bars, shear strain rates $\dot{\varepsilon}$ in s^{-1}.

The first two of the laws given above can be expressed in the form of a power flow law suggested earlier by Perutz (1950). The latter is

$$\dot{\varepsilon} = B\tau^n, \tag{2.3.2}$$

where B and n are constants. The quantity n ranges from 1 to 5 (Lliboutry and Duval 1985; a more recent review of the proposals for the power flow law exponent n has been given by Weertman 1983). The flow of ice seems to be due to the simultaneously operation of two processes: grain boundary creep at low stresses and intracrystalline gliding at high stresses (Meier 1960). Anisotropy may also occur (Lliboutry 1983).

In any application, power flow laws represent great mathematical difficulties. Fortunately, it turns out that the flow law of plasticity theory is in many cases quite adequate. An exposé of the latter may be found in the book of Hill (1950).

2.3.4 Physics of Frozen Ground

2.3.4.1 General Remarks

Frost in the ground represents a powerful geomorphic agent. The theory of how water freezes in the ground is by no means very well understood as of yet: what happens is not only that water simply freezes in situ, thereby expanding and "busting" the pores, but flow occurs concurrently which leads to the buildup of ice lenses and similar features ("ice segregation"). The problem has recently been discussed at a conference in Sapporo (Kinosita and Fukuda 1985).

The theoretical problems, thus, concern the destruction of rocks by freezing water ("cryogenic weathering") and the problem of frost heave.

2.3.4.2 Cryogenic Weathering

If water invades surface cracks and pores of rocks and freezes therein, the latter become very quickly destroyed. Generally, it had been assumed that this is due to the fact that ice occupies a greater volume than the same mass of water which would result in tremendous pressures being built up within the invaded rock (McGreevy 1981).

However, it has been noted by Walder and Hallett (1985, 1986) that the above assumption is at odds with experiences with the breakdown of porous building materials due to freezing water: the damage in these materials is related to the movements of water (Hoekstra 1969) in the pores and corresponding growth of ice bodies within them. A mathematical theory (Walder and Hallett, loc. cit.) of ice growth in rock cracks was based on principles of fracture mechanics and porous media physics, leading to frost damage rates which are comparable with empirical data.

The matter does not seem to be quite as simple, however: there are discrepancies between the experimental results obtained in the laboratory and the field observations in nature (McGreevy and Whalley 1985): experimental moisture regimes tend to promote degrees of saturation which exceed those likely to be attained in nature.

In any case, frost shattering seems to be the result of intense freeze-thaw cycles (Lautridou and Ozouf 1982; Coutard and Mücher 1985), the upper temperature limit for this phenomenon to occur being -3 to $-5\,°C$ (McGreevy and Whalley 1982). The frequency of such cycles seems to be too low in actuality, however, so that hydration shattering by adsorption-desorption of water molecules (French 1981) and other physical weathering processes (cf. Sect. 3.2.3) may have to be advocated as well. In shales, conditions are even more complicated, since clay swelling occurs (Mugridge and Young 1983).

Thus, the problem of cryogenic weathering seems to be still largely unresolved (McGreevy 1981; French 1986, 1987).

2.3.4.3 Frost Heaving

It turns out that in many instances, the actual displacements caused by water freezing in the surface material of the Earth are much greater than can be explained simply by assuming that the volume expansion is solely due to the volume increase undergone by the water contained in the material in question when it becomes ice. This phenomenon is particularly, well known from the frost-heaving experienced under many circumstances (cf., e.g., Taber 1924; Ruckli 1950; Jumikis 1955; Penner 1956; Balduzzi 1959).

The characteristic feature of frost heaving lies in the fact that layers of ice (often called ice lenses) may form within a water-saturated porous medium, without the water freezing in the adjacent pores. An ice lens keeps growing by the addition of water which is being drawn from the porous medium. There are two theories of this process of which the writer is aware. The first is due to Jackson and Chalmers (1958) based on nucleation theory, and the second is due to Gold (1958), based on capillary equilibrium. Some thermodynamic relationships for soils near the freezing point have also been published by Low et al. (1968). A lengthy discussion of the theories available has also been published by Martynov (1959).

We shall discuss first the theory of Jackson and Chalmers. Accordingly, one must take the interfacial tensions between water, ice, and the rock into account. One has generally

$$\sigma_{LB} = \sigma_{SB} + \sigma_{LS} \cos \alpha, \tag{2.3.3}$$

where σ_{SB}, σ_{LB} and σ_{SL} are the interfacial tensions (interfacial energies) between the solid-rock, liquid-rock and solid-liquid interfaces, respectively; α is the contact angle.

According to nucleation theory, the initiation of solidification can occur in a capillary of radius r at a temperature T* given by

$$T^* = T_E + \frac{\sigma_{LS} T_E \cos \alpha}{Lr} \tag{2.3.4}$$

where T_E is the usual equilibrium temperature and L the latent heat. It should be noted that, for $\cos \alpha$ negative, T* will lie below T_E. This case occurs in the water-ice system where $\alpha = 180°$.

In a porous medium which may be regarded as an assemblage of capillaries, freezing will not be initiated until the temperature T* is reached. Once this has happened, a nucleus will form which will rapidly grow to an ice lens, since T* is below the equilibrium freezing temperature T_E, i.e., since the water in the system is supercooled. Water will be drawn from wherever it is available to let the ice lens grow. Thus, volume expansions much greater than those corresponding to the freezing of the water originally contained in the pores may occur.

The theory of Gold to explain the occurrence of frost heaving is similar to that of Jackson and Chalmers, but the role of the freezing temperature depression due to the prevention of nucleation is replaced by a direct shifting of the freezing temperature due to capillary forces. It turns out that parts of the ice water interface in the pores which are convex toward the water with a large radius of curvature will grow more rapidly than parts with a smaller radius of curvature. Thus, a picture as shown in Fig. 15 will result. This represents frost heaving.

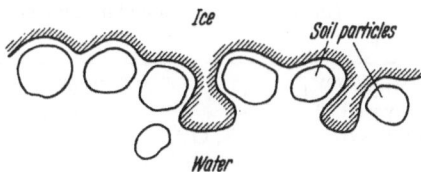

Fig. 15. Ice-water interface in a porous medium. (After Gold 1958)

2.4 Physics of the Atmosphere

2.4.1 Introduction

The study of the atmosphere is the subject of the science of *meteorology*. Many pertinent treatises exist to which the reader is referred for details; in the present context we shall only give a short account of those meteorological results that are of importance with regard to geomorphology. Our main concern will be with climatic effects near the ground, i.e., with "micrometeorology", as this is of utmost geomorphological significance. Comprehensive treatises on this subject have been written by Sutton (1953) and by Geiger (1950).

2.4.2 Statics of the Atmosphere

2.4.2.1 General Remarks

We start with the equilibrium conditions obtaining in the atmosphere.

In this connection, one has first of all to review the structure and the chemical composition of the gaseous envelope of the globe.

Then, we shall review the physical principles that govern the atmospheric equilibrium.

2.4.2.2 Structure and Composition of the Atmosphere

The structure of the gaseous envelope of the Earth is a layered one.

The first 18 km of it are called the troposphere, in which the temperature decreases with altitude from about 290 to 210 K. The troposphere is bounded by the tropopause, beyond which the temperature is first constant at about 210 K and then begins to increase again to about 280 K, which is reached at a height of ca. 50 km. This region is called the stratosphere. Beyond this height, the temperature drops (mesosphere) to an altitude of about 80 km, where it falls below 200 K and the ionosphere begins. In the latter, the temperature rises rapidly to reach more than 900 K at 300 km altitude. Only the troposphere is of geomorphic significance.

The chemical composition of the atmosphere is approximately that shown in Table 3.

2.4.2.3 Equilibrium of the Atmosphere

For most practical purposes, the air forming the atmosphere can be regarded as an ideal gas. We have then the following well-known equation of state:

Table 3. Chemical composition of the atmosphere (volume per cent). (After Paneth 1933)

Nitrogen(N_2)	78.08%	Argon	0.93%
Oxygen(O_2)	20.95%	Carbon dioxide	0.03%
Others (Ne, Kr, He, X, O_3)	0.01%		

$$\frac{p}{\rho} = RT, \tag{2.4.1}$$

where p is the pressure, ρ the density, T the absolute temperature (in Kelvin) and R the universal gas constant divided by the molecular weight of the gas. For dry air we have $R \sim 2.87 \times 10^6 \, cm^2 \, s^{-2} \, deg^{-1}$. Since in a column of fluid in the gravity field (with acceleration g) the condition of static equilibrium yields

$$\frac{dp}{dz} = -\rho g \tag{2.4.2}$$

(z being a coordinate which is counted positive upward), one obtains the following expression for the pressure distribution

$$\frac{dp}{dz} = -g \frac{p}{RT} \tag{2.4.3}$$

and hence

$$\text{lognat } p = -\int_0^z \frac{g}{RT(z)} dz + \text{lognat } p_0 \tag{2.4.4}$$

$$p = p_0 \exp\left\{ -\int_0^z \frac{g}{RT(z)} dz \right\}; \tag{2.4.5}$$

similarly, one obtains for the density distribution

$$\rho = \rho_0 \exp\left\{ -\int_0^z \frac{g}{RT(z)} dz \right\}. \tag{2.4.6}$$

This shows that it is possible to determine theoretically the pressure and the density if the variation of temperature with height is known.

This distribution of temperature with height, in the near to ground layer, which alone is of geomorphological significance, can assume a wide variety of patterns. Stability is attained in the atmosphere if the temperature gradient is below a characteristic value. The latter can be calculated as follows. Assume that a volume of air is displaced upward through the distance dz. This causes a pressure decrease by the amount

$$dp = -g\rho dz = -g \frac{p}{RT} dz. \tag{2.4.7}$$

However, owing to the pressure change, the air will undergo an adiabatic temperature change given by (see, e.g., Planck 1945, p. 63)

$$dT = T \frac{dp}{p} \left(\frac{\gamma - 1}{\gamma} \right),$$ (2.4.8)

with $\gamma \cong 1.41$ for dry air. If the air column is to be stable, this temperature change must produce a temperature which is not higher than that already present in the column. Thus, we have in the limiting case

$$\frac{dp}{p} = -g \frac{dz}{RT} = \frac{dT}{T} \left(\frac{\gamma}{\gamma - 1} \right)$$ (2.4.9)

or

$$\frac{dT}{dz} = -\frac{g}{R} \frac{\gamma - 1}{\gamma},$$ (2.4.10)

which, for air, represents a gradient of roughly $-1\,°C$ per $100\,m$. This is the critical pressure gradient ("dry adiabatic lapse gradient"); if the temperature change is less than (in absolute value) or equal to this gradient, the air column in question will be stable.

The above discussion refers to dry air only. For moist air, suitable modifications can be made which cause a change in the various constants occurring in the equations.

In effect, the molecular weight of water vapor is only 18/29 of that of dry air; hence R is larger for moist than for dry air. Under the same pressure and temperature conditions, the density of moist air is therefore less than that for dry air. Thus, the adiabatic lapse gradient becomes only equal to about $-0.6\,°C$ per $100\,m$ for moist air, as compared to ca. $-1.0\,°C$ per $100\,m$ for dry air.

2.4.3 Dynamics of the Atmosphere

2.4.3.1 General Remarks

The atmosphere at rest obviously can have little effect upon the formation of the Earth's surface. We thus have to consider the *dynamics* of the atmosphere.

The air above the ground forms a complicated thermomechanical system. The basic equations of motion, which include the Navier–Stokes equations for a viscous fluid and the thermodynamic equations for a real gas, can be written down, but it turns out that solutions thereof can be obtained only in the simplest of cases.

In principle, these cases concern quasistatic flow in the atmosphere, which is represented by geostrophic flow on the one hand and by boundary layer flow on the other.

Finally, a few remarks will be made on turbulent flow in the atmosphere.

2.4.3.2 Geostrophic Flow

In geostrophic flow, the vertical acceleration is ignored in the motion, the horizontal component of the angular velocity vector of the Earth is neglected, and the friction is also neglected.

Then the flow is subject essentially to two forces. One is the Coriolis force F_c (per unit mass) which is always normal to the (horizontal) wind velocity vector, to the right in the Northern hemisphere and to the left in the Southern. Its magnitude is

$$F_c = 2\omega v \sin \varphi, \tag{2.4.11}$$

where ω is the angular velocity of the Earth, v the wind speed, and φ the latitude. The other force is caused by horizontal pressure differences; denoting the pressure gradient normal to the isobaric lines by $\partial p/\partial n$, the force per unit mass (F_p) in question is

$$F_p = \frac{1}{\rho} \frac{\partial p}{\partial n}, \tag{2.4.12}$$

where ρ is the density of the air.

In quasistatic equilibrium, the two forces must balance each other, which shows that the geostrophic wind will blow parallel to the isobaric lines with a velocity

$$v = \frac{1}{2\rho\omega \sin \varphi} \frac{\partial p}{\partial n}. \tag{2.4.13}$$

In particular cases of the quasistatic theory, it becomes possible to regard the air flow as potential flow. The pertinent equations have been solved by Pockels (1901) for the flow around a thin mountain. A discussion of the stream-line pattern obtained yields the result that the highest horizontal velocity occurs at the top of the mountain and that the maximum vertical velocity is obtained at the middle of its windward side.

2.4.3.3 Boundary Layer Theory

The assumption of a frictional force leads to the consideration of the effect of viscosity on quasistatic or stationary flow. As with all viscous fluids, viscosity has the effect that a laminar boundary layer must exist near a solid surface. It is possible to set up the equations of motion for this boundary layer. This has been done by Prandtl (1926) in his investigations of turbulence (cf. Sect. 2.2.2.2). In the present context, we are interested only in the *laminar* boundary layer, which corresponds to the laminar sublayer in the theory of turbulent flow. Prandtl obtained his "boundary layer equations" by writing down the pertinent equations for a viscous fluid (cf. Sect. 2.1.2) in two dimensions (x, z) and consistently neglecting terms of an order higher than the first.

Blasius (1908) has integrated Prandtl's boundary layer equations numerically for a particular case ($u = w = 0$ for $z = 0$ and $u = v$ for $x = 0$ and $x = \infty$, $v =$ kinematic viscosity) and obtained an expression for the ground friction σ_0:

$$\sigma_0 = \frac{0.332 \rho v^2}{\sqrt{vx/v}}. \tag{2.4.14}$$

2.4.3.4 Turbulent Flow in the Atmosphere

The theory of quasistatic flow in the atmosphere, as discussed in the last section, has relatively little importance with regard to geomorphological effects. It turns out that the air flow near the ground has to be *turbulent* if geomorphological changes are to be caused by it.

The theory of turbulent air flow near the surface of the ground is entirely analogous to the theory of turbulent flow of water near a surface. The general concepts of the statistical theory of turbulence, as outlined in Sect. 2.2.2, can therefore be directly applied to the flow of air. With regard to the velocity distribution near the ground, as will be discussed in more detail in Sect. 8.2.2.2, the theory is identical to that of turbulent water flow in open channels (Sect. 4.2). The details, therefore, will be discussed later in their proper context.

In all geomorphological applications, the origin of the motion of the wind or of the turbulence is of little importance, as long as its structure is known. Questions of general meteorology will therefore not be discussed here.

2.5 Problems of Climate

2.5.1 The Notion of Climate

In a first approximation, the climate of an area is determined by (1) the average temperature, (2) the average number of hours of sunshine per day, and (3) the average daily (or annual) rainfall. This first approximation may be too crude, inasmuch as *averages* are only adequate in unimodal distributions. If pronounced seasons occur in a locality during the course of a year, these averages must be given for each distinct season, usually two, viz. summer and winter (or dry and wet) seasons. The climate, in turn, governs the kind and intensity of the exogenic processes that can occur.

On the basis of the corresponding averages, the climates of the world can be divided into three large classes:

1. Cold climates, which are characterized by the fact that frost may occur at any day of the year. They are found in polar regions and elsewhere at high altitudes.
2. Moderate climates, which are characterized by seasonal changes. Thus, the distributions of temperature, hours of sunshine per day, and rainfall rate are basically bimodal. The averages must be given for the winter and summer seasons separately. Moderate climates are found in the middle latitudes.
3. Hot climates, which are characterized by the fact that the temperature is such that frost never occurs. Under such conditions, the rainfall rate is the next most important variable in determining the type of landscape that evolves, ranging from rain forest to desert. Hot climates are found in low-latitude regions at low altitudes.

The distribution of climates has not always been the same as it is today. It has been mentioned several times already that, during the Pleistocene (as well as during other geological periods) there were periods of extensive glaciation. A corollary to the glaciations are fluctuations of sea level ("eustatic changes"), because large amounts of water are withdrawn from the sea during such periods to be immobilized in ice caps: sea level fluctuations are the reciprocal of changes in ice volumes (Matthews 1972). Other consequences occur in the middle latitudes.

2.5.2 Climate Change: Evidence

2.5.2.1 General Remarks

A naive person tends to take the world he experiences as permanent and stationary. Thus, to-day's distribution of climates is taken as unchangeable; plans by developers and allocators of resources are based upon the immutability of such factors as annual rainfall, average temperature, etc.

However, already during a single modern human lifespan disturbing changes can be observed: glaciers retreat, desertification (Sahel zone!) expands, sea levels (Venice!) rise.

The changes are even more obvious when the geological record is examined. In fact, it turns out that today's climate is, in a geological context, rather unusual: it is too cool. Normally, i.e., during the vast majority of geological time, the climate was "hot" or "moderate (in the sense as defined in Sect. 2.5.1) right up to the poles. The climate in Earth history has recently been reviewed by Drake et al. (1982) and the corresponding geological record by Crowley (1983).

The most important fact is, as has already been mentioned, that the "normal" climate of the whole Earth has generally been moderate to hot, i.e., globally about 5 °C warmer than at present (Kreutzbach and Gallimore 1989). This "normal" climate was interrupted several times by colder periods ("ice ages"), each of which lasted may be a few 100 000 years. Each ice age itself consisted of a series of fluctuations, in which extensive glaciations alternated with warmer "interglacial stages". The last ice age occurred in the Pleistocene (ending about 13 000 years ago); it is not clear whether it has truly ended or whether we are at the present time in an interglacial stage of this ice age. During an ice age, the temperature is globally 6 °C, on the continents 10 °C colder than at present.

Earlier ice ages occurred in the late Paleozoic and in the Precambrian.

The evidence for the statements made above will now be presented.

2.5.2.2 Holocene Climatic Trends

We begin with a discussion of climatic trends that can be detected to have occurred since the end of the last glaciation at about 13 000 years ago. The period in question is commonly referred to as Post-Pleistocene or "Holocene". The specific periods in question are the "present" (or "modern" period) which includes the last 100 years or so, the "historic period" going back to about 4000 years before the present (B.P.), and the period from 4000 to 13 000 years B.P.

Indications of modern climatic trends can be obtained, for instance, from simple measurements of marine temperature fluctuations (Folland et al. 1984). There does not seem to be a long-term trend indicated from this source during the last 130 years. Using temperature values from meteorological *land* stations, Hansen and Lebedeff (1987) found a global warming trend of 0.5–0.7 °C for the past century. In addition, there seem to be oscillatory cycles (Kukla et al. 1977; Burroughs 1980; Lamb 1980) superposed on the long-term trends. The spectral structure of the cycles is difficult to ascertain. The same observation is also made in specific areas, such as central Europe (Grabau 1987; Wallen 1986). Climatic oscillations express themselves in the El Nino phenomenon (Philander 1983), glacier fluctuations (Allison 1981), and in erosion/desertification phenomena (Leopold 1976; Courel 1983). The latter may or may not have been influenced by human activity (Toya et al. 1985; Levi, 1990).

Turning to larger time intervals, such as can properly be called "historic" (the last 4000–5000 years), one can first of all refer to the written record (Tollner 1974; Ingram et al. 1978). This is, however, somewhat subject to the views of the chronists.

The scientific method of choice for ascertaining historic climates is dendroclimatological: trees grow faster under warm than under cold conditions; tree rings are therefore wider during warm periods. If the wood is dated (e.g., by the ^{14}C method), a dated climatic record can be obtained. Studies along these lines were made, e.g., by Lamarche (1978), by Cropper (1982), by Furrer and Holzhauser (1984). and by Jacoby (1989). Other methods were presented at a symposium in Stockholm (Mörner and Karlen 1984) and by Kashiwaya et al. (1988).

Going back even further in time, to the end of the ice age (Pleistocene), it may be noted that, in addition to the methods already mentioned above, paleogeomorphological features have been used to identify climates, such as the former size of lakes and the distribution of fossil dunes (Goudie 1982; Besler 1983), or vegetation studies (Brakenridge 1980; White and Mathewes 1982).

The result of all these studies is that, about 13 000 years ago (end of the ice age), a long-term warming trend set in until a climatic optimum was reached about 4000 years ago, after which (around the birth of Christ), there was a rather sudden drop; since then, the temperature dropped again slowly to a small minimum (little ice age) around 1700 A.D., since which time there has been a warming trend. Superposed on these long-term trends there have been oscillations whose spectral composition is not well understood (Fairbridge 1968, see p. 531). Therefore, it is not possible to make predictions for the future: the general warming trend may have been arrested during the past 30 years or so, or this may be the effect of a superposed downswing oscillation.

2.5.2.3 Pleistocene Climate

As "Pleistocene" one denotes the timespan between roughly 1.6 million years and 10 000 years B.P. This time-span is characterized by a series of glacial stages (extreme extensions of the ice-covered areas of the world), interspersed with warmer periods (interglacial stages). The last four of the glacial stages are particularly well defined; they are identifiable in Europe as well as in North America, where they have received different names (derived from the localities which were reached during the greatest advance of the corresponding ice sheets) as shown in Table 4 (cf. Fairbridge 1968, p. 923). Thus, between the maxima of

Table 4. Ice ages in the Northern Hemisphere (cf. e.g., Fairbridge 1968, p. 923)

Absolute dates kyears B.P.	North American name	European name
0–67	Wisconsian	Würm
67–128	Interglacial stage	
128–180	Illinoisian	Riss
180–230	Interglacial stage	
230–300	Kansan	Mindel
300–330	Interglacial stage	
330–470	Nebraskan	Günz

sequential Pleistocene ice ages there were intervals of ca. 100 000 years; the actual maxima themselves lasted only a few hundred years. Further glacial and interglacial stages have been postulated in Europe for the time prior to the Günz glaciation, but these are doubtful.

The evidence for the existence of Pleistocene glaciations is first of all morphological, particularly for the last (Würm or Wisconsin) stage, because features caused by them have remained in many places to this day (cf. Sect. 1.7.3). In this connection, it should be noted that not only typically "glacial" features (such as moraines, erratic blocks, etc.) indicate the corresponding climatic conditions, but also traces of pluvial episodes (Strick et al. 1982; Spaulding and Graumlich 1986; Pokras and Mix 1987), types of paleosols (Schubert 1988), paleo-erosion patterns (Colinvaux 1979; Warren 1985), sea level changes (cf. Sect. 2.5.2.5), and faunal extinctions (Martin and Klein 1984).

More quantitative paleoclimatic research methods have been reviewed by Fairbridge 1972, Hecht et al. (1979) and Mahaney (1981). Thus, faunal analyses of deep sea sediments (Wollin et al. 1971; Berger et al. 1983) and, particularly, isotope studies (mainly the $^{18}O/^{16}O$ ratio, which is related to temperature) have been used (cf., e.g., Lorius et al. 1985; Mix and Ruddiman 1985). The temperature oscillations during the Pleistocene have thereby been confirmed; notably, it has been shown that at least four main glaciation stages are correlated with corresponding *global* climatic conditions.

2.5.2.4 Pre-Pleistocene Climates

The pre-Pleistocene climates of the globe are characterized by generally uniformly moderate to warm conditions. Only at the end of the Paleozoic and much earlier in the Precambrian is there some evidence for the occurrence of (geologically speaking) brief ice ages.

The cooling trend leading up to the Pleistocene ice age seems to have begun about 3.5 (Mercer at al. 1975; Keigwin and Thunell 1979) or 2.6 (Louberre and Moss 1986) million years ago. Before that, in the late Tertiary, the climate was "moderate". This in itself was a deterioration ("Oligocene Deterioration") of the climate prevailing in the early Tertiary, when it was warm (Wolfe 1978; Hubbard and Boulter 1983). Most of these results are based on paleobiological (foraminifers, phanerogamic leaves, and pollen) studies.

In the Cretaceous, the global climate had generally been assumed as uniformly hot (Thompson and Barron 1981; Donn 1982), a view which has recently been challenged by Kemper (1986), who claimed that the *generally* hot conditions of the Mesozoic were interrupted by some cold periods during the Cretaceous as well as earlier on in the Jurassic. As evidence, Kemper adduced studies of composition of carbonate sediments. The oscillation periods during cold spells seem to be similar to those of the global advances and retreats of the glaciers in the Pleistocene.

A definite ice age, however, has been identified in the late Paleozoic (Spjeldnaes 1978; Dickins 1984); at least two more in the Precambrian (Spjeldnaes 1978), one with two pulses in the late Proterozoic (Kroener 1977; Crittenden et al. 1983; from an analysis of deposits which are evidently of glacial origin), another in the early Proterozoic (Gowganda formation of Ontario by a study of major element chemistry of lutites: Nesbitt and Young 1982). The glacial periods, in a geological sense, are therefore indeed brief, few, and far between throughout the major portion of the Earth's history. The planet Mars, in fact, may have had a similar climate as the earth in the first billion years of its life (McKay and Stoker 1989).

2.5.2.5 Eustatic Changes

Coastal and submarine geomorphology is much affected by changes of the relative sea level ("eustatic changes") with regard to the land (Mörner 1986). These eustatic changes are in a small measure the result of endogenic processes, such as sea floor spreading (Flemming and Roberts 1973), geoid changes (Mörner 1976), intraplate stresses (Cloetingh et al. 1985; Cloetingh 1986), or vertical crustal movements (Tjia 1975; Rosenberg 1970; Peltier et al. 1978). In the main, however, they are due to climatic phenomena: sea level is low during cold climatic periods (Friedman and Sanders 1970): as noted, sea level is essentially the mirror image of the extent of ice accumulations on the globe because the water immobilized in ice caps is missing in the ocean. Thus, sea level is low during ice ages and lowest during the principal glaciation stages.

Sea level changes are evident already from very short-term recent observations. Thus, tide-level records (Donn et al. 1967) indicate an upward trend along most of the world's coastlines in this century (Newman and Fairbridge 1986): the mean rise has been more than 1 mm/a (Lambeck and Nakiboglu 1984) at least since 1890, at which time there seems to have been a global low point (Fairbridge and Krebs 1962). In the most recent time (the last 30 years), the rise appears to have accelerated globally to 5 mm/a (Rosen 1978): one of the most notorious cases is occurring in Venice (Caputo 1971). The pattern varies somewhat if considered ocean by ocean. Thus, even reversals have been noted, e.g., a lowering of sea level by 3.9 mm/a in Denmark (Christiansen et al. 1985). This seems to be connected with ("isostatic") vertical movements which occur also elsewhere in Scandinavia (Mörner 1973).

A general rise of sea level can be followed back throughout the historic period (2000 a) in many areas of the world, notably in the Mediterranean (Caputo and Pieri 1976) and North America (Pirazzoli 1977). However, emergence (lowering of sea level) prevails around the Arctic, Indian, and Pacific oceans. These facts are attributed to postglacial isostatic uplift and other tectonic phenomena (Flemming 1969; Pirazzoli 1977).

When referring to earlier periods, historic records are no longer available and a special methodology for the study of sea level fluctuations based on dated (^{14}C) plant successions, ancient beach deposits, and sediments in shallow water zones has been developed (Tooley 1985, 1986).

Thus, going back towards the end of the last ice age, we note that one has to contend with two opposing effects: isostatic uplift of the land being freed from the load of the ice, and the rise of sea level due to the addition of the meltwater to the ocean. Thus, for most "stable" (in a tectonic sense) areas of the world, a rise of sea level is postulated for the last 6000–10 000 a (Shepard and Suess 1956; Coleman and Smith 1964; Andrews et al. 1973, and many others). Referring to such past sea levels, Walcott (1972) has constructed postglacial rebound curves for isostatically rising areas. However, this general picture has been challenged by Clark et al. (1978), who noted that the sea level rise in the postglacial period was by no means uniform, even when disregarding obviously deglaciated areas. See levels around Australia seem to have been sinking in the period in question (Chappel 1983). The increase in water mass in the ocean does not necessarily seem to cause a rise in sea level, as its weight causes a deformation of the substratum (as far as the mantle) and of the geoid as well, which has generally been neglected.

The most thoroughly investigated period with regard to eustatic changes is that relating to the succession of Pleistocene glaciation stages, particularly the last (Würm/Wisconsin) one. Data have been collected particularly by Shepard (1963) and by Milliman and Emery (1968), who found a general lowering of sea level from 35 000s to 17 000 B.P., prior to the more or less general rise observed thereafter (as described above). Thus a low point may have been reached at about 17 000 B.P. On the general trends, saw-like fluctuations appear to have been superposed (Kazanskiy 1985).

Regarding eustatic changes connected with the earlier glaciation stages, i.e., those prior to the Würm/Wisconsin glaciations, it is generally proposed that there have been oscillations with periods of about 100 000 superposed upon a general rise of sea levels, leading to a "staircase rise" during the past 400 000 years (Morrow 1986). The superposed oscillations have been inferred to be linked to the cycles of glaciation. These inferences are based on the study of shelf carbonate successions in all parts of the world.

Going back into the Tertiary and the Mesozoic, we note that an excellent summary on the subject of sea levels during that time has been published by Steckler (1984). Accordingly, most authors agree on a eustatic high in the late Cretaceous. Thus, there would have been a steady fall of sea level from that time onward up to the Pliocene, and a rise from the early Cretaceous to the late Cretaceous. In Jurassic, sea level was fairly constant at about 30 m below the present level. Some authors (mainly Vail et al. 1977) assume large fluctuations (+ 200 m and more) superposed on the general trend. Hsü (1985) goes as far as to postulate a dry Mediterranean in the mid-Miocene.

Eustatic fluctuations and transgressive events seem to have occurred also at various times in the Paleozoic (e.g., Johnson et al. 1985; Saunders and Ramsbottom 1986) and at earlier apochs. However, for an understanding of today's geomorphology, evidently only the latest events (i.e., those of the Pleistocene) are of primary importance.

2.5.3 Climate Change: Theory

2.5.3.1 General Possibilities

We have seen that there is no doubt that the climate has changed drastically during the Earth's history: at the very least, it must be considered as established that there have been relatively brief ice-age periods (in the Pleistocene, in the Palezoic, and in earlier epochs) interspersed in long periods of moderate to warm climates.

The reason for the occurrence of the observed large climatic swings is still largely a matter of speculation (cf., e.g., Hunt 1984). The general possibilities for seeking causes of the climatic changes have been reviewed in books by Cornwall (1970), Ponte (1976), Lockwood (1979), and Drake et al. (1982). Reviewing the literature, one can state that the theories that have been advanced have sought the causes either in astronomical or else in endogenic effects. A combination (feedback) between these two types of effects has also been advocated. We shall discuss the various proposals in turn.

2.5.3.2 Astronomical Causes

a) *Irradiation Fluctuations.* The best-known astronomical hypothesis of the origin of ice ages is that based upon irradiation fluctuations which had been postulated by Milankovitch (1941). This theory is based on the fact that, because of the varying gravitational effects of the Moon, Sun, and Planets upon the Earth (due to the different relative position of these heavenly bodies), the position of the equinoxes with respect to the perihelion, the obliquity of the plane of the Earth's ecliptic, and the eccentricity of the Earth's orbit around the Sun vary in an oscillatory fashion with periods of roughly 21 000, 41 000 and 97 000 years, respectively. In the middle and high latitudes, this leads to a variation of insolation with a period of 41 000 years, in the low latitudes to a variation of insolation with a period of 21 000 years. The total insolation over one year stays approximately constant, as summer and winter effects vary in an opposite manner. However, it may be argued that a low summer insolation will generally favor the origination of an ice age, since the snow that has fallen in winter tends to remain permanently. In comparison with this, the extremities of cold reached in winter seem to be of little importance (this had been pointed out earlier by Köppen and Wegener (1924). Of greatest importance are the insolation fluctuations at middle and high latitudes; at these latitudes, they have a periodicity, as noted above, of 41 000 years. These could trigger climatic changes by orbital forcing of a nonlinear climatic oscillator (Broecker 1966; Le Truet and Ghil 1983). The period of 41 000 years for climatic changes is of the right order of magnitude in comparison with the periodicity of glacial and interglacial stages during the Pleistocene. Thus, Milankovitch fits the known record of glacial and interglacial stages with the insolation cycle.

The "Milankovitch theory" has been heavily criticized, but no fundamental argument against it has ever been produced. A recent review of the state of the art has been given by Berger et al. 1984 and by Berger (1988). It is clear that the insolation effect cannot be the only one that is present, simply because there were no ice ages during long periods of geological history.

b) *Solar Emission Theory.* Another astronomical hypothesis of the origin of ice ages is the solar emission theory, which assumes periodic fluctuations in solar radiation. It has been advanced, for instance, by Fritz (1951), Schulman (1951), Öpik (1953, 1968), Willet (1953), Simpson (1940), and Charvatova (1989). As with orbital forcing, solar emission fluctuations could trigger flip-flops between different quasistable states of the (nonlinear) climatic oscillator (Oerlemans 1980).

It is quite clear that the hypothesis of solar emission fluctuations cannot be tested easily. It is known that small emission fluctuations with short periods, such as the sunspot cycle of 11 years or the 179-year cycle of the solar inertial motion (Fairbridge and Shirley 1987) do indeed occur, but nothing definite is known with regard to the necessary long period of 40 000 years for an explanation of the succession of the Pleistocene glaciations, nor indeed with regard to possible emission "lows" hundreds of millions of years apart in order to explain the origin of the ice ages in the Pleistocene and the Paleozoic, with normal "warm" periods in between.

2.5.3.3 Endogenic Causes

a) *Volcanism.* The first endogenic cause we shall discuss is volcanism. A review of the hypothesis of the volcanic origin of ice ages has, for instance, been given by Flint (1947, 1957). Accordingly, volcanic activity may pollute the atmosphere with a large mass of volcanic dust which is an obstacle to the incoming short-wave radiation from the Sun, but does not interfere much with the outgoing long-wave radiation. Indeed, the effects of recent volcanic eruptions on climate variations has been substantial (cf. Bray 1979; Sear and Kelly 1980; Porter 1981; Kelly and Sear 1984). Thus, the presence of volcanic dust could cause an ice age. The rhythm of ice ages would then reflect the rhythm of volcanism. However, it is difficult to see why the volcanic activity of the Earth should be a rhythmic process.

b) *Submarine Tectonism.* Some Russian investigators, notbaly Saks et al. (1955) advocated that submarine tectonics would be able to change the oceanic water circulation in such a fashion that an effective heat exchange between the seas of the world would be prevented; this could produce an ice age. However, the rhythm of ice ages is again difficult to explain in this manner.

c) *Polar Wandering Theory.* It is well known that it has been postulated that the axis of rotation of the Earth may have had different positions with regard to the continents at various times during geological history. This phenomenon has been called "polar wandering". Ewing and Donn (1956, 1958) and Ewing (1960) argue that every time the axis of rotation is such that one of the poles falls into a (nearly) enclosed sea (such as presently into the Arctic Ocean), a series of ice ages will be the result. The reason for this is that an enclosed sea at either pole, according to Ewing and Donn, represents an unstable system. If the polar sea is frozen over, the supply of moisture in the polar areas is limited and existing glaciers recede. This, in turn, leads to a rise of sea level so that circulation in the oceans becomes such that a ready exchange of water and heat takes place between the polar and warmer seas. At this instant, the polar sea will thaw out. However, with an open polar sea, a ready supply of moisture is present which will accumulate in high latitudes as snow and give rise to glaciations. The process continues until so much snow is locked up in land areas that the sea level is sufficiently lowered to prevent an effective interchange of water between the polar and warm seas. At that stage, the polar sea suddenly freezes over and a new interglacial cycle statrs.

Instead of polar wandering, continental drift (also in its modern plate-tectonic version) has been invoked to have caused the formation of enclosed high-latitude seas which would trigger the flip-flop mechanism represented by glacial-interglacial cycles (Donn and Shaw 1977; Beckinsale 1980).

The "pluvial" flip-flop theories described above have been criticized on the grounds that the temporal correspondence between the decrease of glaciation and the icing up of the Arctic ocean has by no means been established (Schwarzbach 1960; Colinvaux 1964; 1970). Furthermore, Cox (1968) pointed out that, according to paleomagnetic evidence, the polar axis has been in its present position for an interval at least ten times longer than the interval of recurrent glaciations.

A modification of the "pluvial" flip-flop theory has been proposed by Wilson (1964), who assumed that the trigger is not provided by the freezing and unfreezing of the Arctic Ocean, but by the buildup and decrease of the Antarctic ice sheet.

d) *Greenhouse Theory.* As of late, the problem of the effect of the content of carbon dioxide and water vapor in the atmosphere upon climate has become of great interest. The relative abundance of these gases would have been due to purely endogenic causes before the advent of man. Recently, however, the natural balances could have been upset by the burning of fossil fuels (Gosselink 1958).

As noted, the possible effect of the relative carbon dioxide content in the atmosphere has become a very modern topic, because of fears of the anthropogenic impact (burning of fossil fuels) on the climate. There is a huge literature on the matter, some of it with a definitely sensational and catastrophist slant. There have been two recent serious reviews on the problem, one by Degens et al. (1984) and one by Ellsaesser et al. (1986). Accordingly, the impact of an (artificial) increase of carbon dioxide on climate is not at all clear: it could be one way or another. Inasmuch as a global warming trend seems to have been observed during the past decades, and at the same time the carbon dioxide content of the atmosphere has increased, most researchers assume a causal connection between the two. However, the warming trend need not or not only be due to increased CO_2.

2.5.3.4 Combined Theories

a) *The Volcanic Combined Theory.* As noted above, the volcanic theory is by itself unable to account for the periodicity of the stages in an ice age. It has therefore been suggested that increased volcanism may have been able to produce a lowering of the over-all temperature of the world so that one of the astronomical causes considered above may have been able to imprint oscillations thereupon of the required periodicity. During long periods of the Earth's history, the volcanism may have been too weak to lower the temperature sufficiently to cause an ice age; this would account for the absence of glaciations during much of the Earth's evolution.

b) *Solar Topographic Theory.* A more general formulation of a relationship between ice ages and topography may be obtained by not restricting oneself to volcanism as a temperature-depressing agent. It stands to reason that the atmospheric turbulence is greater during periods of high topographic relief than during periods of low relief. One might reason that the increased turbulence leads to increased cloud formation and thus to an increased albedo of the Earth. Increased albedo, in turn, will have the effect of reflecting more of the Sun's radiation away from the Earth and therefore lead to lowered temperatures. A connection between climate and orogenesis had, in fact, been postulated long ago by geologists (cf., e.g., Ramsay 1910). A review of the above ideas has been given by Emiliani and Geiss (1959). Accordingly, the two most recent sequences of ice ages coincide with the two most recent periods of increased orogenetic activity: the Pleistocene period and the Paleozoic period. The topographic changes produce a general lowering of the Earth's surface temperature, and some astronomical cause connected with solar irradiation or solar emission impresses the required periodicity.

The initial changes in albedo could also have been caused by effects other than topographic ones, such as changes in solar radiation input, atmospheric composition etc. The feedback mechanism described above would then be triggered immediately in consequence (cf., e.g., Ramanathan and Coakley 1978; Koerner 1980; Oerlemans 1981; Burrett (1982).

c) *An Eclectic Theory.* A theory in which many of the possible causes of ice ages discussed above have been combined, has been proposed by Fairbridge (1961). Accordingly, polar wandering would put the Earth's poles into the Arctic Ocean and into the Antarctic Continent, Tertiary tectonism would close the Tethys Sea and cause high mountains (with attendant effects on the amtopsheric circulation), the ice buildup in Antarctica would increase the Earth's albedo, Milankovitch-type irradiation fluctuations would introduce cyclic changes in the heat available, solar radiation fluctuations would also be active, and the combination of all these factors would produce the characteristics of the present ice age.

Similar "combined" theories involving many factors have been proposed on various occasions. Thus changes in ocean temperature and carbonate productivity (Ruddiman and McIntyre 1981), changes in patterns of ocean currents (Kvasov and Verbitsky 1981; Chalikov and Verbitsky 1984), and a concurrence of about seven different variables (Beaty 1978), as well as of other components of the climate system (Warren 1982) have all been invoked as the causes of climatic changes. Finally, life ("Gaia Hypothesis") may have been involved (prior to the appearance of Man) in influencing the Earth's climate (Kirchner 1989).

d) *Stochastic Fluctuations.* In addition to the possible specific causes of climate changes mentioned above, it has been considered that nondeterministic random fluctuations of the climate system could trigger the swings (Kavvas and Delleur 1976; Lorenz 1976; Kominz and Pisias 1979). Once triggered, a causality chain of atmospheric-oceanic anomalies would ensue (Hirst and Hastenrath 1983) leading eventually to extreme climatic conditions.

e) *Conclusion.* One cannot decide to date which is the correct theory of the origin of ice ages. However, it is seen that there are several, presently equally possible, hypotheses. We shall concern ourselves here mainly with the effects of ice action upon geomorphology and refer the reader interested in the details of ice ages theories to the papers listed in the references.

References

Allison, I. (ed.): Sea level, ice and climatic change; Proc. Canberra Symp., Dec. 1979, Washington: IAHS (1981)
Ambach, W. and H. Eisner: Polarforschung 55(2), 71 (1985)
Ambach, W. and H. Eisner: Cold Regions. Sci. Technol. 13, 1 (1986)
Andrews, J.T., C.A.M. King and M. Stuiver: Geol. Mijnbouw 52(1), 1 (1973)
Balduzzi, F.: Experimentelle Untersuchungen über den Bodenfrost. Zürich: Mitt. Vers. Anst. Wass. Erdbau, 44, 1 (1959)
Ballard, G.E.H. and R.W. McGraw: Publ. Int. Assoc. Sci. Hydrol. 69, 160 (1966)
Barr, D.I.H.: Proc. Inst. Civ. Eng. (2) 69, 555 (1980)
Bass, J.: C.R. Acad. Sci. Paris 228, 228 (1949)
Batchelor, G.K.: The theory of homogeneous turbulence. Cambridge: University Press (1953)
Beaty, C.B.: Am. Sci. 66(4), 452 (1978)
Beckinsale, R.P.: Acta Geol. Acad. Sci. Hung. 23, 229 (1980)
Berger, A.: Rev. Geoph. 26(4), 624 (1988)
Berger, A. et al. (eds): Milankovitch and climate. Dordrecht: Reidel (1984)
Berger, W.H., R.C. Finkel and J.S. Killingeley: Nature 303, 231 (1983)
Bernal, J.D.: Nature (Lond.), 181, 380 (1958)
Besler, H.: Sasqua Intern. Symp. Swaziland, p. 445 (1983)
Blasius, H.: Z. Math. Phys. 56, 4 (1908)
Brakenridge, G.R.: Nature 283, 655 (1980)
Bray, J.R.: Nature 282, 603 (1979)
Broecker, W.S.: Science 151, 299 (1966)
Brown, R.L.: J. Glaciol. 25(91), 99 (1980)
Burrett, C.F.: Nature 296, 54 (1982)
Burroughs, W.J.: Weather 35(6), 156 (1980)
Butkovich, T.R.: Q. Colo. School Mines 54(3), 349 (1959)
Caputo, M.: Atti Acad. Sci. Ferrara 49, 1 (1971)
Caputo, M. and L. Pieri: J. Geophys. Res. 81(33), 5787 (1976)
Chalikov, D.V. and M.Ya. Verbitsky: Nature 308, 609 (1984)
Chappell, J.: Nature 302, 406 (1983)
Charvatova, I.: Stud. geoph. et Geod. Praha 33, 230 (1989)
Christiansen, C., J.T. Möller and J. Neilsen: Eiszeitalter und Gegenwart. 35, 89 (1985)
Clark, J.A., W.E. Farrell and W.R. Peltier: Quat. Res. 9, 265 (1978)
Clarke, G.K.C., U. Nitsan and W.S.B. Paterson: Rev. Geoph. Space Phys. 15(2), 135 (1977)
Cloetingh, S.: Geology 14, 617 (1986)
Cloetingh, S., H. McQueen and K. Lambeck: Earth and planet. Sci. Lett. 75, 157 (1985)
Colbeck, S.C.: Water Resour. Res. 11(2), 261 (1975)
Colbeck, S.C.: Water Resour. Res. 12(3), 523 (1976)
Colbeck, S.C.: J. Geoph. Res. 88(C9), 5475 (1983)
Coleman, J.M. and W.G. Smith: Bull. Geol. Soc. Am. 75, 833 (1964)

Colinvaux, A.P.: Science 145, 707 (1964)
Colinvaux, A.P.: Nature 278, 399 (1979)
Cornwall, I.: Ice ages, their nature and effects. London: Baker, (1970)
Courel, M.F.: Sahel Study. Paris: IBM France (1983)
Coutard, J.P. and H.J. Mücher: Earth Surf. Proc. and Landf. 10, 309 (1985)
Cox, A.: Meteorol. Monogr. 8(30), 112 (1968)
Crittenden, M.D., N. Christie-Blick and P.K. Link: Bull. Geol. Soc. Am. 94, 437 (1983)
Cropper, J.P.: Arct. Alp. Res. 14(3), 223 (1982)
Crowley, T.J.: Rev. Geoph. Space Phys. 21(4), 828 (1983)
Darcy, H.: Les fontaines publiques de las ville de Dijon. Paris: Dalmont (1856)
Degens, E.T., S. Kempe and A. Spitzky: In: Handbook of environmental chemistry (ed. O. Hutzinger, 1c,
   127. Berlin Heidelberg New York, Springer (1984)
Dickins, J.M.: BMR J. Aust. Geol. Geoph. 9(2), 163 (1984)
Donn, W.L.: Paleogeogr., Palaeoclimatol., Palaeoecol. 40, 199 (1982)
Donn, W.L. and D.M. Shaw: Bull. Geol. Soc. Am. 88, 390 (1977)
Donn, W.L., J.G. Patulio and D.M. Shaw: In: Research in Geophysics, ed. H. Odishaw, Vol. 2, p. 243,
   Cambridge, Mass.: MIT Press (1967)
Drake, C.L. and 7 others: Climate in earth history. Washington: National Academy Press (1982)
Duval, P. and L. Lliboutry: J. Glaciol. 31(107), 60 (1985)
Ellsaesser, H.W. and 3 others: Rev. Geoph. 24(4), 745 (1986)
Emiliani, D. and J. Geiss: Geol. Rdsch. 46, 576 (1959)
Ewing, M.: J. Alberta Soc. Petrol. Geol. 8, 191 (1960)
Ewing, M. and W.L. Donn: Science 123, 1061 (1956)
Ewing, M. and W.L. Donn: Science 127, 1259 (1958)
Fairbridge, R.W.: Ann. New York Acad. Sci. 95, 542 (1961)
Fairbridge, R.W.: In: Encyclopedia of Geomorphology, New York ed. R.W. Fairbridge. Reinhold
   (1968)
Fairbridge, R.W.: Quat. Res. 2(3), 283 (1972)
Fairbridge, R.W. and O.A. Krebs: Geoph. J.R.A.S. 6(4), 532 (1962)
Fairbridge, R.W. and J.H. Shirley: Solar Phys. 110, 191 (1987)
Flemming, N.C.: Geol. Soc. Am. Spec. Pap. 109, 1 (1969)
Flemming, N.C. and D.G. Roberts: Nature 243, 19 (1973)
Flint, R.F.: Glacial geology and the Pleistocene epoch, New York: Wiley, (1947)
Flint, R.F.: Glacial and Pleistocene geology, New York: Wiley (1957)
Folland, C.K., D.E. Parker and F.E. Kates: Nature 310, 670 (1984)
Fox, D.J. and D.K. Lilly: Rev. Geoph. Space Phys. 10(1), 51 (1972)
French, H.M.: Progr. Phys. Geogr. 5, 267 (1981)
French, H.M.: Progr. Phys. Geogr. 10, 535 (1986)
French, H.M.: Ecol. Bull. 38, 5 (1987)
Friedman, G.M. and J.E. Sanders: Bull. Geol. Soc. Am. 81, 2457 (1970)
Fritz, S.: In: Compendium of meteorology, ed. T.F. Malone, Amer. Meteorol. Soc., pp. 243–251 (1951)
Furrer, C., and H.P. Holzhauser: Z. Geomorph. Suppl. 50, 117 (1984)
Geiger, R.: Das Klima der bodennahen Luftschicht., 3rd edn., Braunschweig (1950)
Glen, J.W.: J. Glaciol. 2, 111 (1952)
Gold, L.W.: Bull. Highway Res. 168, 65 (1958)
Gold, L.W.: Can. J. Phys. 38, 1137 (1960)
Goldstein, J.: Modern developments in fluid dynamics (2 Vols.) Oxford: Univ. Press (1938)
Gosselink, J.G.: Proc. Indiana Acad. Sci. 68, 294 (1958)
Goudie, A.S.: Progr. Phys. Geogr. 6(3), 446 (1982)
Grabau, J.: Geogr. Helv. 42, 35 (1987)
Haefeli, R.: Schweiz. Bauztg. 85(1/2), 3 (1967)
Hansen, J. and S. Lebedeff: J. Geoph. Res. 92 (D11), 13345 (1987)
Hawkes, I. and M. Mellor: J. Glaciol. 11(61), 103 (1972)
Hecht, A.D. and 6 others: Quat. Res. 12, 6 (1979)
Henderson, F.M. and R.A. Wooding: J. Geoph. Res. 68(8), 1521 (1964)
Hill, R.: The mathematical theory of plsticity. Oxford: Clarendon Press (1950)
Hinze, J.O.: Turbulence. New York: McGraw-Hill Book Co. (1959)
Hirst, A.C. and S. Hastenrath: J. Phys. Oceanogr. 13(7), 1146 (1983)

Hoekstra, P.: Proc. Soil Sci. Soc. Am. 33(4), 512 (1969)
Hsü, K.J.: Giornale di Geol. (3) 47 (1–2), 203 (1985)
Hubbard, R.N. and M.C. Boulter: Nature 301, 147 (1983)
Hunt, B.G.: Nature 308, 48 (1984)
Huppert, H.E.: Nature 300, 427 (1982)
Hurst, H.E.: Proc. Inst. Civil. Eng. Pt. I, 519 (1956)
Hutter, K.: Theoretical glaciology. Dordrecht: Reidel (1983)
Ingram, M.J., D.J. Underhill and T.M. Wigley: Nature 276, 329 (1978)
Jackson, K.A. and B. Chalmers: J. Appl. Physics 29, 1178 (1958)
Jacoby, G.C.: IAWA Bull. 10(2), 99 (1989)
Johnson, J.G., G.K. Klapper and C.A. Sandberg: Bull. Geol. Soc. Am. 96, 567 (1985)
Jumikis, A.R.: The frost penetration problem in highway engineering. New Brunswick, N.J.: Rutgers Univ. Press (1955)
Kashiwaya, K., A. Yamamoto and K. Fukuyama: Quat. Res. 30, 12 (1988)
Kavvas, M.L. and J.W. Delleur: Bull. Hydrol. Sci. 21(3), 407 (1976)
Kazanskiy, A.B.: Quat. Res. 24(3), 285 (1985)
Keigwin, L.D. and R.C. Thunell: Nature 282, 294 (1979)
Kelley, P.M. and C.B. Sear: Nature 311, 740 (1984)
Kemper, E.: Das Klima der Kreidezeit. Geol. Jb. 96. Stuttgart: Borntraeger (1986)
Kinosita, S. and M. Fukuda: Ground freezing. Proceedings of the 4th Int. Sympos. Sapporo, 5–7 Aug. 1985. Rotterdam: Balkema, (1985)
Kirchner, J.W.: Rev. Geoph. 27(2), 223 (1989)
Koeppen, W. and A. Wegener: Die Klimate der geologischen Vorzeit. Berlin: Gebr. Borntraeger (1924)
Koerner, R.M.: Quat. Res. 13, 153 (1980)
Kominz, M.A. and N.G. Pisias: Science 204, 171 (1979)
Kreutzbach, J.E. and R.G. Gallimore: J. Geoph. Res. 94(D3), 3341 (1989)
Kroener, A.: J. Geol. 85, 289 (1977)
Kukla, G.J. and 8 others: Nature 270, 573 (1977)
Kvasov, D.D. and M.Ya. Verbitsky: Quat. Res. 15, 1 (1981)
Lamarche, V.C.: Nature 276, 334 (1978)
Lamb, H.: Hydrodynamics. 6th edn., London: Cambridge Univ. Press (1932)
Lamb, H.: Hydrodynamics. New York: Dover Publications (1945)
Lamb, H.H.: Pure and Applied Geoph. 119(3), 628 (1980)
Lambeck, K. and S.M. Nakiboglu: Geoph. Res. Let. 11(10), 959 (1984)
Lang, T.E. and J.D. Dent: Revs. Geoph. Space Phys. 20(1), 21 (1982)
Lautridou, J.P. and J.C. Ozouf: Prog. Phys. Geogr. 6(2), 215 (1982)
Leopold, L.B.: Quat. Res. 6, 557 (1976)
Le Treut, H. and M. Ghil: J. Geoph. Res. 88(C6), 5167 (1983)
Levi, B.G.: Physics Today 43(2), 17 (1990)
Lliboutry, L.: Houille Blanche, Vol. Spec. 5, 489 (1970)
Lliboutry, L.: J. Glaciol. 10(58), 15 (1971)
Lliboutry, L.: C.R. Acad. Paris 297, 54 (1983)
Lliboutry, L.: Very slow flows of solids: basics of modeling in geodynamics and glaciology. The Hague: Nijhoff (1987)
Lliboutry, L. and P. Duval: Ann. Geoph. 3(2), 207 (1985)
Lockwood, J.G.: Causes of climate. London: Arnold (1979)
Lorenz, E.N.: Quat. Res. 6, 495 (1976)
Lorius, C. and 6 others: Nature 316, 591 (1985)
Louberre, P. and K. Moss: Bull. Geol. Soc. Am. 97, 818 (1986)
Low, P.F., P. Hoekstra and D.M. Anderson: Water Resour. Res. 4, 379, 541 (1968)
Mahaney, W.C. (ed.): Quaternary paleoclimate. Norwich: Geobooks, (1981)
Mandelbrot, B.B. and J.R. Wallis: Water Resour. Res. 5(2), 321 (1969)
Martin, P.S. and R.G. Klein (eds.): Quaternary extinctions, Tucson: Univ. Arizona Press (1984)
Martynov, G.A.: Osnovy geokryologii, Moscow: Acad. Sci. (1959)
Matthews, R.K.: Quat. Res. 2, 368 (1972)
McClung, D.M.: J. Glaciol. 19 (81), 101 (1977)
McGreevy, J.P.: Progr. Phys. Geogr. 5(1), 56 (1981)
McGreevy, J.P. and W.B. Whalley: Arct. Alp. Res. 14(2), 157 (1982)

McGreevy, J.P. and W.B. Whalley: Arct. Alp. Res. 17(3), 337 (1985)
McKay, C.P. and C.R. Stoker: Rev. Geoph. 27(2), 189 (1989)
Meier, M.F.: U.S. Geol. Surv. Prof. Pap. 351, 1 (1960)
Mercer, J.H. and 3 others: In: Quaternary studies. Wellington: Royal Soc. New Zealand p. 223 (1975)
Milankovitch, M.: Kanon der Erdbestrahlung und seine Anwendung auf das Eiszeitenproblem. ed.
    Spec. Acad. Roy. Serbe Tome 133, Belgrad (1941)
Milliman, J.D. and K.O. Emery: Science 162(8), 1121 (1968)
Mix, A.C. and W.F. Ruddiman: Quat. Sci. Rev. 4, 59 (1985)
Mörner, N.A.: Palaeogeogr., Palaeoclimatol., Palaeoecol. 13, 1 (1973)
Mörner, N.A.: J. Geol. 84(2), 123 (1976)
Mörner, N.A.: J. Coastal Res. Sep. Is. 1, 49 (1986)
Mörner, N.A. and W. Karlen: Climatic changes on a yearly to millenial basis. Geological, historical and
    instrumental recods. Proc. 2nd Nordic Symp. Stockholm, May 16–20, 1983. Dordrecht: Reidel
    (1984)
Morrow, D.W.: Bull. Can. Petrol. Geol. 34(2), 284 (1986)
Mugridge, S.J. and H.R. Young: Can. J. Earth Sci. 20(4), 568 (1983)
Mulhearn, P.J.: J. Fluid Mech. 71(4), 801 (1975)
Narita, H.: J. Glaciol. 26(94), 275 (1980)
Nesbitt, H.W. and G.M. Young: Nature 299, 715 (1982)
Newman, W.S. and R.W. Fairbridge: Nature 320, 319 (1986)
Nordin, C.F., R.S. McQuivey and J.M. Mejia: Water Resour. Res. 8(6), 1480 (1972)
Oerlemans, J.: Quat. Res. 14, 349 (1980)
Oerlemans, J.: Quat. Res. 15, 77 (1981)
Öpik, E.J.: A climatological and astronomical interpretation of the ice ages and of the past variations of
    terrestrial climate, Armagh. Obs. Contr. No. 9 (1953)
Öpik, E.J.: Irish Astron. J. 8, 153 (1968)
Pai, S.I.: Viscous flow theory (2 Vols.). New York: Nostrand (1957)
Palm, E. and M. Tweitereid: J. Geoph. Res. 84 (C2), 745 (1979)
Paneth, F.A.: Sci. J. Roy. Coll. Sci. 6, 120 (1933)
Paterson, W.S.B.: The physics of glaciers. Oxford: Pergamon (1969)
Peltier, W.R., W.E. Farrell and J.A. Clark: Tectonophysics 50, 81 (1978)
Penner, E.: Bull. Highway Res. 135, 109 (1956)
Penner, E.: Can. Geotech. J. 23(3), 334 (1986)
Perutz, M.F.: Observatory 70, 64 (1950)
Philander, S.G.H.: Nature 302, 295 (1983)
Pirazzoli, P.A.: Z. Geomorph. 21(3), 284 (1977)
Planck, M.: Treatise on thermodynamics. 3rd edn., New York: Dover Publ.-Co. (1945)
Pockels, F.C.: Ann. Physik 4(4), 459 (1901)
Pokras, E.M. and A.C. Mix: Nature 326, 486 (1987)
Ponte, L.: The cooling, Englewood Cliffs N.J.: Prentice-Hall (1976)
Porter, S.C.: Nature 291, 139 (1981)
Prandtl, L.: Über die ausgebildete Turbulenz. Trans. 2nd Int. Congr. Appl. Mech., Zürich., p. 62 (1926)
Ramanathan, V. and J.A. Coakley: Rev. Geoph. Space Phys. 16(4), 465 (1978)
Ramsay, W.: Forh. Ofversight of Finska Vet. Soc. 52 (1910)
Rayleigh, Lord: Sci. Papers 2, 258 (1883)
Rosen, P.S.: Mar. Geol. 26 (12) M7 (1978)
Rosenberg, G.D.: Bull. Geol. Soc. Am. 81, 525 (1970)
Ruckli, R.: Der Frost im Baugrund. Vienna: Springer, 1950
Ruddiman, W.F. and A. McIntyre: Quat. Res. 16, 125 (1981)
Saks, V.N., N.A. Belov and N.N. Lapina: Priroda 7, 13 (1955)
Salm, B.: Rev. Geoph. Space Phys. 20(1), 1 (1982)
Saunders, W.B. and W.H.C. Ramsbottom: Geology 14, 208 (1986)
Scheidegger, A.E.: Geofis. Pura Appl. 33, 111 (1956)
Scheidegger, A.E.: The physics of flow through porous media. 3rd edn. Toronoto: Univ. Press (1974)
Scheidegger, A.E.: Physical aspects of natural catastrophes, Amsterdam: Elsevier (1975)
Scheideger, A.E.: Principles of geodynamics, 3rd edn., Berlin Heidelberg New York: Springer (1982)
Schubert, C.: Interciencia 13(3), 128 (1988)

Schulman, E.: In: Malone, T.F. (ed.): Compendium of meteorology, ed. T.F. Malone. Am. Meteorol. Soc., pp. 1024–1029 (1951)

Schwarzbach, M.: Z. Dtsch. Geol. Ges. 112, 309 (1960)

Sear, C.B. and P.M. Kelley: Nature 285, 533 (1980)

Shepard, F.P.: In: Essays in marine geology in honour of K.O. Emery, Los Angeles: U. Calif. Press p. 1 (1963)

Shepard, F.P. and H.E. Suess: Science 123, 1082 (1956)

Shumskiy, P.A.: Principles of structural glaciology. Engl. trans. by D. Kraus. New York: Dover (1964)

Simpson, G.C.: Proc. Linnean Soc. Lond. 152, 190 (1940)

Sommerfeld, R.A.: J. Geoph. Res. 79(23), 3353 (1974)

Spaulding, W.G. and L.J. Graumilch: Nature 320, 441 (1986)

Spjeldnaess, N.: In: On climate changes and related problems. Ed. K. Frydendahl. Proc. Nordic. Symp. Copenhagen: Danish Met. Inst. p. 76 (1978)

Steckler, M.: In: Patterns of change in earth evolution, eds. H.D. Holland and A.F. Trendall, p. 103, Berlin Heidelberg New York: Springer (1984)

Strick, M.R. and 3 others: Nature 295, 105 (1982)

Sutton, O.G.: Micrometeorology, New York: McGraw Hill (1953)

Taber, S.: J. Geol. 37, 428 (1924)

Terzaghi, K.: S.-ber. Akad. Wiss. Wien, Math.-Natw. Kl. Abt. IIa, 132, 105 (1923)

Thompson, S.L. and E.J. Barron: J. Geol. 89(2), 143 (1981)

Tjia, H.D.: Z. Geomorph. Suppl. 22, 57 (1975)

Tollner, H.: Alpenver. Jb. 99, 101 (1974)

Tooley, M.J.: Progr. Phys. Geogr. 9(1), 113 (1985)

Tooley, M.J.: Progr. Phys. Geogr. 10(1), 120 (1986)

Toya, H., K. Takeuchi and H. Ohmori (eds): Studies of environmental changes due to human activities in the semi-arid regions of Australia. Tokyo: Dept. Geography (1985)

Vail, P.R. and 7 others: Am. Assoc. Petrol. Geol. Mem. 26, 49 (1977)

Walcott, R.I.: Quat. Res. 2, 1 (1972)

Walder, J. and B. Hallett: Bull. Geol. Soc. Am. 96(3), 336 (1985)

Walder, J. and B. Hallett: Arct. Alp. Res. 18(1), 27 (1986)

Wallen, C.C.: Geogr. Ann. 68A, 245 (1986)

Warren, A.: Progr. Phys. Geogr. 9(3), 434 (1985)

Warren, S.G.: Climatic Change 4(4), 329 (1982)

Weertman, J.: Annu. Rev. Earth Planet. Sci. 11, 215 (1983)

White, J.M. and R.W. Mathewes: Can. J. Earth Sci. 19(3), 555 (1982)

Willet, H.C.: In: Climate change, ed. H. Shapley. p. 51–71. Cambridge: Harvard University Press (1953)

Wilson, A.T.: Nature 201, 147 (1964)

Wolfe, J.A.: Am. Sci. 66(6), 695 (1978)

Wollin, G., D.B. Ericson and M. Ewing: in: The Late Cenozoic glacial ages, ed. K. Turekian. New Haven: Yale University Press, p. 199 (1971)

3 Mechanics of Slope Formation

3.1 Principles

Any cursory inspection of the shape of the surface of Earth shows that *slopes* are the basic constituents of many features of interest. We therefore start the main part of our treatise on theoretical geomorphology with a description of what is known regarding the theory of slope evolution. In this, it should be noted that the present chapter deals only with specific physical effects. As of recently, slopes have been treated as *systems* in analogy with whole landscapes. Thus, thermodynamic analogs and process-response models have been proposed for the explanation of slope evolution. However, such approaches will be relegated to Chapt. 5 on geomorphological system theory.

The exogenic deformation of any slope starts with the *reduction* (i.e., decay) of the constituent material. This reduction may be chemical (corrosion), physical, or even biological. The various possibilities will be discussed in Sect. 3.2.

After the material on a slope has been loosened up, further development takes place by the *removal* of the loose pieces from their original position. This removal may occur *spontaneously* (Sect. 3.3) or it may be due to the action of *various agents*. The agents that are able to remove material from a slope will be discussed in Sect. 3.4. The combined effect of all these agents upon a slope produces *slope denudation*. It is here that mathematical analysis has been most widely employed. The section in question (3.5) will therefore be the most interesting one in the present Chap. (3) for the theorist.

3.2 Reduction of Rocks

3.2.1 Basic Statements

The effect of exogenic agents upon the shaping of the Earth's surface is mostly a destructive one: features built up by endogenic processes are worn down and destroyed.

The destructive action of the exogenic forces begins with the reduction or weathering of rocks. By "reduction" we mean the breaking up of the solid material of the Earth's surface into small particles which are subsequently susceptible to removal by a variety of transporting agents.

The processes that bring about the reduction of rocks may be of diverse natures. Reviews of the subject matter in book form have been given by Carroll (1970) and by Yatsu (1988).

Most effective are probably chemical processes which alter the composition of the rocks so that actual corrosion is the result (Sect. 3.2.2). Next in line are physical processes (Sect. 3.2.3). Finally, rock reduction can also be brought about by plants and animals (Sect. 3.2.4).

We shall discuss the various modes of rock reduction in turn below.

3.2.2 Chemical Weathering

3.2.2.1 General Remarks

One of the most important causes of the decomposition of materials of the Earth's crust is chemical weatheirng. This type of weathering always involves some reaction of water with the rock material. The geological effect of chemical weathering is basically to loosen up the rock so that it can be further attacked by physical agents. Morphogenesis due to weathering is particularly pronounced in the tropics (Modenesi 1983). In some instances, chemical weathering can produce directly largescale geomorphological effects such as the karst phenomena, which are solely due to the dissolution of limestones.

The chemistry of weathering was the subject of a NATO conference in 1984 (Drever 1985). With regard to weathering patterns, McGreevy (1985a) distinguishes between four types of rocks: granite, basalt, sandstone, and calcareous materials. An interesting study of weathering *rates* as functions of time has been made by Colman (1981). Accordingly, most rock-weathering rates (with the exception of the dissolution of limestone) decrease through time. The explanation of this fact appears to be the inhibiting role of residues played in the weathering process.

We shall now discuss typical decomposition processes in detail.

3.2.2.2 Weathering of Igneous Rocks

Igneous rocks are generally a mixture of crystalline and amorphous forms of silicium compounds.

Some of the widest-spread types of igneous rocks are granites and granitoid rocks. The chemical decomposition of such rocks is mainly governed by the breakdown of the feldspar contained therein into clay minerals ("kaolinitization"; Heydemann 1966; Berner and Holdren 1977, 1979): clay formation occurs when slightly carbonated water comes into contact with plagioclase and similar minerals; the sodium (or in other cases: calcium or potassium) is thereby removed from the feldspars and only the silicium and aluminum are left in the end product. The latter is clay. In this process, potassium-rich granitoid rocks are more resistant to weathering than their less potassic cousins. This reflects the fact that orthoclase is more stable against weathering influences than calcic plagioclase (Pye 1986).

Kaolinitization also plays a major role in the decay of basaltic materials, such as basaltic glass and tephra (Hay and Jones 1972).

Other chemical reactions contributing to the weathering of igneous rocks involve the oxidation and chemical hydration of many minerals. The phenomenon of serpentinization belongs in this category (Bakker 1965; Nossin and Levelt 1967).

In addition, silica can also be dissolved by water, particularly at pH values above 9 (Siever 1962; Martin 1987), although much less so than carbonates. Nevertheless, karstic (or better: pseudokarstic) geomorphic features are created in this way in landscapes with a crystalline base (Watson and Pye 1986). The corollary to silica dissolution is the formation of hydrous silica coatings in igneous-volcanic terrains such as Hawaii (Farr and Adams 1984).

3.2.2.3 The Dissolution of Limestone

It is well known that water can dissolve limestone (Augustithis 1983). However, pure water has relatively little effect; the speed of the reaction is greatly enhanced if salts or carbon dioxide are present in the water. The effect of the presence of salts will be treated in the next section (3.2.2.4); we consider here only "pure" water in contact with the atmosphere, i.e., water containing CO_2. In this case, the calcium carbonate is slowly transformed into calcium bicarbonate and removed (Ingle-Smith and Mead 1962; Buhmann and Dreybrodt 1985). The kinetics of the process has been studied in some detail by Weyl (1958). Accordingly, the reactions taking place are the following

1. $CaCO_3 = Ca^{2+} + CO_3^{2-}$
2. $CO_3^{2-} + CO_{2(solution)} + H_2O = 2HCO_3^-$
3. $CO_{2(gas)} + H_2O_{(liquid)} = CO_{2(solution)}.$

Weyl (1958) investigated the kinetics of each of these reactions separately and came to the conclusion that none is slow enough to materially limit the dissolution of limestone by flowing water. In other words, the amount of limestone dissolved is solely determined by the amount of water coming into contact with it and by the prevailing conditions of pressure, temperature, etc. Hence, the rate of limestone dissolution is influenced by the diffusivity of solute from the vicinity of the contact surface into the liquid stream. The diffusivity factor D in question may be taken as equal to

$$D = 2 \times 10^{-5} \, cm^2 \, s^{-1}. \tag{3.2.1}$$

The dissolution of limestone is therefore described by a diffusivity equation with mass transport

$$D \, lap \, C - v \, grad \, C = \frac{\partial C}{\partial t} \tag{3.2.2}$$

where C is the concentration of solute, v the local flow velocity and D, as indicated above, the diffusivity factor. As noted above, the dissolution of limestone at the wall is very rapid so that the boundary condition can be formulated by stating that the solution must be saturated at the wall.

Weyl has solved the above Eq. (3.2.2) (steady state case) for a fissure of width d which contains a fluid flowing laminarly with an average velocity $\bar{v}$ parallel to its walls; he found for the distance L at which the bulk of the fluid is 90% saturated

if $\bar{v}d/D \gg 1$ $L = 0.304 \, d^2/D$

If $\bar{v}d/D > 1$ $L = d.$ (3.2.3)

For $\bar{v} = 1$ cm/s, $d = 1$ mm, this works out to

$\qquad$ L = 1.52 meters. $\hfill$ (3.2.4)

Weyl's theory is valid only for very fine fissures in which one may assume laminar flow. In limestone caverns the flow of water is presumably turbulent so that the rate of mass transfer out of the vicinity of a wall is increased many times. At any rate, Weyl concludes from his investigations that water inside rock is essentially in chemical equilibrium with its surroundings. The solution reaction occurs in times which are short compared with the times during which the water remains in the rock.

Thus, the leaching out of a cavern is entirely determined by the amount of water that percolates through it and by the solubility of the limestone in the water. There is no danger that water will percolate through a cavern without becoming completely saturated while doing so.

3.2.2.4 Effect of Salts

If salts are present in the water, the dissolution of limestone becomes very complicated, as complexing reactions may occur. This process is of some importance in connection with the development of coasts. Thus, morphological field studies have yielded the result that corrosion of beach and reef rock is most pronounced in the intertidal zone, i.e., in the zone between high and low tide. The surface of the rock becomes very irregular with many small depressions which are separated by narrow ridges. The depressions, or basins, remain filled by sea water at low tide and the corrosion taking place therein is very pronounced (Fairbridge 1952; Mac Fadyen 1930; Guilcher 1958; Smith 1941).

It is, of course, well known that water is able to dissolve calcium carbonate, so that it would appear at first glance that there is little difficulty in explaining the observed corrosion. However, it has been noted that in most instances, the basin water in which the dissolution of the calcium carbonate is supposed to take place is, in fact, supersaturated with that substance. Therefore, the corrosion of limestone by sea water in the manner supposed above poses somewhat of a puzzle (Williams 1949; Kuenen 1950).

The problem was analysed by Revelle and Emery 1957, who discussed various attempts at an explanation which were based upon phenomena which are not chemical. However, they came to the conclusion that none of these explanations is satisfactory and that the cause of the rock corrosion is, in fact, chemical. It appears that the mechanism of dissolution of calcium carbonate under the presence of other salts (sodium chloride etc.), and possibly organic matter, is not very well understood. Therefore, the possibility exists of complexing reactions and hydration taking place which might explain the high rate of corrosion observed in intertidal limestone rocks (Garrels et al. 1961; Hodgkin 1964; Coleman 1966). An alternative opinion (Matthews 1974) is that the dissolution on coasts is not primarily effected by sea water at all, but rather by meteoric (fresh) water containing only carbon dioxide.

Other special types of coastal features, such as honeycomb weathering of sandstone exposures, have been ascribed to gypsum (McGreevy 1985b) or halite

(Mustoe 1982), crystal growth in the pores causing mechanical disaggregation of the rock.

Quite generally, salt action can have an effect on silicate rocks as well as on carbonate ones. Thus, salts are able to produce granular disintegration of granites, evidently also because of sulfate (or other) crystal growth in the pores (Kwaad 1970).

3.2.3 Physical Rock Reduction Processes

3.2.3.1 General Remarks

The physical processes causing rock reduction fall into a variety of categories.

Thus, the physical drag of blowing wind or flowing water may be sufficient to separate loosely coherent substances into small particles, causing manifest "reduction" of these substances. If the physical drag is exercised by debris passing over a substratum, it causes "corrasion" of the latter (Sect. 3.2.3.2).

Spattering action of raindrops and rainwash (Sect. 3.2.3.3) represents another type of physical rock reduction. The impact of rain may loosen up soil particles and ready them for further transportation.

A peculiar form of reduction of rocks is due to cavitation in flowing water (Sect. 3.2.3.4).

Reduction of rocks may also occur due to the presence of tectonic stresses which cause mechanical displacements involving the destruction of the material involved (Sect. 3.2.3.5).

Finally, other physical processes causing rock reduction will be discussed in Sect. 3.2.3.6. These involve mainly temperature effects. In this, it should be noted, however, that the effect of freezing and thawing ("cryogenic weathering") has already been discussed in the chapter on the mechanics of ice (Sect. 2.3.4.2).

3.2.3.2 Drag, Contrition, and Corrasion

Some materials of which the surface of the Earth is composed are so loose as to be almost without cohesion. In such cases, the physical drag of water or wind streaming over these materials is by itself sufficient to separate the individual particles. This induces reduction of these materials. The details of the drag theory will be treated in connection with the corresponding geomorphic processes, i.e., for water in Sect. 4.4.3 and for wind in Sect. 8.2.3.

In addition, particles being carried along by wind or water become smaller during their journey because of frictional forces which cause contrition of the material.

Thus, it stands to reason that the cange in weight dW of a given particle is proportional to its weight W and the distance dL through which it travels; hence (with α a proportionality constant)

$$dW = -\alpha W dL \tag{3.2.5}$$

or

$$W = \text{const } e^{-\alpha L}. \tag{3.2.6}$$

The last equation is known as Sternberg's (1875) formula.

The assumption that pebbles become contriturated during their downstream journey is beyond doubt. Experiments to study the contrition have been reported, e.g., by Rayleigh (1942, 1944), who was mostly interested in the shape that pebbles adopt during their contrition. For this purpose, he made artificial pebbles of chalk and contriturated them by placing them in a metal box together with various kinds of abrasive particles (such as hexagon nuts). The box was kept in slow rotation. The result was that, if a spherical shape of pebbles was used to begin with, the shape did not change very much during the experiment.

In his second paper, Lord Rayleigh repeated the experiments with a block of marble being contriturated by steel fragments. A typical result shows that the edges become rounded. Somewhat different results are obtained if the contrition is achieved in an artificial pothole by rotating water in a vessel containing pebbles, forcing the latter to rub against the walls; the result of this type of experiment leads to an elliptical shape of the pebbles. Occasionally, a slightly *concave* shape is encountered. Lord Rayleigh suggested a qualitative explanation for this; the concavity is the result of the pebbles being nicked by other similar pebbles. The nicking process is more pronounced in the center than at the edges of a pebble (where nicking merely acts to round the corners), which gives rise to the concave shape.

If the physical drag is exercised by debris moving over a substtratum it has been termed *corrasion*. This occurs, e.g., on mountain sides where rubble moves over rock strata below. The physical action of the rubble upon the rock is destructive as material is being abraded and the rock thus is being reduced.

3.2.3.3 Splattering of Drops and Rainwash

A notable denudational effect, particulary in soils, is caused by the impact and splattering of raindrops, with subsequent rainwash ("splash and wash"). This phenomenon has been mainly studied by agricultural engineers, as it touches on the problem of soil loss on agricultural lands.

A classic review of the subject matter was given by Smith and Wischmeier (1962).

The mechanics of loosening soil by raindrop impact involves two processes: first, particles are detached by direct impact and then they are dislodged a small distance by the splash. To this is added afterwards the transport of the particles by "wash".

Unfortunately, the mechanism of the impact of raindrops upon soil is very complicated and it appears almost hopeless to try to construct a satisfactory theory thereof, although Esin and Dmitriev (1975) tried to set up a complete mechanical model, which contains, however, many simplifying assumptions. Most of the investigations, therefore, have been directed toward establishing empirical correlations between the amount of soil loosened and several of the variables that one might assume to be important.

The first such variable which might come to mind is the energy of the raindrops. Most experimental studies attempt to establish universal curves purporting to show the raindrop energy versus the amount of soil loosened; such studies have been reported, for instance, by Woodburn (1948), Ekern (1951, 1953), Young and Wiersma (1973), Stocking (1978), and Poesen (1985). It was found essential to specify the drop shape at the time of impact in order to come up with unequivocal relationships.

A different type of correlation has been sought by Rose (1960), who plotted the amount of soil removed against the momentum (rather than the energy) of the rain. The correlation thus obtained is nonlinear.

The momentum change (per unit area and time) is equal to the stress acting on the soil due to the rain. Attempts at measuring such stresses directly (using transducers) were made by Imeson et al. (1981) and by Ghadiri and Payne (1979, 1981). The latter authors found 0.2–0.4 MPa lasting less than 1 ms in one case and 2–6 MPa lasting 50 μs in another.

The above two hypotheses, viz. (1) that the erosive action (per unit time) of rain is proportional to the energy expended by the rain (per unit time) and (2) that the erosive action of the rain (per unit time) is proportional to the momentum of the rain (per unit time) can be formulated mathematically as follows (Scheidegger 1963)

$$1. \quad Q = Cq\rho v^2, \tag{3.2.7}$$

$$2. \quad Q = Cq\rho v, \tag{3.2.8}$$

where Q is the rate of erosion (g/cm^2 s), q the rate of rainfall (cm/s), v the terminal velocity of the rainfall (cm/s), ρ the mass density (g/cm^3) and C a constant.

The above results are altered by the emergence of additional effects. The soil erosion can be slowed down by the formation of surface crusts (McIntyre 1958) which are harder than the material below. The formation of such crusts is due to the washing-in of fine particles into the pores of the soil and to surface compaction of the latter. Similarly, if the rain is strong enough for a surface film of water to form upon the soil, the impact of the raindrops will no longer be directly affecting the soil below, but set up turbulent motions in the water film already present. This will again cause soil erosion but of a different type than that considered above. A study of this process has been reported by Kuron and Steinmetz (1957).

Because of the difficulty inherent in setting up simple correlations, empirical "soil loss" equations have been established. The classic such equation is due to Wischmeier (1960). It is

$$A = R \, KLSCP, \tag{3.2.9}$$

where A is the computed average annual soil loss (mass per area) from a specific field under specific rainfall and specific agricultural use, R is the rainfall factor (erosive potential of the average annual rainfall in the locality), K is the erodibility of the soil, L and S are topographic factors (referring to length and gradient), C is a cropping management factor and P an erosion control practice factor. For any particular area, the corresponding factors have to be established empirically. Wischmeier (1960) himself gave a list of values for these factors referring to typical cases; a great number of other investigations leading to empirical soil loss estimates have been reported in the literature. A summary thereof is contained in the doctoral thesis of Schmidt (1979).

A final process in soil reduction by rain is the transport by wash. This occurs partly by sheetflow, partly by rill formation. The denudation effected thereby will be treated in Sect. 3.4.4.

3.2.3.4 Cavitation

A peculiar form of denudation can occur owing to *cavitation* in rapidly moving water. Cavitation, i.e., the formation of bubbles, takes place if the hydrodynamic forces in the fluid are so great that the local pressure becomes smaller than the vapor pressure. Cavitation is known to be of great significance in hydraulic machinery, where highly deleterious effects are known to be produced by it, particularly on turbine vanes and similar structures. The cavitation phenomenon has therefore been extensively studied by mechanical engineers. Bibliographies of these studies have been compiled, for instance, by Eisenberg (1950) and by Shal'nev (1956). The destructive action of cavitation is probably due to the shock waves created when the bubbles collapse.

Since cavitation is of such great significance in eroding metal surfaces of engineering equipment, it stands to reason that it also is of importance with regard to geomorphological phenomena. The geological effects that may be ascribed to cavitation have been discussed by Hjulström (1935), Barnes (1953, 1956) and Ball (1976).

Accordingly, bubble formation and therewith geological action of cavitation will occur primarily during the rapid flow of water over a slope or within a river channel. One can give an estimate of the velocity required for water flow in order to produce cavitation based upon Bernoulli's equation

$$\frac{v_1^2}{2g} + \frac{p_1}{\rho g} + z_1 = \frac{v_2^2}{2g} + \frac{p_2}{\rho g} + z_2. \tag{3.2.10}$$

This equation must be valid along any streamline if one neglects energy dissipation due to turbulence or bottom friction. In (3.2.10) v_1 is the flow velocity, p_1 the pressure, z_1 the height of the stream line above an (arbitrary) datum level and v_2, p_2, z_2, are the corresponding values at the cavitation point. In the latter, one has p_2 equal to the vapor pressure (p_D) of the water. Let us now consider (with Hjulmström) a case where the water at one point is at rest ($v_1 = 0$) and the corresponding pressure is the atmospheric pressure ($p_1 = p_{atm}$). These conditions will apply to the surface element of a puddle of water. If we neglect gravity, cavitation will occur if v_2 reaches the value

$$v_2 = \sqrt{\frac{2(p_{atm} - p_D)}{\rho}}. \tag{3.2.11}$$

From this, Hjulström calculated that v_2 must equal 14.3 m/s for $p_{atm} = 760$ mmHg at 0 °C. A similar calculation was made by Barnes, who considered streamlines in which the maximum flow velocity v_2 is $2v_1$, v_1 denoting the initial flow velocity on the streamline. Again taking $p_1 = P_{atm}$ as initial pressure, he then calculated the initial flow velocity v_1 required to produce cavitation and obtained from (3.2.10), discounting gravity:

$$v_1 = 8.1 \text{ m/s.} \tag{3.2.12}$$

The result of the above investigation is that the flow velocity of water has to be rather high in order to produce cavitation. Such flow velocities can be realized in waterfalls and in rapids.

3.2.3.5 Stress Weathering

A further physical process to be considered here is weathering due to the action of stresses: every rock mass is subject to a tectonic stress field; thus the formation of joints in such masses is a ubiquitous phenomenon (Scheidegger 1985).

Jointing is particularly frequent in regions of high stress, especially if not only tectonic, but also self-gravitational stresses are involved. This occurs at notches, at the foot of rock walls, and on mountain peaks, leading to typical decay patterns of such features (Gerber and Scheidegger 1969). On surfaces, whole sheets of material may be split off, representing the phenomenon of exfoliation (cf. the discussion in Scheidegger 1982, p. 333). Similarly, undercutting by rivers causes the sheeting-off of sandstone cliffs (Robinson 1970).

Apart from such large-scale features, smaller-scale disintegration patterns may also be produced by stresses. In granite, this process may proceed to the formation of grus and thereby to the destruction of the rock (Whalley et al. 1982; Folk and Patton 1982; Lajtai 1986); in basalt, spheroidal weathering is initiated (Jocelyn 1972). Last but not least, the rock destruction is fault zones by stresses (e.g., mylonitization) is well known (see any textbook on physical geology, e.g., Holmes 1944).

A peculiar type of "weathering" involving stresses is caused by clay swelling: upon contact with water, high internal pressures are created within materials containing clay. In consequence of these pressures, the latter swell up. The pressures depend on the mineralogical composition of the materials (Madsen 1979; Kovari et al. 1981).

3.2.3.6 Other Physical Effects

In addition to the physical effects mentioned above as causing weathering there are other such effects.

Notable amongst these are temperature effects. In this connection, "cryogenic weathering" has already been discussed in Sect. 2.3.4.2.

At the other end of the temperature scale, rock destruction is caused by the heat from forest fires. A recent discussion of the problem has been given by Ollier and Ash (1983), who reviewed much of the earlier literature. Their paper contains pertinent curves showing the behavior of the material constants (thermal expansion, thermal diffusivity, and compressive strength) mainly of granite as a function of temperature. From this, actual patterns of spalling and cracking are explained as a result of the combination of fire behavior, rock shape, and composition.

Not only fires, but even the lesser diurnal temperature changes in a hot desert climate are significant in inducing "desert weathering". Thus, Kerr et al. (1984) made a study of the behavior of rocks under desert conditions. It is seen that thermal fracturing of rocks (Marovelli et al. 1966) and dirt cracking (Ollier 1965) are significant phenomena in deserts.

3.2.4 Biological Effects

Finally, it should be noted that life also may cause the reduction of rocks. Burrowing animals, such as rodents, moles (Wilson and Smart 1984), and earthworms may make the rock susceptible to destruction, and plants (including algae; cf. Kobluk and Kahle 1978) may push shoots into rock cracks so as to cause their widening. Various forms of life also affect the chemistry of the environment (e.g., fungi causing the addition of carbon dioxide to the water) which then will exhibit a different chemical action upon the rocks than it would otherwise. Mostly, however, life in the form of vegetation acts as a protecting agent, inhibiting the speed of denudation (Dieckmann et al. 1985) that would take place without it or even cause the building-up rather than the destruction of rocks ("biokarst"; cf. Viles 1984).

Other biological effects occur in the formation of some surface crusts such as in badlands by the activity of blue-green algae (Finlayson et al. 1987) and in deserts by mixotrophic manganese oxidizing bacteria (Dorn and Oberlander 1982) which are instrumental in the buildup of desert varnish: the latter is composed of clay minerals, iron and manganese oxides, and hydroxides, with the manganese compounds being most characteristic. The latter are the result of the activity of the bacteria mentioned above.

3.3 Spontaneous Mass Movements on Slopes

3.3.1 Taxonomy of Movements

Turning now to the spontaneous mass movements on slopes, we note that these can be classified according to the material that is involved and according to the velocity with which they take place. In addition, it is of importance how deep the layers are that are affected. The basic taxonomy has already been discussed in Section 1.2.5.

In a discussion of mass movements on slopes, it should be noted that these are part of the normal, long-range landscape development. We have seen in Sect. 1.5.2 that a landscape constitutes a dynamic equilibrium between endogenic buildup and exogenic destruction: the mass movements are part of the destruction process.

Inasmuch as mass movements constitute geological hazards to life and property, the literature dealing with them is extremely large. The writer has summarized much of this literature in his book on natural catastrophes (Scheidegger 1975); he has also published several reviews updating the book (Scheidegger 1984, 1987, 1988). In the present book, we will summarize the work that is of importance with regard to geomorphology.

3.3.2 Rheology of Slope Materials

3.3.2.1 General Remarks

The materials that form the land surface of the Earth, such as soil and rock, are in the solid state of aggregation. Our aim is to understand the deformations that they may undergo. The general investigation of the deformation mechanisms in "solid" materials is the subject of the science of "rheology". From a study of this science it ensues that a solid material has generally three "ranges" of behavior: an elastic range, a creep range, and a failure range.

A general review of the mechanics of deformation, as it is of interest to Earth scientists, has been given by the writer in great detail in his book on geodynamics (Scheidegger 1982; cf. p. 133 ff. therein). Thus, for the discussion of fundamental questions, the reader is referred to the cited monograph.

3.3.2.2 The Elastic Range

For small stress changes, earth materials (and particularly rocks) behave elastically. This implies a linear, reversible relation between the components of the tensors of stress and strain.

In an isotropic solid, the elastic behavior is charactrized by *two* elastic parameters (e.g., Hooke's modulus and Poisson's ratio). A still classic compilation of the values of these parameters for various earth materials has been published by Birch (1966).

3.3.2.3 Creep Range

If very small (changes of) stresses are exceeded in earth materials, the latter respond by slow irreversible deformations. This behavior has been called creep.

Attempts at describing creep by linear (differential) equations have generally been unsatisfactory (Scheidegger 1982). The most satisfactory creep equation to date is a nonlinear one due to Lomnitz (1956):

$$\varepsilon = \frac{\sigma}{\mu}[1 + q \ln(1 + bt)], \tag{3.3.1}$$

where ε is the strain (increment), σ the stress (change) and t time. The remaining symbols (μ, q, b) represent constants of the material in question.

Other proposals for creep laws made in the literature imply power relations of the type

$$\dot{\varepsilon} = \text{const} \left(\frac{\sigma}{\mu}\right)^n \tag{3.3.2}$$

(the dot denotes time differentiation, n the power law exponent), of which Tsenn and Carter (1987) have reviewed the possible versions.

Special conditions obtain in sands and clays. We have already mentioned the occurrence of swelling pressure in clay (Sect. 2.3.4.2). In addition, the "creep" of clays (and sands) does not take place by some rheological creep law, but in direct response to the attainment of a shearing limit (Coulomb 1776) in the stresses. The shearing limit, s, is satisfactorily expressed by Coulomb's empirical equation, which reads (see Terzaghi 1943)

$$s = \sigma \tan \Phi + c, \tag{3.3.3}$$

where c is indicative of the cohesion of the material, σ signifies the normal stress, and Φ is commonly called the angle of internal friction.

The limiting shear stress condition (3.3.3) is best represented in a Mohr diagram (cf. Scheidegger 1982; p. 129) where it appears as two lines. If one normal stress is given, there are two possible values for the other so as to attain the shear limit; they are called the active and the passive Rankine states (cf. Scheidegger 1982: Fig. 68 p. 145). The limiting shear [cf Eq. (3.3.3)] is attained in elements oriented in a direction subtending an angle of $\pm (45° + \Phi/2)$ and $\pm (45° - \Phi/2)$ with the given normal stress direction.

Regarding the cohesion, there are two extreme cases of substances which may be considered here. The first occurs if $c = 0$, so that Eq. (3.3.3) yields

$$s = \sigma \tan \Phi. \tag{3.3.4}$$

This corresponds to cohesionless materials, such as sands or piles of gravel. However, even sands may have a little cohesion, particularly if they are moist. If one forms a pile with *completely dry* sand, the material will slide and not come to rest until the angle of inclination of the slopes becomes equal to a certain angle which is called the *angle of repose*. This angle of repose β is equal to the angle of internal friction if the sand is in its *loosest state*. For loose, dry sand the angle of repose is independent of the height of the slope

$$\beta = \Phi \tag{3.3.5}$$

independently of the height of the slope, which is what was to be demonstrated.

For compacted sand or packed gravel the above relationship is no longer true because the formula for friction [Eq. (3.3.4)] does not hold for packed materials. The resistance to shearing motion in this case depends on the packing. It is then possible for two angles of repose to exist: one which corresponds to the slope angle at which sliding masses come to rest, and another, larger one, which corresponds to the slope at which stacked masses begin to slide.

As noted, the above remarks refer to cohesionless substances. The other extreme case, mentioned earlier, is obtained if in Eq. (3.3.3) Φ is set equal to zero. This leads to the classical theory of plasticity (cf. Hill 1950). Clays show a behavior which fits this kind of description very well.

The above exposition refers only to dry materials. In a water-saturated assemblage of grains, the porewater pressure may effect a change in the fundamental relationship (3.3.3). It then becomes, in view of the general "Terzaghi hypothesis" (cf. Sect. 2.2.4.2).

$$s = c + (p_t - p_w) \tan \Phi, \tag{3.3.6}$$

where c is the cohesion as before, p_t is the total earth pressure (due to the weight of the earth material and the water contained therein), p_w is the porewater pressure ($= \rho g H$ with ρ density of the water, g the gravity acceleration and H the "hydraulic head"), and Φ, as usual, the angle of internal friction.

3.3.2.4 Failure Range

The failure behavior of earth materials has been studied by innumerable investigators: indeed, failure is a problem of greatest importance for the design and execution of many engineering works. Unfortunately, the whole subject matter is still only very poorly understood. This fact is reflected by the very number of publications: the latter represent, in essence, nothing but a compilation of special cases and experiences, without the emergence of generally applicable laws (cf. the discussion by Scheidegger 1982).

Fortunately, for geomorphological purposes, some time-honored general rules are quite sufficient to explain many evidently stress-caused landscape features. Of these rules, the hypothesis of Coulomb (1776), as modified by Mohr (1928), is very useful. It states: for an isotropic medium fracturing under the action of three unequal principal stresses, the surface of fracture is parallel to the direction of the intermediate principal stress and inclined at an angle $\varphi \leqq 45°$ (30° is a good average) toward the maximum principal pressure.

Furthermore, it is assumed that the fracture occurs when the maximum shearing stress reaches a limiting value. For the latter, Wuerker (1956) has made a compilation in tables which is still very useful. Further values have been proposed by Labuz et al. (1985) for granite, Laqueche et al. (1986) for slate schist, and Tschierske (1982), as well as Müller-Vonmoos et al. (1985) for clays. The simple description thus obtained is at least sufficient for explaining many specific geomorphological features.

3.3.3 Stability of Slopes

3.3.3.1 General Remarks

The stability of a slope bank depends on the condition that neither the creep nor the failure threshold of the material is reached.

Inasmuch as it is likely in *soils* that some type of *creep* (including plastic yielding) is initiated, whereas in *rocks* outright *fracture* phenomena are most likely to occur, these two types of materials have to be treated separately.

3.3.3.2 Stability of Soil Slopes

As noted, soil slopes fail if a creep threshold is reached by the stresses. The most common procedure is to consider a plastic yield threshold, corresponding to Eq. (3.3.3). Inasmuch as soils are generally porous, one should also consider the possible effect of (artesian) porewater pressure and thus replace Eq. (3.3.3) by Eq. (3.3.6).

Thus, let us consider a slope bank of slope angle β and height h consisting of some material having certain values for the parameters c and Φ. The question is: what are the conditions for such a slope to become unstable?

A review of the possible methods to obtain answers to the above question has been given by Carillo (1942), who observed that an earth bank may become

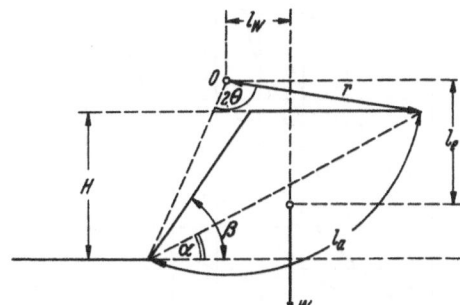

Fig. 16. Slope failure along a critical toe circle

unstable in two ways: slope failure and base failure. Base failure is caused by failure in the supporting base of the slope, whereas slope failure represents a collapse of the slope itself. We are concerned here only with slope failure, since it bears out the possible geomorphological effects sufficiently accurately. The standard analysis today assumes that a slope bank develops a curved sliding surface, which, in a first approximation, can be considered as cylindrical with a circular cross-section (Petterson 1916). This leads to the picture shown in Fig. 16.

The potential circle of sliding is given by the chord angle α and the center angle 2Θ. The equations for stability are then determined by formulating the equilibrium conditions for the moments around the point 0 (in Fig. 16). Assuming the angle of friction to be zero, the only force opposing sliding is that due to the cohesion c. The equilibrium condition is then (Terzaghi 1943; see p. 155)

$$Wl_w - cl_a r = 0, \tag{3.3.7}$$

where l_a is the length of the arc. One can solve this for c which yields

$$c = W\frac{l_w}{rl_a}. \tag{3.3.8}$$

The last equation can be regarded as an expression for the cohesion c required to maintain stability. One can insert the various geometrical quantities in Fig. 16 into it and one then obtains an expression of the form

$$c = \rho gh \frac{1}{f(\alpha, \beta, \Theta)}, \tag{3.3.9}$$

where h, β are the height and slope angle of the slope, respectively, and f denotes a certain function of α, β, Θ. Failure in a given slope would occur around such a circle that c is a minimum, i.e.,

$$\frac{\partial c}{\partial \alpha} = \frac{\partial c}{\partial \Theta} = 0, \tag{3.3.10}$$

since β is fixed. This equation can be solved anaytically and one thus obtains the "critical toe circle" for the failure of a slope. For each critical toe circle, one has a connection between c and h; hence one can determine the critical height h_c for any

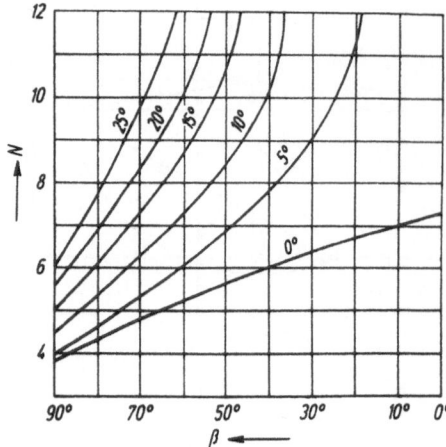

Fig. 17. Stability factors as a function of slope angle for various values of Φ

value of the cohesion c and slope angle β. The connection has the form

$$h_c = \frac{c}{\rho g} N,$$ (3.3.11)

where N is a *stability factor*, depending on the slope angle β only. The calculations sketched above have been carried out by Fellenius (1927); his results are represented in the graph shown in Fig. 17. The constants α and Θ of the corresponding toe circles are shown in Fig. 18.

Fellenius also carried out calculations for a friction angle $\Phi \neq 0$. The stability factor N_Φ tends to infinity for $\beta \rightarrow \Phi$; i.e., as long as the slope angle β is smaller than Φ, the slope can be made arbitrarily high. The results of the calculations of Fellenius are also shown in Fig. 17.

The above standard theory has been modified by assuming different shapes of sliding surfaces (cylinders with elliptical, logarithmic, etc. cross-sections). Many of the attempts have been summarized in the writer's works on catastrophes

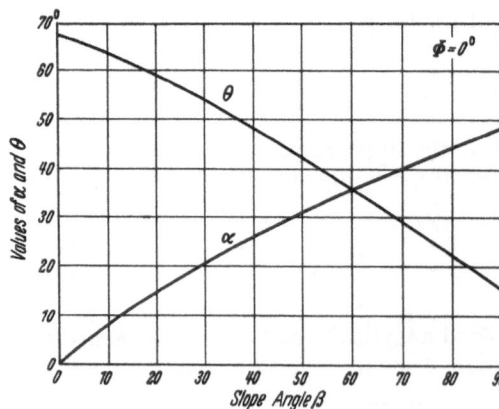

Fig. 18. The angles α and Θ of the critical toe circle for $\Phi = 0$. After Fellenius (1927)

(Scheidegger 1975, 1984, 1988). As of late, finite element calculations have been used in this connection (Toki et al. 1985; Correia 1988). An entirely novel approach to the problem has been made by the introduction or probabilistic features. Whereas conventional slope stability investigations have been performed using two-dimensional limiting equilibrium analysis, probabilistic methods use the failure probability for every slice considered on a slope bank and then apply the principle of maximum entropy for making the individual probabilities of failure all equal (Read and Harr 1988).

Turning to influences on slope stability of effects other than self-gravitational ones, we note that the neotectonic stress field represents one such case. Thus, it has been noted that in many slides, the mean displacement direction coincides with one of the principal neotectonic stress directions (cf. the summary of this work by Scheidegger and Ai 1986): the presence of a horizontal tectonic pressure normal to the crest of a slope bank evidently destabilizes the latter (cf. Ai and Miao 1987, for mechanical estimates).

As mentioned above, the presence of porewater pressure also destabilizes a slope, in conformity with Eq. (3.3.6). This is the explanation for the observation that landslides are triggered usually during and by heavy rainfalls (e.g., Iseda and Tanabashi 1986) and by a rise in groundwater pressure (Okunishi and Okimura 1987). Such a rise, due to the impounding of an artificial lake at the foot of Mount Toc, seems to have triggered the notorious slide at Vajont (cf. the discussion by Scheidegger 1984) in 1963.

Finally, it may be noted that earthquakes can have a destabilizing effect on slope banks: during a seismic event, the inertial forces (D'Alembert's principle) have to be added to the static ones; the sum of the two may well exceed the strength limit of the material.

3.3.3.3 Stability of Rock Slopes

Turning now to rock slopes, we note that the stability condition refers to a fracture or sliding threshold. Very often, the latter occurs on preexisting joints which can become reactivated. In the latter case, it is a frictional process which is initiated: a simple equilibrium calculation will tell whether the instability threshold is exceeded or not (Bjerrum and Jørstad 1968). The problem has been reviewed, e.g., by Scheidegger (1975). Since that time, computers have been used to assist the calculations (Carrara et al. 1977; Leung and Kheok 1987), particularly if nonlinear strength criteria are applied (Madhav and Poorooshasb 1986; Zhang and Chen 1987).

The consequence of a rock slope failure is a rock fall. The basic laws applicable to falling rocks are simply those of freely falling bodies. In addition, saltation (jumping) and rolling may occur (Scheidegger 1975). On uneven ground, it is again helpful to use computers to follow the dynamics occurring at each impact (Bozzolo et al. 1988).

A specific type of geomorphological decay can be observed in consequence of the instability of exposed rock slopes. It is well known that a tectonic stress field is present everywhere. In view of Mohr's fracture condition (cf. Sect. 3.3.2.4), this causes decay patterns referring to peaks and walls. Inasmuch as the corresponding theory refers to the *endogenic* stress field, it has been presented in the writer's book on geodynamics (Scheidegger 1982; cf. Sect. 7.42 therein).

3.3.4 Rapid Mass Movements on Slopes

3.3.4.1 General Remarks

As noted on several previous occasions, landslides represent spontaneous rapid mass movements on slopes. The landslide phenomenon evidently involves basically three distinct aspects: the triggering, the motion itself, and the long-range end result which can be landscape-forming.

Landslides are of great importance to mankind. In this context, they have been studied by engineers (cf., e.g., the book by Scheidegger 1975; also a symposium arranged by Yang and Yamaguchi 1987). In the present study, we will mainly be concerned with the geomorphological effects of the slides.

3.3.4.2 Initiation of Landslides

In a review of the possible mechanical "causes" of landslides, one has to distinguish between long-range and immediate effects. As noted earlier (Sect. 1.5.2), landslides are part and parcel of the normal long-range landscape development: the ongoing endogenic uplift *has* to be compensated by downhill mass movements. Furthermore, the neotectonic forces predesign the future failure surfaces so that the orientation of the slides is quite in conformity with the orientation of the neotectonic stress field (Scheidegger and Ai 1986; Ai and Miao 1987).

Nevertheless, the fact remains that slides occur on slopes that have been stable for a long time, sometimes for many thousands of years. Thus, if a slide occurs, there must have been an agent which caused a decrease of the stability of the slope so as to make it collapse. In every landslide, there must therefore be a direct mechanical cause. A good review of the possible mechanical causes has been given by Terzaghi (1950). Accordingly, one has to distinguish between *external* and *internal* causes of slides. The external causes create an increase of the shearing stresses in the slope. They include, e.g., the undercutting of the slope by a river, the effect of an earthquake shock, and the deposition of material at the upper edge of the slope. Internal causes are those that create a decrease of the shearing resistance of the material. This may occur without a visible change of the conditions affecting the slope (hence their name) and may be due to an increase of the porewater pressure or to a progressive decrease of the cohesion of the slope material.

Terzaghi (1950) discusses the various possibilities as follows.

a) *Increase of Slope Steepness.* A river undercutting a slope bank will generally cause the slope to collapse. This is simply due to an overall increase in slope angle which eventually may reach such a value that the stability criterion is no longer satisfied. A similar situation occurs if material is deposited at the upper edge of the slope, but this is rare under natural conditions although it is common enough in bad engineering practice. As soon as the stability condition is no longer satisfied, a slide will occur.

b) *Earthquakes.* Another external cause of landslides may be represented by earthquakes (Seed 1967; Harp 1986; Svoboda 1986). The equilibrium condition for a slope has been given in Eq. (3.3.7). One can define a "safety factor" G for the slope by forming the quotient of resisting moment and driving moment (for an explanation of the symbols, cf. Fig. 16:

$$G = \frac{cl_a r}{Wl_w}. \tag{3.3.12}$$

A slide will occur if $G < 1$ in conformity with the equilibrium condition (3.3.7).

An earthquake produces a horizontal acceleration g_e so that the equilibrium condition now reads: (cf. Fig. 16)

$$Wl_w - cl_a r + \frac{W}{g} g_e l_e = 0. \tag{3.3.13}$$

Defining again a safety factor as the quotient between resisting moment and driving moment, we have during the earthquake:

$$G' = \frac{cl_a r}{W(l_w + l_e g_e/g)}. \tag{3.3.14}$$

A slide will occur if during the earthquake $G' < 1$. For catastrophic earthquakes, g_e may reach $\frac{1}{2}g$. Thus the lowering of the safety factor evident in formula (3.3.14) explains how landslides may be caused by an earthquake.

c) *Pore Water Pressure.* Turning now to the *internal causes* of landslides, we note that the most common such cause is an increase in pore-water pressure (piezometric landslides). In a water-saturated assemblage of grains, the relation for the shearing resistance (to failure) of a water-filled earth material is given by Eq.(3.3.6).

If one again calculates the "safety factor" G (following the procedure leading to (3.3.12), one has to replace c in (3.3.12) by s from (3.3.6) and obtains

$$G = [c + (p_c - p_w) \tan \Phi] \frac{l_a r}{Wl_w}. \tag{3.3.15}$$

As usual, a slide occurs if, along the critical toe circle, the quantity G becomes smaller than 1. It is obvious that this is possible if the pore-water pressure p is sufficiently increased. A more sophisticated treatment of the problem has been presented by Nago and Maeno (1987). An increase in porewater pressure may occur by the formation of a lake at the bottom of a slope or by soaking of the ground by rainstorms (Schumm and Chorley 1964).

d) *Other Internal Causes.* A mounting porewater pressure is the most common cause of an internal change of the shear resistance of the material forming a slope. However, we have already stated above that structural changes may gradually occur in the slope material. These structural changes affect the cohesion of the material and may be due to chemical and physical weathering. In a similar category belongs the phenomenon of "spontaneous liquefaction" which may occur due to the change of the arrangement of the grains in water-logged fine sand or coarse silt (McCrone 1960; Bazant 1966; Kent 1966). In all these cases, the shearing resistance is lowered to such an extent that the safety factor [Eq.(3.3.12)] becomes smaller than 1, and hence a slide results. Sometimes, the slow progressive changes are confined to the base upon which the slope rests; this too will lead to a slide.

Finally, an internal cause is also represented by the progressive development of a fracture within the material: as the failure arises, the stress state in the material is changed and this may produce new and more critical stress concentrations which create further instabilities and fractures in the material. The stress relocations can thus accelerate their spread, leading to a sudden collapse (cf. also Erismann 1988).

3.3.4.3 Mechanics of Landslides

Once a slope becomes unstable, it will collapse. The collapse must be thought to take place along the critical toe circle, inasmuch as the latter is an acceptable approximation to the surface of sliding (cf. Sect. 3.3.3.2). The final result of a slide is shown schematically in Fig. 19.

The actual mechanics of the sliding motion has generally been described by a "frictional model", using a ficticious coefficient of friction which is different for each slide. In a model based on friction, the average coefficient of friction, f, is simply equal to the total vertical height, h, of the path of the landslide divided by the total reach, x (cf. Fig. 20; after Scheidegger 1973), provided the slide started from rest:

$$f = h/x = \tan \alpha. \tag{3.3.16}$$

It is a matter of experience that for large, catastrophic landslides, the coeffcient of friction becomes progressively smaller as the volume V of the sliding mass

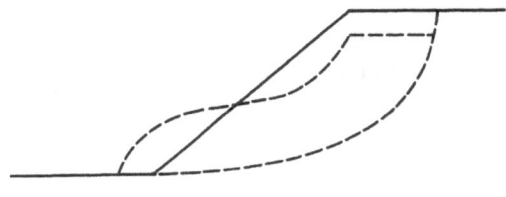

Fig. 19. Schematic drawing of the final result of a landslide. (After Terzaghi 1950)

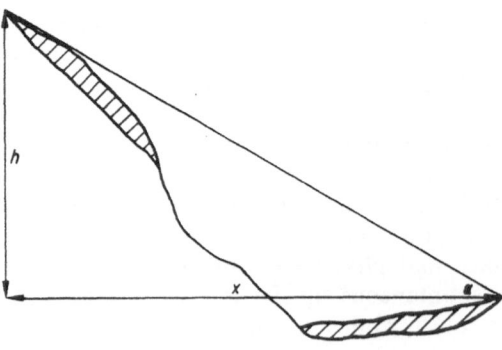

Fig. 20. Geometry of a slide. (After Scheidegger 1973)

increases. Scheidegger (1973) showed (entirely empirically) that the best correlation is obtained if $\log f$ is plotted against $\log V$

$$\log f = a \, \log V + b \tag{3.3.17}$$

with (V in cubic meters; logarithms to the basis 10)

$a = -0.15666$
$b = 0.62419.$

The empirical correlation of Scheidegger (1973) gives no explanation for the observed low coefficients of friction. In fact, the exact reason for this observation is not at all clear as of yet. Proposals include the idea that the slides ride on a cushion of trapped air (Shreve 1966) or on vaporized pore-water (Habib 1975), fused rock (Erismann 1979), dispersive fluidization (Hsü 1975), and others (see the review by Scheidegger 1984).

The frictional model discussed above has been very successful for describing the general dynamics of landslides. More specific investigations following the details of the sliding motion have also been made (Iverson 1986a, b; Hutter and Savage 1988), but these are generally unnecessary if only the final geomorphological effects are of interest.

3.3.4.4 Geomorphological Effects of Landslides

A single landslide has evidently a specific local effect in a landscape. Cumulatively, recurring landslides can form the character of a landscape.

Thus, the general hump-slump form typical of slides (cf. Fig. 19) can be seen to recur in many landslide-prone areas. Traces of floods may be found inasmuch as rivers have been dammed up by landslide masses which have subsequently been broken through, releasing the water suddenly. This is particularly evident in seismically active regions, where frequent landslides are earthquake-induced (Johnson 1984; Sorriso-Salvo 1986).

Slides are a form of exogenic denudation agent. In plan, they are more prone to occur in "hollows" than on "noses", so that a topography of hollows and noses can develop (Iida and Okunishi 1983).

Peculiar landslide-caused features have also been observed in coastal zones (Prior and Renwick 1980) where slides cause steep-walled, cirque-type depressions opening to the sea in soft materials.

3.3.5 Slow Spontaneous Mass Movements

3.3.5.1 General Remarks

The phenomena of mass movement that were discussed above are characterized by a rapid downhill movement of earth masses in the form of slides. However, mass movement may also occur in the form of *slow flowage*. Such slow flowage is almost imperceptible if it is not observed over a long period of time; it consists in a slow downslope movement of soil or rock.

There are several types of slow flowage that may be discerned. These depend to an extent on the climate that is prevalent. In temperate and tropical climates, one encounters rock creep, talus creep, and soil creep; in niveal climates, one has solifluction.

3.3.5.2 Rock Creep

Even in hard rock, such as in schists (Huder 1976) or in crystalline metamorphics (Bonzanigo 1988), it is often possible to recognize mass creep that reaches depths of hundreds of meters. At the head of slopes, such movements appear as "mountain fracture", in the middle as "slumps", and at the bottom as "valley closures". Frequently, the material breaks up into plate-like fragments which remain relatively intact and move individually.

A general theory of the problem has been given by Brückl and Scheidegger (1972), who have shown that reasonable sliding velocities (i.e., up to 1 m/a) can be obtained if a logarithmic or generalized Newtonian creep law is assumed for the moving mass.

3.3.5.3 Talus Creep: Scree

The fall and accumulation of rock fragments at the foot of a wall leads to the formation of talus slopes that consist of scree (cf. Sect. 1.2.3). In its equilibrium position, a scree slope will have a slope angle equal to the angle of repose (Sect. 3.3.2.3) of the material. Typical values (29–40°) for various natural scree slopes were given by Piwowar (1903), Chandler (1973), and Knoblich (1975). It is then of some interest to investigate the motion that occurs in such slopes of scree.

The most natural and oldest theory is that the scree motion is caused by temperature changes. The model is illustrated by Fig. 21 (Scheidegger 1961b). Thus, we consider a long, straight slope of slope angle β. We shall assume that all the material is constrained from direct movement parallel to the slope. It is then possible to look at the effect of the daily temperature changes in the following

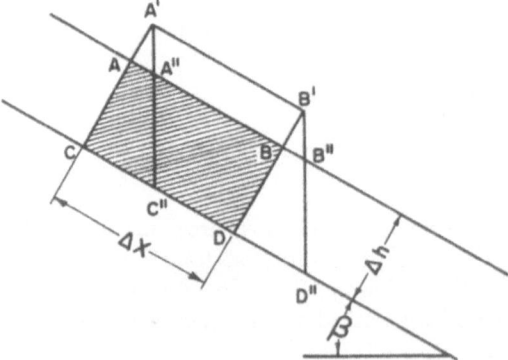

Fig. 21. Geometry of constrained thermal expansion. (After Scheidegger 1961b)

manner: During the heating-up period each infinitesimal block of material ABCD (cf. Fig. 21) will expand to the shape A'B'CD. This must be so if it be assumed that no sideways expansion can take place. During the cooling period, the particles will rearrange themselves and the shrinkage of any block A'B'C"D" will then the result in the form A"B"C"D" in Fig. 21. Thus, during one cycle, a net transport of surface material by the distance AA" takes place. During this whole process no net sideways expansion (i.e., in the x-direction) occurs in correspondence with our assumption.

A detailed calculation (Scheidegger 1961b) shows that one obtains indeed yearly net downhill motion rates of the required magnitude. Similar calculations were made by Harrison and Herbest (1977) and by Yair and De Ploey (1979).

Unfortunately, observations in nature do not bear out the correctness of the above model. By applying strips of paint to the rocks at various elevations and revisiting these after some years, it was noted that no creep occurs at all, but that some of the rocks had moved in a statistical fashion by substantial distances. Cumulatively, of course, the phenomenology of "steady" downhill motion is thereby created (Carniel and Scheidegger 1976; Luckman 1978; Gardner 1979; Perez 1988).

Thus, the motion on scree slopes evidently takes place in the form of "miniature landslips" (Gerber and Scheidegger 1966): Small landslips occur in the layer (having a cohesion c and an angle of internal friction Φ) of scree of (minimum) thickness D and average slope angle α; h_c is the critical height (cf. Sect. 3.3.3.2) of the slope bank that is stable in the scree material with slope angle β, and 2θ is the central angle of the critical toe circle for the landslip (Sect. 3.3.3.2) that is just possible. Inspecting the geometry of Fig. 22, one can calculate the critical thickness D_c

$$D_c = \frac{h_c(1 - \cos \theta)}{2 \sin \alpha \sin \theta}.$$
(3.3.18)

Using Eq. (3.3.11) and the curves of Fellenius (see Fig. 17) one has a connection between all the characteristic quantities. Thus, a scree thickness $D > D_c$ is unstable, a thickness $D < D_c$ is stable. Material will therefore accumulate until $D = D_c$ and then will begin to slide by means of miniature landslides.

Furthermore, it has been noted that most scree slopes are concave upward. Since in a cohesionless material the slope angle should be equal to the angle of repose which is independent of the height of the slope (cf. Sect. 3.3.2.3), one is faced with the problem of explaining this fact.

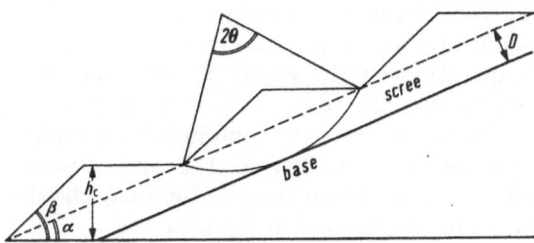

Fig. 22. Geometry of a scree slope. (After Gerber and Scheidegger 1966)

Machatschek (1952) ventured the opinion that the concavity of slopes of screes is due to the grading of material which is generally observed thereupon. However, it appears that the angle of repose, which would generally be believed to determine the slope angle, does not depend on the size of the screes, but rather on their angularity. A somewhat more convincing explanation has been advanced by Sharpe (1938), who noted that there are, in fact, *two* angles of repose: one at which the material begins to slide, and another at which it is being deposited. The former angle is always greater than the latter. If the fomation of the slope is due to talus creep, one would expect that, at the top, the slope angle is close to the angle of repose for incipient sliding and, at the bottom, close to the angle of repose for deposition. This would at least partially explain the observed concavity, but no numerical comparisons are available. It would appear, however, that the two angles of repose do not differ sufficiently to account for all of the observed change in steepness.

A considerably greater difference in angle than that accounted for above may be caused by the fact that the packing of the screes becomes looser the further the material has moved from its source. We have noted in Sect. 3.3.2.3 that the customary angle of repose in a cohesionless material is reached for the loosest possible packing only. If the packing is denser than the loosest one possible, a slope of greatly increased slope angle is attained. This increase may be very substantial and can easily account for any increase of slope angle with height that may be observed.

3.3.5.4 Spontaneous Soil Creep

Turning now to soft materials, we note that, in this instance, one generally speaks of "soil" creep. The laws governing the displacement mechanisms are the various rheological creep laws.

The very shallow (50 cm) types of movements are referred to as "skin creep" or "surface soil creep" (Lewis 1976; Göbel 1977; Mills 1983). The motion is spatially continuous, but may be locally nonuniform so that small terracettes result with wave lengths of the order of 1 m. The actual cause of the formation of such terracettes is, in fact, still a matter of controversy. Scheidegger (1983) has seen in their genesis the operation of an instability principle of general systems theory (cf. Sect. 5.2.3). The terracettes are certainly also enhanced by grazing animals (Higgins 1982), but animals cannot be the sole cause, since hummocky terrains are also found in regions without animals (Hamann 1984, 1985). Some hummocks may be cryogenic in origin (Bolster 1985).

The more familiar types of soil creep affect layers of some tens of meters of thickness. These phenomena are particularly spectacular in argillaceous materials, where they can be chronic for centuries. Like many mass movements, they show active and quiescent phases: short periods (1–2 years) of activity alternate with long periods (8–10 years) of quiescence (Gerber and Scheidegger 1984). The activity periods are evidently caused by the high availability of water (e.g., heavy precipitation in winter combined with rapid melting of the snow). In an active phase, the actual triggering of the sliding is caused by the immediate precipitation, which acts on the first day at the top of the slide, after 3–6 days in its middle part and after 10–12 days at the foot.

The actual mechanics of soil creep is analogous to that of ice in glaciers, at least if the latter is treated by the method of plasticity theory. Ice flow is, in its turn, very similar to the theory of the Rankine state of the slow downhill creep of scree (Sect. 3.3.5.3): the driving force is gravity. As has been shown in Sect. 3.3.2.3, the material can flow in either an active or a passive Rankine state. The formulas are similar to those in glacier flow and will be discussed in detail in that connection (Sect. 7.3.3.2).

In addition, the porewater pressure in the sliding material is a further variable which, if it is excessively high, can contribute to the mobility of the material (Hutchinson 1986; Vulliet and Hutter 1988).

3.3.5.5 Niveal Solifluction

In cold climates, one encounters niveal solifluction: frost heaving occurs in a direction normal to a slope, settling upon thawing in the vertical direction so that during each freezing-thawing cycle a net motion of the soil in the downslope direction is generated. Williams (1957, 1959) has envisaged the mechanism of niveal solifluction in a manner as shown in Fig. 23. Qualitatively, it was invoked for the first time, it appears, for the explanation of frost-induced soil creep by Davison (1889). The line A–B need not necessarily be straight. The increments Δl of downslope movement (cf. Fig. 23) are proportional to the increment of frost heave ΔE and to the tangent of the slope; hence

$$\Delta l = \tan \alpha \, \Delta E. \tag{3.3.19}$$

However, the increment of frost heave ΔE is proportional to the time taken for Δx

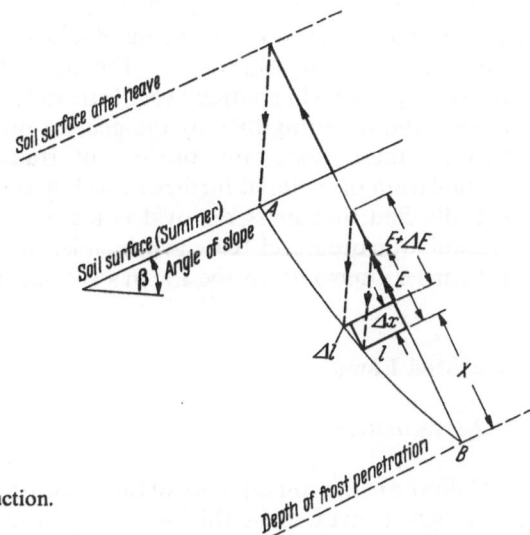

Fig. 23. Details of frost action in solifluction. (After Williams 1957)

to freeze times the water flowing in during that time, Thus:

$$\Delta E = f(x)\Delta x. \tag{3.3.20}$$

where $f(x)$ is the quotient of water inflow rate over rate of freezing. Hence, by integration

$$l(x) = \tan \beta \int_0^x f(x)\,dx. \tag{3.3.21}$$

This yields the equation of the curve AB. Because of the conservation of mass, $f(x)$ must equal the amount of excess water present in the soil at the level x. It can thus be measured. Hence, the shape of the curve AB can be calculated in any practical case.

A somewhat more elaborate theory of essentially the same model of solifluction has been given by Kirkby (1967). The mechanism of solifluction envisaged above was verified in the laboratory by Coutard and Mücher (1985). Quite generally, the theory seems to be supported by the facts.

3.4 External Transporting Agents

3.4.1 Basic Statements

Mass displacements on slopes do not only occur spontaneously, but also in consequence of the action of external transporting agents: these agents can be water or snow/ice. In the present context, we consider only water-caused transport; the problem of transport by snow or ice will be relegated to Chap. 7, which deals with niveal geomorphology.

Mass displacements of the type considered here are characterized by the phenomenon that a "flow" (carrying mud or debris) is flowing over a "substratum". In argillaceous materials, such flows are caused by the thixotropy of the clay, leading to mud flows. When the material carried by the water is unconsolidated rock, one speaks of "debris flows". Finally, water may rush as a free surface sheet down a slope, carrying initially very little solid material, but leading to local dense gullying and initiating thereby the general process of slope retreat.

In all these cases, the process of transport entails erosion as well as accumulation of material further downhill: on a steep part of the slope, one will generally find that mass removal is taking place, whereas on a less steep part accumulation occurs (cf. "catena principle" discussed in Sect. 1.2.4). A good review of the mass transport on the Earth's surface has been given by Statham (1977).

3.4.2 Mud Flows

3.4.2.1 Definition

Mud flows are the consequence of the thixotropy of clayey materials. Their surface is strongly convex; their thickness may reach 30 m, and their mean velocities

10 m/s. Mud flows are intermittent events (Strömquist 1985) that follow preexisting gulleys in spurts (Aulitzky 1982): local blockages may bring them temporarily to a halt.

3.4.2.2 Mechanism

The mechanical explanation of mud flows has generally been based on river flow formulas (cf. Sect. 4.2.2) such as the Manning (1890) relation [Eq. (4.2.12)]. Based on experimental investigations, Goldin and Lubashevsky (1966) found for the Manning parameter n

$$n = 1/5.15 \qquad (3.4.1)$$

so that the velocity-equation (v in m/s) becomes

$$v = 5.15 H^{2/3} S^{1/4}, \qquad (3.4.2)$$

where H is the thickness and S the slope (tangent of slope angle) of the mud flow.

3.4.3 Debris Flows

3.4.3.1 Definition

When the material carried by the water is unconsolidated rock, one speaks of "debris flows" (cf., e.g., Breitfuss and Scheidegger 1974). For the activation of such flows, two conditions have to be satisfied: in the first place, sufficient debris has to be piled up in the gulleys so that there is material in place that can be moved; in the second place, a heavy storm must occur which activates the debris.

3.4.3.2 Mechanism

The actual mechanics of debris flows is again based in the Manning (1890) formula (cf. 4.2.12); with the pertinent empirical parameters the latter becomes now

$$v = 3.15 H^{1/6} d^{1/3} \sqrt{\frac{\rho_s - \rho_m}{\rho_m}}, \qquad (3.4.3)$$

where v (m/s) is again the velocity and H the thickness of the flow; furthermore, d(m) is the average grain diameter, ρ_s the density of the rock debris and ρ_m the density of the flow mixture (Goldin and Lubashevsky 1966).

For the mobility of the rock-water mixture, Bagnold (1954) has given an explanation by postulating the existence of a "dispersive" stress which is caused either by the collision between the individual rock fragments (inertial flow regime) or by an interaction between the rock fragments and the water (macroviscous regime). An estimate of the numerical values required by Bagnold's (1954) theory shows that the latter are entirely reasonable.

3.4.3.3 Large Boulders

Matters have to be modified somewhat if the rock debris moved are large.

Thus, in this case, the mass movements are caused by water drag on the individual fragments rather than by movement in a flow mixture. This applies, e.g., to granite gruss (De Ploey and Moeyersons 1975), or to rocks (of dimensions $8 \times 3 \times 4$ cm) on playas in Nevada which even left tracks of their motion (Wehmeier 1986). Using the usual formulas describing the scouring force in rivers (Sect. 4.4.3), Wehmeier (1986) showed that water velocities of from 0.5 to 2.0 m/s are capable of producing the sliding motion of the blocks that was inferred from observed tracks.

3.4.4 Slope Development by Free Flow of Water

3.4.4.1 General Remarks

Slopes are affected not only by intrinsic and induced mass movements, but also by the direct action of water (and of snow; however, the effect of the latter will be treated in Chap. 7 on niveal landforms).

First, there are the geomorphological effects of the rainwash (cf. Sect. 3.2.3.3). On a large scale, rainwash can pass over into sheet floods; the latter can also originate in some catastrophic event such as the emptying of a lake.

Second, the free flow of water further initiates slope development by the formation of rills and gulleys. Upon an integral treatment of all water-caused effects, models of slope development referring to this base can be set up. Finally, it is desirable to investigate the possibilities of establishing scaling rules for making scale model experiments.

3.4.4.2 Water Drag on Slopes

The physical effects of raindrops on soil has been discussed in Sect. 3.2.3.3 in the context with physical denudation processes. We investigate now the geomorphological effects of rain on slopes.

The main effect is the inducement of a net material loss of a slope bank and thereby its recession. Interestingly enough, rain splash can also cause upslope soil movements (Poesen 1986), particularly under certain wind conditions. The normal effect, however, is runoff creep, temporarily of some 20 cm/h, downhill (Moeyersons 1975; De Ploey et al. 1976). Soil loss rates in a semiarid climate have been found to be around $1-6$ g/m^2 per year (Le Roux and Roos 1986), higher in wetter climates. There are hillslope gradient-particle size relations (Abrahams et al. 1985) as local equilibria are established. The soil detachment rate decreases as the runoff depth increases (Torri et al. 1987b) since the detachment power of raindrops is partially dispersed by a preexisting water layer.

3.4.4.3 Sheet Flow and Sheet Floods

The runoff on a slope may eventually take on the aspect of an areal sheet *flow* if the precipitation suffices to maintain a steady water film. Under such conditions, a stable equilibrium profile is maintained (Trofimov and Moskovkin 1976). The theory is based on a laminar flow regime on gentle slopes (Savat 1977). The erosion is caused by drag on individual ground particles (Tödten 1976; Bryan 1979; Band 1985). The areal erosion seems to be caused mainly by bed load progression in individual concentrated water streams (Seuffert 1981), not by a regular process across the profile. Concentrated surface flows tend to become unstable (Trowbridge 1987).

When the water runoff on a slope becomes very large and intermittent, one speaks of sheet *floods*. Such floods may be caused by the breaking of a blockage or by a slide into an existing lake. In such cases, an almost instantaneous flood is created. In profile, a nonstationary gravitational solitary water wave moves along a shallow body of water. The basic mathematical formulation of the problem was given in two dimensions by De Saint-Venant (1871); the basic equation is

$$\frac{\partial u}{\partial t} + u\frac{\partial u}{\partial x} + g\frac{\partial h}{\partial x} + g(S_e - S_s), \tag{3.4.4}$$

where u is the mean velocity at the time t at the point x along the slope, g the gravity acceleration and S_e the slope of the energy line and S_s that of the slope. The equation (3.4.4), in fact, is nothing but the Eulerian equation for waves in shallow water [Eq. (6.2.23)], but with a sloping ground and water surface; its deduction is given in Sect. 6.2.2.2. The theory can be extended to floods in channels where the problem may be of engineering interest (Scheidegger 1975).

In cases of large floods, the geomorphic effects are drastic. The scablands of Washington State seem to have been largely formed by a flood caused by the sudden breaking of an ice dam on glacial Lake Missoula between 13 500 and 16 000 years B.P. (Baker and Nummedal 1978); a similar cause may be advocated for the genesis of some landscapes in the Yukon (Clarke and Mathews 1981). Large flood events may not be as rare as had commonly been thought, as evidenced by paleohydrologic analysis of slack water sediments (Costa 1983; Baker et al. 1983, 1988; Baker 1983, 1984; Clarke 1986; Mayer and Nash 1987).

3.4.4.4 Slope Development by Rill Formation

The slope recession occurs microscopically generally by the formation of little rivulets and gulleys in which material is carried away. The manner in which this might happen has, for instance, been discussed by Beaty (1959); reviews on the subject matter have been given by Bryan (1987) and Jones (1987).

Gullying is naturally most pronounced in soft materials. Most studies that have been reported refer to specific geographic locations (e.g., Flood 1981) or to specific conditions (such as mine tailings; e.g., Haigh 1978, 1979) and are entirely phenomenological, albeit quantitative empirical initiation thresholds may be arrived at in this fashion (De Ploey 1983; Govers 1985; Torri et al. 1987a). Gullying can also, more rarely, develop in more resistant rock (Twidale and Campbell 1986).

Of particular interest are studies of the evolution of gulley *systems*. For this purpose, Mosley (1972a, b) has proposed a random walk model such as had been done by the writer in connection with river nets in general (cf. Sect. 5.5.2.2 of this book).

3.4.4.5 Alluvial Fan Formation

The opposite to rill formation is alluvial fan buildup. The latter is due to the loss of carrying capacity of the water owing to a corresponding loss of velocity (Zollinger 1983). There is a feedback effect because the buildup of the fan decreases the slope and therewith the velocity of the water (Mizutani 1986). The genesis of fans may not be due to a single mechanism: deposits from debris flows, transitional flows, and stream flows may alternate (Beaty 1970, 1974; Wells and Harvey 1987). The intermittency of deposition mechanisms is particularly evident in deserts and semideserts where rain storms and therefore mass transport by water are rare events amongst the wind-caused deposits (Frostick and Reid 1982; Dorn 1988).

It is difficult to describe the genesis of an alluvial fan theoretically, and therefore semi-empirical studies have been made. Bull (1964) and Denny (1965) came up with the following relationship between the fan area A_f and the drainage-basin area A_d

$$A_f = cA_d^n, \tag{3.4.5}$$

where c and n are constants; Hooke (1968) found for the mean value of n

$$n = 0.90. \tag{3.4.6}$$

The study by Hooke (1968) suggested that the relationship (3.4.5) may be due to the establishment of a steady state between deposition and erosion in a fan. An interesting model of this quasi-steady state has been suggested by Price (1974), who based it on the assumption of a random walk for determining the pattern of deposition. Each flow event was taken as a stochastic process so that the fan buildup occurs in a random fashion. General relationships cannot be obtained in this way, but simulating the processes on a computer produces patterns that correspond to the natural fan structure.

3.4.4.6 Integral Slope Development by Water Flow

The general relationships between carrying capacity and velocity of water that have already been alluded to in the last section (3.4.4.5) in connection with the formation of alluvial fans, can also be used for a discussion of the overall integral slope development caused by water flow (Scheidegger 1959).

Thus, let us assume that sediment-carrying water hits a flat part of ground with a constant velocity v_1. The water carries with it a certain amount of material c per unit volume. Due to its being slowed down, it will start building up a slope on the originally flat ground. It is clear that the mass-carrying capacity of the water increases with the velocity v, and hence one may set as a first approximation

$$v \sim c \tag{3.4.7}$$

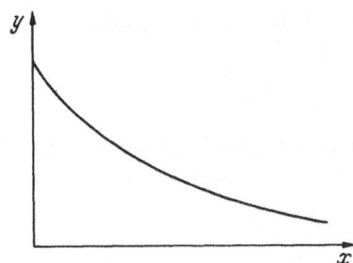

Fig. 24. General geometry of a slope

for a *particular body of water*. Furthermore, we also assume that the velocity of the water is proportional to the instantaneous slope S over which it travels

$$v \sim S. \tag{3.4.8}$$

The last assumption implies that we neglect the effects of momentum which, in turn, means that the layer of water carrying the mass is not very thick (Horton et al. 1934).

We now introduce coordinates x (horizontal), y (vertical, upward) and t (time), and assume that the mass-carrying water hits the originally flat surface at $x = 0$, with a constant velocity v_1. The geometrical layout is shown in Fig. 24. In the slope that will develop, the increase in height is due to the loss of mass from the water; hence

$$\frac{\partial y}{\partial t} \sim -\frac{\partial(vc)}{\partial x}. \tag{3.4.9}$$

Equation (3.4.8) yields, in terms of x and y

$$v \sim \frac{\partial y}{\partial x}. \tag{3.4.10}$$

Thus

$$\frac{\partial y}{\partial t} \sim -\frac{\partial v^2}{\partial x} \sim -\frac{\partial}{\partial x}\left(\frac{\partial y}{\partial x}\right)^2. \tag{3.4.11}$$

Differentiating with regard to x

$$\frac{\partial^2 y}{\partial x\, \partial t} \sim -\frac{\partial^2}{\partial x^2}\left(\frac{\partial y}{\partial x}\right)^2 \tag{3.4.12}$$

and setting

$$\zeta = -\frac{\partial y}{\partial x} \tag{3.4.13}$$

yields

$$\frac{\partial \zeta}{\partial t} \sim \frac{\partial^2 \zeta^2}{\partial x^2} \tag{3.4.14}$$

or, written as an equality

$$\frac{\partial \zeta}{\partial t} = a \frac{\partial^2 \zeta^2}{\partial x^2}, \qquad (3.4.15)$$

where a is some constant. The above differential equation has to be solved for the following boundary condition implicit in our assumptions:

$$\zeta(t, 0) = \zeta_1 = \text{const} \qquad (3.4.16)$$

and the initial condition

$$\zeta(0, x) = \zeta_0 = 0, \qquad (3.4.17)$$

since ζ is proportional to the velocity.

The differential equation (3.4.15) is nonlinear and is thus difficult to solve. Green and Wilts (1952) have suggested a way to linearize the equation so that it will read:

$$\frac{\partial \zeta^2}{\partial t} = a' \frac{\partial^2 \zeta^2}{\partial x^2}, \qquad (3.4.18)$$

where a' is a new constant depending on the old constant a and on an intermediate value ζ' between ζ_0 and ζ_1:

$$a' = a/\zeta'. \qquad (3.4.19)$$

Since we are interested only in general features of the solution, it will be sufficient to deal with the linearized equation. The latter can easily be solved; the required solution is:

$$\zeta^2 = \zeta_1^2 \left(1 - \text{erf} \frac{x}{\sqrt{4a't}} \right) \equiv \zeta_1^2 \, \text{erfc} \frac{x}{\sqrt{4a't}}. \qquad (3.4.20)$$

Hence, we obtain for the slope

$$S \cong - \zeta_1 \sqrt{\text{erfc} \frac{x}{\sqrt{4a't}}}, \qquad (3.4.21)$$

for the velocity

$$v \sim v_1 \sqrt{\text{erfc} \frac{x}{\sqrt{4a't}}}, \qquad (3.4.22)$$

and for the height of the accumulation

$$y \cong - \int_{\infty}^{x} \sqrt{\text{erfc} \frac{x}{\sqrt{4a't}}} \, dx. \qquad (3.4.23)$$

Since the error function complement behaves for large x just like e^{-x^2}/x (see, e.g., Prange 1943) and its square root just like $e^{-(1/2)x^2} x^{-1/2}$, one does not have to fear any difficulties with regard to the convergence of the integral. The shape of the $y = y(x)$ curve is thus exactly as was anticipated in Fig. 24.

The last equation obtained (3.4.23) thus gives a description of what accumulative slopes should look like: they should be essentially concave. Gradually, with time, they should approach a horizontal level.

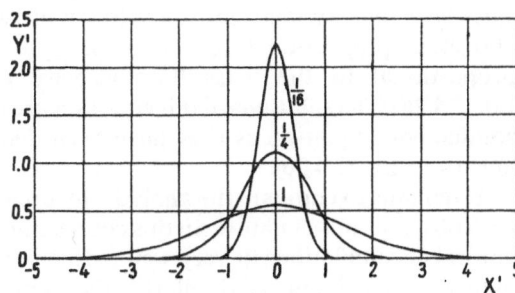

Fig. 25. Solutions of the diffusivity equation

It may be noted that the above theory can also be extended to that part of a slope where material transport effects a removal of mass. In this instance, it leads to a case of *slope recession*. This can be demonstrated as follows.

The differential Eq. (3.4.18), being a diffusivity equation, has solutions of the form (cf. Fig. 25)

$$Y' = \zeta^2 = \frac{\text{const}}{\sqrt{t}} \exp\left\{-\frac{x^2}{\sqrt{4at}}\right\}. \tag{3.4.24}$$

Thus, the slope itself has the form

$$-y' = S = \frac{\text{const}}{\sqrt[4]{t}} \exp\left\{-\frac{x^2}{4\sqrt{at}}\right\}, \tag{3.4.25}$$

which, qualitatively, leads to the picture shown in Fig. 26. The figure indicates that a lowering of the slope is taking place on one side of the axis of symmetry, a rising on the other side. The equation of the slope, being given by the negative integral of the expression (3.4.25), is always an error function complement. The transport of mass thus effects erosion on one side of the center of the slope and accretion on the other side.

Some of the difficulties inherent in the above discussion are due to the nonlinear character of the basic Eq. (3.4.15). Although the latter can be linearized, as shown above, it may also be possible to justify physically the assumption of a *linear* differential equation in the first place, viz. of

$$\frac{\partial y}{\partial t} = a \frac{\partial^2 y}{\partial x^2}. \tag{3.4.26}$$

This, in fact, has been proposed by Culling (1960), Hirano (1975), Yesin and Skorkin (1970), and Trofimov and Moskovkin (1984a) by assuming (a) that the

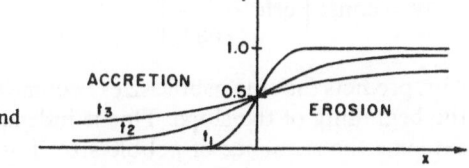

Fig. 26. Slope development by accumulation and erosion. (After Scheidegger 1961a)

mass flow velocity (and not the velocity of the mass-carrying water as assumed above) is proportional to the slope $\partial y/\partial x$ and, (b) that the deposition is proportional to the mass flow velocity gradient. This immediately yields Eq. (3.4.26). The solutions of this equation are the well-known solutions of the heat conductivity equation as they have been discussed above in the analysis of the linearized Eq. (3.4.26).

It remains to compare the analytically predicted slope shapes of accumulations with observations in nature. In this connection, we note that the slopes observed in natural accumulations fit very well indeed with those postulated in the analytical theory outlined above. Similarly, the concave nature of the slopes of most volcanoes also fits the above analytical theory. Schumm (1962) has shown that pediment profiles also fit the above characteristics.

The universality of the concavity of accumulative slopes has led Strahler (1952) to postulate a concave shape a priori from which it would then be possible to calculate the change of shape with time. Strahler (1952) took as fundamental equation (as an a priori assumption) for slopes caused by accumulation (especially river beds)

$$y = Ae^{-kx}. \tag{3.4.27}$$

He then introduced the assumption that the rate at which the slope is lowered at a given point is proportional to the slope at that point. This immediately leads to an exponential lowering of the slope with time; one ends up with

$$y = Ae^{-k_1 x - k_2 t}. \tag{3.4.28}$$

It should be noted, however, that the above procedure does not really predict what the shape of a slope should be (since this has been introduced as an a priori assumption). Furthermore, it is doubtful whether the rate of lowering of the slope can indeed be regarded as proportional to the slope. If the flow of the water is in dynamic equilibrium on the slope, this is certainly not true. The theory given earlier would therefore appear to be preferable.

One can now try to investigate the sequence of the pebble sizes found in natural slopes in terms of the analytical theory of the preceding paragraphs. Remarks with regard to the observed grading have been made in Chap. 1.

In order to explain the observed grading of pebble sizes, one can proceed as follows. It is often assumed that the weight W of the pebbles that can just be moved by flowing water is proportional to the sixth power of the velocity v of the water:

$$W = \text{const } v^6. \tag{3.4.29}$$

Reasons for this assumption will be given later (in Sect. 4.4.3). Combining (3.4.29) with the velocity formula (3.4.22) yields

$$W = \text{const} \left[\text{erfc} \frac{x}{\sqrt{4a't}} \right]^3. \tag{3.4.30}$$

This predicts that the pebble sizes become smaller the farther away one moves from the beginning of the slope. This is indeed confirmed in nature, as it immediately explains the sequence of pebble sizes found in flood plains and alluvial fans.

3.4.4.7 Scaling of Water Erosion Processes on Slopes

Whenever it is difficult to describe a process analytically, recourse may be taken to scale model experiments. In order to set up such scale model experiments, the appropriate dynamic similarity conditions must be observed. For the erosional processes caused by water flowing over a slope, these conditions have been deduced by Scheidegger (1963).

The scaling relations are developed simply by postulating that the fundamental equations be valid in the prototype as well as in the model. The fundamental equations are first written down in terms of dimensionless variables; requiring then that the equations be indentical in prototype and model yields the required scaling relations (Focken 1953).

Of the slope development agents discussed above, we turn first to raindrop action. Of the possibilities discussed in Sect. 3.2.3.3, the hypothesis referring to the significance of the *momentum* rather than of the energy in rain action will be used here, as implied in Eq. 3.2.7. In that case, the constant C is a measure of the "ruggedness" r (cm/s) of the soil, i.e., a measure of its resistance to erosion. The first fundamental equation of raindrop action becomes therefore:

$$Q = \frac{1}{r}q\rho v, \tag{3.4.31}$$

where all the symbols have been defined in Sect. 3.2.3.3. Because of turbulent drag, the terminal velocity v of the raindrops is approximately proportional to $a^{1/2}$, where a is their diameter (see Eq. (4.4.11)) thus:

$$Q = K\frac{1}{r}a^{1/2}. \tag{3.4.32}$$

Here ρ, the density of the water, has been incorporated into the constant K which has the dimension $g\,cm^{-5/2}\,s^{-1}$.

Now, the last Eq. (3.4.22) must be written in terms of dimensionless variables, following the general scheme. In order to do this, the gravity acceleration g has to be introduced as a reference parameter. We can then set (bars denoting dimensionless quantities)

$$\bar{Q} = Qg^{1/2}/(Kr), \tag{3.4.33}$$

$$\bar{q} = q/r, \tag{3.4.34}$$

$$\bar{a} = ag/r^2. \tag{3.4.35}$$

The fundamental equation is then indeed dimensionless:

$$\bar{Q} = \bar{q}\bar{a}^{1/2} \tag{3.4.36}$$

and the scaling relations (α_i denoting the *ratio* of the variable i in prototype and model) are

$$\alpha_Q = \alpha_K\alpha_r/\alpha_g^{1/2}, \tag{3.4.37}$$

$$\alpha_q = \alpha_r, \tag{3.4.38}$$

$$\alpha_q = \alpha_r^2/\alpha_g. \tag{3.4.39}$$

Next, we turn to spontaneous mass transport. We have to go to the fundamental equation given as 3.3.11. The transformation to make this dimensionless is (see Sect. 3.3.3.2 for the definition of the symbols involved)

$$\bar{h}_c = h_c \rho g/c, \tag{3.4.40}$$

$$\bar{N}_s = N_s(\beta). \tag{3.4.41}$$

Then, indeed, (3.3.11) becomes

$$\bar{h}_c = \bar{N}_s \tag{3.4.42}$$

and we obtain the scaling conditions:

$$\alpha_{h_c} = \alpha_c/(\alpha_\rho \alpha_g) \tag{3.4.43}$$

$$\alpha_{N_s} = 1. \tag{3.4.44}$$

Finally, the slope development by water erosion was given by which may be written as follows:

$$\frac{\partial y}{\partial t} = -c\frac{\partial}{\partial x}\left(\frac{\partial y}{\partial x}\right)^2, \tag{3.4.45}$$

where c is a constant and the other symbols have been defined in Sect. 3.4.4.6. This constant is connected with the erodibility of the material over which the water flows and with the quantity of water that is available. The latter is best measured as the water volume that flows per unit time across a slope profile of unit width. We shall denote this quantity by Q; the dimensions are obviously L^2T^{-1}. The relation is then

$$c = Q/r, \tag{3.4.46}$$

where r *defines* the resistance to erosion of the material on the slope, the quantity r is dimensionless (note, however, that Q and r are not the same quantities as in the previous paragraph).

Substituting this value of c in the basic equation results in

$$\frac{\partial y}{\partial t} = -\frac{Q}{r}\frac{\partial}{\partial x}\left(\frac{\partial y}{\partial x}\right)^2. \tag{3.4.47}$$

It may be of some interest to compare slopes with different slope angles β. In this case, one has to differentiate the fundamental differential equation with regard to x

$$\frac{\partial^2 y}{\partial t\,\partial x} = -\frac{Q}{r}\frac{\partial^2}{\partial x^2}\left(\frac{\partial y}{\partial x}\right)^2; \tag{3.4.48}$$

then it is possible to introduce as a new (dimensionless) variable (instead of y):

$$\zeta = \tan\beta = \frac{\partial y}{\partial x} \tag{3.4.49}$$

so that one obtains

$$\frac{\partial\zeta}{\partial t} = -\frac{Q}{r}\frac{\partial^2\zeta^2}{\partial x^2}. \tag{3.4.50}$$

Assuming a slope bank of height H and slope angle β_0 (so that $\zeta_0 = \tan\beta_0$), the complete system of equations defining the process is then

$$\frac{\partial \zeta}{\partial t} = -\frac{Q}{r}\frac{\partial^2 \zeta^2}{\partial x^2},\tag{3.4.51}$$

$$\zeta(t=0) = \zeta_0 \quad \text{for} \quad 0 \leqslant x \leqslant H/\zeta_0 \tag{3.4.52}$$
$$ = 0 \quad \text{otherwise.}$$

This system of equations can be made dimensionless by the transformation

$$\bar{t} = tQ\zeta_0^3/(rH^2),\tag{3.4.53}$$
$$\bar{x} = \zeta_0 x/H,\tag{3.4.54}$$
$$\bar{\zeta} = \zeta/\zeta_0.\tag{3.4.55}$$

Then,

$$\frac{\partial \bar{\zeta}}{\partial \bar{t}} = -\frac{\partial^2 \bar{\zeta}^2}{\partial \bar{x}^2},\tag{3.4.56}$$

$$\zeta(t=0) = 1 \quad \text{for} \quad 0 < \bar{x} < 1 \tag{3.4.57}$$
$$ = 0 \quad \text{otherwise.}$$

Thus, the dynamic similarity conditions are

$$\alpha_t = \alpha_r \alpha_H^2/(\alpha_Q \alpha^3_{\zeta_0}),\tag{3.4.58}$$
$$\alpha_x = \alpha_H/\alpha_{\zeta_0},\tag{3.4.59}$$
$$\alpha_\zeta = \alpha_{\zeta_0}.\tag{3.4.60}$$

On the basis of scaling relationships, experiments have been carried out successfully, e.g., by Haigh (1981) and by Torri and Sfalanga (1986).

3.5 Mathematical Models of Denudation

3.5.1 Introduction

The various agents of mass removal and transport have a combined effect upon the development of a slope which is commonly called "denudation". Inasmuch as many agents are involved which lead to rather complicated "microscopic" aspects, it is convenient to treat the development of a slope bank in a macroscopic fashion by simply considering its overall cross-section.

The development of such overall cross-sections has been phenomenologically studied and the observed patterns have been classified by many people. Thus, Takeshita (1964) and Schumm (1966) discussed various possibilities, distinguishing between mountain, hill, and valley slopes; and Macar (1966) edited a volume containing many papers on the evolution of slopes mostly concerned with phenomenological development patterns. Furthermore, Ahnert (1978) and Seuffert et al. (1984) gave reviews of research trends in morphostructural analysis, and Louis (1977) classified slope types phenomenologically.

In the present Section (3.5), we shall consider only specific deterministic mathematical models of slope profile evolution. The approach based on systems theory (including process-response models) will be relegated to Chap. 5 which deals with the systems theory of landscapes in general. On the indicated specific deterministic models, reviews exist by Kirkby (1976) and by Trofimov and Moskovkin (1983). Much of the work reported here, however, is the writer's own.

3.5.2 Slope with a Rocky Core

3.5.2.1 General Remarks

The first group of models which we shall discuss are those referring to the recession of a rocky slope; it is assumed that debris comes off the slope which accumulates at the bottom , forming a pile of screes which is leaning against the original slope. The slope angle α of the pile (screes angle) will be determined by the geometrical properties of the weathered material according to the conditions established in Sect. 3.3.2.3.

As the pile of debris grows, it will cover up more and more of the original slope and thereby protect it from further denudation. Underneath the pile, the originally straight slope will evolve into a curved one, an occurrence which had already been noted by Fisher (1866) in the last century. The problem now is to determine more exactly the shape of the curve. There are, then, various assumptions possible regarding the manner of weathering which will now be presented.

3.5.2.2 Parallel Rectilinear Slope Recession

The first model of slope recession which we shall consider is the following:

1. the rock is homogeneous so that its reduction by weathering proceeds at an equal rate at all points of the slope that are exposed;
2. all material is instantly removed due to mass transport so that the exposure to the reducing elements is the same everywhere.

The above assumptions imply that the denudation proceeds at an equal rate everywhere on the slope. It is easy to see that this leads immediately to parallel slope recession. If the slope is rectilinear to begin with, then it will remain so and one has the case of parallel rectilinear slope recession.

If the debris piles up against the original slope, a picture as shown in Fig. 27 will be the result.

An attempt to study this case has been made by Lehmann (1933), who envisaged the following model of a receding slope (cf. Fig. 27): a steep slope of angle β is bounded by two horizontal planes which are the (vertical) distance h apart. Each individual amount weathering off the slope will pile up at the bottom, the pile forming a screes angle α. Some of the material, however, may get lost during the transfer, or else its volume might increase since the density of the debris may be less than that of the slope material. Hence one has to set

$$V_R/V_D = 1 - c; \qquad\qquad\qquad (3.5.1)$$

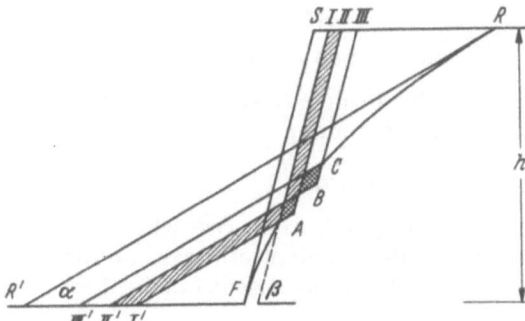

Fig. 27. Building up of a pile of débris at the bottom of a steep slope. (After Lehmann 1933)

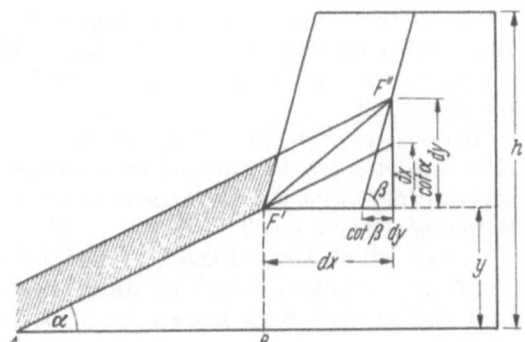

Fig. 28. Infinitesimal increase of the heap of débris at the bottom of a slope. (After Lehmann 1933)

where V_R is the volume of slope material removed and V_D the corresponding volume debris piling up at the bottom of the slope. Expressing the condition (3.5.2.1) for each infinitesimal amount of slope recession, one obtains (from a inspection of Fig. 28:

$$(dx - dy \cot \beta)(h - y) = (1 - c)\left(dy - \frac{dx}{\cot \alpha}\right) y \cot \alpha \qquad (3.5.2)$$

or

$$\frac{dx}{dy} = \frac{h \cot \beta + (\cot \alpha - c \cot \alpha - \cot \beta) y}{h - cy}. \qquad (3.5.3)$$

This is a differential equation which may easily be integrated. One obtains (choosing as a suitable boundary condition $y = 0$ for $x = 0$):

$$x = k(l + m) \operatorname{lognat} \frac{m}{m - y} - ky \qquad (3.5.4)$$

with

$$m = h/c, \qquad (3.5.5)$$

$$k = \frac{(1 - c) \cot \alpha}{c} - \frac{\cot \beta}{c}, \qquad (3.5.6)$$

$$l = \frac{h \cot \beta}{\cot \alpha - \cot \beta - c \cot \beta}. \qquad (3.5.7)$$

In order to investigate the meaning of these equations, it is convenient to consider two limiting cases. If one sets $c = 0$, $\beta = 90°$, one has

$$y = \sqrt{2hx \tan \alpha}, \qquad (3.5.8)$$

which corresponds to the case where all the debris are piled up against the slope; the latter, then, underneath takes on the shape of a parabola.

A second limiting case is obtained by setting $c = -\infty$. One then obtains

$$y = x \tan \alpha. \qquad (3.5.9)$$

This corresponds to the case where all the debris disappears somehow. It indicates that, under these circumstances, the slope will become rectilinear and that the slope angle will become equal to the screes angle. This result, in fact, has been enounced as a geomorphological law by Bakker and Le Heux (1952). The latter authors stated it as follows:

"In every type of weathering-removal recession of steep walls, a denudation slope with a rectilinear cross-profile and a constant slope angle equalling the screes angle in nature will be formed, provided no or hardly any screes are deposited on the terrace at the foot of the initial wall".

Bakker and Le Heux (1952) also provided physiographical examples to show that their law is indeed obeyed in nature. The end result of the process envisaged by them had, in fact, been known for some time as "Richter's (1901) slope of denudation".

The above discussion deals only with the denudation of a slope bounded by two plateaux. One can extend the theory to the weathering of a symmetrical crest, to that of a plateau of finite width, and to a combination of the two (Van Dijk and Le Heux 1952). Thus, suppose that one starts with a symmetrical plateau of finite width. Eventually, a stage will be reached where the two weathering edges meet at the top (Fig. 2.9a). The further progress will be by symmetrical crest recession, but without a plateau being present at the level RQ (Fig. 29). Thus, the development is supposed to be somewhat as it is shown in Fig. 29b.

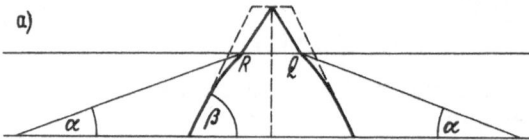

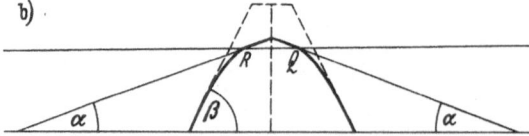

Fig. 29. Two successive stages of the weathering of a symmetrical plateau of finite width. (Explanation see text). (After Van Dijk and LeHeux 1952)

3.5.2.3 Central Rectilinear Slope Recession

The theory of parallel rectilinear slope recession discussed above has certain unsatisfactory aspects. It stands to reason that the amount of denudation at the top is greater than the amount of denudation at the bottom and that one should take such variations into account. This has been done by Bakker and Le Heux (1947) in their postulate of *central* rectilinear slope recession.

The postulate's basic to central rectilinear slope recession can best be seen by inspecting Fig. 30, representing the case of a steep slope of slope angle β that is bounded by two plateaux which are the (vertical) distance h apart. During each infinitesimal time interval, the cross-sectional area weathering off the slope is bounded by two straight lines that always meet at the focal point F which is the original foot point of the slope. As in Sect. 3.5.2.2, the debris then accumulates at the bottom of the slope forming a pile of screes with screes angle α and thereby protects part of the slope from further denudation. During the denudation process, the slope angle β changes.

In order to deduce the relevant differential equation for the slope underneath the debris, we investigate the geometry of an infinitesimal change in some detail as shown in Fig. 31.

The solution is

1. for $c \neq \frac{1}{2}$

$$x = ay - (a - b_0)y \left[\frac{h^2 + (1 - 2c)y^2}{h^2} \right]^{(c-1)/(1-2c)}$$
(3.5.10)

2. for $c = \frac{1}{2}$

$$x = ay - (a - b_0)y \exp\left\{ -\frac{y^2}{2h^2} \right\},$$
(3.5.11)

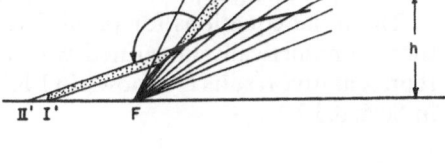

Fig. 30. Central rectilinear slope recession. (After Bakker and Le Heux 1947)

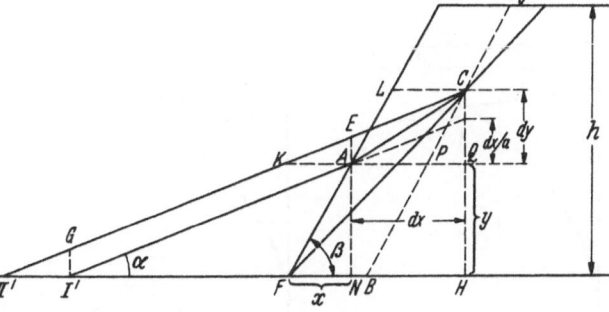

Fig. 31. Diagram for the deduction of the differential equation of central rectilinear recession of a plateau. (After Bakker and Le Heux 1947)

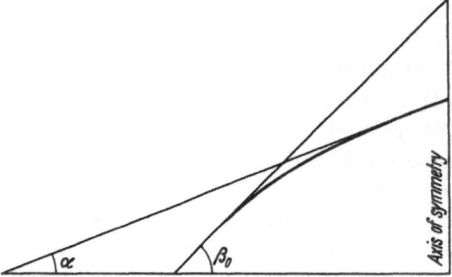

Fig. 32. Central rectilinear recession of a symmetrical crest for $\beta_0 = 45°$, $\alpha = 22°$, $a = 2.5$, $c = 0$. (After Bakker and Le Heux 1950)

with $a = \cot \alpha$, $b = \cot \beta$, Furthermore, b_0 is the cotangent of the initial slope angle and the boundary conditions have been chosen so that $y = 0$ for $x = 0$.

As in the case of parallel slope recession, the above theory has been extended to receding (symmetrical) crests (Bakker and Le Heux 1950). Using the same geometrical layout as in Fig. 29 (showing a symmetrical crest), but expressing the condition of *central* slope recession, one immediately ends up with the differential equation

$$y(1 - c)(a\,dy - dx) = \frac{k^2 - x^2}{2x^2}(y\,dx - x\,dy). \tag{3.5.12}$$

A particular solution for this case is shown in Fig. 32.

3.5.2.4 Nonparallel Slope Banks

The theory of parallel slope recession was extended by Gerber and Scheidegger (1973) to the denudation of vertical rocky walls whose top is not horizontal, but is inclined at an angle γ with the horizontal.

The usual equations for parallel rectilinear slope recession (Sect. 3.5.2.2) can then be numerically integrated with the appropriate boundary conditions; two representative results are shown in Fig. 33. The symbols have the same meaning as in Sect. 3.5.2.2.

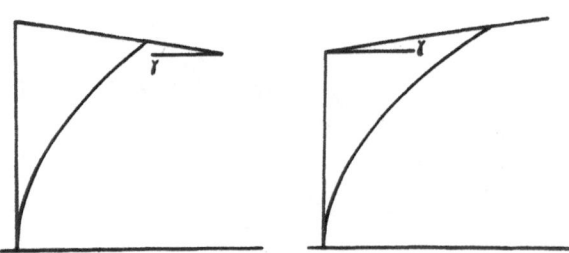

Fig. 33. Two representative coro curves for nonparallel slope banks; $\gamma = \pm 10°$, $\alpha = 35°$, $c = 0.2$. (After Gerber and Scheidegger 1973)

3.5.3 Variations of Exposure

3.5.3.1 General Remarks

The models of denudation reviewed above are concerned with the shape of a rocky core beneath a pile of debris. However, it is also of interest to investigate the development of a slope that is exposed to the elements in a nonuniform fashion.

Indeed, in the theory of central rectilinear slope recession, an attempt was made to take variations of the rapidity of denudation into account. However, this was done by a postulate regarding the geometry of the slope without justifying it by a proper model of mass transport. One might therefore want to analyze systematically the effect of various assumptions regarding the rapidity of denudation upon the shape of the slope. At the same time, one might entertain the hope that the various shapes encountered in nature could be obtained directly from appropriate postulates regarding the process of denudation.

This will be done in the subsequent sections (after Scheidegger 1961a).

3.5.3.2 Linear Theory

Assuming a homogeneous slope material, variations of the rapidity of denudation will be due to variations of exposure of the rock to the elements of the weather. One might think of the following possibilities:

Case 1. The denudation is independent of the slope and proceeds at an equal rate at any exposed portions of the slope.

Case 2. The denudation is proportional to the height of the point under consideration above a certain base level. This could be justified by the observation that in certain areas, precipitation increases with height.

Case 3. The denudation is proportional to the steepness of the slope. This could be justified by noting that the weathering is due to the exposure of the slope. The steeper the slope, the faster will the debris be removed. Thus, the steeper slopes will generally be more exposed than less steep ones.

Let us investigate how the denudation affects the shape of slopes in the above three cases. In order to do this, we assume that the lowering of the slope per unit time at any given point is proportional to a constant (case 1), to the height of the slope (case 2) or to the slope (case 3). Thus, denoting the height by y, the location by x, we have (cf. Fig. 34)

$$\frac{\partial y}{\partial t} = -\text{const } \Phi \tag{3.5.13}$$

with

case 1: $\Phi = 1$ $\tag{3.5.14}$

case 2: $\Phi = y$ $\tag{3.5.15}$

case 3: $\Phi = \partial y / \partial x.$ $\tag{3.5.16}$

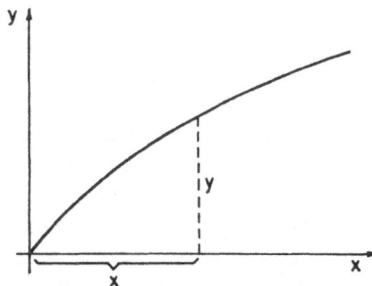

Fig. 34. General geometry of a slope

It is obviously always possible to change the time scale in such a fashion that the constant in Eq. (3.5.13) can be set equal to 1. Thus, one has a partial differential equation to solve; the shape of the original slope represents the arbitrary function that enters into the solution of every partial differential equation. In the three cases under consideration, the solution is very easily obtained.

Case 1. The differential equation is

$$\frac{\partial y}{\partial t} = -1 \tag{3.5.17}$$

with the initial condition $y = f_0(x)$. The solution is

$$y = f_0(x) - t. \tag{3.5.18}$$

This represents the case of equal slope recession. The slope retains its shape and simply moves downward. If the slope is rectilinear to begin with, then one has *parallel slope recession* downward.

Case 2. The differential equation is

$$\frac{\partial y}{\partial t} = -y \tag{3.5.19}$$

with the same initial condition $y_0 = f_0(x)$. The solution is

$$y = f_0(x)e^{-t}. \tag{3.5.20}$$

At any time, all slope heights, therefore, are reduced proportionately. If the slope is rectilinear to begin with, then this indeed represents *central slope recession*. The latter has thus been given a clear physical justification.

Case 3. The differential equation is

$$\frac{\partial y}{\partial t} = -\frac{\partial y}{\partial x}. \tag{3.5.21}$$

Using the usual initial condition $y_0 = f_0(x)$, the solution is

$$y = f_0(x - t). \tag{3.5.22}$$

This solution signifies that any given slope profile will wander to the right with time. If the slope is rectilinear, this means *parallel slope recession* as in case 1. One has thus the interesting fact that parallel rectilinear slope recession can occur in case 1 as well as in case 3.

The basic shape of the slope remains unaltered in all three cases treated above. The hope that a variation of exposure would change the slope shapes is therefore not fulfilled in the above mathematical models. In order to obtain such changes, one still has to take recourse to the idea of building up and afterwards destroying piles of screes.

In conclusion of this section on linear models, it may be remarked that Hirano (1968) was able to achieve more realistic results by combining various models that are still linear in the derivatives, although additional functions of the variables themselves are introduced. Nevertheless, it is the opinion of the present writer that the nonlinear models to be discussed in the next section are preferable.

3.5.3.3 Basic Nonlinear Theory

The mathematical models of slope development discussed so far appeal very much to the imagination because of their basic simplicity. However, precisely because of the latter, some important conditions obtaining in nature have been neglected and the calculated slope profiles appear therefore as far too simple.

A serious oversimplification has been made in the models when the *vertical* lowering of the slopes was set proportional to some expression which was either a constant, equal to y, or equal to $\partial y/\partial x$. One really should allow for the fact that weathering acts *normal* to the slope so that the vertical lowering is then represented by the vertical effect of the weathering action (the latter being taken as proportional to a constant, y, or $\partial y/\partial x$ according to the case under consideration) which is directed *normally* against the slope. From an inspection of Fig. 35, which shows the geometrical layout of the weathering action, one can seen that the slope development is then represented by the differential equation

$$\frac{\partial y}{\partial t} = - \sqrt{1 + \left(\frac{\partial y}{\partial x}\right)^2} \, \Phi, \qquad\qquad (3.5.23)$$

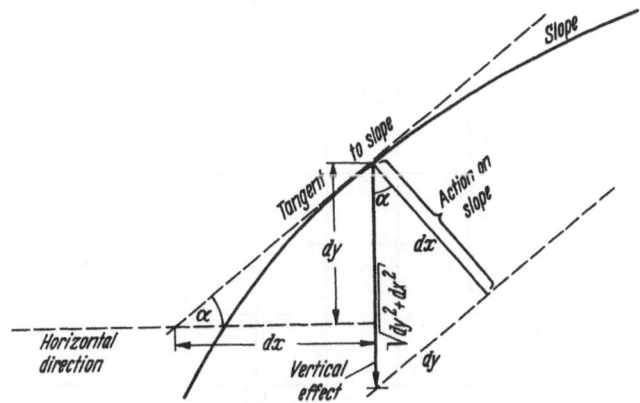

Fig. 35. Vertical effect of weathering action normal to the slope. (Scheidegger 1961a)

where Φ is again given by one of the expressions (3.5.14–3.5.16) corresponding to the three possible cases under consideration.

The improved "new" differential equation (3.5.23) of slope development differs from the old one in a very fundamental regard: it is nonlinear. Easy solutions of the new equation can therefore no longer be obtained. Although analytical solutions are possible (using a Legendre transformation for case 2 and the theory of characteristics for cases 1 and 3; cf. Mitin and Trofimov 1964; Trofimov 1966; and Luke 1972 and 1976), one ends up with a system of equations which can be written down in a reasonably simple form only for the case of an infinitely long straight slope as initial condition. The latter has not much relation to conditions in nature. If one wishes to have actual numerical results that have a visualizable meaning, it is best to solve the equations directly by means of an electronic computer. As always with nonlinear hyperbolic partial differential equations, the choice of the steps in the approximation procedure is critical. The steps for Δx and Δt have to be consistent with the domain of influence defined by the net of characteristics (see, e.g., Collatz 1951) but this is merely a necessary, not a sufficient condition for achieving stability for the solution.

In all cases considered, the development of a slope step was studied (in profile). The original height of the step was assumed as equal to 0.5 (arbitrary) scale units of y, the original slope at one end of the slope as equal to 2. The coördinate x varies from 0 to 1 in 100 steps. The original slope, thus, has the shape shown in Fig. 36. Then, the procedure adopted in the individual cases was as follows.

Case 1. The differential equation is

$$\frac{\partial y}{\partial t} = -\sqrt{1 + \left(\frac{\partial y}{\partial x}\right)^2},\tag{3.5.24}$$

which was approximated by the following difference equation

$$(y_{t_{m+1}} - y_{t_m})\bigg|_n = -\sqrt{1 + \left(\frac{y_n - y_{n-1}}{x_n - x_{n-1}}\right)^2}\,\bigg|_m (t_{m+1} - t_m).\tag{3.5.25}$$

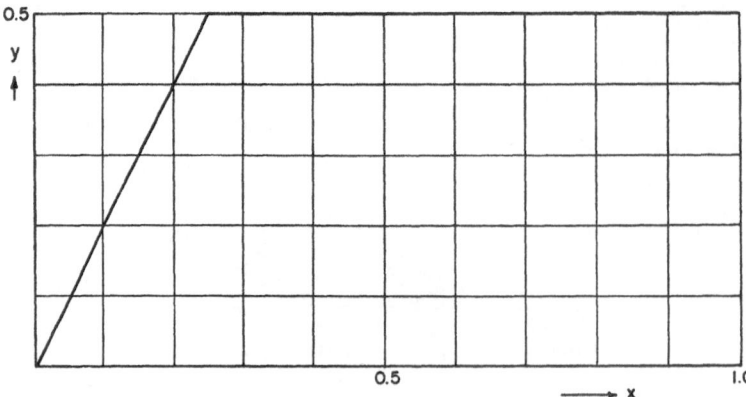

Fig. 36. Original slope step

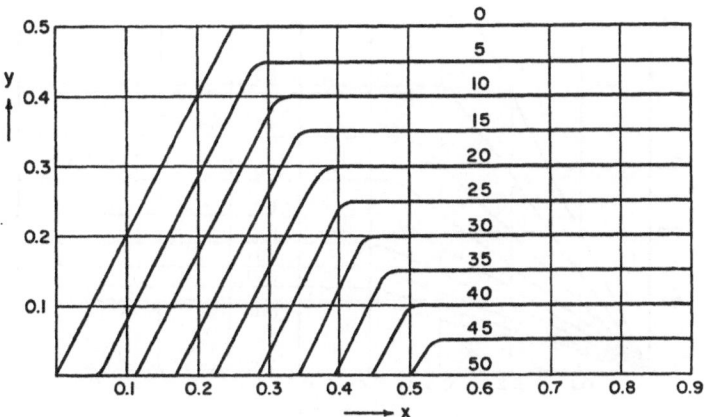

Fig. 37. Slope recession in case 1 of the nonlinear theory. (Scheidegger 1961a)

The result of carrying out the approximation procedure is shown in Fig. 37. The convex curvature at the top edge is due to computer rounding inaccuracies.

It is evident that there is a difference if the present case be compared with the analogous one of the linear theory. The recession is now no longer straight downward, but partly sideways.

Case 2. The differential equation is

$$\frac{\partial y}{\partial t} = -y \sqrt{1 + \left(\frac{\partial y}{\partial x}\right)^2}. \tag{3.5.26}$$

The difference equation approximating this is

$$y_{t_{m+1}} - y_{t_m} \bigg|_n = -\sqrt{1 + \left(\frac{y_n - y_{n-1}}{x_n - x_{n-1}}\right)^2} \, y_n \bigg|_m (t_{m+1} - t_m). \tag{3.5.27}$$

The results obtained by this approximation procedure are shown graphically in Fig. 38. The convex curvature at the top edge of the slope is again due to computer rounding inaccuracies. It is seen that the slope recession is now no longer "central", as was the case in the linear theory.

Case 3. The differential equation is

$$\frac{\partial y}{\partial t} = -\frac{\partial y}{\partial x} \sqrt{1 + \left(\frac{\partial y}{\partial x}\right)^2}. \tag{3.5.28}$$

This is approximated by the difference equation

$$y_{t_{m+1}} - y_{t_m} \bigg|_n = -\frac{y_n - y_{n-1}}{x_n - x_{n-1}} \sqrt{1 + \left(\frac{y_n - y_{n-1}}{x_n - x_{n-1}}\right)^2} \bigg|_m (t_{m+1} - t_m). \tag{3.5.29}$$

The results of the calculation are shown in Fig. 39.

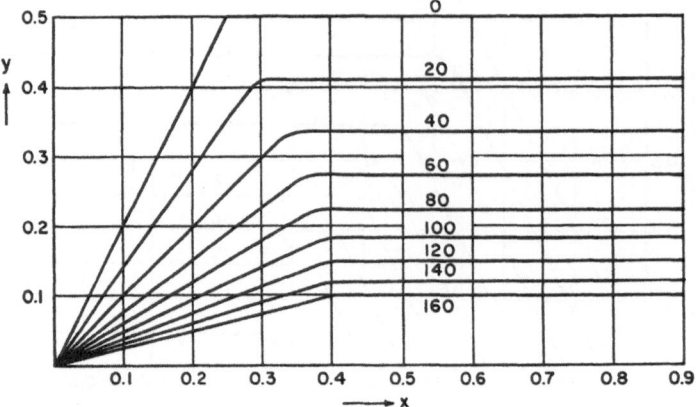

Fig. 38. Slope recession in case 2 of the nonlinear theory. (Scheidegger 1961a)

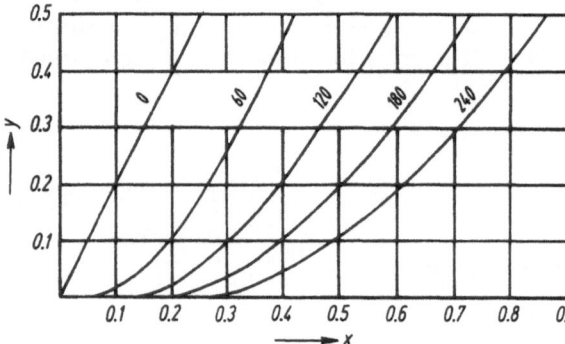

Fig. 39. Slope recession in case 3 of the nonlinear theory. (After Scheidegger 1986)

The results obtained in the present case are eminently reasonable. The originally straight slope eats its way into the bank. The toe becomes very broad, the head remains sharp and thus the slope assumes a concave overall appearance. Simultaneously, the average inclination becomes less and less, and as time progresses, the steep bank will eventually yield to a very gentle slope.

Of the three models discussed, the last is thus the most reasonable one. However, comparison of an actual slope in nature will, in every case, nail down the physical conditions that produced it.

3.5.3.4 Modifications of the Basic Nonlinear Theory

The non-linear theory discussed above can evidently be modified by choosing functions Φ in Eq. (3.5.23) which are different from those given in the preceding sections.

Thus, Yesin (1968) discussed various possibilities. In particular, he considered the expression

$$\Phi = \frac{0.26}{y + 7},$$ (3.5.30)

which he applied to underwater slopes.

Another possibility was considered by Takeshita (1963) and by Young (1963), who both assumed the proportionality (s is the arc length)

$$\Phi \sim \frac{d(\partial y/\partial x)}{ds},$$ (3.5.31)

which implies that the "weathering" action normal to the slope is proportional to the curvature of the slope. Solutions of the basic nonlinear slope equation (3.5.23), using the expression (3.5.31) for Φ have been calculated; we present an example of the results of Takeshita (1963) in Fig. 40.

The physical justification for the choice of Φ given by (3.5.31) is that it follows from the hypothesis that the normal lowering or raising of the slope is proportional to the change dM/ds in down-hill mass flux M; the mass flux M, in turn, is assumed to be proportional to the steepness of the slope. Contrary to the hypothesis in the preceding sections, where all material "eroded" from the slope was assumed to be taken off the slope, Eq. (3.5.31) is the result of the postulate of complete mass balance of the material moving along the slope. These conditions may be thought to be applicable under conditions of soil creep, solifluction etc.

Another modification of the nonlinear theory has been suggested by Mizutani (1970) who reduced the nonlinear basic equation (3.5.23) by nonlinear transformations of the variables to a linear diffusivity equation for which he gave solutions for a variety of boundary conditions.

Aronsson (1973) has shown that the solution of Eq. (3.5.23) can also be formulated in terms of a minimum principle.

Finally, Armstrong (1976) extended the computer approach to three dimensions. He obtained particularly impressive computer graphs of the sequences of the genesis of three-dimensional land forms.

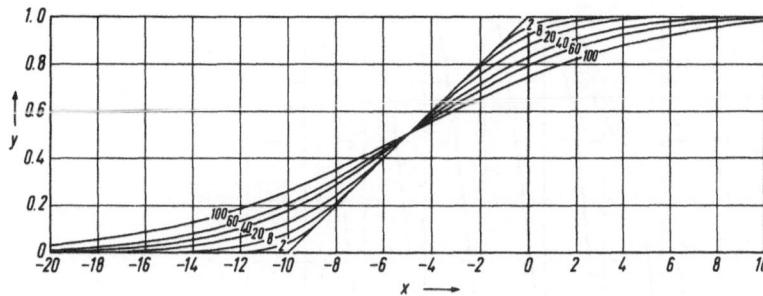

Fig. 40. Time-evolution of a slope bank where Φ is given by Eq. (3.5.31) (arbitrary units). (After Takeshita 1963)

3.5.4 Nonuniformly Exposed Slopes

3.5.4.1 General Remarks

The models discussed above lend themselves immediately to the investigation of many special cases. The writer has himself investigated many such cases by varying the initial and boundary conditions during the computer solution of the basic equations. (Ahnert (1971) also has set up a very flexible program, where not only the boundary conditions, but also the basic denudation equation can be changed at will. In this fashion, slope sequences for nonuniform conditions (on account of exogenic agents, lithology, climate, etc.) can be obtained.

3.5.4.2 Slope Development by Undercutting

The first case which we shall investigate corresponds to slope recession in consequence of an undercutting river, as envisaged in Crickmay's (1960) unequal activity theory (cf. Sect. 1.5.1.4). In this, we assume that river undercutting as well as slope denudation due to direct action of erosive agents on the whole surface of the slope proceed simultaneously. This case, in fact, lends itself easily to a mathematical investigation (Scheidegger 1960). We assume again that the action of the erosive forces is described by Eq. (3.5.23) with $\Phi = \partial y/\partial x$. At the same time, we assume that a river is cutting away at the bottom of the slope. Its action is accounted for by the assumption that the river is able to carry away a certain amount of material (per unit length) m per unit time. This can be done very easily by introducing into the computer program for the solution of equation (3.5.23) a correction accounting for the action of the river after very time step. This is achieved simply by calculating the integral $\int y dx$ from zero to a point X so that

$$\int_0^{} y dx \leqslant m\Delta t \qquad (3.5.32)$$

and setting y for all

$$x \leqslant X \qquad (3.5.33)$$

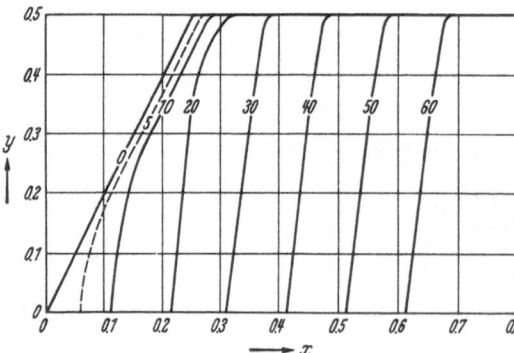

Fig. 41. Slope development with an undercutting river. (Scheidegger 1960)

equal to zero. The calculation, for an initial slope bank of the usual type as shown in Fig. 36, was carried out on an internally programmed computer for

$$m = 0.5 \qquad\qquad\qquad (3.5.34)$$

and some 120 iterations. The result is shown graphically in Fig. 41.

An inspection of the results presented above shows that the action of a river fundamentally changes the development pattern of a slope. The slope now becomes steeper as time goes on and reaches asymptotically an inclination determined by the rapidity of the two types of erosion that are involved. The development in its latter stages is essentially a parallel slope recession.

It thus turns out that, if an undercutting river be involved in the development of a slope, *parallel slope recession* is the ultimate outcome. This is indeed what is observed in nature in the case of undecutting rivers (Schipull 1978).

Incidentally, essentially parallel slope recession is also the outcome if other types of basic denudation equations are used. Thus, Trofimov and Moskovkin (1984b) have applied the boundary conditions corresponding to a steadily undercutting river to their diffusion model (cf. Sect. 3.4.4.6) and also obtained asymptotically parallel slope recession.

3.5.4.3 Deposition of Debris as Aprons

The above slope-development models assume that all debris is completely carried away from the slope during its development, which is evidently an over-simplification of what occurs in nature. Thus, we shall now investigate the case where all the debris from the lowering of the slope is deposited in a horizontal apron in front of the slope bank. This case may occur if a V-shaped valley is being filled up by the debris (after Scheidegger 1962).

The above case can be investigated on the basis of Eq. (3.5.28) where now, however, the conditions are arranged in the numerical solution that the total volume of material which is taken off the slope during each time step is calculated

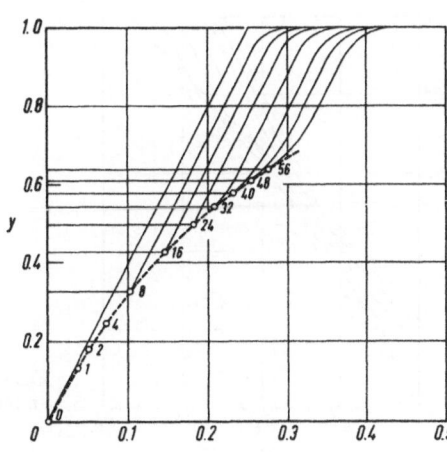

Fig. 42. Denudation of a valley side, the debris filling in the valley. (After Scheidegger 1962)

and then spread horizontally between the zero abscissa and the slope. The results of this calculation, starting with a straight original slope bank (see Fig. 36), are shown in Fig. 42. As the valley fills up, part of the slope becomes covered by debris and is therefore protected from further attack by the weathering agents. This may give rise to a "rocky core" of the final slope, similarly as this was invisaged in Sect. 3.5.2.2. The rocky core forms an envelope to the various stages of the receding slope; it is shown as a dotted curve in Fig. 42.

The above model can be further modified by assuming that the debris piles up at the foot of the slope, forming an inclined apron corresponding to the angle of repose of the material. Starting again with a straight slope bank (Fig. 36) as before, it was assumed that the tangent of the angle of repose was $\frac{1}{2}$ of the tangent of the slope angle of the original slope bank. Then, it is assumed that the total volume of material taken off the slope during each time step is spread over an inclined strip at the foot of the slope. The results of the calculations are shown in Fig. 43. It is noted that the debris will again protect a rocky core beneath the slope from further weathering; it is shown as a dotted line in Fig. 43. The calculations cannot be carried on indefinitely, inasmuch as the apron eventually becomes tangent to the slope, corresponding to Richer's slope of denudation (cf. Sect. 3.5.2.2).

Comparisons of models of the above type with field measurements have been reported by Parsons (1976), who found reasonable agreement.

3.5.4.4 Influence of Lithological Variations

Another important feature which can easily be taken into account in the solution of Eq. (3.5.28) is the influence of lithological variations in the slope. For doing this, one can simply introduce a function a(y) into Eq. (3.5.28) so that it reads

$$\frac{\partial y}{\partial t} = - a(y) \frac{\partial y}{\partial x} \sqrt{1 + \left(\frac{\partial y}{\partial x}\right)^2}.$$

(3.5.35)

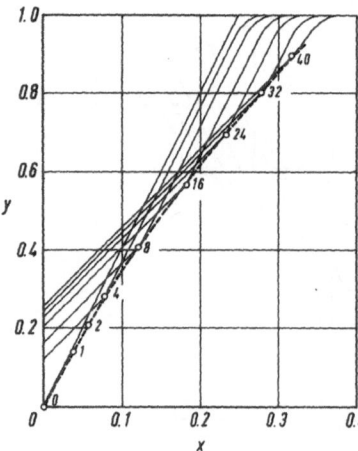

Fig. 43. Denudation of a slope bank, the debris forming an inclined apron. (After Scheidegger 1962)

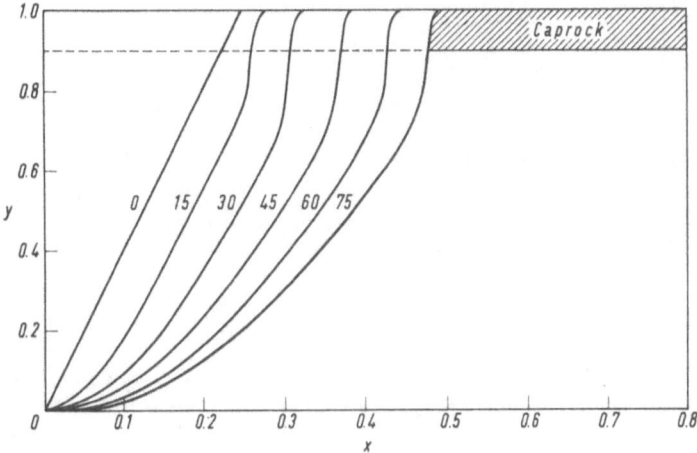

Fig. 44. Graph showing slope with caprock. (After Scheidegger 1964)

Then, choosing suitable values for a(y) models the influence of lithological conditions (Scheidegger 1964).

Thus, starting with a slope bank of the form shown by the zero-time line in Fig. 44, one can choose

$$a = 1.0 \quad \text{for} \quad 0 \leqslant y \leqslant 0.9$$
$$a = 0.1 \quad \text{for} \quad 0.9 \leqslant y \leqslant 1.0. \tag{3.5.36}$$

This represents a slope with a cap rock which is ten times more resistant to erosion than the parts below. The solution is shown in Fig. 44. The calculated profiles are very good representations of mesa-type structures found in nature.

The influence of a horizontal resistant layer can be calculated by setting.

$$a = 0.1 \quad \text{for } 0.4 \leqslant y \leqslant 0.5$$
$$a = 1.0 \quad \text{for all other values of y.} \tag{3.5.37}$$

The solution obtained under these conditions is shown in Fig. 45.

The reverse condition to that considered above is that of a slope with a horizontal soft layer. It is assumed that

$$a = 1.0 \quad \text{for } 0.4 \leqslant y \leqslant 0.5$$
$$a = 0.1 \quad \text{otherwise,} \tag{3.5.38}$$

and the solution of Eq. (3.5.35) is recalculated for these values for a(y). The results are shown in Fig. 46.

The condition of a slope with a soft bottom is obtained if it is assumed that

$$a = 0.1 \quad \text{for } 0.1 \leqslant y$$
$$a = 1.0 \quad \text{for } 0 \leqslant y \leqslant 0.1. \tag{3.5.39}$$

The results of the calculations are shown in Fig. 47.

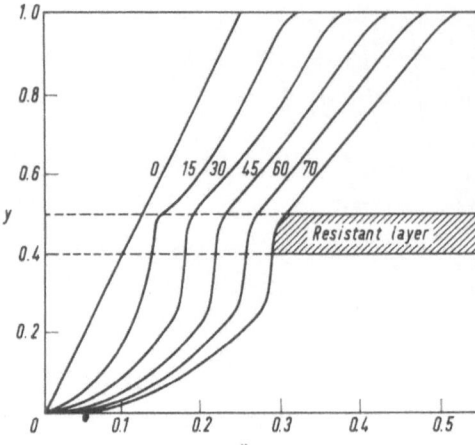

Fig. 45. Graph showing slope with resistant layer. (After Scheidegger 1964)

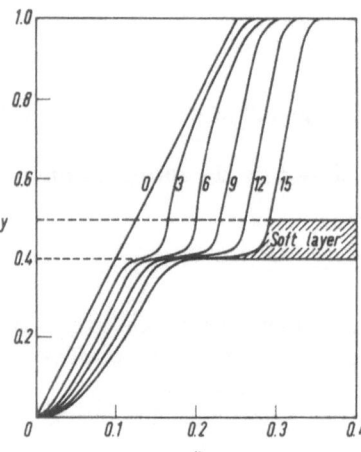

Fig. 46. Graph showing slope with soft layer. (After Scheidegger 1964)

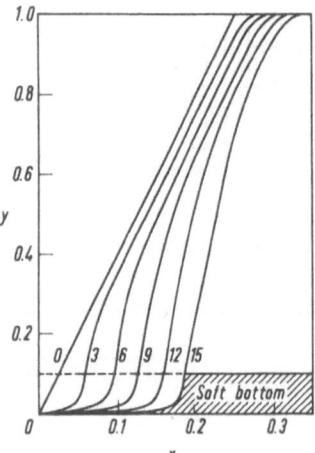

Fig. 47. Graph showing slope with soft bottom. (After Scheidegger 1964)

An inspection of the various figures presented here shows that the calculated models correspond to observed conditions very well (cf. Twidale and Milnes 1983).

3.5.4.5 Climatic Effects

Finally, it may be noted that climate can also have an effect on slope development inasmuch as it may vary even on a single slope, e.g., with altitude.

In essence, the nonuniform exposure of a slope to the elements can be taken care of by considering quasi-uniform climatic *zones*: the glacial, the periglacial, the humid, the pediment zone, etc. In the various zones, the mechanisms of denudation may be quite different: ice abrasion, solifluction, water erosion, etc.

It is therefore not possible to presence integral mathematical models for the climatic effects, but one has to combine different processes that are defined by the climatological conditions and that are representative for specific cases (Gossmann 1970, 1976).

3.5.5 Endogenic Effects in Slope Development

3.5.5.1 General Remarks

The various mathematical models of slope development that have been discussed do not take any endogenic movements into account. They thus fit into the Davisian concept of a geomorphic cycle: it is assumed that an original slope bank is some how created by a diastrophic process and that for ever thereafter the denudation proceeds at a steady pace. If one wishes to introduce Penckian ideas which postulate that endogenic and exogenic geodynamic processes occur *simultaneously*, then endogenic movements have to be superimposed upon the exogenic development patterns.

In other words, we shall now study the modifications that are required in the various models of slope recession discussed earlier, if endogenic effects are assumed to occur simultaneously with exogenic phenomena.

3.5.5.2 Surface Action and Endogenic Effects

The models of slope development due to surface action that have been discussed in Sect. 3.5.3.3 lend themselves easily to a modification so as to describe external effects, simply by introducing an additional function F into the basic differential equation (3.5.23). The latter then becomes (Scheidegger 1962)

$$\frac{\partial y}{\partial t} = -\sqrt{1 + \left(\frac{\partial y}{\partial t}\right)^2}\,\Phi + F. \tag{3.5.40}$$

It is at once apparent that there exist many possibilities for the choice of F. We assume that F is a function of x and y and thus set

$$F = F(x, y). \tag{3.5.41}$$

In the present context, two cases can easily be investigated. In the first case one may assume an endogenic *decrease* of the slope and therefore one may set

$$F = - \text{const } y. \tag{3.5.42}$$

Only the possibility

$$\Phi = \frac{\partial y}{\partial x} \tag{3.5.43}$$

deserves analysis because the corresponding model of weathering appears to be the most reasonable one. Using (3.5.42) would presumably give a good picture of a body of mass (such as a rapidly thrown-up volcanic island) sinking due to its tendency to achieve isostasy. The speed of sinking is then proportional to the height of the mass above a certain base level; this is expressed by Eq. (3.5.42).

The above case was solved on an electronic computer, using a procedure analogous to that employed in Sect. 3.5.3.3. The constant in Eq. (3.5.42) was chosen as equal to 16; the results obtained in this manner are shown graphically in Fig. 48. The second case investigated corresponds to that discussed above, but with a reversed sign; thus the constant was taken as equal to − 16. This yields a slope whose height is endogenically increasing; the rate of increase is proportional to the height already reached. This may perhaps correspond to conditions obtaining in recent orogenetic belts that are still active. The results obtained are shown graphically in Fig. 49.

It appears from the results obtained above that the superposition of an endogenic displacement does not materially affect the character of the slope that will develop. An originally straight slope bank will become concave at the toe; the convex curvature at the top is due to computer rounding inaccuracies.

Similar calculations as those reported above have also been made by Devdariani (1966), but using a much simpler model for the "erosion", viz. one of the type considered in Sect. 3.4.4.6. Thus, Devdariani had to solve a nonhomogeneous diffusivity equation rather than an equation of the type of (3.5.40). The result is a diffusive type of "damping" of the tectonic motion, as might be expected. Hirano (1967), also based calculations on similar assumptions.

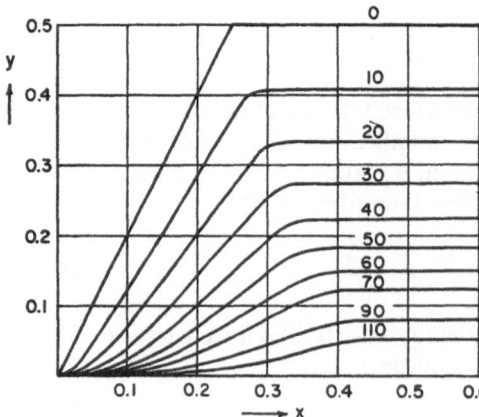

Fig. 48. Development of an endogenically decreasing slope. (Scheidegger 1962)

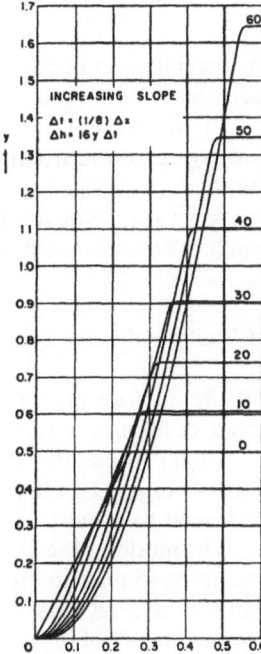

Fig. 49. Development of an endogenically increasing slope

3.5.5.3 Sideways Erosion and Endogenic Movements

It remains to analyze the effect of endogenic movements if, as postulated by Crickmay (1960; see Sect. 1.5.1.4) slope recession is assumed to be due to the sideways erosion of rivers at the bottom of such slopes.

It would appear that in this case, the endogenic movements would not affect the slope recession in its pattern. On a slope consisting of gravel or debris, an essentially parallel slope recession will always be maintained no matter whether there is or is not an endogenic movement occurring at the time. The same is true if the slope recession occurs by successive landslides.

The general patterns of receding slopes should therefore be independent of endogenic movements, if these patterns are due to thesideways erosion of rivers.

3.5.6 Conclusion of Slope Theory

3.5.6.1 General Remarks

In conclusion of the theory of slope development, we discuss some special problems and then give an evaluation of the range of applicability of the various models referring to slope development.

3.5.6.2 Slope Effects of Groundwater Flow

A special case occurs in slope development if groundwater is of importance: erosion occurs by outflow (sapping) of groundwater or by piping by soil water.

Piping is the internal washing out of hollows within the slope, sapping causes seepage erosion; it can occur as artesian sapping or as spring sapping (Higgins 1984).

As of yet, however, few theoretical investigations seem to have been made on such problems. Most work is entirely phenomenological (La Fleur 1984).

3.5.6.3 Evaluation of Slope Development Theories

In the theories of slope recession presented above, little has been said regarding their domain of application.

Under slopes, one may understand mountain sides, river banks, or even actual valley slopes. In all these instances, different slope recession theories must be assumed to apply.

The models of slope recession discussed in this section on "denudation" may be assumed to refer to any mountain side or river bank. In highly cohesive material, such as rock, it may be assumed that the direct action of external agents on the whole surface of the slope is small, so that the observed development will be either that envisaged by Bakker et al. (see Sect. 3.5.2) which occurs under a scree, or else that envisaged by Crickmay (Sect. 3.5.4) which is due to rivers cutting away at the bottom. The direct slope recession caused by action of eroding agents upon the whole surface of the slope will probably most commonly be observed in materials of little cohesion such as clay, shale, and incompletely consolidated sandstone. Actual valley slopes may be expected to have developed in conformity with the erosion and deposition caused by flowing water (Sect. 3.4.4). The latter also may act on mountain sides and shape the slopes involved if sheet flooding is common in the area.

Considering the theoretical effects of endogenic movements upon the development of slopes, we note that an inspection of Figs. 48 and 49 shows at once that there is no support of Penck's ideas regarding a characteristic form of *waxing* and *waning* slopes (cf. Sect. 1.2.4): all slopes, notwithstanding the endogenic movements, are essentially concave upward. There is no indication from theory that slopes ever become convex.

On the other hand, we have noted that the action of a laterally eroding river will produce essentially parallel slope recession, regardless of endogenic movements. Before a dynamic equilibrium is attained, the slope may be convex. This appears to support the principle of unequal activity outlined in Sect. 1.5.1.4, just as the latter is supported by the mathematical slope development theories which did make no allowance for endogenic movements. Gossmann (1970) indeed maintains that slopes in nature are subject to parallel slope recession and normally do not become flatter.

Thus, it appears that the theoretically postulated slope types have actually been observed in nature. In natural slopes, the various agents must be assumed to occur all at the same time. Thus, undercutting by rivers may cause landslides whose faces then may be directly affected by surface action of slope-changing agents. Therefore, landscapes characteristic of the various processes discussed above have been found, depending on which agent has been the most powerful one (Ahnert 1970; Gossmann 1981).

References

Abrahams, A.D., A.J. Parsons and P.J. Hirsch: J. Geol. 93(3), 347 (1985)
Ahnert, F.: Z. Geomorph. Suppl. 9, 88 (1970)
Ahnert, F.: Occ. Pap. Geogr., Univ. Maryland 1, 1 (1971)
Ahnert, F.: Die Erde 109, 49 (1978)
Ai, N.S. and T.D. Miao: Catena Suppl. 10, 21 (1987)
Armstrong, A.: Z. Geomorph. Suppl. 25, 20 (1976)
Aronsson, G.: A new method for calculating the recession of slopes. Uppsala: Univ. Dept. Phys. Geogr. (1973)
Augustithis, S.S. (ed.): Leaching and diffusion in rocks and their weathering products. Athens: Theophrastus, Pub. (1983)
Aulitzky, H.: Mitt. Forstl. Bundesversuchsanst. Wien, 144, 243 (1982)
Bagnold, R.A.: Proc. R. Soc. Lond. A 225, 49 (1954)
Baker, V.R.: Quat. Stud. Pol. 4, 19 (1983)
Baker, V.R.: Mem. Can. Soc. Petrol. Geol. 10, 87 (1984)
Baker, V.R. and D. Nummedal: The channeled scanbland. Washington: Nasa (1978)
Baker, V.R. and 3 others: Spec. Pub. Int. Ass. Sediment. 6, 229 (1983)
Baker, V.R., R.C. Kochel and P.C. Patton (eds.): Flood geomorphology. New York: Wiley (1988)
Bakker, J.P.: Z. Geomorph. Suppl. 1, 69 (1965)
Bakker, J.P. and J.W.N. Le Heux: Proc. K. Akad. Wet. Amsterdam 50, 959, 1154 (1947)
Bakker, J.P. and J.W.N. Le Heux: Proc. K. Akad. Wet. Amsterdam 53, 1073, 1364 (1950)
Bakker, J.P. and J.W.N. Le Heux: Proc. K. Akad. Wet. Amsterdam 55, 399, 554 (1952)
Ball, J.W.: J. Hydraul. Div. ASCE 102 (HY9) 1283 (1976)
Band, L.: Catena 12, 281 (1985)
Barnes, H.L.: Bull. Geol. Soc. Am. 64, 1392 (1953)
Barnes, H.L.: Am. J. Sci. 254, 493 (1956)
Bazant, J.: Proc. 3rd World. Conf. Earthqu. Eng. 1, 16 (1966)
Beaty, C.B.: Bull. Geol. Soc. Am. 70, 1479 (1959)
Beaty, C.B.: Am. J. Sci. 268 (1), 50 (1970)
Beaty, C.B.: Z. Geomorph. Suppl. 21, 39 (1974)
Berner, R.A. and G.R. Holdren: Geology 5(6), 369 (1977)
Berner, R.A. and G.R. Holdren: Geochim. Cosmochim. Acta 43(8) 1161 and 1173 (1979)
Birch, F.P.: Mem. Geol. Soc. Am. 97, 97 (1966)
Bjerrum, L. and A.F. Jorstad: Publ. Nor. Geotek. Inst. 79, 1 (1968)
Bolster, S.J.S.: Swansea Georgr. 22, 94 (1985)
Bonzanigo, L.: Proc. 5th Internat. Symp. Landslides Lausanne ed. C. Bonnard 2, 1315 (1988)
Bozzolo, D., R. Pamini and K. Hutter: Proc. 5th Intern. Symp. Land-slides, Lausanne, ed. C. Bonnard 1, 555 (1988)
Breitfuss, G. and A.E. Scheidegger: Ann. Geofis. Roma 27, 47 (1974)
Brückl, E. and A.E. Scheidegger: Rock Mech. 4, 237 (1972)
Bryan, R.B.: Earth Surf. Proc. 4, 43 (1979)
Bryan, R.B.: Catena Suppl. 8, 1 (1987)
Buhmann, D. and W. Dreybrodt: Chem. Geol. 48 (1–4), 189 (1985)
Bull, W.B.: U.S. Geol. Surv. Prof. Pap. 437A, A1 (1964)
Carillo, N.: Investigation on stability of slopes and foundations. D. Sc. Thesis, Cambridge, Mass.: Harvard Univ. (1942)
Carniel, P. and A.E. Scheidegger: Riv Ital. Geofis. 3 (1–2), 82 (1976)
Carrara, A., E.P. Carratelli and L. Merenda: Z. Geomorph. 21(2), 187 (1977)
Carroll, D.: Weathering. London: Plenum (1970)

Chandler, R.J.: J. Geol. 81(1), 1 (1973)

Clarke, G.K.C.: Can. J. Earth Sci. 23, 859 (1986)

Clarke, G.K.C. and W.H. Mathews: Can J. Earth Sci. 18(9), 1452 (1981)

Coleman, J.M. et al.: Bull. Geol. Soc. Am. 77, 205 (1966)

Collatz, L.: Numerische Behandlung von Differentialgleichungen. Berlin Göttingen Heidelberg: Springer (1951)

Colman, S.M.: Quat. Res. 15, 250 (1981)

Correia, R.M.: Proc. 5th Intern. Symp. Landslides, Lausanne, ed. C. Bonnard 1, 595 (1988)

Costa, J.E.: Bull. Geol. Soc. Am. 94, 986 (1983)

Coulomb, C.A.: Mem. Math. Phys. Acad. Sci. Paris 7: 343 (1776)

Coutard, J.P. and H.J. Mücher: Earth Surf. Proc. Landf. 10, 309 (1985)

Crickmay, C.J.: J. Geol. 68, 377 (1960)

Culling, W.E.H.: J. Geol. 68, 336 (1960)

Davison, C.A.: Geol. Mag. (3), 26, 255 (1889)

Denny, C.S.: U.S. Geol. Surv. Prof. Pap. 466 (1965)

De Ploey, J.: Z. Geomorph. Suppl. 46, 15 (1983)

De Ploey, J. and J. Moeyersons: Catena 2, 275 (1975)

De Ploey, J., J. Savat and J. Moeyersons: Earth Surf. Proc. 1, 151 (1976)

De Saint-Venant, B.: C.R. Acad. Paris 1871, 73 (1871)

Devariani, A.S.: Izv. Akad. Nauk SSSR. Ser. Geogr. (3), 7 (1966)

Dieckmann, H. and 3 others: 'Geoökodynamik 6, 121 (1985)

Dorn, R.I.: Nat. Geogr. Res. 4, 56 (1988)

Dorn, R.I. and T.M. Oberlander: Progr. Phys. Geogr. 6(3), 317 (1982)

Drever, J.I. (ed.): The chemistry of weathering. Dordrecht: Reidel (1985)

Eisenberg, P.: On the mechanism and prevention of cavitation. D.W. Taylor Model Basin: Navy Dept. (1950)

Ekern, P.C.: Proc. Soil. Sci. Soc. Am. 15, 7 (1951)

Ekern, P.C.: Agric. Eng. 34, 23 (1953)

Erismann, T.: Rock Mech. 12, 15 (1979)

Erismann, T.: Z. Geomorph. 32(3), 257 (1988)

Esin, N.V. and V.D. Dmitriev: Geomorfologiya (Akad. Nauk SSSR), 1975 (4), 68 (1975)

Fairbridge, R.W.: Proc. 7th Pac. Sci. Cong. 3, 347 (1952)

Farr, T.G. and J.B. Adams: Bull. Geol. Soc. Am. 95, 1077 (1984)

Fellenius, W.: Erdstatische Berechnungen, Berlin: W. Ernst & Sohn, (1927)

Finlayson, B.L., J. Gerits and B. Van Wesemael: Catena 14, 131 (1987)

Fisher, O.: Geol. Mag. 3, 334 (1866)

Flood, R.D.: Sedimentology 28, 511 (1981)

Focken, C.M.: Dimensional methods and their application, London: Arnold (1953)

Folk, R.L. and E.B. Patton: Z. Geomorph. 26(1), 17 (1982)

Frostick, L.E. and I. Reid: Z. Geomorph. Suppl. 44, 53 (1982)

Gardner, J.S.: Geomorph. 23(1), 45 (1979)

Garrels, A.M. et al: Am. J. Sci. 259, 24 (1961)

Gerber, E.K. and A.E. Scheidegger: Geograph. Helv. 21(1), 20 (1966)

Gerber, E.K. and A.E. Scheidegger: Eclogae Geol. Helv. 62(2), 401 (1969)

Gerber, E.K. and A.E. Scheidegger: Peterm. Geogr. Mitt. 117(1), 23 (1973)

Gerber, E.K. and A.E. Scheidegger: Vierteljahrsschr. Naturforsch. Ges. Zür. 129(3), 294 (1984)

Ghadiri, H. and D. Payne: In: Soil Physical properties and crop production in the tropics, ed. R. Lal and D.J. Greenland (New York: Wiley) p. 95 (1979)

Ghadiri, H. and D. Payne: J. Soil Sci. 32(1), 41 (1981)

Göbel, P.: Catena 3, 387 (1977)

Goldin, B.M. and S.L. Lubashevsky: Sov. Hydrol. 2, 179 (1966)

Gossmann, H.: Würzburger Geogr. Arb. 31, 1 (1970)

Gossmann, H.: Z. Geomorph. Suppl. 25, 72 (1976)

Gossmann, H.: Geoökodynamik 2, 205 (1981)

Govers, G.: Catena 12, 35 (1985)

Green, L. and C.H. Wilts: Proc. 1st U.S. Nat. Congr. Appl. Mech., p. 777 (1952)

Guilcher, A.: Coastal and submarine geomorphology; transl. from French. London: Methuen & Co. (1958)

Habib, P.: Rock Mech. 7, 193 (1975)
Haigh, M.J.: Oklahoma Geol. Notes 38(3), 87 (1978)
Haigh, M.J.: Earth Surf. Proc. 4, 183 (1979)
Haigh, M.J.: In: Catchment experiments in fluvial geomorphology, ed. T. Burt and D. Walling: Norwich: Geobooks. P. 247 (1981)
Hamann, C.: Berliner Geogr. Abh. 36, 69 (1984)
Hamann, C.: Salzburger Geogr. Abh. 10, 1 (1985)
Harp, E.L.: Geol. Appl. e Idrogeol. (Bari) 21(2), 1559 (1986)
Harrison, J.C. and K. Herbst: Geophys. Res. Lett. 4(11), 535 (1977)
Hay, R.L. and B.F. Jones: Bull. Geol. Soc. Am. 83, 317 (1972)
Heydemann, A.: Geochim. Cosmochim. Acta 30, 995 (1966)
Higgins, C.G.: Z. Geomorph. 26(4), 459 (1982)
Higgins, C.G.: In: Groundwater as a geomorphic agent, ed. R.G. La Fleur. Boston: Allen & Unwin, p. 18 (1984)
Hill, R.: The mathematical theory of plasticity. Oxford: Clarendon (1950)
Hirano, M.: Chikyu Kagaku 21(5), 27 (1967)
Hirano, M.: J. Geosci. Osaka City Univ. 11(2), 13 (1968)
Hirano, M.: J. Geol. 83, 113 (1975)
Hjulström, F.: Bull. Geol. Inst. Uppsala 25, 221 (1935)
Hodgkin, E.P.: Z. Geomorph. 8, 385 (1964)
Holmes, A.: Principles of physical geology. London: Nelson (1944)
Hooke, R.L.: Am. J. Sci. 266, 609 (1968)
Horton, R.E., H.R. Leach and R. Van Vliet: Trans. Am. Geoph. Union 15, 393 (1934)
Hsü, K.J.: Bull. Geol. Soc. Am. 86, 129 (1975)
Huder, J.: Mitt. Inst. Grundbau ETH Zürich 107, 125 (1976)
Hutchinson, J.N.: Can. Geotech. J. 23(2), 115 (1986)
Hutter, K. and S.B. Savage: Proc. 5th Int. Symp. Landslides, Lausanne, ed. C. Bonnard 1, 691 (1988)
Iida, T. and K. Okunishi: Z. Geomorph. Suppl. 46, 67 (1983)
Imeson, A.C., R. Vis and E. De Water: Catena 8, 83 (1981)
Ingle-Smith, D. and D. Mead: Proc. Univ. Bristol Speleolog. Soc. 9, 188 (1962)
Iseda, T. and Y. Tanabashi: Nat. Disaster Sci. 8(1), 55 (1986)
Iverson, R.M.: J. Geol. 94(1), 1 (1986a)
Iverson, R.M.: J. Geol. 94(3), 349 (1986b)
Jocelyn, J.: Nature (Phys. Sci.) 240(98), 39 (1972)
Johnson, P.G.: Z. Geomorph. 28(2), 235 (1984)
Jones, J.A.A.: Progr. Phys. Geogr. 11(2), 207 (1987)
Kent, P.E.: J. Geol. 74, 79 (1966)
Kerr, A. and 3 others: Geology 12(5), 306 (1984)
Kirkby, M.J.: J. Geol. 75, 359 (1967)
Kirkby, M.J.: Z. Geomorph. Suppl. 25, 1 (1976)
Knoblich, K.: Catena 2, 1 (1975)
Kobluk, D.R. and C.F. Kahle: Bull. Can. Petrol. Geol. 26(3), 362 (1978)
Kovari, K., F.T. Madsen and C. Amstad: Proc. Int. Symp. Weak Rock Tokyo 21–24 Sept. 1981, 1019 (1981)
Kuenen, P.H.: Marine geology, New York: J. Wiley & Sons (1950)
Kuron, H. and H.J. Steinmetz: C.R. Assemb. Gen. Toronto, Assoc. Hydrol. Sci. 1, 115 (1957)
Kwaad, F.J.: Pub Fysisch Geogr. Bodenkund. Lab. 16, 67 (1970)
La Fleur, R.G. (ed.): Groundwater as a geomorphic agent. Boston: Allen & Unwin (1984)
Labuz, J.F., S.P. Shah and C.H. Dowding: Int. J. Rock Mech. Min. Sci. 22(2), 85 (1985)
Lajtai, E.Z.: Rock Mech. 19(2), 71 (1986)
Laqueche, H., A. Rousseau and G. Valentin: Int. J. Rock Mech. Min. Sci. 23(5), 347 (1986)
Lehmann, O.: Vierteljahrsschr. Naturforsch. Ges. Zür. 78, 83 (1933)
Le Roux, J.S. and Z.N. Roos: Z. Geomorph. 30(4), 477 (1986)
Leung, C.F. and S.C. Kheok: Rock Mech. 20, 111 (1987)
Lewis, L.A.: Z. Geomorph. Suppl. 25, 132 (1976)
Lomnitz, C.: J. Geol. 64, 473 (1956)
Louis, H.: Z. Geomorph. Suppl. 28, 1 (1977)
Luckman, B.H.: Z. Geomorph. Suppl. 29, 117 (1978)

Luke, J.C.: J. Geoph. Res. 77(14), 2460 (1972)
Luke, J.C.: Z. Geomorph. Suppl. 25, 114 (1976)
Macar, P.: L'evolution des versants. Symp. Int. Geomorphologie, Liege (1966)
Mac Fadyen, G.A.: Geog. J. 75, 24 (1930)
Machatschek, F.: Geomorphologie, 5th edn. Leipzig: Teubner (1952)
Madhav, M.R. and H.B. Poorooshasb: Geol. Appl. Idrogeol. (Bari) 21(2), 139 (1986)
Madsen, F.T.: Mitt. Inst. Grundbau ETH Zürich 114, 1 (1979)
Manning, R.: Trans. Inst. Civ. Eng. Ireland 20, 161 (1890)
Marovelli, R., T.S. Chen and K.F. Veith: Trans. Soc. Mining Eng. 235, 1 (1966)
Martin, C.: Z. Geomorph. 31(1), 73 (1987)
Matthews, R.K.: Spec. Publ. Soc. Econom. Paleont. Min. 18, 234 (1974)
Mayer, L. and Nash, D. (ed.): Catastrophic flooding. London: Allen & Unwin (1987)
Mc Crone, A.W.: J. Sediment Petrol. 36, 270 (1960)
Mc Greevy, J.P.: Earth Surf. Proc. Landforms 10, 125 (1985a)
Mc Greevy, J.P.: Earth Surf. Proc. Landforms 10, 509 (1985b)
Mc Intyre, D.S.: Soil Sci. 85, 185 (1958)
Mills, H.H.: Geogr. Ann. 65A (3–4), 255 (1983)
Mitin, A.V. and A.M. Trofimov: Uch. Zap. Kazansk. Gos. Un-ta 124(4), 112 (1964)
Mizutani, T.: Geogr. Rept. Tokyo Metrop. Univ. 5, 49 (1970)
Mizutani, T.: Geogr. Rept. Tokyo Metrop. Univ. 21, 251 (1986)
Modenesi, M.C.: Catena 10, 237 (1983)
Moeyersons, J.: Catena 2, 289 (1975)
Mohr, O.: Abhandlungen aus dem Gebiete der technischen Mechanik. 3. Aufl. Berlin: Wilh. Ernst &
    Sohn (1928)
Mosley, M.P.: East Midland Geogr. 5(5), 235 (1972a)
Mosley, M.P.: Ann. Assoc. Am. Geogr. 62(4), 655 (1972b)
Müller-Vonmoos, M., P. Honold and G. Kahr: Mitt. Inst. Grundbau, ETH Zürich 128, 1 (1985)
Mustoe, G.E.: Bull. Geol. Soc. Am. 93, 108 (1982)
Nago, H. and S. Maeno: Nat. Disaster Sci. 9(1), 23 (1987)
Nossin, J.J. and T.W.M. Levelt: Z. Geomorph. 11, 14 (1967)
Okunishi, K. and T. Okimura: In: Slope stability, ed. M.G. Anderson and K.S. Richards, Wiley, p. 265
    (1987)
Ollier, C.D.: Aust. J. Sci. 27, 236 (1965)
Ollier, C.D. and J.E. Ash: Z. Geomorph. 27(3), 363 (1983)
Parsons, A.: Z. Geomorph. Suppl. 25, 145 (1976)
Perez, F.L.: Z. Geomorph. 32(1), 77 (1988)
Petterson, K.E.: Tekn. Tidskr. 46, 289 (1916)
Piwowar, A.: Vierteljahrsschr. Naturforsch Ges. Zürich 48, 337 (1903)
Poesen, J.: Z. Geomorph. 29(2), 193 (1985)
Poesen, J.: Z. Geomorph. Suppl. 58, 81 (1986)
Prange, G.: Vorlesungen über Differential- und Integralrechnung. Berlin: Springer (1943)
Price, W.E.: Water Resour. Res. 10(2), 263 (1974)
Prior, D.B. and W.H. Renwick: Z. Geomorph. Suppl. 34, 63 (1980)
Pye, K.: Catena 13, 47 (1986)
Rayleigh, Lord: Proc. R. Soc. (Lond.) A181, 107 (1942)
Rayleigh, Lord: Proc. R. Soc. (Lond.) A182, 321 (1944)
Read, J.R. and M.E. Harr: Proc. 5th Int. Symp. Landslides, Lausanne, ed. C. Bonnard 1, 749 (1988)
Revelle, R. and K.O. Emery: U.S. Geol. Surv. Prof. Pap. 260-T, 699 (1957)
Richter, E.: Peterum. Geograph. Mitt., Ergänzungsbd. 24, 1 (1901)
Roberts, W.A.: Icarus 5, 459 (1966)
Robinson, E.S.: Bull. Geol. Soc. Am. 81, 2799 (1970)
Rose, C.W.: Soil Sci. 89, 28 (1960)
Savat, J.: Earth Surf. Proc. 2, 125 (1977)
Scheidegger, A.E.: Geofis. Pura e Appl. 56, 58 (1951)
Scheidegger, A.E.: Geol. u. Bauw. 25, 3 (1959)
Scheidegger, A.E.: J. Alberta Soc. Petrol. Geol. 8(7), 202 (1960)
Scheidegger, A.E.: Bull. Geol. Soc. Am. 72, 37 (1961a)
Scheidegger, A.E.: J. Alberta Soc. Petrol. Geol. 9(4), 131 (1961b)

Scheidegger, A.E.: Geoph. J. R. Astronom. Soc. 7, 40 (1962)
Scheidegger, A.E.: Geofis. Pura Appl. 56, 58 (1963)
Scheidegger, A.E.: U.S. Geol. Survey Circ. 485 (1964)
Scheidegger, A.E.: Rock Mech. 5, 231 (1973)
Scheidegger, A.E.: The physics of flow through porous media. Third edition. Toronto: University of Toronto Press, (1974)
Scheidegger, A.E.: Physical aspects of natual catastrophes. Amsterdam: Elsevier, (1975)
Scheidegger, A.E.: Principles of geodynamics, 3rd. edn. Berlin Heidelberg New York: Springer (1982)
Scheidegger, A.E.: Z. Geomorph. 27(1), 1 (1983)
Scheidegger, A.E.: Earth Sci. Rev. 21, 225 (1984)
Scheidegger, A.E.: Geoph. Surv. 70, 259 (1985)
Scheidegger, A.E.: Z. Geomorph. 30, 257 (1986)
Scheidegger, A.E.: Wildbach- und Lawinenverbau 51, 3 (1987)
Scheidegger, A.E.: In: Natural and man-made hazards, ed. M.I. El-Sabh & T.S. Murty. Dordrecht: Reidel p. 21 (1988)
Scheidegger, A.E. and N.S. Ai: Tectonophysics 126, 285 (1986)
Schipull, K.: Z. Geomorph. Suppl. 30, 93 (1978)
Schmidt, R.G.: Probleme der Erfassung, Quantifizierung von Ausmaß von Prozessen der aktuellen Bodenerosion (Abspülung) auf Ackerflächen. Dr. Diss. Univ. Basel (1979)
Schumm, S.A.: Bull. Geol. Soc. Am. 73, 719 (1962)
Schumm, S.A.: J. Geol. Educ. 14, 98 (1966)
Schumm, S.A. and R.J. Chorley: Am. J. Sci. 262, 1041 (1964)
Seed, H.B.: J. Soil Mech. Found. Div. ASCE 93(SM4), 299 (1967)
Seuffert, O.: Geoökodynamik 2, 141 (1981)
Seuffert, O. and 4 others: Z. Geomorph. Suppl. 50, 31 (1984)
Shal'nev, K.K.: Izv. Akad. Nauk SSSR, Otd. Tekh. Nauk 1956, No. 1, 3 (1956)
Sharpe, C.F.S.: Landslides and related phenomena. New York: Columbia Univ. Press (1938)
Shreve, R.L.: Science 154 (3757), 1639 (1966)
Siever, R.: J. Geol. 79, 127 (1962)
Smith, C.L.: J. Mar. Biol. Assoc. 25, 234 (1941)
Smith, D.D. and W.J. Wischmeier: Adv. Agron. 14, 109 (1962)
Sorriso-Salvo, M.: Geol. Appl Idrogeol. (Bari) 21(2), 291 (1986)
Statham, I.: Earth surface sediment transport. Oxford: Clarendon Press (1977)
Sternberg, H.U.: Z. Bauw. 1875, 483 (1875)
Stocking, M.A.: Z. Geomorph. Suppl. 29, 141 (1978)
Strahler, A.: Bull. Geol. Soc. Am. 63, 923 (1952)
Strömquist, L.: Z. Geomorph. 29(2), 129 (1985)
Svoboda, B.: Geol. Appl. Idrogeol. (Bari) 21(2), 69 (1986)
Takeshita, K.: Bull. Fukuoka-Ken Forest Exp. Stn. 16, 115 (1963)
Takeshita, K.: Bull. Fukuoka-Ken Forest Eyp. Stn. 17, 1 (1964)
Terzaghi, K.: Theoretical soil mechanics New York: J. Wiley & Sons (1943)
Terzaghi, K.: Geol Soc. Am. Engineering Geology (Berkey) Volume p. 83, New York: Geol. Soc. Am. (1950)
Tödten, H.: Z. Geomorph. Suppl. 25, 89 (1976)
Toki, K., F. Miura and Y. Oguni: Earthqu. Eng. and Struct. Dyn. 13(2), 151 (1985)
Torri, D. and M. Sfalanga: Dev. Environ. Model. 10, 161 (1986)
Torri, D., M. Sfalanga and G. Chisci: Catena Suppl. 8, 97 (1987a)
Torri, D., M. Sfalanga and M. Del Sette: Catena 14, 149 (1987b)
Trofimov, A.M.: Izv. Vsesoy. Geograf. Ob-va 98(2), 166 (1966)
Trofimov, A.M. and V:M. Moskovkin: Z. Geomorph. Suppl. 25, 105 (1976)
Trofimov, A.M. and V.M. Moskovkin: Matematicheskoe modelirovanie y geomofologii sklonov. Kazan: Izd. Kaz. Univ. (1983)
Trofimov, A.M. and V.M. Moskovkin: Earth Surf. Proc. and Landforms 9, 435 (1984a)
Trofimov, A.M. and V.M. Moskovkin: Z. Geomorph. 28(1), 71 (1984b)
Trowbridge, J.H.: J. Geoph. Res. 92 (C9), 9523 (1987)
Tschierske, N.: Rock Mech. Suppl. 12, 89 (1982)
Tsenn, M.C. and N.L. Carter: Tectonophysics 136, 1 (1987)
Twidale, C.R. and E.M. Campbell: Z. Geomorph. 30(1), 35 (1986)

Twidale, C.R. and A.R. Milnes: Z. Geomorph. 27(3), 343 (1983)
Van Dijk, W. and J.W.N. Le Heux: Proc. K. Akad. Wet. Amsterdam B55, 115, 123 (1952)
Viles, H.A.: Progr. Phys. Geogr. 8(4), 523 (1984)
Vulliet, L. and K. Hutter: Geotechnique 38(2), 199 (1988)
Watson, A. and K. Pye: Z. Geomorph. 29(3), 285 (1986)
Wehmeier, E.: Catena 13, 197 (1986)
Wells, S.G. and A.M. Harvey: Bull. Geol. Soc. Am. 98, 182 (1987)
Weyl, P.K.: J. Geol. 66, 163 (1958)
Weyl, P.K.: Geochim. Cosmochim. Acta 17, 214 (1959)
Whalley, W.B., G.R. Douglas and J.P. McGreevy: Z. Geomorph. 26(1) 33 (1982)
Williams, J.E.: Geogr. Rev. 39, 129 (1949)
Williams, P.J.: Geogr. J. 123, 42 (1957)
Williams, P.J.: Am. J. Sci. 257, 481 (1959)
Wilson, C.M. and P.L. Smart: Catena 11, 145 (1984)
Wischmeier, W.H.: Proc. Soil. Sci. Soc. Am. 24(4), 322 (1960)
Woodburn, R.: Agric. Eng. 29, 154 (1948)
Wuerker, R.G.: Pap. Petroleum Branche AIME 663-G, 1 (1956)
Yair, A. and J. De Ploey: Catena 6, 245 (1979)
Yang, X.Q. et al. and S. Yamaguchi et al. (eds.): The China-Japan field workshop on landslide. Lanzhou:
    Gansu Soc. on Landslides (1987)
Yatsu, E.: The nature of weathering, an introduction. Tokyo: Maruzen (1988)
Yesin, N.V.: Izv. Akad. Nauk SSSR, Ser. Geogr. No. 3, 126 (1968)
Yesin, N.V. and N.A. Skorkin: Vestn. Moskovsk. Univ. Geogr. 1970 (3) 56 (1970)
Young, A.: Nachr. Akad. Wiss. Göttingen, II. Math.-Phys. Kl. (5) 45 (1963)
Young, R.A. and J.L. Wiersma: Water Resour. Res. 9(6), 1629 (1973)
Zhang, X.J. and W.F. Chen: Int. J. Numer. Analyt. Meth. Geomech 11(1), 103 (1987)
Zollinger, F.: Die Vorgänge in einem Geschiebeablagerungsplatz. Diss. ETH Zürich (1983)

4 Theory of River Action

4.1 General Remarks

On the land areas of the world, rivers are undoubtedly some of the most important geomorphological agents. Rivers act geomorphologically in fundamentally two fashions: first by interaction with their bed, i.e., downwards (this type of action is usually referred to as river bed process) and second, by interaction with their banks (sideways erosion). It is the purpose of this chapter (4) to present the theories of these processes.

Much of the pertinent information is available in corresponding textbooks which will be referred to as required (e.g., Dingman 1984). In addition, annual reviews on the subject matter are being published (Gregory 1985; Walling 1986, 1987; Richards 1987a, 1988).

4.2 Linear Flow in Open Channels

4.2.1 General Principles

In order to understand the mechanics of river action properly, it is first of all necessary to acquaint oneself with the fundamentals of open channel flow.

Flow in rivers is basically turbulent, but reasonable approximations can often be obtained by considering laminar flow, treating the ever-present turbulence as a "perturbation". Thus, turning first to (frictionless) laminar flow, we note that it is characterized by the existence of stream lines. The flow along each streamline is determined by the well-known Bernoulli equation:

$$H = z + \frac{p}{\rho g} + \frac{v^2}{2g} = const, \qquad (4.2.1)$$

where z is the vertical co-ordinate, p the pressure, ρ the fluid density, v the flow velocity, and g the gravity acceleration. The Bernoulli equation is an expression of the principle of conservation of energy. H represents the energy content at the point under consideration expressed as a height (hydraulic head).

It is often convenient to write the Bernoulli equation for the bottom streamline in a stream of depth h; then, assuming static pressure distribution, (4.2.1) becomes (with $z = 0$):

$$H = h + \frac{v^2}{2g}. \qquad (4.2.2)$$

Disregarding the change of velocity with depth, and introducing into the above equation for v the value of the average velocity

$$v = \frac{Q}{A} \tag{4.2.3}$$

(Q denoting the volume flow rate and A the cross-sectional area), we obtain

$$H = h + \frac{Q^2}{2gA^2}. \tag{4.2.4}$$

Again, the quantity H represents the energy content expressed as a height above the river bottom. If we denote the distance along the river by s, then H(s) defines a line which has been called *energy line*.

Let us assume that the channel is rectangular of width b. Then (4.2.4) becomes

$$H = h + \frac{Q^2}{2gb^2h^2} \tag{4.2.5}$$

or:

$$Q = \sqrt{(H - h)2gb^2h^2}. \tag{4.2.6}$$

From this equation it is at once obvious that if H (the energy) and Q (the volume flow) be given, there are two possible water depths h (and corresponding velocities v) with which the flow may occur, provided, of course, that H be large enough. The faster one of these flows is termed *shooting* (or "*supercritical*") flow, the other *streaming* (or "*subcritical*") flow. One can show that, in streaming flow, the flow velocity v is always less than the shallow water wave velocity u

$$v < u \equiv \sqrt{gh} \tag{4.2.7}$$

[for u, cf. Eq. (6.2.28)]; in shooting flow, the reverse is true:

$$v > u \equiv \sqrt{gh}, \tag{4.2.8}$$

if

$$v = u, \tag{4.2.9}$$

the flow is termed *critical*.

It is possible to build the whole structure of flow theory upon the Bernoulli equation. However, as noted above, it turns out that most river flow is turbulent with corresponding velocity fluctuations.

The study of turbulent flow in open channels is of particular interest with regard to a solution of the problem as to how sediment is being transported in rivers. There are several monographs bearing upon the subject, the most pertinent having been written by Velikanov (1954), Schmidt (1957), and Chow (1959; update by Shen and Yen 1984).

In the present context, we shall study only (quasi-) stationary flow in open channels. If "stationary" flows are under investigation, one has to average out all the velocity fluctuations that characterize the turbulence. Moreover, only the variation of average flow with *height* above the river bed is usually considered, which represents somewhat of an oversimplification. In Fig. 50 we show the

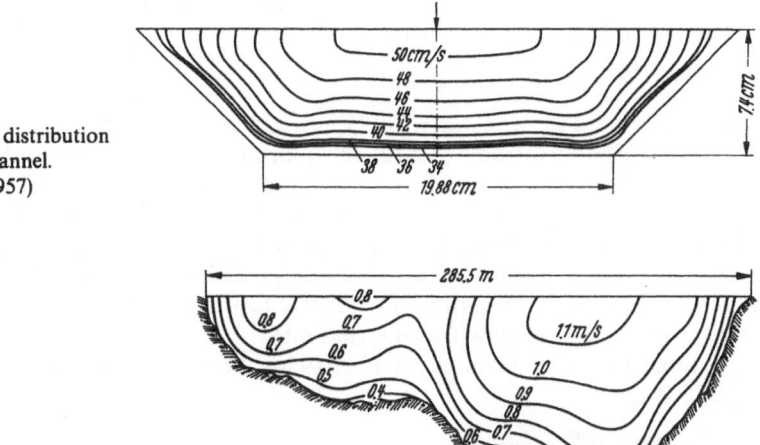

Fig. 50. Velocity distribution
in a prismatic channel.
(After Schmidt 1957)

Fig. 51. Velocity distribution in
a river. (After Schmidt 1957)

average flow velocities as they have been measured in a regular channel. Whereas
in this case there is a fairly regular velocity distribution over the cross-section, this
is no longer the case in a natural channel (see Fig. 51). Nevertheless, without the
assumption of a certain regularity of the velocity distribution, it is almost
impossible to arrive at any theory at all. The above remarks will serve to bear out
some of the limitations of the investigations that will follow.

4.2.2 Empirical Formulas

In our discussion of flow in open channels, we shall first turn towards empirical
flow formulas. Pertinent investigations of this problem have been made for at least
200 years, so that the fundamental patterns of the relationships have been known
for a long time. The formulas have been deduced from tests in the field as well as in
the laboratory.

The resulting formulas are generally quoted in the form

$$v = \text{const } h_m^a S^b, \tag{4.2.10}$$

where v is the average flow velocity in the channel, h_m its mean depth, and S the bed
slope. The quantities a and b, as well as the "const", are empirical constants.
Instead of h_m, it is often convenient to introduce a hydraulic radius R, defined as
follows:

$$R = A/P, \tag{4.2.11}$$

where A denotes the cross-sectional area and P the wetted perimeter of the section
under consideration. For large rivers, R evidently becomes equal to h_m. With a
slight change of constants, it is usually possible to write the empirical formulas to
be discussed here, in terms of either h_m or R.

Various values have been suggested in the literature for the constants a and b occurring in (4.2.10). A well-known relationship is of the form

$$v = R^{2/3} S^{1/2}/n, \qquad (4.2.12)$$

which is called the *Manning (1890) formula*.

In the Manning formula (4.2.12), the quantity n is a constant (the "roughness" of the channel), whose value varies from river stretch to river stretch. A catalog of representative values of n, with many colored pictures illustrating the river reaches in question, was published by Barnes (1967). Other investigations of the friction factor have been made by Martinec (1967), Limerinos (1969), Smith and Yates (1975) and Haque and Mahmood (1983). Accordingly, the values of n found may vary between 0.029 and 0.075 metric units.

A more elaborate form of Eq. (4.2.10) has been suggested by Bessrebrennikov (1958). It is

$$v^n = \text{const } h^m S, \qquad (4.2.13)$$

where n and m are supposedly connected in such a fashion that

$$m + n = 3. \qquad (4.2.14)$$

Bessrebrennikov (1958) quotes investigations bearing out that, in weedy channels, n equals 1, in nonweedy channels, n equals 2.

Another well-known formula for the average flow velocity v in a channel has been proposed by Chézy (in an unpublished report; see Herschel 1896). Chézy's formula is also essentially empirical, but one can give a somewhat rational deduction of it. Thus, let us write down the force balance equation for a slug of water flowing downstream. In order to do this, we consider a section of a stream of length L, cross-section A, wetted perimeter P, hydraulic radius R [defined by (4.2.11)] and slope S. The forces parallel to the current then are:
a) from the weight of the water

$$F_w = AL\rho gS, \qquad (4.2.15)$$

where ρ is the density of the water and g the gravity acceleration;
b) the frictional force

$$F_R = \sigma_m LP, \qquad (4.2.16)$$

where σ_m is the tractive force per unit surface (commonly referred to as *drag*). A reasonable assumption for the drag is (which can be justified because each obstacle offers quadratic resistance to the flow; see Eq. (2.2.6)]

$$\sigma_m = C^2 v^2, \qquad (4.2.17)$$

where C is called *Chézy's coefficient*. In flow that is essentially uniform, the sum of the forces acting upon a slug of water must be zero. Hence

$$AL\rho gS = \sigma_m LP = C^2 v^2 LP, \qquad (4.2.18)$$

whence we obtain Chézy's equation

$$v = \frac{1}{C} \sqrt{S\rho g \frac{A}{P}} = \frac{1}{C} \sqrt{S\rho gR}. \qquad (4.2.19)$$

Furthermore, we have

$$\sigma_m = \rho g \frac{A}{P} S,$$ (4.2.20)

or, if P is large (h = depth)

$$\sigma_m = \rho g h S.$$ (4.2.21)

This is the basic formula of the *drag theory*.

Experimental evidence seems to show that C is, in fact, not independent of the hydraulic radius. Remembering Manning's formula (4.2.12) we can write:

$$c \equiv \frac{1}{C} = kR^{1/6} \approx kh^{1/6}$$ (4.2.22)

where the last approximation holds for wide rivers. The Manning formula, and hence also Eq. (4.2.22), are based upon observational data. Tables giving values for the constant k for various cases are availble (Schmidt 1957) and can also be calculated from values of the Manning n mentioned above.

A different presentation of Chézy's formula is the so-called Darcy-Weisbach equation which reads (cf. Chow 1959, see p. 8 therein)

$$v = K\sqrt{hS},$$ (4.2.23)

where the factor K contains everything which was written out explicitly in (4.2.19). This factor is commonly written as follows

$$K = \sqrt{\frac{8g}{f}},$$ (4.2.24)

where g is the gravity acceleration and f the Darcy-Weisbach friction factor. If the appropriate dependence of f on h is inserted, one ends up with the Chézy or Manning equation.

It has been pointed out by Francis (1957) that formulas of the type of (4.2.10) can be shown to be approximations to the theoretical formulas (logarithmic laws) which we shall deduce in the next section (4.2.3).

4.2.3 Turbulent Flow in Rigid Channels

We consider next the mechanics of turbulent flow in open channels with a fixed, immobile, rough bottom.

Quite generally, it can be stated that the turbulent resistance is caused by the interaction of the roughness elements with their own separation vortices (Bertschler 1972). Experimental investigations of the pressure fluctuations near the boundaries have been published (Mulhearn 1976). The reaction to these in the fluid leads to an internal friction therein and therewith to a velocity decrease with distance from the boundary.

A pertinent theory for the above phenomena has been developed by Keulegan (1938) in analogy with investigations of the theory of flow in pipes made by Karman (1930), Nikuradse (1932), and others around 1930. Accordingly, the expression of Prandtl (1926), see Sect. 2.2.2.2) for the turbulent shear stress σ at any point in a fluid moving past a solid wall is [cf. Eq. (2.2.9)]

$$\sqrt{\sigma/\rho} = l\, d\bar{u}/dy, \tag{4.2.25}$$

where ρ is the density of the fluid, $\bar{u}$ the (time-averaged) velocity in question, y the distance from the wall, and l is the turbulent mixing length. The last equation can also be written

$$u_* \equiv \sqrt{\frac{\sigma_m}{\rho}} = l\frac{d\bar{u}}{dy}\sqrt{\frac{\sigma_m}{\sigma}}, \tag{4.2.26}$$

where σ_m is the maximum shear stress which occurs at the wall (i.e., for $y = 0$). The abbreviation u_* is frequently used; the quantity it denotes is often called *shear velocity*.

Dimensional analysis then yields

$$l = -k\frac{d\bar{u}/dy}{d^2\bar{u}/dy^2}, \tag{4.2.27}$$

where k is Karman's (1930) universal dimensionless constant of turbulence (equal to roughly 0.4). Substituting the expression for l into that for u_* yields

$$u_* = -k\frac{(d\bar{u}/dy)^2}{d^2\bar{u}/dy^2}\sqrt{\frac{\sigma_m}{\sigma}}. \tag{4.2.28}$$

For small values of y, the square root approaches 1 and one ends up with

$$u_* = -k\frac{(\bar{u}')^2}{\bar{u}''}, \tag{4.2.29}$$

This can be integrated to yield

$$\frac{\bar{u}}{u_*} = \frac{1}{k}\,\text{lognat}\,\frac{y}{y_0}, \tag{4.2.30}$$

where y_0 is a constant of integration ("displacement height"); the latter depends on the detailed geometry of the roughness elements (Jackson 1981). Equation (4.2.30) is Karman's (1930) law of velocity distribution in the neighbourhood of a solid wall.

If the surface of the wall is smooth, y_0 will depend solely on u_* and v, the latter denoting the kinematic viscosity. Dimensional analysis then yields

$$\frac{y_0 u_*}{v} = m, \tag{4.2.31}$$

where m is a constant (generalized Reynolds number). Hence

$$\frac{\bar{u}}{u_*} = a_s + \frac{1}{k}\,\text{lognat}\,\frac{y u_*}{v}. \tag{4.2.32}$$

Experimental evidence yields (Keulegan 1938)

$$a_s = 5.5. \tag{4.2.33}$$

If the surface is rough, then the "constant" m must depend on the height k_s of the surface roughness. Dimensional analysis yields

$$\frac{y_0 u_*}{v} = f\left(\frac{k_s u_*}{v}\right). \tag{4.2.34}$$

Again using experimental data, one obtains for water:

$$\frac{\bar{u}}{u_*} = 8.5 + \frac{1}{k}\text{lognat}\left(\frac{y}{k_s}\right). \tag{4.2.35}$$

In order to obtain expressions for the average flow velocity in the channel, the above expressions must be integrated from δ to R where R is again a hydraulic radius

$$R = \frac{A}{P} \tag{4.2.36}$$

(with A = cross-sectional area and P the wetted perimeter) and δ the thickness of the laminar sublayer. This thickness δ, from dimensional reasoning, must be proportional to v/u_*; Keulegan (1938) quotes the following relationship as determined from experiments:

$$\delta = 11.5\frac{v}{u_*}. \tag{4.2.37}$$

One then obtains for the average channel velocity v (after Keulegan 1938)
a) for smooth channels

$$\frac{v}{u_*} = 3.5 + 5.75 \log_{10}\left(\frac{R u_*}{v}\right) \tag{4.2.38}$$

b) for rough channels

$$\frac{v}{u_*} = 6.25 + 5.75 \log_{10}\left(\frac{R}{k_s}\right). \tag{4.2.39}$$

The expressions for smooth and rough channels can be taken together if one writes

$$\frac{v}{u_*} = 5.75 \log_{10}\left(12.27\frac{Rx}{k_s}\right) \equiv 5.75 \log_{10}\left(12.27\frac{R}{\Delta}\right), \tag{4.2.40}$$

where x is a correction factor shown in Fig. 52 and Δ is simply

$$\Delta = \frac{k_s}{x}. \tag{4.2.41}$$

As noted in Sect. 4.2.2, it can be shown that the Chézy formula (4.2.9) is an approximation to Eq. (4.2.40). In addition, it may be noted that the logarithmic laws quoted above have also been deduced from dimensional considerations (Wooding et al. 1973).

The equations of Keulegan have been found to describe circumstances in alluvial channels reasonably well, providing that a mathematical correction is

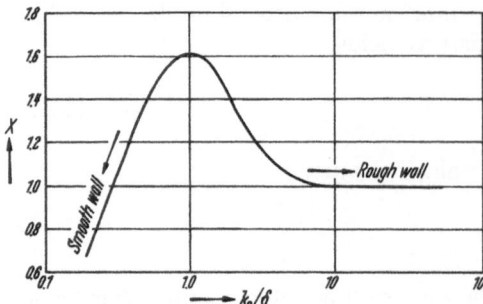

Fig. 52. Correction factor in open channel flow

made for sand bars and that certain channel conditions exist (Burkham and Dawdy 1976). Otherwise, discrepancies may arise (Taylor 1939).

The above theories refer to flow in fixed (immobile) channels. If there is an interaction with the bed, one is led to the theory of channel formation (Sect. 4.6 of this book).

4.2.4 Nonuniform Flow

4.2.4.1 General Remarks

The formulas deduced thus far refer to uniform flow is space and time. Thus, the cross-section of the flow has been assumed as constant along the channel and it has been assumed that there is no variation of the (mean) flow parameters at any one point. It is of interest to investigate the changes in the formulas that are necessary if the flow is nonuniform.

4.2.4.2 Spatially Nonuniform Flow

This problem has been treated by many authors, for instance by Ovsepyan (1955) and in a book by Rouse (1950). A physically most satisfactory deduction of the relevant formulas has been given by Liu (1958) based upon the Bernoulli equation with turbulent energy dissipation treated as a perturbation. Accordingly, we envisage the geometry of the flow as shown in Fig. 53. The theory can best be

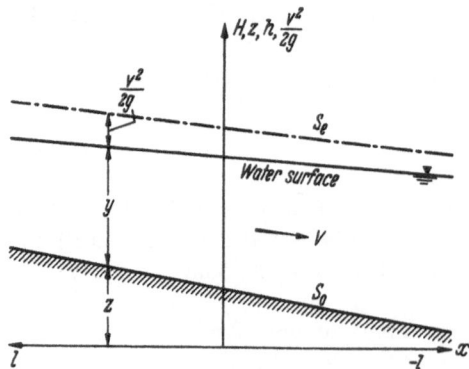

Fig. 53. Geometry of nonuniform flow. (After Liu 1958)

represented if we introduce the hydraulic head H at every section (located by giving x); the former is given by

$$H = \frac{v^2}{2g} + h + z, \qquad (4.2.42)$$

where v is the average velocity corresponding to the section under consideration, and h is the depth of the river for that section.

Differentiating (4.2.42) with regard to $l(= -x)$ yields

$$\frac{dH}{dl} = \frac{d}{dl}\left(\frac{v^2}{2g}\right) + \frac{dh}{dl} + \frac{dz}{dl} \qquad (4.2.43)$$

The slope of the energy line is denoted by S_e, the slope of the bottom by S_0. Thus

$$S_e = \frac{d}{dl}\left(\frac{v^2}{2g}\right) + \frac{dh}{dl} + S_0. \qquad (4.2.44)$$

Introducing x instead of l, we have

$$-S_e = \frac{d}{dx}\frac{v^2}{2g} + \frac{dh}{dx} - S_0. \qquad (4.2.45)$$

According to the Chézy relation (4.2.19), the velocity for a wide channel ($R \sim h$) and *constant slope* is a function of the depth:

$$v^2 = \text{const } h. \qquad (4.2.46)$$

Differentiating this with regard to x and substituting it into (4.2.45) yields

$$\frac{dh}{dx} = \frac{S_0 - S_e}{1 + K}, \qquad (4.2.47)$$

where K is some constant. This is a differential equation for the variation of the river depth.

4.2.4.3 Temporally Nonuniform Flow

The theory given above assumes that the flow in the open channel is (in the mean) temporally stable. However, it is possible that surges develop downstream with alternating high and low (mean) flow velocities.

The theory treating such phenomena has turned out to be very difficult. In fact, what is involved is the mathematical treatment of long waves in shallow water: attempts have been based on the method of characteristics. Stability criteria for open channel flows are then obtained on the basis of a Taylor series expansion of the flow equations in the neighbourhood of the first characteristic (Dracos and Glenne 1967).

The most extreme "temporally nonuniform flows" are the flash-sheet floods already mentioned in Sect. 3.4.4.3.

4.3 Three-Dimensional Flow in Open Channels

4.3.1 The Problem

Up to now, we have regarded a river as an essentially linear feature with the (mean) velocity vector at any one point in its cross-section directed strictly parallel to the (mean) direction of the river channel.

However, in actuality there are always velocity components at right angles to the (mean) river flow; these are particularly significant in river bends and in junctions.

4.3.2 Hydraulics in River Bends

4.3.2.1 General Remarks

Much information on water movement in curved channels may be found in the general textbooks on open channel hydraulics already mentioned in Sect. 4.2.1. In addition, Wittmann and Böss (1938) wrote a monograph on water and bed load movement in curved river reaches and Rosovskiy (1957) as well as Ananyan (1957) published books on water movement in a curved channel. References to specific investigations will be given in their proper context.

As a very crude approximation, the flow in curved channels can be described as a two-dimensional process. This leads to the discussion of "primary" currents in river bends (Sect. 4.3.2.2). This is generally done by treating the two-dimensional flow as potential flow, the (turbulent) energy dissipation being considered as a perturbation. Whereas the potential flow type of an approximation might lead to acceptable results in *straight* river reaches, this is not the case in *curved* reaches. In the latter case, helicoidal cross-currents ("secondary" currents) are of great importance. Some aspects and a few of the standard theories for the explanation of such secondary currents will be discussd in Sect. 4.3.2.3. However, it will also be shown that these "standard" explanations are really inadequate. Therefore, in Sect. 4.3.2.4, we shall present some attempts to explain the existence of the helocoidal cross-currents in terms of more basic principles.

Finally, we shall discuss (Sect. 4.3.2.5) some shock phenomena that may occur in very rapid flow around corners in curved channels.

4.3.2.2 Primary Currents in River Bends

As mentioned above, the flow in a river, in a first approximation, may be treated as two-dimensional, and, furthermore, the energy dissipation may be neglected. We have noted in the section on general principles of flow in open channels (Sect. 4.2.1) that frictionless laminar flow is characterized by the existence of stream lines. If the flow is also irrotational, then it can be represented by the introduction of a velocity potential Φ. It is well known that, if one restricts oneself to two dimensions, the stream lines can be represented as the equipotential curves of a stream function Ψ

which is connected with the velocity potential Φ by the Cauchy-Riemann differential equations. Both Φ as well as Ψ are subject to the Laplace equation:

$$\text{lap } \Psi = \text{lap } \Phi = 0. \tag{4.3.1}$$

Because of the Cauchy-Riemann differential equations, the absolute value of the velocity vector is equal to the absolute value of the gradient of Ψ as well as of Φ.

The velocity field derived from a potential is often called the field of "primary currents" in a river. These primary currents may be considered as a first approximation of river flow.

Let us calculate the velocity in a circular river. In this case, the stream lines must be circular, hence the stream function may be assumed to be a function of the distance r from the center of the circle only. The Laplace equation for Ψ yields in polar coordinates:

$$\frac{\partial^2 \Psi}{\partial r^2} + \frac{1}{r} \frac{\partial \Psi}{\partial r} = 0. \tag{4.3.2}$$

However, $\partial \Psi / \partial r = v$, where v denotes the absolute value of the velocity. Hence

$$\frac{dv}{dr} + \frac{1}{r} v = 0 \tag{4.3.3}$$

or

$$v = \frac{\text{const}}{r}. \tag{4.3.4}$$

Thus, the primary-current velocity is greatest on the *inside* of the curved river stretch, smallest on the outside. The stream lines are concentric circles, the lines of equal velocity potentials are radial lines.

The above simple pattern of stream lines is somewhat distorted if the river does not form a complete circle.

The decrease of velocity from the inner to the outer bank has the effect that a superelevation of water occurs at the outer bank. If one assumes that the energy is constant and equal to H along any one radius in a curve, then the water depth can be calculated by means of the Bernoulli equation (4.2.2). Thus

$$h = H - \frac{v^2}{2g}, \tag{4.3.5}$$

which yields with (4.3.4):

$$h = H - \text{const} \frac{1}{2gr^2}. \tag{4.3.6}$$

The last formula gives the water height (above a datum) as a function of r. One can also set up similar formulas for velocity distributions that are more complicated.

4.3.2.3 Elementary Theory of Secondary Currents in River Bends

It has been known for a long time (Thomson 1876) that potential flow does not describe the flow around river bends correctly. In addition to the primary currents, there are closed cross-currents which have been called secondary flows. A summary of the observations of such cross-currents has been given by Braden (1959); further observations have been reported by Bathurst et al. (1977), Thorne and Hey (1979), Nakagawa and Nezu (1984) and Falcon (1984). An inexpensive model to study such currents has been described by Tanner (1982).

There have been many attempts to give a rational theoryy of the secondary currents in river bends. Some general discussions of the phenomenon have been published by Prus-Chacinski (1958) and by Shapiro (1958). Our present concern will be only with helicoidal currents in *streaming* flow (cf. Sect. 4.2.1); in *shooting* flow additional complications occur since *cross-waves* will arise which are analogous to shock waves. This will be discussed later (Sect. 4.3.2.5).

We shall now give a review of the commonly quoted elementary theories of secondary currents in curved open channels.

One of the earliest theories of secondary currents has been proposed by Boussinesq (1877) who started from an analogy of flow in rivers with flow in pipes. Using some semi-empirical formulas for pipe flow, he postulated corresponding formulas for river flow. Thus, according to Boussinesq, the loss of head in a straight river (per unit length) is given by (v is the velocity, h the depth of the river)

$$\Delta H_1 = \beta \frac{v^2}{h}, \tag{4.3.7}$$

where β represents some coefficient. In a curved channel, Boussinesq (1877) postulates an additional loss (α is another coefficient, R the radius of the curve, and b is the width of the river):

$$\Delta H_2 = \frac{\alpha}{h} \sqrt{\frac{b}{R}} v^2 \tag{4.3.8}$$

and hence

$$\frac{\Delta H_2}{\Delta H_1} = \frac{\alpha}{\beta} \sqrt{\frac{a}{R}}. \tag{4.3.9}$$

The *total* loss in a curved river is then

$$\Delta H = \Delta H_1 + \Delta H_2 = \beta \frac{v^2}{h} + \frac{\alpha}{h} \sqrt{\frac{b}{R}} v^2 = \frac{v^2}{h} \left(\beta + a \sqrt{\frac{b}{R}} \right). \tag{4.3.10}$$

Boussinesq then reasons that in a river composed of straight and curved stretches, the head loss per unit distance must be the same. Hence

$$\frac{\beta}{h_{straight}} v^2 = \frac{v^2}{h_{curved}} \left(\beta + \alpha \sqrt{\frac{b}{R}} \right) \tag{4.3.11}$$

or

$$h_{curved} = h_{straight} \left(1 + \frac{\alpha}{\beta} \sqrt{\frac{b}{R}} \right). \tag{4.3.12}$$

This gives a correlation between the depth of water in straight and in curved stretches of the river. It is obvious from the structure of the formula that the depth increases if the radius of curvature decreases. Boussinesq concludes that the deeper the channel, the more effective must be the cross-currents.

The Boussinesq formulas are semi-empirical because they are based upon a rather tenuous analogy with empirical formulas found for curved pipes. It would be desirable to devise a more analytical theory describing secondary currents in a curved channel.

Such an analytical theory may be based on the hypothesis that the flow in a river may be described by a velocity potential (cf. Sect. 4.3.2). According to the potential flow theory, the flow is fastest at the inner (convex) side of the river (see Sect. 4.3.2.2). However, such a distribution of flow cannot exist if the drag force at the bottom of the river is considered. The latter will cause the velocity to decrease from the surface to the bottom in a vertical column of water. The individual fluid particles moving with their individual velocities v [cf. Eq. (4.3.4)] are forced around their circular paths with a radius of curvature r by a radial pressure drop of dp/dr (centrifugal force) across the channel which is given by the following equation:

$$dp/dr = \rho v^2/r, \tag{4.3.13}$$

where ρ is the density of the water. In a slow-moving river, the radial pressure gradient dp/dr will not change along a vertical line, and hence its value is too great to keep the slow-moving bottom fluid (which is retarded by the drag) on the same curvature as the fast-moving fluid particles at the surface. Hence, the fluid at the bottom of the river is forced inward onto paths with stronger curvature. This gives rise to secondary currents which, if superimposed upon the mean flow, will create a helicoidal flow pattern as shown in Fig. 54.

The above explanation of secondary flow in curved channels is the standard one found in the textbooks. It seems to be due to Einstein (1926). However, the basic assumption, viz. that river flow is fundamentally irrotational, so that it can be described by potential flow theory, is certainly not tenable. It is for this reason that the above explanation of the onset of secondary currents in river bends is scarcely acceptable.

4.3.2.4 Basic Theory of Helicoidal Flows

The discussion presented thus far leaves one with the task of finding an acceptable explanation for the existence of the cross-currents in river bends. It would be

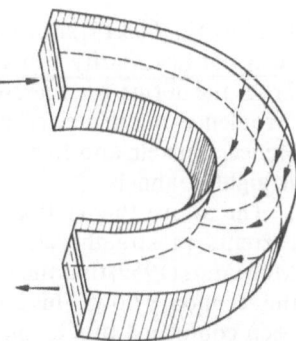

Fig. 54. Effect of secondary currents to produce helicoidal flow in a river bend

desirable to find such an explanation in terms of basic principles, viz. in terms of the Navier-Stokes equation and in terms of the statistical theory of turbulence. A clue to how to do this might be pesented by the fact that secondary currents have been observed not only in curved channels, but also in straight ones (see, e.g., Murota 1960). Therefore, a basic instability in turbulent flow might be involved (Parker et al. 1982). The best known such theory is due to Einstein and Li (1958).

Accordingly, the existence of secondary currents in straight channels can be explained if one starts from the formulation of the Navier-Stokes equations in terms of the vorticity (ξ, η, ζ)

$$\xi = \frac{\partial w}{\partial y} - \frac{\partial v}{\partial z} \text{ etc.} \tag{4.3.14}$$

with

$$\frac{\partial \xi}{\partial x} + \frac{\partial \eta}{\partial y} + \frac{\partial \zeta}{\partial z} = 0. \tag{4.3.15}$$

Here xyz are Cartesian co-ordinates, uvw are the velocity components. The Navier-Stokes equations then read as follows (cf. Lamb 1945, p. 578)

$$\frac{D\xi}{Dt} = \xi \frac{\partial u}{\partial x} + \eta \frac{\partial u}{\partial y} + \nu \text{ lap } \xi \text{ etc.}, \tag{4.3.16}$$

where D/Dt is the total time derivative moving with the fluid and ν is the kinematic viscosity. One now introduces the concept of mean and fluctuating velocity components

$$u = \bar{u} + u', \tag{4.3.17}$$

where

$$\overline{u'} = \overline{\frac{\partial u'}{\partial t}} = 0. \tag{4.3.18}$$

This is introduced into the Navier-Stokes equations, the average is taken, and one then ends up with

$$\frac{\overline{\partial \xi}}{\partial t} = \frac{\partial^2}{\partial y \partial z} (\overline{v'^2} - \overline{w'^2}) + \frac{\partial^2}{\partial z^2} (\overline{v' w'}) - \frac{\partial^2}{\partial y^2} (\overline{v' w'}). \tag{4.3.19}$$

The last Eq. (4.3.19) shows that in turbulent flow, the x-component of the vorticity need not necessarily vanish and therewith that secondary currents are possible. True, the terms on the right hand side of (4.3.19) add up to zero in *isotropic* turbulence, but in a river, particularly near its bed, the turbulence is not isotropic. Hence, Einstein and Li conclude that secondary currents must be expected even in straight channels.

The above theory does not have a bearing upon the *initiation* of secondary currents in straight channels. However, it has been shown by Delleur and McManus (1959) that the boundary shear in straight open channel flow can initiate the secondary flow. This is certainly also the case in curved flows, a view which has been confirmed by Gessner (1973) and by Bridge and Jarvis (1977).

A very interesting analysis of the origin of cross-currents based upon the basic equations of turbulent flow, has also been made by Ananyan (1957), who actually carried out calculations of the turbulent instability for flow in curved open channels. Ananyan started with a set of fundamental flow equations which are based upon Reynolds' (1894) equations for the turbulent stresses

$$\rho\left(\frac{\partial \bar{u}_i}{\partial t} + \bar{u}_j \frac{\partial \bar{u}_i}{\partial x_j}\right) = \rho F_i + \frac{\partial}{\partial x_j}(\tau_{ij} - \rho \overline{u_i' u_j'}). \tag{4.3.20}$$

Here, the x_i are the co-ordinates, the $\bar{u}_i$ are the average velocities, the u_i' are the velocity fluctuations, the τ_{ij} are the average hydrodynamic stresses,

$$\sigma_{ij} = \rho \overline{u_i' u_j'} \tag{4.3.21}$$

are the turbulent stresses, F_j is the external volume force and ρ is the fluid density. The viscosity term has been neglected because it is insignificant in relation to the turbulent stresses.

Ananyan then introduced a heuristic assumption for the turbulent streses, viz.

$$\sigma_{ij} = \varepsilon'(x_i)\left(\frac{\partial \bar{u}_i}{\partial x_j} + \frac{\partial \bar{u}_j}{\partial x_i}\right), \tag{4.3.22}$$

where $\varepsilon'(x_i)$ is the eddy viscosity (cf. Sect. 2.2.2.3). Following Prandtl (1926) the latter may be expressed as follows [see Eq. (2.2.11)]

$$\varepsilon'(z) = \rho l^2 \left|\frac{d\bar{u}_x}{dz}\right|, \tag{4.3.23}$$

where x is the horizontal, z the vertical co-ordinate and l is Prandtl's mixing length [cf. Eq. (4.2.27)]:

$$l = -k \frac{d\bar{u}_x/dz}{d^2\bar{u}_x/dz^2}. \tag{4.3.24}$$

If all the above equations are taken together, if the continuity equation

$$\text{div}\,\bar{u} = 0 \tag{4.3.25}$$

is added, and if the boundary conditions are specified, then one has a problem which is, in principle, determined. Ananyan wrote the equations in cylindrical coordinates, he specified the boundary conditions pertaining to a circular river bend, and showed that there are solutions which represent cross-circulations.

Ananyan's treatment of the cross-circulation problem in river bends is the most satisfactory one available todate. Difficulties still exist in the introduction of Prandtl's heuristic theory into the fundamental equations of Reynolds. Furthermore, in obtaining his final solutions, Ananyan had to make a series of approximations. It is not certain whether the solutions of Ananyan are unique and, furthermore, what the fundamental influence of the channel curvature is.

Because of the difficulty in describing secondary flows analytically, methods of dimensional analysis and experimental correlations have also been tried. A lengthy reivew of such investigations has been published by Kolar (1956). As usual with such investigations, these do not explain the physical facts but only make them more accessible to observation.

4.3.2.5 Shooting Flow Around Corners

Peculiar phenomena occur in shooting (supercritical) flow around river bends. In shooting flow, the wave propagation velocity u, with which disturbances travel, is smaller than the flow velocity v. The behavior of such a flow in a curved channel has, for instance, been analyzed by Knapp (1951).

Thus, let us consider the beginning of a channel bend containing shooting flow (see Fig. 55). If the flow is smooth in the straight section above the bend, the first disturbance due to the bend will start at the points A and A_1. At A, a piling up of water will be created, at A_1, a lowering of the level. The disturbance travels with the wave velocity u, and hence it cannot make itself felt above the point B (see Fig. 55) where

$$\sin \beta = \frac{ut}{vt} = \frac{u}{v} = \frac{\sqrt{gh}}{v}. \tag{4.3.26}$$

Here, the shallow water formula (6.2.28) has been used for the wave velocity; t denotes some arbitrary time and h the river depth. The disturbances will reach the opposite bank at the points M and M_1 which lie approximately on the radius 0C. For the angle Θ_0 determining the position of this radius, one has (see Fig. 55)

$$\Theta_0 = \arctan \frac{b}{\left(r + \dfrac{b}{2}\right)\tan \beta}, \tag{4.3.27}$$

where b denotes the width of the river. The disturbances which arrive at the banks at the radius with angle Θ_0 will again start further disturbances of the same nature, and hence a pattern of *cross-waves* will be the result which have a wavelength λ of

$$\lambda = 2r\Theta_0. \tag{4.3.28}$$

The disturbances being reflected from the banks of the channel have a scouring effect.

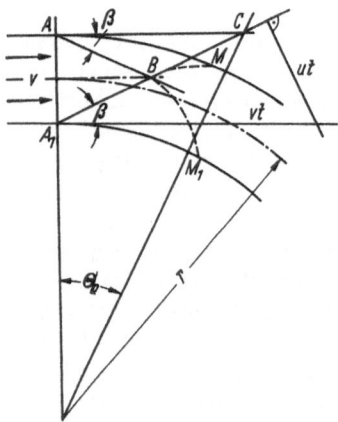

Fig. 55. Geometry of shooting flow at the beginning of a river bend

It is obvious that the propagation phenomena of disturbances in shooting flow are analogous to the corresponding phenomena in gas dynamics. Particular problems arise when $v = u$ ("critical" flow) where shock fronts, analogous to the "sonic boom" in gases, can arise. The geomorphological effects of shooting flow thus consist in a high stressing of the banks.

4.3.3 Hydraulics of Junctions

4.3.3.1 General Remarks

The consideration of three-dimensional flow patterns is important not only in river bends, but also in the vicinity of junctions. The latter include confluences as well as bifurcations.

4.3.3.2 Hydraulics of Confluences

Special conditions apply at the confluence of two rivers. A review of recent work on this subject has been published by Bomer (1988); accordingly, there are, in fact, not many meaningful studies in existence.

Physically meaningful models of the hydraulics of river junctions must be based upon the principles of fluid mechanics. The pertinent equations are, however, extremely complicated (Chow 1959 see p. 512 therein). Thus Ertel (1965) made an early attempt at treating the problem by linearizing the equations of continuity and motion, which is justified if it is assumed that the junction causes only small perturbations in the flow regimes in the three branches involved.

A more general approach has been tried by Mosley (1976), who based his analysis of the problem on the equations of continuity, energy, and momentum, the latter being integrated forms of the equations of motion. The momentum equation can be used for the descrption of bulk properties of the flow and can thus give an overall view of junction hydraulics. (Mosley 1976). For the confluences of two tributaries 1 and 2 the momentum equation has the form (bold face types denote vectors)

$$\sum F = \Delta M = M_3 - (M_1 + M_2) = \rho[Q_3 V_3 - (Q_1 V_1 + Q_2 V_2)] \qquad (4.3.29)$$

where $\sum F$ is the summation of the forces acting on a specific volume of fluid in the junction, ΔM is the momentum change from the upstream to the downstream boundary of the volume considered, ρ is the density of water, Q is the discharge and V the (mean) velocity in the channel; the indices refer to the river links in question.

A similar equation can be written down for the energy change ΔE from the upstream domain to the down-stream domain:

$$\Delta E = \rho g[Q_1(d_1 + V_1^2/2g) + Q_2(d_2 + V_2^2/2g)$$
$$- (Q_1 + Q_2)(d_3 + V_3^2/2g)], \qquad (4.3.30)$$

where g is the gravity acceleration and d is mean channel depth.

Inasmuch as V, M, F etc. are vectors, the confluence angles enter into the above equations as independent variables, and the two equations permit calculation of the (mean) hydraulic conditions in the vicinity of the junction of fixed channels. In fact, the confluence angle is not an independent variable in natural channels with movable sides, but questions of this sort belong to the Section 4.6 on the interaction of river flow with its confines.

4.3.3.3 Hydraulics of Bifurcations

Bifurcations are the opposite to confluences. They are of less importance than the latter. True bifurcations between drainage systems are very rare, but braidings occur on alluvial fans and in flood plains. Artificial bifurcations have been constructed in connection with channeling of water in hydroelectric systems.

For the latter case, Ertel (1963) has studied a model where a "bifurcation" in form of a connecting channel exists between two rivers with different discharge rates. He has shown that unstable oscillations are induced in the connecting channel.

A more pertinent study of bifurcation hydraulics has been reported by Sobey and Drazin (1986), based on solutions of the Navier-Stokes equations. These authors demonstrate that stable flow exists at low Reynolds numbers, which becomes unstable at larger Reynolds numbers.

Again, these studies refer to fixed, given channels. The mutual interaction of the river flow with its confines will be treated later in this book.

4.4 Forces of Fluids on Particles

4.4.1 General Remarks

In many instances, the surface of the Earth consists of loose particles. This is particularly the case in alluvial areas, on beaches and in river beds. The morphology of such surfaces is conditioned by the interaction of water with the individual particles.

Therefore, our next task is to study the forces that act on individual particles immersed in a fluid. First, we shall study the gravity force with its attendant effect on the settling velocity of particles in water. Then we shall turn to the scouring effect of flowing water on river bed particles, and finally we shall study the modes by which flowing water can actually lift particle from a river bed to bring them into suspension.

4.4.2 Gravity Force: Settling Velocity

4.4.2.1 Principles

If we consider a particle suspended in a liquid, we note that amongst all the forces acting upon it, the best known is the gravity force. It is due to the gravity acceleration and is equal to the underwater weight W of the particle given by (making use of Archimedes' principle):

$$W = Vg(\delta - \rho). \tag{4.4.1}$$

Here, V is the volume of the particle, δ is the density of the particle, ρ the density of the fluid, and g is the gravity acceleration.

Owing to the gravity force, a submerged particle has the tendency to settle towards the bottom, regardless of whether the liquid be at rest or whether it be flowing. During the settling process, the resistance that the particle has to overcome may be caused either by viscous drag or by turbulence, depending on the relative velocity of fluid and particle. We shall investigate these possibilities in turn.

4.4.2.2 Viscous Drag

In case the resistance is due to viscous forces, the settling velocity can easily be calculated from Stokes' law. As is well known, the resistance F_R of a sphere of radius a to flowing liquid is given by

$$F_R = 6\pi a \eta v, \tag{4.4.2}$$

where η is the viscosity of the fluid and v the relative velocity between fluid and sphere. Equation (4.4.2) is modified if there is a boundary wall in the vicinity (Lee and Leal 1980). Setting (4.4.2) equal to the underwater weight W (4.4.2) of the sphere

$$W = \tfrac{4}{3}\pi a^3 g(\delta - \rho)$$

yields the following settling velocity of the particle

$$v = \frac{2a^2(\delta - \rho)g}{9 \quad \eta}. \tag{4.4.3}$$

If the particles are no spherical, one has to introduce a shape factor κ, and one then has

$$F_R = 6\pi\kappa\eta v. \tag{4.4.4}$$

The length κ can be calculated theoretically for various simple shapes. In the case of a circular disc of radius c moving broadside on, one has

$$\kappa = 0.85\,c \tag{4.4.5}$$

and in the case of a disc moving edgewise (cf. Lamb 1945, p. 605)

$$\kappa = 0.556\,c. \tag{4.4.6}$$

Other shapes have been considered by Komar (1980: cylinders) and by John and Goyal (1975: flocculation particles).

The viscous drag formulas are good for particles settling in water on the Earth of a radius up to about 0.114 mm. For larger particles, the formulas given in Sects 4.4.2.3 and 4.4.2.4 have to be used.

The gravity and inertial forces acting upon a spherical particle in a moving fluid can be combined and an equation of motion can be deduced. This has been done by Tchen (1947), who started with a discussion of the problem of slow motion of a spherical particle under the influence of gravity in a fluid at rest. The latter problem had been studied previously by Basset (1888), Boussinesq (1903), and Oseen (1927). Oseen arrived at the following equation of motion in the direction of the axis Oy (+y being directed vertically upward):

$$\frac{4\pi a^3}{3}\delta\dot{v} = -\frac{2\pi a^3}{3}\rho\dot{v}^2 - 6\pi\eta a\left\{v + \frac{a}{\sqrt{\pi v}}\int_{t_0}^{t}dt_1\frac{\dot{v}(t_1)}{\sqrt{t-t_1}}\right\} - \frac{4\pi a^3}{3}(\delta-\rho)g. \quad (4.4.7)$$

In this equation, $v(t)$ is the velocity of the particle, ρ, δ are density of fluid and particle, respectively, a is the radius of the (spherical) particle, g is the gravity acceleration and η, v are viscosity and kinematic viscosity, respectively. In deriving Oseen's equation, it had been assumed that the particle as well as the fluid had been at rest until $t = t_0$. It will usually be expedient to asume $t_0 = -\infty$.

Tchen now rewrote the Oseen equation in terms of a particle whose velocity is $(v - u)$ and endowed the entire system (particle plus fluid) with a uniform rectilinear velocity $u(t)$. The result of performing this operation is

$$\frac{4\pi a^3}{3}\delta\dot{v} = \frac{4\pi a^3}{3}\delta\dot{u} - \frac{2\pi a^3}{3}\rho(\dot{v}-\dot{u}) - 6\pi\eta a(v-u)$$

$$- 6\pi\eta\frac{a^2}{\sqrt{\pi v}}\int_{-\infty}^{t}\frac{\dot{v}(t_1)-\dot{u}(t_1)}{\sqrt{t-t_1}}dt_1 - \frac{4\pi a^3}{3}g(\delta-\rho). \quad (4.4.8)$$

This is Tchen's (1947) equation.

4.4.2.3 Turbulent Drag

We now direct our attention to the settling velocities of particles where the resistance is due to turbulent energy dissipation. Turbulence occurs if the relative velocity between fluid and particle exceeds a critical value given by

$$Re = \frac{2\rho v a}{\eta} \gtrless Re_{crit}, \quad (4.4.9)$$

where Re is commonly called Reynolds number. As usual, ρ denotes the density of the fluid, v the relative velocity, a the radius of the particle, and η the viscosity of the fluid.

Turbulent settling has mostly been studied by meterologists, since it is the mode by which raindrops fall from the sky.

The settling velocity in purely turbulent flow can be calculated in the same fashion as that for laminar flow by setting the underwater weight of the particle (cf. 4.4.1) equal to the well-known expression for turbulent drag: (cf. 2.2.7)

$$F_R = C_D \pi a^2 \frac{\rho v^2}{2} \quad (4.4.10)$$

[cf. also Eq. (4.4.25)], where C_D is some drag coefficient (see Leliavsky, 1955, p. 36) equal to about 0.79 for spherical particles. We then obtain

$$C_D \pi a^2 \frac{\rho v^2}{2} = \frac{4}{3} \pi a^3 g(\delta - \rho),$$

or for the settling velocity:

$$v = \sqrt{\frac{8}{3} \frac{ga(\delta - \rho)}{\rho C_D}}. \tag{4.4.11}$$

The nonlinearity of the drag force causes an assembly of settling particles to undergo diffusion (Murray 1970).

4.4.2.4 Intermediate Flow Regime

Shifrin (1958) has given a general formula for the settling rate of a sphere in a fluid, which encompasses both the turbulent and the viscous settling formulas into one expression.

Shifrin assumes that, for low Reynolds numbers, the factor C_D is a function of the Reynolds number

$$C_D = f(Re). \tag{4.4.12}$$

Then, setting the drag equal to the underwater weight, one obtains the following relation under steady-state conditions (Shifrin 1951)

$$\alpha a^3 = \frac{1}{24} Re^2 f(Re) \tag{4.4.13}$$

with

$$\alpha = \frac{4}{9} \frac{\rho(\delta - \rho)g}{\eta^2}, \tag{4.4.14}$$

where $(\delta - \rho)$ is the difference between the density of the sphere and that of the fluid. We then set

$$F(Re) \equiv \frac{1}{24} Re^2 f(Re). \tag{4.4.15}$$

The determination of the settling velocity (or of the corresponding Reynolds number) is then accomplished by a solution of the equation

$$F(Re) - \alpha a^3 = 0. \tag{4.4.16}$$

Denoting the inverse function of F by Φ

$$\Phi \equiv F^{-1}$$

we can write the solution of (4.4.16) as follows

$$Re = \Phi(\alpha a^3). \tag{4.4.17}$$

Thus, we have for the settling velocity

$$v = \frac{\eta}{2\rho a} \Phi(\alpha a^3). \tag{4.4.18}$$

This is Shifrin's general formula for the settling velocity which is valid for the turbulent and laminar flow regimes alike. This can be verified as follows. The function f(Re) in Eq. (4.4.12) has been discussed by Goldstein (1938), who gave a series development for small values of Re. Using Goldstein's series leads to the following expression of Φ for small Re:

$$\Phi(F) = F\left\{1 - \frac{3}{16}F + \frac{109}{1280}F^2 - \frac{1031}{20480}F^3 + \cdots\right\} \tag{4.4.19}$$

and hence one obtains

$$v = \frac{\eta}{2\rho a}\alpha a^3 = \frac{2}{9}\frac{(\delta - \rho)g}{\eta}a^2, \tag{4.4.20}$$

which is indeed Stokes' formula (4.2.10). For large Re, C_D is constant; thus

$$F(Re) = \frac{C_D}{24}Re^2 \tag{4.4.21}$$

$$\Phi = \sqrt{24F/C_D}. \tag{4.4.22}$$

Hence

$$v = \frac{\eta}{2\rho a}\sqrt{\frac{24\alpha a^3}{C_D}}$$

$$= \sqrt{\frac{8(\delta - \rho)ga}{3\rho C_D}}, \tag{4.4.23}$$

which is indeed the formula found earlier (4.4.11). Thus, Shifrin's general formula for the settling velocity of a spherical particle indeed yields the correct limiting values for high and low Reynolds numbers and hence it stands to reason that it correctly describes the intermediate flow regime, too.

The above investigations refer to the settling velocity of a single particle. If there are many particles present, it may be expected that an interaction effect occurs. The interaction between several particles following each other whilst they are settling has received some attention in the literature (cf. Happel and Pfeiffer 1960; Happel and Brenner 1965). Generally, it is found that two particles will fall faster than one. Experiments seem to bear out this prediction.

4.4.2.5 Empirical Relations

The above exposition shows that the mechanics of settling of smooth spherical particles is well understood, both in the viscous and in the turbulent flow regime. This, however, is not the case for irregularly shaped grains. Attempts at introducing various types of shape factors have not been entirely successful.

In any case, there are no theoretical explanations of the effect of grain shape on settling rates. Such studies as have been reported are basically empirical.

A corresponding investigation has been made by Komar and Reimers (1978), using scale model experiments with ellipsoidal pebbles settling in glycerine. These authors obtained general relationships of the dependence of the drag coefficient C_D

[Eq. (4.4.10)] on the Reynolds number (cf. 4.2.25) and on a shape factor S_C defined as follows (Corey 1949):

$$S_C = D_s/(D_i D_m)^{1/2}, \tag{4.4.24}$$

where D_m is the major, D_i the intermediate, and D_s the minor diameter of the pebble.

Similar studies were made by Nasser (1978) and by Dietrich (1982). The latter author produced an empirical equation that accounts (in terms of four dimensionless parameters) for the effects of size, density, and roundness of the particles on the settling mechanism. Introducing a further shape parameter (the Powers 1953, roundness index), Goossens (1987) produced a set of master curves giving the dependence of C_D on the Corey shape factor, the Powers index, and the Reynolds number.

4.4.3 Scouring Force

4.4.3.1 General Remarks

Another type of force of flowing fluids whose existence is intuitively well known is the scouring force. It acts on the bottom of a stream bed.

The general aspects of the problem have been described, for instance, by Christofoletti (1976a). The scour may act on particles at the bottom which are already loose, tending to entrain them. Alternatively, scour can also act on a rocky substratum attacking the latter.

We shall discuss these possibilities in their turn.

4.4.3.2 Grain Scour

The simplest model of the scouring force acting on grains can be obtained by regarding each particle as sitting alone at the bottom of an otherwise smooth container with the fluid streaming by it. It is well known that such a particle offers a resistance to the fluid that varies with the square of the velocity (provided the Reynolds number is high enough), as this is implicit in the momentum transfer theory (cf. Sect. 2.2.2.2). By the principle of action and reaction this is also the force experienced by the particle. Thus, the scouring force F_s acting upon the particle can be expressed as follows

$$F_s = C_D \frac{\pi}{4} d^2 \frac{1}{2} \rho_F v^2. \tag{4.4.25}$$

In this equation, d is the diameter of the particle, ρ_F is the density of the fluid, v is the fluid velocity at the particle level, and C_D is a drag coefficient, depending on the size of the particle and on the Reynolds number. Expression (4.4.25) is the well-known expression for turbulent drag [cf. Eq. (4.4.10)]. As noted in Sect. 4.4.2.3, in the case of the particle being spherical, the drag coefficient is given by $C_D = 0.79$ at high Reynolds numbers.

The scouring force will be able to *move* a (spherical) particle if it is large enough to overcome the frictional resistance F_f

$$F_f = \varepsilon(\rho_s - \rho_F)g\tfrac{1}{6}\pi d^3, \qquad (4.2.26)$$

where ε is the coefficient of friction, ρ_s the density of the particle, g the gravity acceleration and the other symbols have the previously define meaning.

Equating the two forces and solving for v yields an expression for the critical velocity v_{cr} of the stream which will just be able to move bottom particles. It is:

$$v_{cr}^2 = \frac{4g}{3}\frac{\varepsilon(\rho_s - \rho_F)}{C_D\rho_F}d. \qquad (4.4.27)$$

The square of the critical velocity thus turns out to be proportional to a linear dimension of the particle. However, the latter is roughly proportional to the cube root of its weight W, and hence we can write

$$v_{cr} \sim W^{1/6}, \qquad (4.4.28)$$

This is correlation (*Brahms' equation*) which had been found from empirical data over 200 years ago.

Somewhat more sophisticated models of the forces acting on single grains (e.g., lying on a *rough* rather than on a smooth bed) have been discussed by Nakagawa and Tsujimoto (1981).

The above theory is obviously somewhat oversimplified. More generally, in newer investigations, it is therefore assumed that it is a critical value for the bottom tractive force σ_m (also called drag force) introduced in Sect. 4.2.2 which starts the bottom particles moving. This critical drag may be related to the mechanics in the turbulent and laminar boundary layer in rough channels. Microscopically, the phenomena taking place are not entirely clear; it may be that grain entrainment results from the interaction between fluid elements within an eddy and the sediment grains (cf. Sutherland 1967). Phenomenologically, Shields (1936) postulated a general relation for the critical drag force $\sigma_m = \sigma_{crit}$

$$\frac{\sigma_{crit}}{\rho_s - \rho_w} = f\left(\frac{u_* d}{\nu}\right), \qquad (4.4.29)$$

where ρ_s is the sediment density ρ_w the water density, u_* the shear velocity [cf. Eq. (4.2.26)], d the grain diameter ν the kinematic viscosity of the water and f denotes a universal function.

The above theoretical models are obviously very primitive. Unfortunately, it has been very difficult to account mathematically for the processes involving many grains on a bed; thus, one has turned to experimental investigations. The best known of these is due to Hjulström (1935), who established empirically critical drag curves separating the flow regions in the fluid where entrainment, deposition, and steady transportation occur. (see Fig. 56).

Other types of experiments have been directed at establishing the pertinent values of σ_{crit} [cf. Eq. (4.4.29)] for various types of sedimentary particles. These measurements, originating from various laboratories, have been collected by Leliavsky (1955) and the result is shown in Fig. 57. The figure bears out a definite correlation (the circles are the values measured by various laboratories). From

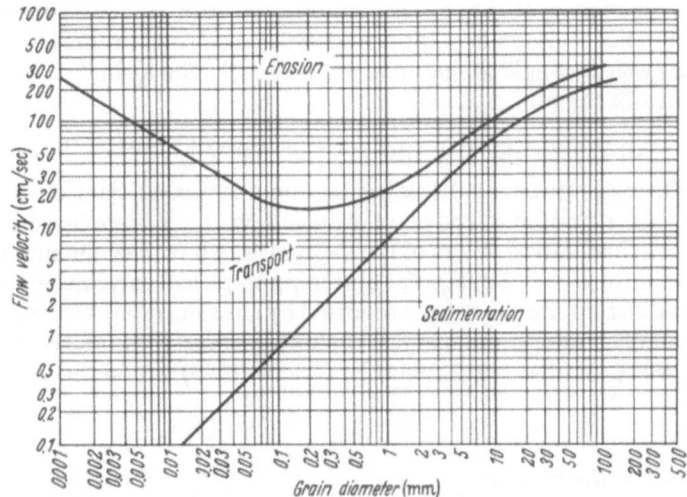

Fig. 56. Hjulstrom's (1935) critical drag curves, separating the flow regions where entrainment, deposition, and steady carriage occurs

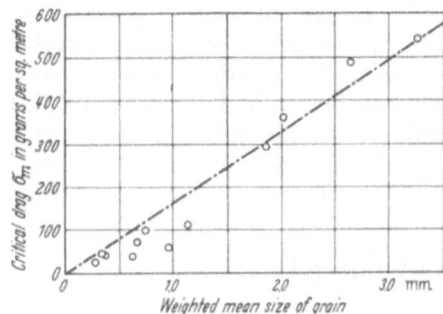

Fig. 57. Chart showing critical drag (ordinate) in g/m² as function of grain diameter (abscissa) in mm compared with experimental results (circles) from various laboratories. (After Leliavsky 1955)

Leliavsky's graph, one takes the following formula for the critical drag force

$$\sigma_{\text{crit}} = \text{const} \cdot d, \qquad (4.4.30)$$

where d is the particle diameter. A variety of formulas of this type has been suggested in the literature, most of which, if dimensionally correct, can be reduced to the form (being a simplified version of 4.4.29; cf. Carter 1950)

$$\frac{\sigma_{\text{crit}}}{g(\rho_s - \rho_F)d} = A, \qquad (4.4.31)$$

where again σ_s and ρ_F are the densities of sediment and fluid, respectively, g is the gravity acceleration, d the particle diameter and A is a dimensionless parameter. The latter may be taken as a function of d/D, where D is the thickness of the laminar sublayer.

Sundborg (1956b) has generalized the above models somewhat by introducing modifications to allow for *cohesion* between sand grains to exist and also to allow for a *slope* (angle α) that may not be small. The formula corresponding to (4.4.30) then becomes

$$\sigma_{crit} = \text{const}_1[gd(\rho_s - \rho_F)(\varepsilon \cos \alpha - \sin \alpha) + \text{const}_2 \cdot C], \qquad (4.4.32)$$

where C represents the cohesion and ε is a coefficient of friction.

As noted, the above relations have to be confirmed and completed by experimentation. Most of the experiments have been made on beds consisting of uniform grains. Attempts at obtaining empirical correlations for mixtures of grains of different sizes have been reported by Gessler (1965). Other empirical relations for establishing entrainment thresholds have been proposed by Bohlen (1976), Carling (1983), Andrews (1983, 1984), Brayshaw (1985), Mersi and Graf (1985), and Rakoczi (1987a).

4.4.3.3 Rock Scour

If the particles on the stream bed are not only subject to cohesion, but are actually cemented together, the scouring process becomes one of physical rock reduction. Some of the possibilities for the latter to occur have already been discussed in Sect. 3.23, notably that referring to cavitation (Sect. 3.2.3.4).

In the present context, however, it is worthy of note that the forces on the individual cemented grains can also be considered: one has to take the flow in the interstices between the grains into account. A rather involved theory of these effects has been set up by Rehbinder (1977a, b). The matter is important in connection with the technology of cutting slots into rocks by means of water jets. In geomorphology, instances of this type could occur in rock beds beneath waterfalls, as this is also the case with cavitation.

4.4.4 Lifting Force

The forces acting on particles contained in a flowing fluid discussed so far are not able to explain the lifting and transportation of sediment. The scouring force might at best provide for an explanation of the inception of motion of bottom sediment, but for an explanation of the well-known transportation of sedimentary particles in suspension, a *lifting* force is required.

It has long been recognized that *turbulence* must play a fundamental role for this lifting force. In this instance, turbulence, in sediment transportation, plays a similar role as in the dynamics of flight. A semitheoretical discussion of this fact has been given, for instance, by Leighly (1934). General discussions of the statistical theory of turbulence in connection with the lifting force have been presented by Herczynski and Pienkowska (1980) and by Sumer and Müller (1983). An analytical theory has been provided by Friedlander (1957).

Accordingly, each particle is assumed to be contained in an *eddy* of fluid, large compared with the former, but still small compared with the dimensions of the container of the fluid. Let the velocity of the eddy be u_F and its acceleration a_F. Correspondingly, the velocity and acceleration of the particle are u_P and a_P.

The force balance equation, neglecting gravity, for the particle can now be written down. If the Reynolds number

$$Re = d_P \sqrt{\overline{u_R^2}} \rho_F / \eta \tag{4.4.33}$$

is less than 1 (where d_P is the particle diameter, furthermore

$$u_R = u_F - u_P, \tag{4.4.34}$$

ρ_F is the fluid density and η the fluid viscosity), then the resistance F_f is given by Stokes' law which may be written as follows, to allow for the various shapes of the particles [cf. (4.4.4)]

$$F_f = -f(u_P - u_F). \tag{4.4.35}$$

In addition to this resistance, the particle will be subject to a force F_b because of the pressure gradient in its vicinity. If the Reynolds number is high, then the Stokesian force can be neglected and one has (by Newton's law of motion; p is pressure):

$$\frac{\partial p}{\partial x} = \rho_F a_F. \tag{4.4.36}$$

The force F_b is then given by (m_F being the mass of the displaced fluid)

$$F_b = m_F a_F \tag{4.4.37}$$

so that the force balance equation yields (m_P being the mass of the particle)

$$m_P a_P = m_F a_F - f(u_P - u_F) \tag{4.4.38}$$

or

$$\frac{du_P}{dt} + \beta u_P = \beta u_F + \gamma \frac{du_F}{dt}, \tag{4.4.39}$$

where β is some coefficient (depending on f and the particle inertia) and

$$\gamma = \rho_F / \rho_P. \tag{4.4.40}$$

Friedlander now obtains the rate of eddy diffusion of the suspended particles by expressing their mean square displacement $\overline{x_P^2}$ as a function of time. The distance traveled by a particle which at $t = 0$ is at the position $x = 0$ is obtained by integrating equation (4.4.39):

$$x_P + \frac{u_P - u_{PO}}{\beta} = \int_0^t u_F(T) \, dT + \frac{\gamma(u_F - u_{FO})}{\beta}. \tag{4.4.41}$$

Squaring and averaging this and introducing a correlation function

$$R(\Theta) = \frac{\overline{u(t)u(t + \Theta)}}{\overline{u^2}} \tag{4.4.42}$$

yields asymptotically for large t

$$\overline{x_P^2} = 2\overline{u_F^2} t \int_0^{t \to \infty} R(\Theta) \, d\Theta. \tag{4.4.43}$$

This is an expression characteristic of *diffusion*.

There is some question as to what the correlation function should be. Tchen (1947) took it as equal to the correlation function of the fluid and interpreted the last equation as meaning that fluid and particles diffuse at the same rate. However, Friedlander points out that the correlation functions which should be used range between two extremes: the Lagrangian function for small particles able to follow the fluid, and the Eulerian function for particles which remain almost fixed.

In a more sophisticated fashion, some authors have actually also accounted for the rotational (and not only the vertical) motion of the fluid and of the grains (Müller et al. 1971; Hinch and Leal 1976; Tooby et al. 1977; Willets and Murray 1981).

In any case, it always turns out that the existence of diffusion in turbulent flow provides a means for keeping particles in suspension against gravity.

4.5 Sediment Transportation

4.5.1 General Remarks

Sediments in a river are mixtures of particles of various kinds, sizes, and shapes. Mechanically, the "kind" of particle expresses itself in the grain density ρ_s, the size by its largest diameter d and the shape by some measure of ellipticity. Measurements of these properties consist in sieve analyses leading to weight fractions etc. (cf. e.g., Griffith 1961; Adams 1977).

The particles making up the sediment mixture in a river are often subject to Rosin's law, expressible by the equation (Rosin and Rammler 1934)

$$R = 100\,e^{-bx^n}, \tag{4.5.1}$$

which can be directly obtained from probability considerations (Krumbein and Tisdel 1940). In Eq. (4.5.1) R is the weight percent retained on a sieve of mesh x, and b and n are constants for any given material. For the study of sedimentary *structures*, the orientation of grains relative to each other and in space is of importance, but in the present context this is not of much significance.

There are two fundamentally different modes by which sediment can be transported in a river. These modes are called *suspended* sediment transportation and *bottom* sediment transportation. In the former, the particles in transit are in suspension within the river, whereas in the latter they are dragged along at the bottom. The transitional stage between these two modes of sediment transportation has been called *saltation*; in its particles perform jumping motions.

The characteristics of sediment transportation have been described in various places in the literature. Much information is contained in general textbooks on river bed processes (Hjulström 1935; Meinzer 1942; Leliavsky 1955; Jarocki 1957; Velikanov 1955, 1958; Allen 1970; Statham 1977; Yalin 1977; Graf 1984; Middleton and Southard 1984). In addition, the problem has been reviewed by Benedict (1957), Mitchell (1957), Lehmann (1975) Christofoletti (1977) and Garde and Ranga Raju (1977). References to the original investigations, which are basic to the various theories that will be presented below, will be given in their proper context.

In order to characterize sediment transport, it is customary to introduce a Reynolds number Re

$$Re = \frac{u_* d}{v},$$ (4.5.2)

where u_* is the shear velocity [cf. (4.2.26)] in the river, d the particle diameter and v the kinematic viscosity of the water. It then is thought that definite values of Re separate the various modes of sediment transportation from each other.

It turns out that the Reynolds number is only a rather inaccurate criterion for determining what type of sediment transportation can be expected in a river whose velocity and bed composition are known. A more elaborate attempt to arrive at a criterion, based on similarity considerations, has been undertaken by Abduraupov (1958). However, it would appear that criteria regarding the type of sediment transportation based on similarity considerations could never be very satisfactory. A direct solution of this problem should be sought by actually studying the various modes of sediment transportation; the fraction and type of material that will be subject to each mode will then automatically follow.

Unfortunately, a complete analysis of the general problem of sediment transportation has not yet been achieved. We shall, in the next sections, describe the theories that have been proposed for an explanation of suspended and bottom sediment transportation, respectively, and then return to the general questions regarding total sediment load transportation.

4.5.2 Suspended Sediment Transportation

4.5.2.1 Principles

We shall first look at the mechanism of suspended sediment transportation. As outlined earlier, it must be held that it is *turbulence* which keeps the particles in suspension. However, there are two theories with regard to the manner in which this turbulence might act. The first theory assumes that there is an effect acting on the particles which is analogous to diffusion (eddy diffusion), and that the vertical concentration of the particles is therefore subject to a diffusivity equation. This leads to the *diffusivity theory* of suspended sediment transportation. The second theory is based directly upon the dynamics of the supporting stream. This leads to the *gravitational theory* of suspended sediment transportation. Finally, it is of interest to compare the theoretical investigations with phenomenological and experimental observations.

Much of the general information on suspended sediment transportation is contained in the textbooks and reviews on river transport processes mentioned in Sect. 4.5.1. In addition, an entire symposium was devoted specifically to suspended sediment transport; the proceedings provide a review of the state of the art (Bechteler 1986).

4.5.2.2 Diffusivity Theory

Let us discuss first the diffusivity theory. As noted above, the mass exchange in turbulent flow may be regarded as analogous to diffusion (cf. Sect. 4.4.4). We therefore start with a suitable diffusivity equation for the vertical particle density which might be written as follows:

$$\frac{\partial n}{\partial t} = \frac{\partial}{\partial z}\left(\varphi \frac{\partial n}{\partial z}\right). \tag{4.5.3}$$

Here, n is the particle number per unit volume, φ is a suitable diffusivity factor (sometimes called sediment transfer coefficient), t is time and z is the vertical co-ordinate. Every particle, if left alone, will tend to settle downward with the settling velocity w; hence the above diffusivity equation should apply only in a system which is moving downward with the velocity w (assuming that all particles have the same settling velocity). Hence, in a rest system (co-ordinates y, t), the diffusivity equation becomes

$$\frac{\partial n}{\partial t} = \frac{\partial}{\partial y}\left(\varphi \frac{\partial n}{\partial y}\right) + w \frac{\partial n}{\partial y}, \tag{4.5.4}$$

where

$$y = z - wt. \tag{4.5.5}$$

Eq. (4.5.5) is the well-known diffusivity equation with a mass transport term.

Seeking a steady-state solution one obtains (O'Brien 1933)

$$\varphi \frac{\partial n}{\partial y} + nw = 0. \tag{4.5.6}$$

An inspection of the last equation shows that it can be interpreted as a continuity equation: in a steady state, the rate of particle settling (nw) must be exactly balanced by some quantity which must be equal to the rate of particles being transferred upward by the turbulence. Hence, φ must be the turbulent transfer coefficient for the particles.

One could have arrived at the above relationship also directly by writing down the balance equation for the particles (Vanoni 1941)

$$-\varphi \frac{\partial n}{\partial y} = nw, \tag{4.5.7}$$

where φ is *defined* as the transfer coefficient for the particles. The recourse to the diffusivity equation (4.5.4) is then not necessary.

Integrating Eqs. (4.5.6) and (4.5.7) yields (n_0 = particle number at $y = y_0$)

$$\frac{n}{n_0} = e^{-w \int_{y_0}^{y} dy/\varphi}. \tag{4.5.8}$$

Now, the problem is to find an expression for φ. This problem is solved by making the basic assumption that the transfer coefficient for the particles and the transfer coefficient for the momentum are identical. Thus, using (4.5.8) one can write

$$\sigma = \varphi \rho \frac{d\bar{u}}{dy}, \tag{4.5.9}$$

where $\bar{u}(y)$ is the time-averaged velocity at the height y, σ is the turbulent shear stress and ρ is the density of the fluid. It may then be assumed that the stress σ is a linear function of the depth concerned

$$\sigma = \sigma_m \left(1 - \frac{y}{h} \right), \tag{4.5.10}$$

where σ_m signifies the maximum stress which is reached at the bottom (y = 0) and h the total depth of the river. Hence, one obtains

$$\varphi = \frac{\sigma_m}{\rho} \frac{1 - y/h}{d\bar{u}/dy}. \tag{4.5.11}$$

The further discussion of the problem depends on he assumption of the vertical velocity distribution in the stream. Using Karman's (1930) logarithmic law [cf. Eq. (4.2.39)] i.e.,

$$\frac{u_{max} - \bar{u}}{\sqrt{\sigma_m/\rho}} = \frac{1}{k} \text{lognat} \frac{h}{y} \tag{4.5.12}$$

yields

$$\varphi = \sqrt{\frac{\sigma_m}{\rho}} \left(1 - \frac{y}{h} \right) ky$$

$$\frac{n}{n_0} = \left[\frac{h - y}{y} \frac{y_0}{h - y_0} \right]^K \tag{4.5.13}$$

with

$$K = \frac{w}{k} \sqrt{\frac{\rho}{\sigma_m}}. \tag{4.5.14}$$

Here, k is as usual Karman's "universal" constant of turbulence [cf. Eq. (4.2.27)] which is approximately equal to 0.4, and h is the total depth of the stream (Rouse 1937). The above equation applies to a suspended load where the particles are characterized by the settling velocity w. In a mixture of particles, one has to sort out the individual fractions.

The above procedure employs, so to speak, the diffusivity equation on a "macroscopic" basis. Following the stochastic motion of the individual particles, Conover and Matalas (1967) also arrived at a similar relation. In a corresponding fashion, Alonso (1981) and Gani and Todorovic (1987) based theories on random walk models.

From a practical standpoint, it is extremely important to proceed from the above formulas (which explain the distribution of sediment with height in a stream) to a quantitative expression of the carrying capacity of streams with regard to given sediment sizes. An attempt in this direction was made by Einstein (1950), simply by integrating the last equation over the vertical, assuming that in the *horizontal* direction, the speed of the particles is identical to that of the surrounding fluid, corresponding to Eq. (4.5.12). He obtained:

$$q_s = \int_y^h n_y \bar{u} dy$$

$$= \int_y^h n_0 \left[\frac{h-y}{y} \frac{y_0}{h-y_0} \right]^\kappa \left[\sqrt{\frac{\sigma_m}{\rho}} \frac{1}{k} \text{lognat} \frac{y}{h} + u_m \right] dy, \qquad (4.5.15)$$

where q_s is the sought-after sediment flux. This integral cannot be expressed in closed form, it can only be evaluated numerically. For this purpose, tables supplied by Einstein (1950) must be applied. Furthermore, the integral cannot be valid down to the bottom of the bed but must be broken off at the "laminar sublayer" since it becomes infinite.

Thus, an inspection of the above formula immediately shows up the difficulties that are inherent in the method. First of all, the integral is extremely sensitive to the lower limit of integration. Second, the concentration of sediment at the level y_0 enters as a parameter into the formulas. It is therefore impossible to calculate, say, the total amount of sediment suspended in a particular stream: one must always have at least one value that is measured. On such a basis, Willis (1979) made predictions of the suspended-load discharges for flumes.

Finally, it may be remarked that the above spatially essentially one-dimensional (in the y-direction) diffusion theory was extended by Kerssens et al. (1979) to two and by Chiu and Chen (1969) to three dimensions. The diffusion can also be regarded as the result of random walks of the individual particles (Gani and Todorovic 1987).

4.5.2.3 Gravitational Theory

The "diffusivity" theory of suspended sediment transportation is based essentially on the hypothesis that an exchange coefficient exists for the solid particles which is proportional to the exchange coefficient of water in the river. This is a somewhat indirect approach to the sediment transportation problem. It might be considered as preferable if it were possible to write down the equations of motion of the particles and take directly into account the forces exerted by the turbulent river. This type of approach has been called *gravitational theory*; it seems to have been initiated by Velikanov (1944, 1948), whose original approach gave rise to much controversy. A similar theory applicable to small sediment particles and small accelerations in the river was later proposed by Barenblatt (1953). It is also controversial (Velikanov (1954; Barenblatt 1955; Romanenko 1956).

The difficulties in these theories arise from the fact that somehow the equations of motion for the river must be averaged, since the local fluctuating velocity is of very little use in applications. Velikanov (1944) uses a doubtful energy relationship to achieve this, Barenblatt (1953) follows some relationships of Kolmogorov (1942), which are somewhat more reliable, but still open to objections. Velikanov's (1951) theory leads to an exchange equation although its deduction is quite novel. Barenblatt ends up with a formula relating the turbulent energy in the river with the suspended sediment content.

On a more macroscopic basis, Bagnold (1966) has calculated momentum flux balances for *swarms* of particles in shear turbulence. By this means, he could obtain limits for the *total* amount of sediment that can be carried in suspension.

The theories referred to above turn out to be mathematically quite involved. In view of this and in view of the fact that the various controversies have not yet been finally settled, the reader is referred to the original papers for further details.

There is no doubt that a "gravitational" type of theory of suspended sediment transportation will ultimately provide the most satisfactory description of the suspended sediment transport phenomenon. However, the structure of the turbulence in the river must be known accurately to carry this approach to a satisfactory stage. This has not yet been achieved. Thus, the "diffusivity theory" of suspended sediment transportation is still the most useful one to date.

4.5.2.4 Phenomenological and Experimental Studies

Finally, there is some interest in making comparisons between the various theories of suspended sediment transportation with phenomenological and experimental observations.

Thus, in attempts at checking the result [Eq. (4.5.13)] obtained from the diffusivity theory experimentally, Vanoni (1941) found that, in absolute terms, the agreement is not very good. Coleman (1970) also found that the sediment transfer coefficient [cf. Eq. (4.5.6)] is not constant as supposed in the theory, but reaches a maximum at a level equivalent to about 1/5 to 1/3 of the distance from the bed to the water surface. Nevertheless, the general shape of the curves found experimentally agrees reasonably well with that postulated by theory. In fact, Barfield et al. (1969) made predictions of sediment profiles in open channels on the basis of the diffusion theory.

However, it should be noted that there is a considerable spatial variability in the particle-size characteristics in suspended sediments; in addition, rivers also exhibit patterns of temporal variation (Ciet and Tazioli 1976; Walling and Moorehead 1987). Thus it is not only the concentration of the particles which is of importance, but also their size distribution expressing itself in the "mean fall velocity"; the latter cannot be assumed as constant as was implied in the various theories.

4.5.3 The Transportation of Bottom Sediment

4.5.3.1 General Remarks

We investigate now the mechanisms that might be responsible for transporting bottom sediments. Good summaries of this subject have been given, for instance, by Chien (1954) and by Shulits and Hill (1968). There are essentially three types of theories of bottom sediment transportation available which might be called *drag* theories, *statistical* theories, and *semi-empirical physical* theories. In addition, it is of interest to review the experimental and phenomenological evidence. We shall discuss these matters in turn below.

4.5.3.2 Drag Theories

It is to be expected, that the scouring force discussed in Sect. 4.4.3 would cause movement of bottom sediment. Theories of bottom sediment transportation based upon this scouring force are called *drag theories*.

Accordingly, the forces acting upon a bottom particle are:

1. The *gravity force* as given in Eq. (4.5.15), acting vertically downward

$$W = (M_s - M_f)g, \tag{4.5.16}$$

   where M_s denotes the mass of the particle under consideration and M_f the mass of the displaced fluid.

2. The *scouring force* F_s from Eq. (4.4.25):

$$F_s = C_D \frac{\pi}{4} d^2 \frac{1}{2} \rho (v_f - v_s)^2, \tag{4.5.17}$$

   where v_f denotes the fluid velocity and v_s the velocity of the sediment particle. The scouring force acts parallel to the stream bed.

3. The *bottom friction force* due to the friction of the particle at the bottom, acting parallel to the stream bed. This can be represented most easily in terms of a frictional coefficient ε (cf. Sect. 4.4.3)

$$F_f = -\varepsilon W \cos \alpha, \tag{4.5.18}$$

   where α denotes the angle of the stream bed with the horizontal.

According to the drag theory, these are *all* the forces that act upon the sedimentary particles, provided forces due to the acceleration of the flowing water are neglected.

The equation of motion for the particle can be obtained in the usual fashion by equating all the forces acting in the direction of the stream bed to the product of mass and acceleration of the particle:

$$M_s \frac{dv_s}{dt} = C_D \frac{\pi}{8} d^2 \rho (v_f - v_s)^2 - (M_s - M_f)g \sin \alpha - (M_s - M_f)\varepsilon g \cos \alpha. \tag{4.5.19}$$

For the steady-state case one postulates $dv/dt = 0$, and hence one has

$$v_s - v_f = \sqrt{\frac{8g}{C_D \pi d^2 \rho}(M_s - M_f)(\sin \alpha - \varepsilon \cos \alpha)}. \tag{4.5.20}$$

This is the expression for the velocity of a bottom sediment particle, provided, of course, the drag is large enough to overcome the frictional force. The above equation has been deduced by Eagleson et al. (1957) for use in an investigation of bottom sediment on a beach due to shoaling waves, where it has yielded reasonable results.

The principles of the drag theory have been known for a long time. Early advocates of it were Du Boys (1879), Chang (1939), and O'Brien and Rindlaub (1934). The best known of these early formulations is that of Du Boys (1879) based upon a model of sliding layers of bed material being kept in motion by the drag exerted by the fluid so that the shearing force parallel to the bed decreases with distance downward. Thus:

$$q_B = c_B \sigma(\sigma - \sigma_{cr}),$$ (4.5.21)

where q_B is the transport rate (in weight per unit width and time), σ is the drag force, σ_{cr} the critical drag force (cf. Sect. 4.4.3), and c_B is some parameter. The Du Boys formula has been generalized by O'Brien and Rindlaub (1934)

$$q_B = c'_B(\sigma - \sigma_{cr})^m,$$ (4.5.22)

where c'_B and m are two new constants. This has been done by assuming that in a state of dynamic equilibrium, the shearing force must be constant on planes parallel to the bottom.

In a mixture of particles of various sizes, there are different critical drag stresses for each fraction, leading to a different mobility for each particle size. On this basis, Egiazaroff (1965) calculated gradation curves for channel sediments under various discharge rates.

In cohesive soils, additional terms have to be introduced into the drag theory (Sundborg 1956a) which express the cohesion (cf. also Sect. 4.4.3.2).

4.5.3.3 Stochastic Theories

The various versions of the drag theories of bottom sediment transportation have not proven too satisfactory. Different approaches have therefore been sought.

One of these is based on statistical considerations, treating the movements of particles as random walks. It was originally developed by Einstein and Polya (1937) and later summarized and expanded by Einstein (1950, 1968). Modifications were made by Chien (1954), Cheong and Shen (1976), and Bardsley (1981). Reviews were given by Hung and Shen (1971) and Han and He (1980a).

Accordingly, every particle in a river bed has a finite probability to be lifted off the bed (due to turbulent velocity fluctuations) and to be carried a certain distance.

The rate at which particles of a given size are eroded from the stream bed is proportional to the number of such particles present and to the probability of each particle being eroded. If i_b denote the fraction of (resting) bed sediment of the size range considered (characterized by the "diameter" d), then the number of particles per unit area is $i_b/(A_1 d^2)$ where A_1 is a corrective coefficient so that $A_1 d^2$ is the exposed cross-section. If we denote by p the probability of erosion of a particle, and by t_1 the time necessary to replace a bed particle by a similar one (*exchange time*), then the number of particles eroded per unit time and area is

$$N_E = i_b p/(A_1 d^2 t_1).$$ (4.5.23)

The exchange time may be assumed to be proportional to the time necessary for a particle to fall in the fluid through a distance equal to its own diameter; thus (with w = settling velocity; cf. 4.4.11)

$$t_1 = A'_3 \frac{d}{w} = A_3 \sqrt{\frac{d\rho_f}{g(\rho_s - \rho_f)}},$$ (4:5.24)

(where ρ_f is the density of the fluid and ρ_s that of the particle) so that the number N_E

referred to above becomes

$$N_E = \frac{i_b p}{A_1 d^2 A_3} \sqrt{\frac{g(\rho_s - \rho_f)}{d\rho_f}}. \tag{4.5.25}$$

If there be dynamic equilibrium in the stream, then the number of particles eroded per unit time (N_E) must be equal to the number deposited (N_d). The latter number can be calculated as follows.

The rate at which particles of the given size moe through the unit width of the stream is $i_B q_B$, where q_B equals the rate at which the bed moves per unit width and i_B is the fraction of particles of the given size in the *moving* bed. All the particles of the given size (diameter d) perform "jumps" of length $A_L d$. When these particles pass through a particular cross-section of the stream, it is not known where in its "jump" each of the particles is. It must be assumed, therefore, that a particular particle may be deposited anywhere in an area of length $A_L d$ and unit width downstream from the cross-section under consideration. If q_B is measured in dry weight per unit time and width, and if $A_2 d^3$ is the volume of each particle, then

$$N_d = \frac{q_B i_B}{A_L d A_2 d^3 \rho_s g}. \tag{4.5.26}$$

Equating N_E to N_d yields the *bed load equation*:

$$\frac{i_B q_B}{\rho_s A_2 A_L g d^4} = \frac{i_b p}{A_3 A_1 d^2} \sqrt{\frac{g(\rho_s - \rho_f)}{d\rho_f}}. \tag{4.5.27}$$

There must obviously be a connection between the distances travelled by the particles $(A_L d)$ and the probability of erosion p. As long as p is small, A_L is a general constant λ, roughly equal to 100. If p is large, however, then deposition cannot occur on that part of the bed (p) where the lifting force exceeds the particle weight. Thus, only $(1 - p)$ particles (of the unit available) are deposited after traveling a distance λd, p particles are not deposited. Of these $(1 - p)$ are deposited after traveling $2\lambda d$ so that p^2 are not yet deposited, etc. Hence

$$A_L d = \sum_{n=0}^{\infty} (1 - p)p^n(n + 1)\lambda d = \frac{\lambda d}{1 - p}. \tag{4.5.28}$$

Thus, the bed load equation may be written as follows:

$$\frac{p}{1 - p} = A_* \Phi_*. \tag{4.5.29}$$

with

$$A_* = A_1 A_3 / (\lambda A_2) \tag{4.5.30}$$

and

$$\Phi_* = \frac{i_B}{i_b} \left\{ \frac{q_B}{\rho_s g} \sqrt{\frac{\rho_f}{\rho_s - \rho_f}} \sqrt{\frac{1}{gd^3}} \right\} \equiv \frac{i_B}{i_b} \Phi. \tag{4.5.31}$$

Thus, the bed load equation relates the bed velocity (this is essentially what Φ or $\Phi*$ is) to the probability of erosion p.

It remains to express the probability of erosion p in terms of the river flow velocity in order to obtain a direct relationship between flow and bed load transportation. This can be done by taking the dynamic lift L of the water into consideration which acts upon each particle. The probability p, then, is equal to the probability that this lift is greater than the underwater weight W' of the particle. This underwater weight is, as usual:

$$W' = g(\rho_s - \rho_f)A_2 d^3,$$
(4.5.32)

whilst for the lift L we can write, in conformity with the general principles of turbulent action,

$$L = c_L \rho_f \tfrac{1}{2} u^2 A_1 d^2.$$
(4.5.33)

The factor c_L has been found from experiments to be equal to 0.178, and the velocity u occurring in Eq. (4.5.33) must be taken as that velocity which obtains at a level 0.35 X from the "theoretical" bed surface, where

$$X = 0.77\Delta \quad \text{if } \Delta/\delta > 1.80$$
$$X = 1.39\delta \quad \text{if } \Delta/\delta < 1.80.$$
(4.5.34)

For the meaning of the Greek symbols in the above equations, the reader is referred to Sect. 4.2.3.

Equation (4.5.33) represents the average lift. The instantaneous lift is subject to random fluctuations which can be expressed as follows:

$$L_{inst.} = L_{aver.}(1 + \eta),$$
(4.5.35)

where η is some random function. The last equation can be further evaluated by introducing the value for u from the various theories presented earlier, in terms of the (energy) slope S and the hydraulic radius R.

Einstein assumes that η follows a Gaussian distribution where the standard deviation η_0 is a universal constant. Then the probability p is equal to the probability of W'/L to be smaller than unity. What one ends up with is thus a relationship between

$$\Psi = \frac{\rho_s - \rho_f}{\rho_f} \frac{d}{RS}$$
(4.5.36)

(note that Ψ is essentially $1/v^2$ with v equal to the mean river velocity) and Φ_* (or Φ) in which some constants enter. It is

$$1 - \frac{1}{\sqrt{\pi}} \int_{-B_*\Psi - 1/\eta_0}^{B_*\Psi - 1/\eta_0} e^{-t^2} dt = \frac{A_* \Phi}{1 + A_* \Phi}.$$
(4.5.37)

This relationship, in essence, represents a relationship between bedload movement and flow intensity. Einstein's plot of it is shown in Fig. 58. The constant are (as corrected by Chien 1954)

$$A_* = 43.5$$
$$B_* = 1/7$$
$$\eta_0 = 1/2.$$
(4.5.38)

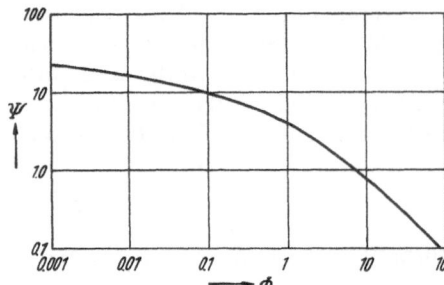

Fig. 58. Einstein's (1950) relationship between Ψ and Φ (note Ψ is essentially $1/v^2$ with v mean river velocity; Φ is bed load velocity)

When using Einstein's plot, it should be noted that there are various parameters in it. First of all there is R, the hydraulic radius of the river. Second, Einstein gives various correction factors to account for the effect of the sandbars, the effect of small particles hiding behind large ones, etc., which may be incorporated in Φ, Ψ. Third, the universal constants A_1, A_2, η_0 are not very well defined. Lee and Jobson (1975) proposed some experimental means for gaining the required knowledge of the basic probability distributions. Thus, although it is in principle possible to calculate the bed load transport of a river from its characteristics, it is in practice very difficult to arrive at the correct numerical values.

4.5.3.4 Physical Semi-Empirical Theories

The third group of bedload transport theories which have gained acceptance are based on fundamental physical principles supplemented by semi-empirical relations.

Some considerations of this type are contained in the work on drag thresholds by Shields (1936) mentioned earlier. A well-known attempt is due to Meyer-Peter and coworkers (Meyer-Peter et al. 1934; Meyer-Peter and Müller 1948). Using Froude's law of similarity as a guide, it may be expected that the quantity Q

$$Q = q_s^{2/3} S/d \tag{4.5.39}$$

should be independent of the scale used. Here, S is the slope, d is the diameter of the particles, and q_s is the specific discharge quantity (mass of water flowing per unit width of bed and per unit time). Meyer-Peter and coworkers plotted the quantity Q against G

$$G = g_s^{2/3}/d, \tag{4.5.40}$$

where g_s is the dry mass of the bedload transported per unit width of bed and per second. The resulting points from various sources all fell near a straight line given by

$$Sq_s^{2/3}/d = a + bg_s^{2/3}/d. \tag{4.5.41}$$

In the SI system, the values of the constants are

$$a = 17 \, kg^{2/3} \, m^{5/3} \, s^{2/3}$$

$$b = 0.4. \tag{4.5.42}$$

Meyer-Peter and Müller (1948) have generalized the above equation from further empirical tests to include the effect of sediment density and also to account for channel roughness. The formula then reads

$$\rho_f R \left(\frac{k_s}{k_r}\right)^{3/2} S = A(\rho_s - \rho_f)d + B\left(\frac{\rho_f}{g}\right)^{1/3}\left(\frac{\rho_s - \rho_f}{\rho_s}g_s\right)^{2/3}. \qquad (4.5.43)$$

Here, all the symbols have the previously defined meaning; in addition, R is the hydraulic radius of the channel, ρ_s, ρ_f are the densities of sediments and fluid, respectively, and g is the gravity acceleration. A and B are constants, and the ratio k_s/k_r is indicative of the roughness of the channel (the ratio decreases with increasing roughness). The modified Meyer-Peter formula (4.5.43) is dimensionally homogeneous so that A and B are dimensionless constants. Their values are (Meyer-Peter and Müller 1948)

A = 0.047

B = 0.25. $\qquad (4.5.44)$

It has been shown by Chien (1954) that the empirical (modified) Meyer-Peter formula, in fact, gives results which are in excellent agreement with the Einstein bed load theory.

Since the Meyer-Peter formula represents, in essence, an embodiment of experimental data, its existence and close agreement with the bed load formula must be considered as a substantiation of Einstein's (1950) theory as far as laboratory investigations are concerned.

A further theory of bed load transport, based on fundamental physical principles, was proposed by Bagnold (1966, 1977), who noted that the only source of energy to move the bed is the stream power Ω defined as follows

$$\Omega = \rho g Q S, \qquad (4.5.45)$$

where Q is the whole discharge of the stream, S the gravity slope, ρ the water density and g the gravity acceleration. The power ω per unit width is then

$$\omega = \frac{\Omega}{\text{flow width}} = \frac{\rho g Q S}{\text{flow width}} = \rho g d S \bar{u}, \qquad (4.5.46)$$

where d is the mean flow depth and $\bar{u}$ the mean (water) flow velocity.

The stream power must in some fashion be connected with the sediment transport rate. With regard to the bottom sediment, we denote the mean transport velocity by $\bar{U}_b$. The corresponding transport rate i_b is

$$i_b = w_b \bar{U}_b, \qquad (4.5.47)$$

where w_b is the underwater weight of the bed load over a unit bed area. The bed load work rate ω_b (power dissipation) is then

$$\omega_b = i_b \tan\alpha, \qquad (4.5.48)$$

where $\tan\alpha$ plays the role of a friction factor ("dynamic bed load friction factor"). The rate of power dissipation in the bedload ω_b is always smaller than the stream power available; thus one can try to set

$$\omega_b = e_b \omega \qquad (4.5.49)$$

or

$$i_b = \omega e_b/\tan\alpha, \tag{4.5.50}$$

where $e_b < 1$ is the "bed load efficiency". Incidentally, a corresponding relationship could also be written down for suspended sediment transport; the *sum* of the powers dissipated by the two transport mechanisms must then always be smaller than ω. Here, however, we are only concerned with the bed load.

The further elucidation of the "coefficients" $\tan\alpha$ and e_b must evidently be done empirically. In fact, the proportionality relation (4.5.50) may be too simple. Bagnold (1977) writes as a possibility

$$i_b \sim (\omega - \omega_0)[(\omega - \omega_0)/\omega_0]^{1/2}, \tag{4.5.51}$$

where ω_0 is a further parameter, a threshold power. There is evidently a large indefiniteness in the basic Bagnold scheme. Discussions of the various versions of the Bagnold theory have been given by Engelund and Fredsoe (1976), Ledder (1979), Bridge (1981), and Hardisty (1983).

Finally, additional attempts along the lines discussed in the present chapter have been published by Schoklitsch (1950), Tanner (1977, leading to estimators), Azia and Prasad (1985, considering stability factors) and Samaga et al. (1986, introducing dimensionless similarity considerations). As with all the attempts under the present heading, there is a large amount of indeterminacy if an attempt is made to establish specific relations.

4.5.3.5 Experimental Investigations

Experimental investigations of bed load transport have been made with several aims in mind.

Thus, studies have been undertaken to observe and ascertain the reality of some of the basic assumptions underlying the various theories. For instance, the existence of entrainment thresholds has been established by Brayshaw (1985), by Ikeda and Iseya (1986) and by Rakoczi (1987b). Next, the actual occurrence of the supposed "saltation" of the grains has been verified by strobe photography (Murphy and Hooshiari 1982), as well as the statistical dispersion of the grains in nature (Beltaos 1982).

Tests of the various theories of bed load transportation have been reported on several occasions. As noted earlier, agreement between laboratory experiments and theoretical predictions is generally reasonably good, mostly because the various constants are based on laboratory experiments. The correlation between these laboratory experiments, however, is a different matter. This problem has been analyzed, e.g., by Yalin (1959), Karolyi (1958), Amin and Murphy (1981), Nakagawa et al. (1982), Fujita and Muramoto (1982a), Smart (1984), and Bathurst et al. (1986).

However, it should be noted that even the basic premises of the usual bed load theories have been questioned: the transport may basically take place by shifting of bottom ripples and dunes (Norgaard 1968; Shteynman 1969; Griffiths and Sutherland 1977; Whiting et al. 1988). In this instance, bedload transport may still be considered as a steady-state process. However, there is also evidence that such transport may also be temporally unstable (Soni 1981), resulting in wavelike oscillatory movement (Hallermeier 1982; Meade 1984; Iseya and Ikeda 1987) or outright surges (Heggen 1986).

In view of these results, it must be held that the present understanding of bed load transport in rivers is still only qualitative, with the numerical (predictor) values being somewhat in doubt.

4.5.4 Total Sediment Transport

4.5.4.1 General Remarks

In order to base the total carrying capacity of a river on fundamental physical principles, it stands to reason that all one would have to do is to combine the relations for suspended load with those for bed load (Shen and Hung 1983); wherein perhaps a further term would have to be added for the load carried in solution (Gunn 1982): the latter may, in fact, be more than half the total load (Corbel 1959).

However, an inspection of the theories of suspended and bottom sediment transportation makes it at once clear that, based upon these theories, it is not possible to predict the amount of sediment transported in a given river with known flow and bed composition. An attempt to do this in the case of Money Creek (Ill. U.S.A.) by Stall et al. (1958) yielded most inconclusive results.

Thus, the effort has been directed at obtaining relationships between total sediment load and river characteristics empirically from observations in nature and in the laboratory. Of interest are also the interrelationships (process-response) between the various parameters.

The literature is large. A review has been given by White et al. (1975) and an extensive bibliography has been published by Brown (1979).

4.5.2.4 Observations in Nature

First of all, it is possible to obtain empirical relationships (usually through some usual regression analysis) between sediment load and river characteristics from observations in nature.

One of the best-known studies of this type has been undertaken by Laursen (1958), who investigated the significance of various parameters and finally plotted a quantity A as follows

$$A = \frac{\bar{c}}{\left(\dfrac{d}{h}\right)^{7/6}\left(\dfrac{\sigma_0'}{\sigma_c} - 1\right)} \tag{4.5.52}$$

against another, B, which is

$$B = \frac{\sqrt{\sigma_m/\rho}}{w}; \tag{4.5.53}$$

in the above formulas, $\bar{c}$ is connected with the sediment concentration in the

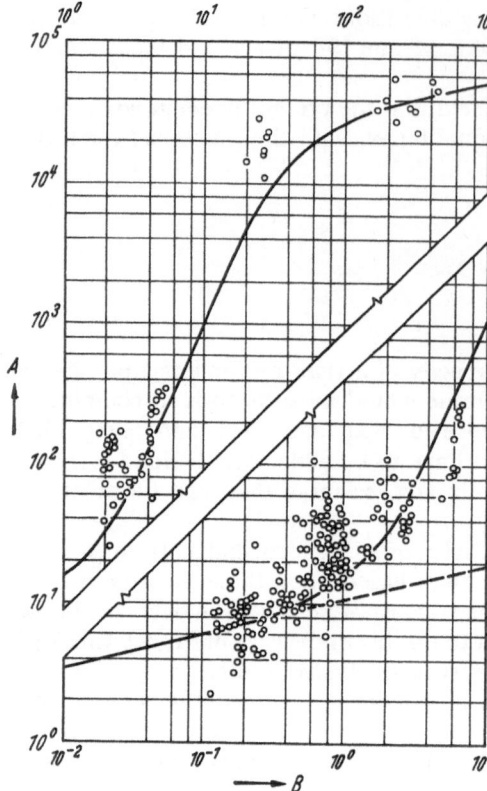

Fig. 59. Correlation between sediment transportation (dotted line: bottom load; solid line: total load) and river characteristics. The measured points are indicated by dots. (After Laursen 1958)

following manner:

$$\bar{c} = 265 \frac{q_s}{q}, \qquad (4.5.54)$$

where q_s is the volume rate of sediment transport, and q is the volume rate of flow. Furthermore, d is the particle diameter, h is the depth of the river, σ_c is the critical drag [cf. Eq. (4.4.30)], σ_m is the boundary shear [cf. Eq. (4.2.17)], ρ is the density of the water, w is the setting velocity of the particles (cf. Sect. 4.4.2), and

$$\sigma_0' = \frac{v^2 d^{1/3}}{30 h^{1/3}}, \qquad (4.5.55)$$

where v is the mean flow velocity of the river. In terms of A and B, Laursen obtained a universal curve which is shown in Fig. 59. In this figure, the dashed line represents the bed load, the solid line the total load. The relationship, according to Laursen, is valid for each size fraction of the bed; in order to obtain the total sediment load composition, an appropriate summation over each fraction has to be performed.

The discussion by Laursen gives only the total load as a function of discharge; however, the discharge changes with time. It is thus necessary to obtain a frequency

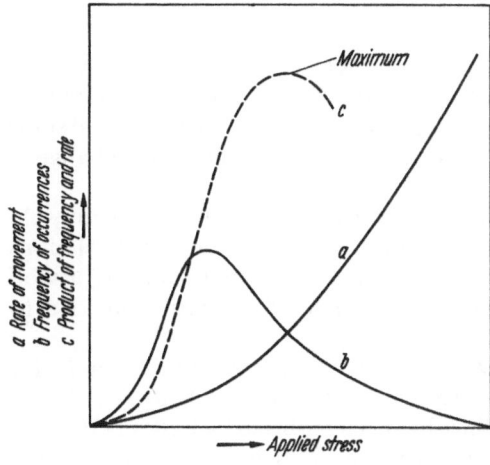

Fig. 60. Relation between transport rate, frequency of occurrence, and effectiveness of an erosional process. (After Wolman and Miller 1960)

curve for discharge values that may be observed in a drainage basin, as it is apparent that the frequency of occurrence of an erosive process is just as important as its magnitude. This has been pointed out, for instance, by Wolman and Miller (1960). The situation is illustrated in Fig. 60. The total of the erosive processes in an area can be determined if one combines the frequency distribution of each discharge rate with the corresponding carrying capacity.

Other empirical relationships from observations in nature have been proposed by Egiazaroff (1967), Flaxman (1972), Williams and Berndt (1972), Christofoletti (1976b), and Griffiths (1979). Such natural relationships were subsequently used to predict the sediment movement behavior in specific types of rivers (Laursen 1958; Ashida et al. 1976, 1981; Wolman 1987). The influence of very peculiar conditions, such as the presence of lichens, has been studied by Gregory (1976b).

4.5.4.3 Laboratory Experiments

Attempts at deducing general relationships between sediment discharge and river parameters have also been made by the use of laboratory experiments.

For this purpose, the appropriate scaling relations have to be observed. Dynamic similarlity can be obtaining by specifying four modeling parameters (Southard et al. 1980): exact Reynolds-Froude modeling of loose sediment transport is then valid and workable. Parker and Anderson (1977) have shown that any pair of resistance and load relations uniquely specify depth discharge and sediment transport rating curves at given slopes and bed particle Reynolds numbers. Yang (1979) has transformed the Bagnold semi-empirical procedure into dimensionless form.

On this basis, flume experiments have been reported by Stein (1965), by Whittaker and Davies (1982), and in a symposium that was held on the subject in Toronto (McColl 1986).

4.5.4.4 Process-Response in Sediment Transport

Inasmuch as any set of landscape phenomena represent, at any one time, a dynamic (quasi-) equilibrium (cf. Sect. 1.5.2.1), a process changing any one of the parameters will entail a response in the other parameters (process-response idea). Thus, if some process changes the river characteristics, the system will respond with a change in sediment transport characteristics.

It is therefore possible to base calculations of the response of the sediment yield to changing flow conditions on the various theoretical and empirical relations presented above. Thus, Sharma and Dickson (1979) related monthly runoff sediment yields to rainfall, and Ribberink (1987) related the sediment load to morphological changes in rivers.

The most efficient procedure is, however, the construction of unit graphs for sediment yield, as this is common practice otherwise for river hydrographs (cf. Scheidegger 1975). A study along these lines has been reported by Rendon-Herrero (1978). The same superposition principle (Sherman 1932) as is valid for creeks is assumed to the valid for sediment discharge. Thus, the following convolution integral

$$S(t) = \sum_0^t Q(\tau) u(t - \tau) \, d\tau \tag{4.5.56}$$

permits one to calculate the (excess over normal) sediment yield S(t) at a point in the river from the (excess over normal) water discharge $Q(\tau)$ at times τ in the river at the point under consideration. The term $u(t - \tau)$ denotes an influence function (corresponding to the unit hydrograph) which must be determined empirically for a given watershed.

4.6 Mutual Interaction of Bed, Flow, and Sediment Transport

4.6.1 General Remarks

Thus far we have discussed how the shape of the river bed affects the flow velocity, and how the flow velocity affects the sediment discharge. However, in reality all these effects react mutually with each other and it is not really permissible to treat them in the sequence indicated above.

Therefore, we shall discuss in the present section (4.6) how the turbulent river acts upon the bed (where it causes bottom ripples) and then how the bed acts upon the sediment transportation pattern. It will turn out, in essence, that every one of these phenomena depends on all the others and that there are various possibilities for the establishment of stationary states. The difficulties encountered in defining these states forces one to introduce the semi-empirical concept of a "graded river", which designates a river which is in dynamic equilibrium, i.e., which neither silts up nor scours. The conditions for a river to be graded will be given.

Subsequently, we shall present the theories that have been advanced for an explanation of the observed river profiles, both longitudinal and transverse.

Finally, we shall give a review of the scaling laws that have been proposed for making model experiments of river bed processes.

General discussions of the subject have been given by Parker (1977), Richards (1977), and Richards and Ferguson (1987).

4.6.2 Longitudinal Instabilities

4.6.2.1 General Description

We noted in Sect. 1.4.2 that the longitudinal course of a river is generally marked by an unstable sequence of riffles and pools. When the riffles are large, they appear as transverse bars or dunes.

In the riffles and pools, the bed load particles appear sorted as to size in some fashion (Ikeda and Iseya 1980: Hirsch and Abrahams 1981; Lisle 1979).

Thus, the first question which we shall analyze in connection with the interaction between the river bed and the flowing water is the origin of the ripples and bars which are commonly found at the bottom of a river bed.

4.6.2.2 Empirical Investigations

The first types of attempts at investigating the course of the development of riffles, dunes, bars, and pools in a river were by empiricism.

Inasmuch as the features mentioned have been observed everywhere in natural rivers, the reality of the phenomenon in *nature* is beyond doubt.

Experiments then, have shown that ripples can also be reproduced in the *laboratory*. The literature is large (cf. e.g. Dzulynski and Simpson 1966a, b; Anketell et al. 1969; Southard and Dingler 1971); much of it has been reviewed in a monograph by Allen (1968). With regard to the larger features represented by dunes, Iseya (1984) has published a lengthy study. Accordingly, the tests bear out the fact that the suspended sediment concentration is greater on the upstream side of a developing dune than on the downstream side. The dune, evidently, has a strong influence on sediment suspension, thus giving rise to a positive feedback tending to accumulate material in a place where some has already been accumulated until an "equilibrium dune stage" is attained (cf. the "traffic-jam theory" in Sect. 4.6.2.4).

4.6.2.3 Direct-Action Mechanics

The most obvious attempt at explaining the interaction mechanism between river flow and the bed is on the basis of the direct local action of the fluid on the individual elements of the bed.

In this instance, the first problem is the question of the action of the moveable river load on the flow. The formulas of open channel flow discussed in Sect. 4.2 refer to channels with an immovable bottom. If the bottom is not fixed, the flow regime in the river is affected.

Early studies of this problem were undertaken by Gilbert (1876), who argued that the energy loss in a flowing river was incurred in two ways, first, in overcoming the hydraulic friction of the flow and, second, in transporting the sediment. On the one hand, this leads directly to the Chézy formula of flow [Eq. (4.2.19)] of which a more sophisticated deduction has been given in Sect. 4.2.2, and on the other hand, it was concluded that a clear stream must flow faster than a comparable one laden with sediment.

Special studies to investigate the validity of the clean channel flow formulas for dirty channels have been undertaken by Liu and Hwang (1959) and by Vanoni and coworkers (Vanoni and Brooks 1957; Vanoni and Nomicos 1960; Brooks 1958). In all these studies it turned out that the formulas for clean channel flow are indeed no longer valid. Liu and Hwang (1959) started with the assumption of a formula of the type of Eq. (4.2.10) and gave empirical correlation curves for the constants occurring therein. Vanoni and Brooks (1957) made special studies to determine the influence of the sediment load upon the discharge. Their data are best represented by the introduction of a friction factor f defined as follows:

$$f = 8(u_*/v)^2, \tag{4.6.1}$$

where u_* is the shear velocity [cf. Eq. (4.2.26)] and v the average velocity in the river. This friction factor does not vary in accordance with the ideas of Gilbert (1876), in that it does not increase with sediment load. In fact, it turned out that there is no single-valued relationship between the velocity and any combination of depth and slope of the river. The results of Vanoni and coworkers in this regard, however, are purely empirical and no theoretical explanation could be given. It also turned out that Karman's "universal constant" (cf. 4.2.27) is not, in fact, a constant but varies with sediment load. Other investigations of the bar resistance on flow have been made by Prestegaard (1983).

The next problem is that of the formation of the ripples by the flow. It is clear that these ripples are the expression of some sort of instability; this can be treated by the methods of system dynamics, as will be shown in the next Section (4.6.2.4). Here we shall look at the direct action on a local basis.

Thus, the formation of the ripples has been considered by Liu (1957) to be the expression of instability between *superimposed stratified flows*. Liu's argument is mainly qualitative and is based upon the fact that, even in turbulent flow, there is a laminar boundary layer or sublayer very close to the boundary. He then treats the river bed as an extreme case of a *density current*: the movable layer containing the bed load of the river is treated as one dense fluid of (almost) infinite viscosity, the laminar sublayer of clear water attached thereto as another. One has thus, in essence, two fluids—the layer containing the bed load and the laminar boundary layer—which are both in laminar motion but with a different velocity. According to investigations reported in Sect. 2.2.3.3, this causes an instability at the interface which could create the ripples.

Liu was not able to substantiate his claims by an actual calculation of the size of the ripples as a function of the size of the bed particles and the flow velocity. His explanation of the bottom ripples in rivers must therefore be regarded as a conjecture. Similar studies were made by Hayashi (1970).

Another theory of ripple formation has been proposed by Anderson (1953), who assumed a resonance effect between the river bed with waves on the surface of the water. The surface waves were assumed as corresponding to those found by the *laminar* flow theory. Anderson's theory is therefore similar to the theory of particle movement on beaches to be discussed in Sect. 6.4.3.3. A similar attempt was made by Kondratyev (1964). However, the application of laminar flow theory to sediment ripple formation in a river is questionable, since it stands to reason that the ripples are intimately connected with the turbulence of the water.

Another approach to the problem of sediment ripple formation has been taken by Velikanov (1957). The latter author assumes that there is a connection between the bottom ripples and the *large-scale turbulence* in a river. Commonly, the statistical theory of homogeneous turbulence is applied to the turbulence which is present in a river only in the high-frequency end of the spectrum. The low-frequency part is not homogenous and it stands to reason that the velocity fluctuations will come into resonance with the walls. Unfortunately, the statistical theory of turbulence has not been developed to such an extent where the question as to the *form* of turbulence could be posed. However, it has been shown by Velikanov and Mikhailov (1950) by direct measurements that the periods of the fluctuation of the turbidities are very close to the periods of the large-scale velocity fluctuations and these in turn to the periods of the sand ripples just when the latter begin to be formed. Other investigations ascribing ripple formation to eddy patterns have been made by Thomas (1979), by Du Toit and Sleath (1981), by Folk (1976) and Jackson (1975).

This leads to the now most common direct-action theory of ripple formation: the latter is assumed to be due to *pulsations* in the bed load transport and river flow rates. The pulsations may be due to external (meteorological) conditions (Jones 1977), to unstable particle size sorting (Iseya and Ikeda 1987), or to positive feedback (Allen 1978; Schumm et al. 1982). In any case, the scouring and deposition effect of the pulsating flow can be directly calculated and leads to dune and ripple formation, e.g., with the bed load entrainment theory (cf. Sect. 4.5.3.4) of Bagnold (Pratt 1973).

4.6.2.4 System Theory of Ripple and Dune Formation

A completely different approach to the problem of ripple, dune, and bar formation is based on the model of an intrisically unstable system: if deviations from uniformity tend to reinforce themselves, riffles and pools will of necessity develop. In quite general terms, such system-dynamic mechanisms have been invoked by Allen (1969a), Kennedy (1969), Smith (1970), Culbertson and Scott (1970), Ledder (1980), Davies and Sutherland (1980), Hecde (1981), and Young and Davis (1987). A very much noted approach to the problem was taken by Langbein and Leopold (1968), who reported that during the downstream movement of sediment, individual particles will tend to impede each other's progress. Thus, once several particles have accidentally come together, they impede the progress of others and thus create a phenomenon analogous to a "traffic jam". These are the dunes and bars.

The reason for the initiation of the instability may simply be sought in the stochastic fluctuations of the flow parameters: thus, Nordin and Algert (1966) showed that the process of ripple formation can be represented by a Markov second-order linear model. Other stochastic models have been proposed by Allen (1976b), Lee and Jobson (1977), and Fredsoe (1981). Alternatively, some specific effects, such as the presence of obstacles (Richardson 1968; Chang 1970; Williams and Kemp 1972), trochoidal waves (Hand 1969), or changes in bed form (Allen 1973) have been invoked for the origin of the instabilities.

Once the instabilities have been originated, their further development can be followed by tracing mathematically various possible feedback mechanisms (Callander 1969; Engelund 1970; Smith 1970; Mercer and Haque 1973; Parker 1975a; Engelund and Fredsoe 1970, 1974; Fredsoe and Engelund 1975; Richards 1978, 1980). More interesting, however, are considerations based on general system theory: rivers minimize their rate of potential energy expenditure per unit mass of water (Yang 1971) or the sums of the variances of power (Cherkauer 1973). This is entirely in the spirit of Langbein and Leopold (1968) mentioned above.

4.6.3 Mechanics of the Formation of Sedimentary Structures

4.6.3.1 General Remarks

The interaction of river flow with the bed leads to the formation of sedimentary structures, notably cross-bedding, flutes, ridges etc. Much of the pertinent literature has been reviewed and discussed by Ikeda (1983) and Zollinger (1983) and, notably, by Allen (1968, 1969, 1984), who wrote some monographs on the subject. We shall discuss some of the specific problems in the sections following.

4.6.3.2 Cross-Bedding

The ripples and dunes mentioned in the last sections give rise to the phenomenon of cross-bedding in sandstones: the total height H of a layer of cross-bedded sediments corresponds to the height of the sandwaves in a stream. In cross-bedded sediments, relationships between mean grain size diameter d and the height H of cross-bedded sediments is of the following type (β is a constant):

$$d = \text{const } H^{\beta}. \tag{4.6.2}$$

This relationship appears to be quite universal. It can be explained theoretically by assuming the following model: Due to the presence of the ripples, the turbulence in the stream is affected; the latter, in turn, is directly related to the carrying capacity (with regard to the grain sizes involved) of the stream. The chain of interactions leads to a relation between H and d.

Thus, from Eq. (4.2.39) we have, taking the height of the surface roughnesses k_s equal to the sand-wave height H:

$$v = K_1 \log\left(\frac{R}{H}\right) + K_2, \tag{4.6.3}$$

where v is the mean flow velocity, R the hydraulic radius of the river and K_1, K_2 are constants. Henceforth R will be assumed as equal to 1 (i.e., the river "depth" is approximately taken as unit length), so that H represents the "relative" roughness; it is a number $H \ll 1$.

An obstacle of height H creates turbulent velocity fluctuations u' of magnitude [cf. Eq. (2.2.18)] neglecting β; after Scheidegger and Potter 1967:

$$\text{const } u'^2/v^2 = H. \tag{4.6.4}$$

Furthermore, the carrying capacity of a river must be constant over the whole rippled bed; an inspection of (4.5.8) shows that for this to be the case, the ratio of the settling velocity w to the turbulent diffusion φ must be constant. However, from Eq. (4.4.3) we have

$$w \sim d^2, \tag{4.6.5}$$

where d is the grain diameter; furthermore, we have

$$\varphi \sim u'^2. \tag{4.6.6}$$

Therefore, we have a constant carrying capacity if

$$d \sim u'^2. \tag{4.6.7}$$

Thus, taking (4.6.3), (4.6.4) and (4.6.7) together, we obtain

$$d \sim H(\log H)^2 + \text{small terms} \sim H^\beta, \tag{4.6.8}$$

where the last proportionality results if the upper bound is taken instead of the logarithm of H. Thus, the empirical relation (4.6.2) has been explained.

The above argument explains the relation between grain size and thickness of cross-beds. The existence of cross-bedding itself is due to the existence of the ripples and dunes whose faces are not parallel to the main bedding. As the dunes wander, layers are added to the dune faces which represent the cross-stratification (e.g., Nottvedt and Kreisa 1987).

4.6.3.3 Other Sedimentary Structures

Apart from cross-beds, a great number of sedimentary structures have been noted (cf. e.g., Potter and Pettijohn 1963). In the present context, only structures built by rivers are of interest.

In this instance, we note that Allen (1971) published a kaleidscopic review of pertinent phenomena. In particular, he has shown that scallops, flutes, furrows, and ridges can be the result of corrsponding vorticity patterns.

In a similar fashion, Dionne and Laverdiere (1972) proposed a whirlpool flow regime for the explanation of an observed vertical cylindrical structure in a thick Quaternary sand and silt deposit near Montreal.

Unfortunately, no actual mechanical theories seem to be available for the explanation of the above features, other than rather schematic conjectures regarding the flow fields necessary to produce them.

4.6.4 Regime Theory

4.6.4.1 General Remarks

From the above discussion of the mutual interaction between a river bed, the total flow, and the sediment transported therein, it becomes obvious that the problem of finding the correct relationship between these phenomena is a very difficult one. People have therefore tried to at least establish the conditions for channel equilibrium empirically. The idea has thus arisen that certain relations between variables characterizing the river may be significant for the latter to be in a regime of dynamic equilibrium. A river in such a dynamic equilibrium regime has been called at grade (Gilbert 1877).

The efforts have been directed, therefore, first of all towards establishing the pertinent relations between the various parameters for a river to be at grade. Once this has been achieved, it is obviously possible to find the response of the remaining parameters if some process changes one of them, so that equilibrium is reestablished. This leads again to the process-response idea which we have already met in connection with sediment transport (Sect. 4.5.4.4).

Many of the problems connected with regime theory have been discussed in the books of Allen (1968, 1984) cited earlier. A summary of the regime theory as used by engineers has been given in monographs by Blench (1957, 1969). Other reviews have been published by Richards (1977, 1987b), West (1978) Chang (1979a) Gregory (1977), Morisawa (1985), and White (1988).

4.6.4.2 Conditions for a Graded River

The conditions for a graded river refer to the hydraulic geometry variables introduced in Sect. 1.4.2. Certain relations between these variables will determine whether a river is at grade or not. Many experiments have been performed, notably, by hydraulic engineers, to determine when a channel is at grade (in regime). Also, innumerable data have been collected from actual canals in order to determine what the values of the external variables are that determine whether a channel is at grade.

Assuming that the kind of material that is being transported and the total discharge are given by nature, there remain three self-adjusting variables which are the depth h, the gradient S, and the width w of the channel. In order to determine the values of these variables in a particular case, one needs obviously three independent relationships between them.

These empirical relationships form the body of what engineers call regime theory of channel formation. It was probably first developed by Lacey (1933) and as noted, was summarized by Blench (1957, 1969). Accordingly, from innumerable factual data, the first equation of regime theory turns out to be

$$v^2/h = B, \tag{4.6.9}$$

where v is the mean velocity of the flow, h is the depth of the water, and B is a bed factor related to the type of bed load carried by the river.

The second equation of regime theory is

$$v^3/w = s, \qquad (4.6.10)$$

where w is that width which multiplied by h yields the cross-sectional area of the flow. The quantity s is a side factor depending on the interaction of the river with the material of which the sides consist.

The third equation of regime theory is

$$\frac{v^2}{ghS} = c\left(\frac{vw}{v}\right)^{1/4}, \qquad (4.6.11)$$

where S is, as usual, the slope, v is the kinematic viscosity and c is a dimensionless constant.

The above three equations are the conditions required by the regime theory for a channel to be at grade. One can manipulate these equations algebraically to obtain "design equations", giving the width, depth, and slope that are required to produce a channel which is to be at grade, in terms of the total discharge Q and the constants c, s, B.

It may be remarked that the first equation of regime theory [i.e., Eq. (4.6.9)] is equivalent to the statement that stable flow is only possible if the Froude number

$$F_r = \frac{v}{\sqrt{gR}} \qquad (4.6.12)$$

(where R is the hydraulic radius of the river, equal to h for wide rivers) is less than a certain critical value. This is a condition of the regime theory which has often been used by geologists. It is then generally said that for Froude numbers below a certain critical value (in the neighborhood of l), one has turbulent flow in which stable conditions are possible; for Froude numbers above the critical value, one has turbulent flow in which stable conditions are no longer possible. Bottom slopes that produce flow at a subcritical Froude number are then termed *mild*, otherwise they are termed *steep*.

The regime theory has been tested not only in India, but also in many other places, such as in Britain (Hey 1986) and Australia (Riley 1976).

The reason for the adjustment of the hydraulic variables so as to correspond to a river at grade is to be sought in the requirement that the variance of the stochastic fluctuations of the variables becomes a minimum (minimum variance principle; cf. Langbein 1964a; Chang 1979b; Trofimov and Moskovkin 1983; Davies and Sutherland 1983). This subject will be further discussed in later sections of this book.

4.6.4.3 Process-Response Theory for Rivers

As noted above, the existence of conditions for a river to be at grade leads to the process-response theory: If a *process* changes some of the variables defining the hydraulic geometry of a river, the others must *respond* by changing in their turn so that the river will again be at grade. The classic formulation of this idea seems to be due to Whitten (1964): specific applications to rivers have been made by Maddock

(1969), Pickup (1976), Mosley (1976), by Richards (1987d), Yu and Wolman (1987), and Baker (1988). In essence, the regime equations yield the required process-response relations (Ferguson 1986). Thus, the variations of channel geometry downstream (with increasing drainage area and therefore increasing discharge) can be explained (Klein 1981a; Carragher et al. 1983). Similarly, Gregory and Park (1974) and Richards (1987b) have predicted the response of channel capacity to the construction of engineering works changing other hydraulic parameters, Orme and Bailey (1971) investigated the response of rivers to changes in bank vegetation; Wilcock (1971), Smith (1974), Miller and Onesti (1979), Soni et al. (1980), Klein (1982) and Sawada et al. (1983) the response to loading with sediment; Gardiner (1981) the response to changing climatic conditions, and Begin et al. (1981) the response to basement level lowering. Many local examples of the operation of the process-response model have been published, such as from the Chandler River in New South Wales (Loughran 1976), from the Mississippi in Louisiana (Burnett and Schumm 1983), from the Canadian Rocky Mountains (Desloges and Gardner 1984), and from the Alberta Badlands (Bryan and Campbell 1986).

Of particular interest is the problem of thresholds, when some parameters change drastically in response to minute changes in some others. This can lead to an actual "metamorphosis" in a river (Schumm 1969; Allen 1976a; Louis 1976; Slinger-Land 1981). The original trigger for such drastic changes can lie in on-going geological (tectonic) processes or in anthropogenic activity. Most of the investigations, however, are entirely qualitative and empirical.

4.6.5 River Bed Profiles

4.6.5.1 Introduction

We have already discussed some aspects of river bed profiles: notably the formation of pools, riffles, and dunes on its bed. These, however, are local features. What concerns us now is the overall aspect of the longitudinal and transverse profiles of rivers.

4.6.5.2 Longitudinal Profiles of a River

We begin with a discussion of the longitudinal profiles of rivers.

In this instance, the first problem is one of deciding on the most appropriate phenomenological description of a "typical" profile. Generally, an exponential law (Sternberg 1875)

$$S = S_0 \exp(-aL), \tag{4.6.13}$$

(where S is the slope, L the distance from some point and a a constant) has been taken as representative. However, other types of profile functions have also been suggested. These have been compared by Duran (1964) with actual river profiles in Colombia and by Shepherd (1985) with river profiles in Texas. The latter author finds that the exponential function is actually appropriate for profiles that end downstream in aggradational areas, but that a logarithmic function is preferable for profiles that terminate upstream at scarps or steep divides; power functions are generally inadequate in spite of the fact that Duran (1964) assumes a parabolic profile for Colombian rivers.

The exponential profile equation may be only an approximation of the solution of a diffusivity equation which is the natural result in direct-action mechanical considerations referring to rivers (Exner 1931). The corresponding argument has already been presented in connection with the discussion of integral slope development by water flow [Sect. 3.4.4.6, and particularly Eq. (3.4.26) in this book]; inasmuch as a river bed is indeed a "slope formed by water flow".

The fact that a diffusivity equation seems to be the correct basis for calculating river bed profiles can also be explained by referring to general system theory: the adjustment of systems affected by random influences and subject to external constraints generally leads to equations of this type. In this vein, Langbein (1964b) has proposed a specific random walk model and has obtained concave profiles by numerical empiricisms.

Another approach to the problem of longitudinal river profiles can be made by starting from the concept of a graded river. Inasmuch as some of the hydraulic parameters change with distance from the source, the slope has to adjust itself in a corresponding way. On this basis, Hormann (1965) has distinguished various profile tracts, Cherkauer (1972) and Knighton (1975) have related slope to the longitudinal variation of drainage area and discharge, and Snow and Slingerland (1987) set up mathematical models for the calculation of longitudinal profiles of graded rivers.

The above considerations allow one at once to use again process-response theories: the river bed slope must adjust itself to any change in other parameters so that the river remains at grade. The response can be to tectonically induced processes (Zuchiewicz 1980) or to an increase in water discharge, decrease in size of bed material, or decrease in bed material discharge (Galay 1983).

Peculiar phenomena occur if there are abrupt changes ("knick points") in the bed slope of a river.

The initiation of knick points can be caused by lateral flow constrictions (Ertel 1964), by periodic flooding (Schick and Magid 1978), or in the form of rapids as relics of geomorphic features (Graf 1979). An extreme case of knick points are waterfalls, generally caused by lithological conditions. Form and process of such falls have been discussed by Young (1985): a classic case is the Niagara Falls (Philbrick 1974): at an escarpment, an erosion-resistant layer of dolostone overlies a soft shale layer below (both Silurian); the shale at the bottom is eroded away by the action of the falls, the resistant layer keeps *breaking* into the void and the falls thus retreat into the escarpment, forming a canyon.

At knick points, the flow is nonuniform (cf. Sect. 4.2.4), being "mild" (in the sense of Sect. 4.6.4.2) and "steep" below. The classical view is that no equilibrium (in the sense of regime theory) is possible below a knick point and that the latter must therefore wander upstream (Sölch 1912; Brush and Wolman 1960) and become obliterated. This view has been challenged, however, by Gerber (1956), who sees in the whole knick point development a selective erosion process.

4.6.5.3 Transverse Profiles of a River

We now come to a discussion of the development of the transverse profile of a stable river.

The first problem in this connection is the identification of the "cross-section" itself. Gregory (1976a) has shown that this identification, e.g., at "bankfull" discharge, is by no means obvious; occurrences of lichens may indicate the limit of the actual "river bed". Knighton (1981) has introduced a series of indices encompassing the shape of a river. Better, of course, is a geometrical curve characterization.

Next is the problem of the theoretical explanation of the development of observed stable river cross-sections. This problem was analyzed long ago by Koechlin (1924), and somewhat more recently in investigations by Ibad-Zade (1952), Pokhsraryan (1957, 1958), Bretting (1958), and Ashida and Sawai (1976). In these investigations, a force-balance equation was written down for a bottom particle being at rest. This yields a relationship between the bottom slope (in profile) and the local bottom velocity. In order to arrive at a differential equation of the profile, it is evident that a further relationship is needed. This additional relationship is provided by some more or less justified assumptions regarding the distribution of either the shearing stress or the bottom velocity with height.

Pokhsraryan bases his connection between slope angle and velocity upon the drag theory (cf. Sect. 4.2.2), but in doing this he allows for the possibility that the bottom velocity, although parallel to the bottom, may have a component in the transverse cross-section because of the possibility of existence of an overall circulation. Furthermore, Pokhsraryan assumes that the drag force vector is not necessarily parallel to the velocity vector, although the general quadratic form of the relationship [cf. Eq. (4.2.17)] is retained.

We shall present here the exposition of Pokhsraryan in some detail.

Thus, taking the general geometry indicated in Fig. 61, where the flow velocity is parallel to the vector W, we have the following forces:

a) the "frontal drag force", which we shall denote here by W

$$W = c_f \rho \frac{\pi d^2}{4} v_{bot}^2, \qquad (4.6.14)$$

where c_f is the "frontal drag coefficient", ρ the fluid density, and d the particle diameter;

b) the "transverse drag force" V (cf. Fig. 61), which is perpendicular to the flow velocity and perpendicular to the profile cross-section

$$V = c_t \rho \frac{\pi d^2}{4} v_{bot}^2, \qquad (4.6.15)$$

where c_t is a "transverse drag coefficient";

c) the underwater weight P of a particle

$$P = g(\delta - \rho) \frac{\pi d^3}{6}, \qquad (4.6.16)$$

where δ is the density of the particle and g the gravity acceleration:

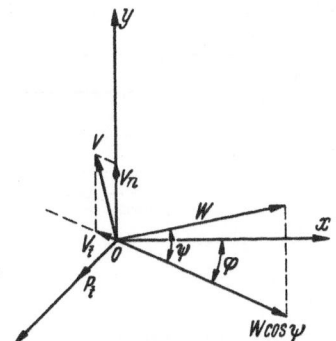

Fig. 61. Geometry of forces at the bottom of a river. (After Pokhsraryan 1957)

d) and finally the frictional force [cf. Eq. (4.4.27)]

$$F = \varepsilon(P_n - V \cos \psi), \qquad (4.6.17)$$

where P_n is the component of P normal to the bed

$$P_n = P \cos \alpha \qquad (4.6.18)$$

(α being the bed slope angle, cf. Fig. 62) and ψ is an angle as shown in Fig. 61. Furthermore, ε is a coefficient of friction.

One can now express the condition of force balance for a particle on the surface of the bed which must obviously be satisfied in a stable channel. This yields

$$[(W \cos \psi - V \sin \psi) \cos \varphi]^2 + [(W \cos \psi - V \sin \psi) \sin \varphi + P_t]^2 = F^2, \qquad (4.6.19)$$

where P_t is the tangential component of the underwater weight:

$$P_t = P \sin \alpha. \qquad (4.6.20)$$

Introducing the individual expressions for the various forces and adding an analog with Chézy's relationship (4.2.19) in the form of ($\bar{v}_m$ at center line)

$$\bar{v} = \bar{v}_m \left(\frac{h}{h_m} \right)^{2/3}, \qquad (4.6.21)$$

the bottom profile can be found theoretically by an integration, for which the following equation is an approximate expression (h' is a relative coordinate,

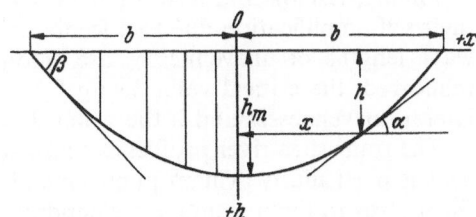

Fig. 62. Geometry of the bottom profile of a river. (After Pokhsraryan 1957)

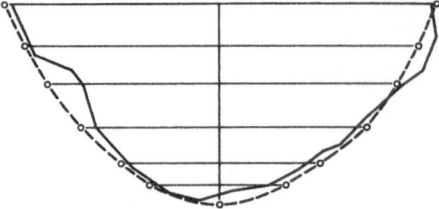

Fig. 63. Comparison of measured values for the cross-section of a river compared with those calculated from Pokhsraryan's theory. (After Pokhsraryan 1957)

varying from 0 to 1)

$$x = m_0 h_m \left[\arccos h' + \frac{1}{4m_0^2} (\arccos h' + h'\sqrt{1 - h'^2}) \right.$$

$$\left. - \frac{1}{64\,m_0^4} (3 \arccos h' + 3h'^2 \sqrt{1 - h'^2} + 2h'^3 \sqrt{1 - h'^2}) \right]. \qquad (4.6.22)$$

A comparison of the calculated values with experimental observations is shown in Fig. 63.

It may also be worth while to make a few remarks regarding Bretting's (1958) treatment of the problem of the cross-section of a stable river. Bretting, essentially, writes down a basic force balance equation which is very similar to that of Pokhsraryan (4.6.19), but he considers from the beginning only the forces parallel to the mean flow; he then introduces a hydrodynamic lifting force L

$$L = c\sigma_{bottom}, \qquad (4.6.23)$$

which corresponds to Pokhsraryan's V (see Eq. (4.16.15)].

The force balance equation, again, yields essentially a relation between v_{bot} and α. In order to calculate the profile, a further relation is needed. Bretting chooses

$$\frac{\sigma_{bottom}}{\sigma_{max}} = \frac{h}{h_{max} \cos \alpha}, \qquad (4.6.24)$$

which replaces (4.6.21) in Pokhsraryan's theory. One duly ends up with a differential equation for the profile. Bretting, in his study, supplies a series of tables giving the numerical results which he obtained. He claims good agreement with tests.

Finally, Ashida and Sawai (1976) investigated lateral bed unevennesses and their self-amplification due to a feedback effect in the tractive force: whilst short wave lengths of unevennesses are smoothed out, long ones tend to become reinforced, the critical value being $L/h \sim 5$, where L is the wave length of the lateral unevennesses and h the water depth.

The transverse river profiles are also subject to process-response phenomena. This is particularly evident in ephemeral channels (Thornes 1980), which adjust themselves to the instantaneous conditions.

4.6.6 Particle Size Profiles

4.6.6.1 General Remarks

It is common knowledge that the sediment particles in an alluvial reach show a gradation of size (Krumbein and Tisdel 1940; Knighton 1982; Gardner 1985). Normally, the finest particles are found farthest away from the source of the carrying water. However, the morphometric indices of the particles may vary in some other way (Ozer et al. 1984; Walling and Moorehead 1987); in particular, a bimodal size distribution has been found by Middleton (1976). Nevertheless, the common observation is a size decrease with distance. There are several possible explanations of this phenomenon.

First, it has been proposed that the particles undergo contrition during their downstream journey so that the finest particles would be those that have traveled farthest.

Second, it has been postulated that the gradation is in fact due to a sorting effect. There are two possible explanations for the occurrence of such a sorting effect. First, the sorting may be due to the fact that only a certain size of particles can remain fixed on any one slope angle of the bed, the finest particles requiring the smallest slope angle. Since the slope angle generally diminishes in rivers if the distance from the source increases, this could account for the observed pebble gradation. Second, the sorting may not be the result of a fixation of particles on a certain slope, but of differential transportation.

The various theories that have been proposed as an explanation of pebble gradation will be discussed below and a section containing an evaluation of these possible explanations will be added.

4.6.6.2 Contrition of Pebbles

The best-known explanation of the gradation of sediment particles in a river is that based upon the contrition of the bed load as it travels along the water course.

The mechanism of contrition has already been discussed in the chapter on physical rock reduction processes (Sect. 3.2.3.2). It leads to an exponential pebble size reduction with distance travelled (cf. Eq. (3.2.6); Sternberg's (1875) formula]. At the same time, the pebbles would become more rounded (cf. Sect. 3.2.3.2), which would conform to the observations.

The contrition mechanism, thus, would on the face of it appear as a very attractive one for the explanation of the size gradation in rivers. It has indeed been verified directly for granitic gravel in the (Texan) Colorado River (Bradley 1970). In other cases, however, it appears as uncertain whether the *amount* of contrition is sufficient to produce the actually observed gradation of bottom sediment in a river. This will be discussed more fully in Sect. 4.6.6.5.

4.6.6.3 Fixation and Selective Erosion

The explanation of pebble gradation given in the last section may not be the correct one, although it undoubtedly yields a qualitatively acceptable result. It is beyond question that particles become contriturated during their downstream journey, but, as noted above, it is not certain whether this phenomenon is large enough to produce the gradations observed.

An alternative explanation of size gradation may be postulated on a similar basis as the explanation of particle size gradation on alluvial slopes (cf. Sect. 3.4.4.6). It is a known fact that, in general, the slope of a river bed declines from its source to its end (cf. Sect. 4.6.5.2). In a river which is in dynamic equilibrium (so that in every one of its reaches just such particles are present that cannot be dragged along by the scouring force), the size gradation of pebbles ought to be given by the same equation (3.4.40) as the size gradation on alluvial slopes. Applied to rivers, taking all constants into the two factors α and β, this equation reads

$$W = \alpha(\text{erfc } \beta L)^3, \tag{4.6.25}$$

where W is the weight of the pebble and L the distance from the source.

In examining this equation, it should be noted that considerable simplifications had been made during the deduction of formula (3.4.30) which cannot be expected to be valid for the case of river flow: particularly Eqs. (3.4.7) and (3.4.8) are in doubt for this case.

A somewhat more heuristic approach to the idea, that it is essentially the slope of the river bed which determines the particle size to be found there, has been provided by Lokhtin (1897). In this approach, it has been assumed that for any single river there exists a constant *coefficient of fixation* C_F defined as follows:

$$C_F = d/S, \tag{4.6.26}$$

where d is the diameter of the pebbles and S the corresponding slope. The reason that this coefficient is assumed to be constant is that, by its definition, it represents the ratio of two forces: (1) of the resistance of a pebble to being moved which is proportional to its weight and hence to d^3 (cf. 4.4.26), and (2) of the scouring force which tends to propel the pebble which is proportional to d^2v^2 (cf. 4.4.25) and, in view of the Chézy relation (4.2.19) to d^2S. If a river is in any sort of dynamic equilibrium, Lokhtin reasons that the ratio of these two forces, and hence the coefficient of fixation C_F, must be constant.

Equation (4.6.26), if the left hand side be assumed as constant, yields a law for the size distribution of pebbles in rivers. Taking again the weight W instead of the diameter d as the criterion of pebble size, Lokhtin's relationship yields:

$$W = \text{const } S^3. \tag{4.6.27}$$

This becomes a relation for the pebble gradation with distance L from the source of the river, if S is inserted as a function of L. Lokhtin (1897) assumed that this would be done empirically. However, we have noted in Sect. 4.6.5.2 that Sternberg postulated a general law for the decrease of slope angle in a graded river which he writes as follows (cf. 4.6.13)

$$S = S_0 e^{-aL}. \tag{4.6.28}$$

Using Sternberg's equation, Lokhtin's formula becomes

$$W = \text{const } e^{-\alpha L}, \tag{4.6.29}$$

where α is a new constant (equal to 3a). It should be noted that this has the same form as the Sternberg formula, but the physical explanation of this equation is totally different.

The argument given above is quite old. In the meantime, some more sophisticated analyses have been made on the same basic premises. Thus, mathematical modeling has been attempted (Thomas and Prasuhn 1977), experimental modeling of selective erosion (Rakoczi 1987a) has been tried and an additional intermediary process—one based on the degree of exposure of particles—has been introduced (Sawai 1987). In addition, Brayshaw (1984) has analyzed the stability of the fixation–erosion process under the presence of occasional large clasts: he found experimentally that the latter act as obstacles to the flow so that cluster bed forms are the result.

4.6.6.4 Differential Transportation

The theory of pebble gradation given above is essentially a heuristic one. It would appear as desirable to set the same basic idea upon a sounder mechanical basis.

This has been attempted by Sundborg (1957) who showed how a sedimentation formula can be deduced in the following fashion. Let $\varphi(w)$ denote the ratio of the number of particles in a certain interval of grain size (characterized by their settling velocity between w, w + dw) that are picked up to the number which are deposited. Furthermore, let s_w denote the average vertical concentration of the corresponding material in the stream. The concentration of that material immediately above the bed may similarly be denoted by $\Phi(w)s_w$. Then, conservation of mass yields

$$h\frac{ds_w}{dt} = -ws_w\Phi(w)[1 - \varphi(w)],\tag{4.6.30}$$

where h is the mean depth of the river. The above equation is valid for any size fraction of particles characterized by the settling velocity w (i.e., the latter assumed as being between w and w + dw). Assuming $\varphi = 0$, the above equation can be integrated to yield

$$s_w(t) = s_w(0)e^{-w\Phi(w)t/h}.\tag{4.6.31}$$

This is Sundborg's formula. If we assume that the mean flow velocity of the river is constant (equal to $\bar{u}$), then we can write

$$s_w(L) = s_w(0)e^{-w\Phi(w)L/(\bar{u}h)},\tag{4.6.32}$$

where L denotes the distance from the source. The quantity s_w represents the concentration still in suspension. The amount of material deposited (per unit time and length) at the distance L from the source is thus

$$K(w, L) = -\frac{ds_w(L)}{dL}\bar{u} = \frac{w\Phi(w)}{h}s_w(0)e^{-w\Phi(w)L/(\bar{u}h)}.\tag{4.6.33}$$

As indicated by the notation, K(w, L) represents the size distribution of particles at the distance L from the source. This equation embodies a decrease of grain size with distance L. This can be shown, e.g., if $s_w(0)$ and $\Phi(w)$ are taken as independent of w. The grain size (characterized by w) which has the maximum frequency of occurrence is calculated by setting

$$\partial K/\partial w = 0. \tag{4.6.34}$$

This yields under the above assumptions

$$w = \frac{\bar{u}h}{\Phi L}. \tag{4.6.35}$$

Since w is proportional (for low settling velocities) to the $\frac{2}{3}$ power of the particle weight W, (cf. 4.4.3) we have

$$W \sim (1/L)^{3/2}. \tag{4.6.36}$$

This does not have the same form as the Sternberg formula although, qualitatively, it also embodies a decrease of particle size with distance from the source. However, it should be noted that the assumption of independence of $s_w(0)$ and $\Phi(w)$ of w constitutes a great oversimplification.

The dispersion (sorting) mechanism as discussed above is the outcome of a mechanically deterministic process. It is also the outcome if stochastic models are used (Todorovic 1975; Han and He 1980b).

The actual occurrence of the sorting mechanism has been verified in laboratory experiments (Jopling 1964; Meland and Norrman 1969). It has also been advocated as causing petrological separations by the flow in the great rivers in the world (Potter 1978) and in secondary mineralization (Best and Brayshaw 1985). In intermittent flows, the gradation produced is also intermittent; the intermittency may be externally induced (flood seasons: Butler 1977) or be due to an internal flow instability (Iseya and Ikeda 1987; Lisle 1979; Bhomik and Demissie 1982; Keller 1971).

4.6.6.5 Evaluation of Theories of Pebble Gradation

It remains to give an evaluation of the various theories of pebble gradation presented above.

Starting with the contrition model, we note that the latter describes the facts accurately enough. A comparison of observational data with Sternberg's formula has been given, for instance, by Leliavsky (1955) for the river Rhine. In this particular case, one obtains a good fit of the data with the contrition theory with

$$\alpha \cong 0.01 \text{ km}^{-1}. \tag{4.6.37}$$

In the case of the Mississippi, the fit is not so good (Shulits 1941).

Thus, it is not at all certain whether the physical postulates used in the deduction of Sternberg's formula, are acceptable, particularly in view of the fact that Lokhtin's theory, which is based upon an entirely different physical model, leads to an identical theoretical result. Whether Sternberg's physical model is acceptable depends on whether the assumption, that the contrition of a pebble is proportional to its size, is true.

Experiments to decide this question were started in the last century by Daubree (1879), who was followed by many investigators thereafter (cf. Krumbein 1941, for

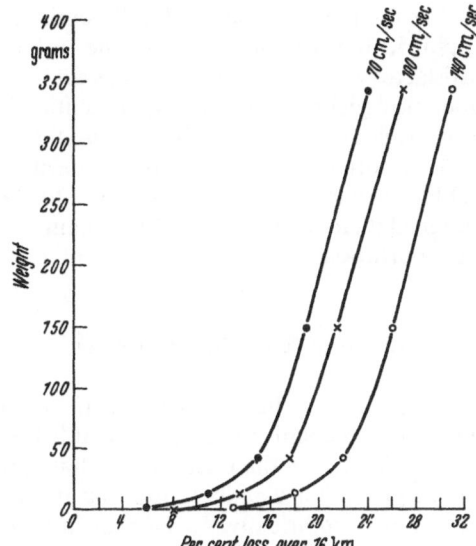

Fig. 64. Relation between percent weight loss and weight for limestone cubes over the equivalent of 16 km travel distance on pebbly concrete floor, as deduced from Kuenen's experiments. (After Kuenen 1956)

a review). However, the experiments performed by these people seem to be inconclusive because the conditions under which they were performed were rather far removed from those prevailing in a natural river. It was not until some investigations of Kuenen (1955, 1956, 1958, 1959) that experimental results on pebble detrition could really be trusted.

Kuenen used a revolving current which drove pebbles over a concrete floor. The abrasion, expressed in distance traveled, was measured. Typical results obtained by Kuenen are shown graphically in Fig. 64.

It is immediately obvious from Fig. 64 that the percent loss of weight is not proportional to the weight of the pebble or boulder. Very small particles (such as quartz sand grains less than 0.5 mm in diameter) do not experience any appreciable loss of weight at all.

The percentage weight loss over 16km corresponding to the Rhine river data [using Eq. (4.6.37)] works out to 16%. This value falls within the range acceptable according to Kuenen's experiments (see Fig. 64) so that the main objection to Sternberg's model is the experimentally observed variation of the relative abrasion with pebble size. However, this objection is serious. Thus, in view of the fact that Lokhtin's model (see Sect. 4.6.6.3) yields the same phenomenological result as Sternberg's, and that there is therefore an alternative explanation of the observed data, Sternberg's model ought to be rejected. No doubt, pebble abrasion plays *some* role in establishing the observed gradation patterns, but it cannot be the sole reason thereof.

Turning now to the second model, viz. to the hypothesis that the observed pebble gradation is the result of a differential fixation rather than that of an abrasion of the pebbles, we note that the theory of Lokhtin yields an acceptable result. A possible objection here lies in the definition of the coefficient of fixation, which may vary widely from river to river.

The final model discussed (differential transportation) represents a much more sophisticated treatment than the others and is therefore physically more satisfactory. Unfortunately, using some simple boundary and initial conditions, it does not yield the correct exponential law of gradation, but in view of the indeterminacy of the coefficients used, this is not too serious.

Therefore the most likely mechanism of pebble gradation in rivers consists of pebbles becoming contriturated due to the action of frictional forces, but being assigned their position along the stream bed by a sorting process due to differential transportation.

4.6.7 Scaling of River Bed Processes

An inspection of the discussion of river bed processes presented up to this point makes one recognize that it is possible only in a very few cases to give an actual physical theory of the phenomena that have been observed. Even if such a theory can be given, it is usually found that it involves many constants which have to be determined empirically. As usual in circumstances where the available theories are inadequate, it is useful to take recourse to model experiments. The problem therefore arises how best to scale river-bed processes.

In order to develop suitable scaling laws, one may assume as a first approximation that *gravity forces* and *turbulent forces* are the only forces that are acting in river bed processes. This leads immediately to the condition that the Froude numbers F_r

$$F_r = \frac{v}{\sqrt{gl}},\tag{4.6.38}$$

(where v is a characteristic velocity, l a characteristic length and g the gravity acceleration) must be equal in the prototype and in the model. One can prove this easily as follows. In order to scale *any* force f, Newton's law of motion (m denotes mass, v velocity, t time)

$$f = m\frac{dv}{dt}\tag{4.6.39}$$

has to be satisfied equally in the model as in the prototype. This implies the following conditions for the scaling ratios of force (α_f), mass (α_m), length (α_l), time (α_t) and velocity (α_v):

$$\alpha_m\alpha_v\alpha_t^{-1} = \alpha_m\alpha_l\alpha_t^{-2} = \alpha_f.\tag{4.6.40}$$

However, the gravity force is

$$f = mg.\tag{4.6.41}$$

Hence

$$\alpha_f = \alpha_m,\tag{4.6.42}$$

which yields

$$\alpha_l = \alpha_t^2.\tag{4.6.43}$$

For the velocity ratio α_v this yields

$$\alpha_v = \alpha_l/\alpha_t = \alpha_l^{1/2}. \tag{4.6.44}$$

Writing

$$\alpha_v = V/v \tag{4.6.45}$$

$$\alpha_1 = L/l, \tag{4.6.46}$$

(where capitals denote variables in the prototype, lower case letters variables in the model) this can be written

$$\frac{v}{\sqrt{gl}} = \frac{V}{\sqrt{gL}} = F_r, \tag{4.6.47}$$

so that one has the result that gravity forces are correctly scaled if the Froude numbers are equal.

In order to scale turbulent forces, one starts with Prandtl's equation (2.2.8)

$$\sigma = \text{const. } \rho l^2 (dv/dy)^2 \tag{4.6.48}$$

(see Sect. 2.2.2.2) for a definition of symbols). This yields the following expression for the various ratios:

$$\alpha_f \alpha_l^{-2} = \alpha_m \alpha_l^{-3} \alpha_l^2 \alpha_v^2 \alpha_l^{-2}. \tag{4.6.49}$$

Inserting this into (4.6.40) leads to

$$\alpha_l^2 \alpha_t^{-2} = \alpha_v^2, \tag{4.6.50}$$

which does not yield a new condition. Hence, equality of the Froude numbers is the only scaling condition that is required to model gravity and turbulence effects correctly. This "Froude condition" can be written as follows:

$$\alpha_v^2 \alpha_l^{-1} = 1. \tag{4.6.51}$$

It turns out that scaling experiments based upon Froude's similarity criterion alone are not always satisfactory. The reason for this is, of course, that gravity and turbulent forces are not the only ones that are present. Thus, in addition to Froude's criterion, other conditions, corresponding to the various sediment transportation formulas, have to be used. Investigations into various possibilities have been made by numerous people. Nazaryan et al. (1959) based their scaling law on a generalized Manning (cf. 4.2.12) relationship

$$v = \beta \left(\frac{h}{d}\right)^n \sqrt{ghS}, \tag{4.6.52}$$

where, as usual, v is the average velocity, h the river depth, d the particle diameter and S the slope. The quantities n and β are empirical constants. This leads immediately to the following scaling law:

$$\alpha_v = \alpha_h^{n+1/2} \alpha_d^{-n} \alpha_S^{1/2}, \tag{4.6.53}$$

where the α's again denote the scaling ratios of the quantities indicated in their indices. In addition, Nazaryan et al. (1959) based further scaling laws upon some morphometric formulas of Velikanov (1955).

Similarly, Einstein and Chien (1956) also used the above generalized Manning formula, but, in addition, based further scaling laws upon the bed load function theory outlined in Sect. 4.5.3.3 (e.g., the condition that the dimensionless functions Φ and Ψ introduced in Sect. 4.5.3.3 remain the same in model and proto-type immediately yields two scaling laws). All in all, Einstein and Chien ended up with nine conditions for ten independent scaling ratios, which were later modified somewhat by Bogardi (1959).

Other attempts at setting up scaling relations for special conditions have been reported by Razumikin (1960), Yalin (1960), Chauvin (1962), Amorocho and Hart (1965), Zeller (1965), Znamenskaya (1966) and Bruun (1966). Reviews have been published by Shen (1979), Alexander (1979), Anderson and Howes (1984), and by Stephenson and Meadows (1986).

As with all model experiments, the latter never *explain* any physical facts, but only make their characteristics accessible to laboratory investigations.

4.7 Planar Aspects of River Flow

4.7.1 Introduction

We have remarked earlier that, in general, rivers do not flow in a straight line. Not only do they curve (meander) back and forth, but, on occasion, they also form braided patterns. Thus, it is essential to investigate the river behavior in plan.

4.7.2 Hydraulic Geometry Theory

4.7.2.1 General Remarks

The first type of analysis of river behavior in plan that has been put forward is based upon the concept of a graded river (Sect. 4.6.4). In a graded river, the hydraulic geometry must conform to certain relations; the idea is that these relations can be fulfilled only if the river has a nonstraight course. This idea also leads to process-response phenomena.

4.7.2.2 Meanders in a Graded River

In any sedimentary plain, the size of the pebbles and the total water discharge of the river presumably would effect that the river is *not* at grade if it were to run in a straight course. Thus, the river would seek itself a crooked course until its bed has been lengthened to such an extent that the river would be at grade. An early expression of this idea is due to Wundt (1949).

Later, Leopold and Miller (1956) showed that there is a relation between stream order and hydraulic geometry which has a bearing on meanders. This relation was tested further by Miller and Onesti (1977). Other "equilibrium configurations" of hydraulic variables in meandering streams have been proposed by Shahjahan (1970), Ferguson (1973a), Abrahams (1982), and Williams (1986).

The above explanation of meanders, appealing though it might appear at first glance, nevertheless cannot be upheld in the light of criticism. For, even though a river be not at grade, there appears to exist no reason whatsoever why it should scour *sideways* in order to rectify the situation. The river might well simply deepen its channel and form a gorge. A different explanation of meanders is therefore required.

Nevertheless, one can make some general remarks upon meander mechanics based on such purely phenomenological terms as a graded river. Thus, Sundborg (1956a) stated that a meander loop must of necessity widen until the radius of curvature has become so large that a further increase would slow down the flow velocity to such an extent that lateral erosion would cease. That such a state will be reached eventually may be ascertained from the principles of the drag theory of fluvial erosion. The shape of the meander loop thus depends on both the flow conditions and on the material constituting the river bed.

The fact, mentioned in Sect. 1.4.4, that the ratio of the radius of curvature R over the channel width w tends to lie around 2 or 3 was explained by Bagnold (1960) by noting that the resistance to flow in a channel falls sharply in the above range of the value of R/w.

Unfortunately, our present knowledge is not sufficient to exploit the above remarks any further.

4.7.2.3 Process-Response Phenomena

It is clear that any "hydraulic equilibrium theory" carries within itself the possibility of a response of the variables to external processes.

Basically, the grading formulas connect the grade (and therewith the length) of a river stretch to other variables; thus they affect the corresponding meanders.

On an entirely natural basis, this means that the meander patterns change downstream along a river course, simply because the pertinent drainage area and therewith the discharge rate becomes progressively bigger (Klein 1981b). Correspondingly, in "underfit" streams the discharge is reduced and the meander patterns change in the opposite way (Chang 1984a).

Of greatest interest are the responses of rivers to artificial (anthropogenic) processes. Thus, channel widening in alluvial stretches (Fujita and Muramoto 1982c) as well as animal grazing on the banks (Petit 1984) was shown to lead to the formation of alternating bars. A rather striking example showing the river channel changes in response to urbanization (dramatic increase of floods) was presented by Leopold (1973) in a presidential address to the Geological Society of America.

4.7.3 Mechanistic Theories of Meandering

4.7.3.1 General Remarks

We come now to the mechanistic theories of meander initiation and formation. These include studies of secondary currents and their effects, the consideration of general flow instabilities, and even the assumption of external (mainly geological) controls.

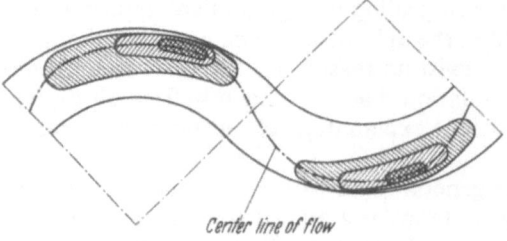

Center line of flow

Fig. 65. Erosion at the outer and deposition at the inner bank in a meander coil. *Darker shades* deeper. (After Leliavsky 1955)

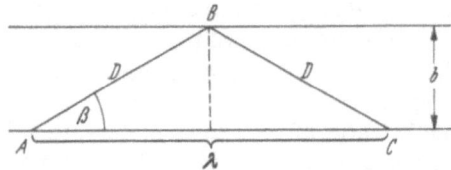

Fig. 66. Geometry of an incipient meander

4.7.3.2 Secondary Flows

According to the theories outlined in Sect. 4.3.2.3, "secondary" currents may be assumed to arise in river bends which, eventually, initiate helicoidal flows (cf. Fig. 54). It stands to reason that such flows will cause erosion at the outer bank of a river bend and deposition at the inner bank (see Fig. 65), simply because there is a tendency to push high-velocity water towards the outer bank. Hence, the process of meandering, once it is initiated, is seen to intensify itself. The existence of this chain of events has been verified and observed in nature (Leliavsky 1955; Tanner 1960; Hickin 1978; Knighton 1984: Thorne et al. 1985).

A theory for the initiation of meanders on this basis has been given by Werner (1951) who based his considerations upon the assumption that the start of the cross-currents might be connected with the propagation of a disturbance across the width b of the river. Referring to Fig. 66, Werner argues that any disturbance originating at the point A will propagate itself to the point B, be reflected there and hit the original bank at the point C. The length AC, then, would correspond to the wave length λ of an incipient meander.

If we denote the propagation velocity of a disturbance along the diagonal D (see Fig. 66) by v_c, we have (see Fig. 66).

$$\cos \beta = \frac{v}{v_c} \qquad (4.7.1)$$

and hence

$$\lambda = 2b \cot \beta = 2b \frac{v/v_c}{\sqrt{1 - v^2/v_c^2}}. \qquad (4.7.2)$$

The velocity v_c is a "critical serpentine velocity"; Werner represents it as follows:

$$v_c = \gamma \sqrt{gh}, \qquad (4.7.3)$$

where γ is an empirical constant ($\leqslant 1$) depending on the silt content and h is the depth of the river (g is as usual the gravity acceleration). Then, we have

$$\lambda = 2b\frac{v}{\sqrt{\gamma^2 gh - v^2}}.$$ (4.7.4)

One can discuss the above meander formula further by introducing some drag formula such as Manning's [cf. Eq. (4.2.12)] into it. It is immediately obvious, however, that the meander length can be real only if

$$v < \gamma\sqrt{gh}.$$ (4.7.5)

The expression $\sqrt{gh}$ is the wave-velocity in the water [see Eq. (6.2.19)]; we have noted that $\gamma \lessgtr 1$, hence it is seen that meanders can form only if the flow velocity is smaller than the wave velocity. The latter condition characterizes streaming flow; in shooting flow, no true meanders are possible.

The theory of Werner (1951) concerns only the initiation of meanders. Naot and Rodi (1982), as well as Chang (1984b), attempted to calculate the integral flow in a meander bend, but this refers only to fixed channels, so that the development and changes of the meander trends are not accounted for. Calculations of the erosive (shear) effects of the secondary currents remain generally on an entirely local level (Gorycki 1973; Hooke 1975; Kikkawa et al. 1976; Dietrich et al. 1979; Chiu and Hsiung 1981; Buch 1982; Odgaard 1982, 1986; Kitanidis and Kennedy 1984; Parker and Andrews 1985; Petit 1987; Furbish 1988).

4.7.3.3 Flow Instabilities

Mechanistic explanations of the meander phenomenon have also been based on general considerations of the occurrence of instabilities in the river flow. Early attempts at such explanations go back to Hickin (1968), who tried to establish thresholds for meander initiation, to Engelund and Skovgaard (1973: also Engelund 1975), who described a linear stability analysis of the interaction between the water flow and the bed, and to Parker (1975b), who did the same thing for meandering of supraglacial melt streams. Furthermore, Blondeaux and Seminara (1985) and Yih (1983) postulated a resonance and a wave phenomenon in river bends.

The most pertinent study of an incipient instability in straight streams that would lead to meander formation, however, was made by Bejan (1982a), who showed analytically that the equilibrium shape of a river bed is a unique sinusoid. Thus, Bejan showed theoretically that a straight, undisturbed river chooses a sinuous course of a precise wavelegth. Setting down the static equilibrium conditions for a straight river (as for a "hose") and passing to the limit of infinitesimally small deviations from the rectilinear shape, Bejan (1982a) found for the (initial) meander wave length λ (w is width, H depth, and v_0 the mean flow velocity):

$$\lambda = \pi w/(2 + gH/v_0^2)^{1/2}.$$ (4.7.6)

For shallow rivers and plane simulation ($gH/v_0^2 \ll 1$) this reduces to

$$\lambda = w\pi/2^{1/2} = 2.22\, w.$$ (4.7.7)

The argument of Bejan thus predicts a universal proportionality between meander wave length and river width which is, actually, strongly supported by observations.

4.7.3.4 External Control

Finally, mechanistic explanations of meander formation have been based on external influences.

Most of the assumed external influences are of endogenic origin (Chang and Toebes 1970). Incised meanders are commonly ascribed to an elevation of the country surrounding a pre-existing meander train (Shepherd 1972). Lithological differences produce dimensional differences in meander trains (Braun 1983); tectonic tilting (Leeder and Alexander 1987) causes asymmetrical meander belts to arise.

Apart from endogenic influences, exogenic ones (in the form of varying climatic conditions) have been advocated as affecting meander forms. Thus, Cogley (1973) suggested that discharges during deglaciation were greater than today, and gave rise to patterns that would be ununderstandable on the basis of present-day hydrology.

4.7.4 Stochastic Theory of Meander Formation

4.7.4.1 General Remarks

A completely different approach to the problem of meander formation is based on statistical considerations: in it, a meander train is assumed to be the result of the stochastic fluctuations of the direction of flow, caused by the random presence of direction-changing obstacles in the river path. This idea can be followed up in several ways: one can seek a "most probable" path for the river between two points, or one can seek the "expectable path" etc. According to the basic probabilistic models chosen, different results are obtained.

The probabilistic approach to the meander problem was initiated by Leopold and Langbein (1966). Subsequently, many further studies were made along this line. A review of them has been given, for instance, by Richards (1986).

4.7.4.2 Most Probable Random Walk

As noted, Leopold and Langbein (1966; see also Langbein and Leopold 1966) proposed that the actual course of a river is the most frequent random walk of the set of random walks under the given geological constraints. It should be noted, however, that this line of reasoning is *not* in conformity with the commonly accepted principles of statistical mechanics (cf. Sect. 5.3.3.3) usually, the expectation value of an observable (the latter may be the "wiggly line" under consideration) is taken as its average over all the configurations of the ensemble in question. In Leopold's and Langbein's meander theory, however, the "characteristic" pattern is taken as that pattern of the observable which occurs in the most probable configuration of the ensemble. This is generally not satisfactory. Nevertheless, some interesting indications can be obtained from a consideration of such most probable random walks in the meander case.

Thus, let us assume that, in Cartesian coordinate system (x, y), a particle starts a random walk at the origin (0, 0). Each step has the (constant) length l; there are a total of j steps after which the particle arrives at the point $J(x_j, y_j)$. The total length given of the walk is $L = jl$. Each step (number i) is a straight line segment, whose angle with the $+x$ axis is φ_i.

One can then ask various types of questions: (a) Assume that the frequency distribution of angles φ_i be $f(\varphi_i)$, what is the most probable path (under the assumed constraints) from 0 to J? (b) Assume that the frequency distribution of the first differences $\Delta\varphi = \varphi_i - \varphi_{i-1}$ is $f(\Delta\varphi)$, what is the most probable path (under the assumed constraints) from 0 to J?

We turn first to the case where the probability distribution $f(\varphi_i)$ of the angles φ_i is given.

Since the random walk, after j steps, must go through the point $x_j y_j$, we have the following constraints in the problem:

$$F \equiv \sum_{i=1}^{j} \frac{L}{j} \cos \varphi_i - x_j = 0, \tag{4.7.8}$$

$$G \equiv \sum_{i=1}^{j} \frac{L}{j} \sin \varphi_i - y_j = 0. \tag{4.7.9}$$

If the probability of finding a particular angle φ_i is $f(\varphi_i)$ (this is assumed to be the same for all steps), the total probability P of a random walk with angles φ_1, $\varphi_2,\ldots,\varphi_j$ is

$$P = f(\varphi_1)f(\varphi_2)\cdots f(\varphi_j). \tag{4.7.10}$$

Let us assume that the probability of f can be represented as follows

$$f = Ce^{-g(\varphi)}, \tag{4.7.11}$$

where C is a normalization constant. Then

$$P = C^j \exp\left(-\sum_{i=1}^{j} g(\varphi_i)\right). \tag{4.7.12}$$

We wish to find the most probable random walk, i.e., $P = \max$ or

$$Q \equiv \log P = \max \tag{4.7.13}$$

under the auxiliary conditions $F = 0$ and $G = 0$. In well-known fashion, this is achieved by maximizing

$$H = Q + \lambda F + \mu G, \tag{4.7.14}$$

where λ and μ are Lagrange multipliers. Thus, we require $\partial H/\partial\varphi_i = 0$ for all i, which yields

$$0 = \frac{\partial H}{\partial\varphi_i} = -g'(\varphi_i) - \lambda \frac{L}{j} \sin \varphi_i + \mu \frac{L}{j} \cos \varphi_i. \tag{4.7.15}$$

This equation has, generally, a series of solutions for φ_i, so that the most probable random walk is a series of legs with certain angles whose sequence cannot further be ascertained. The minimization of H thus does not lead to a definite result: this is an indication that the positioning of the problem (most probable instead of expected state) is not fortunate.

We turn our attention now to the case where $f(\Delta\varphi_i)$ is prescribed; i.e., a certain probability of the change of direction after each step in the random walk is given. This case has been solved by Schelling (1951, 1964). The auxiliary conditions are now

$$F = \cos\Delta\varphi_i + \cos(\Delta\varphi_1 + \Delta\varphi_2) + \cdots + \cos(\Delta\varphi_1 + \cdots\Delta\varphi_j) - x_j = 0$$
$$G = \sin\Delta\varphi_i + \sin(\Delta\varphi_1 + \Delta\varphi_2) + \cdots + \sin(\Delta\varphi_1 + \cdots\Delta\varphi_j) - y_j = 0. \qquad (4.7.16)$$

We again write the frequency distribution as an exponential

$$f(\Delta\varphi) = Ce^{-g(\Delta\varphi)}. \qquad (4.7.17)$$

The probability of a particular random walk is

$$P = f(\Delta\varphi_1)f(\Delta\varphi_2)\cdots f(\Delta\varphi_j), \qquad (4.7.18)$$

which is a maximum if

$$Q = \log P = j\log C - \sum g(\Delta\varphi_i) \qquad (4.7.19)$$

is a maximum. However, one has the auxiliary conditions (4.7.16) which can be taken into account by using Lagrange multipliers, λ, μ, and thus by maximizing the expression

$$H = Q + \lambda F + \mu G. \qquad (4.7.20)$$

This yields

$$\frac{\partial H}{\partial\Delta\varphi_s} = -g'(\Delta\varphi_s) - \lambda\sum_{u=s}^{j}\sin\left(\sum_{i=1}^{u}\Delta\varphi_i\right) + \mu\sum_{u=s}^{j}\cos\left(\sum_{i=1}^{u}\Delta\varphi_i\right) = 0. \qquad (4.7.21)$$

We have

$$\varphi(s) = \Delta\varphi_1 + \Delta\varphi_2 + \cdots\Delta\varphi_s. \qquad (4.7.22)$$

Going to the limit of $l/L = ds \to \infty$, $jl \to L$, s = arc length yields

$$-g'\left(\frac{d\varphi}{ds}\right) - \lambda\int_{u=s}^{L}\sin\varphi(u)\,du + \mu\int_{u=s}^{L}\cos\varphi(u)\,du = 0 \qquad (4.7.23)$$

and after some transformation (differentiation with regard to s, multiplying by $d\varphi/ds$, and integration)

$$\frac{d\varphi}{ds}g'\left(\frac{d\varphi}{ds}\right) - g\left(\frac{d\varphi}{ds}\right) + [\lambda\cos\varphi(s) + \mu\sin\varphi(s)] + \cos 2\omega = 0, \qquad (4.7.24)$$

where ω is a constant of integration. The term in square brackets can be written as $v\cos(\varphi - \alpha)$, where v and α are two new constants; it is clear that setting $\alpha = 0$, $v = 1$ is no restriction of generality. Thus, one has the differential equation

$$\frac{d\varphi}{ds}g'\left(\frac{d\varphi}{ds}\right) - g\left(\frac{d\varphi}{ds}\right) + \cos\varphi + \cos 2\omega = 0. \qquad (4.7.25)$$

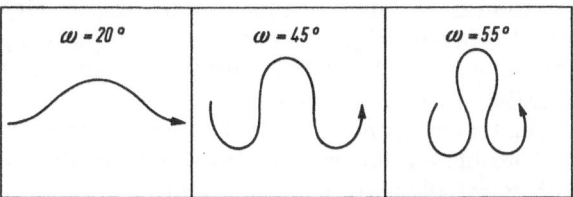

Fig. 67. Gaussian most frequent random walks (After Schelling 1964)

If the distribution f is specified as Gaussian, viz.

$$g = \frac{1}{2}\frac{1}{\sigma^2}\left(\frac{d\varphi}{ds}\right)^2 \qquad (4.7.26)$$

the above differential equation can be solved. One has in parametric form

$$s = \frac{1}{\sigma}\int\frac{d\varphi}{\sqrt{2(-\cos 2\omega + \cos\varphi)}}$$

$$x = \frac{1}{\sigma}\int\frac{\cos\varphi\,d\varphi}{\sqrt{2(-\cos 2\omega + \cos\varphi)}}. \qquad (4.7.27)$$

Examples of these curves, for various values of ω, are shown in Fig. 67

In fact, these curves are solutions of a general family of differential equations in which the rate of change of curvature along the channel is an odd function of path direction and bends are symmetric (Ferguson 1973b).

The curves, as is evident from their inspection, are fairly regular and, while reminiscent of stretches of river meanders, are too regular to describe a whole system of such features. The reason must be sought in the fact that the most frequent are not the expected random walks.

4.7.4.3 Expectable Paths

As noted, the statistical assumptions of the theory presented in the last section (4.7.4.2) lead to a most probable path for a river. This, however, does not conform to the usual procedure of statistical physics in which the *expected* pattern of observables is determined as the average of these observables over an ensemble of possible states. Unfortunately, attempts at setting up such ensembles numerically have not been successful heretofore because of the difficulty of random-generating meander patterns for given boundary conditions.

Thus, empirical approaches have been resorted to.

One of them is due to Thakur and Scheidegger (1970): a necklace chain of a given length L was used to generate theoretical "meanders". Keeping both ends of the chain fixed and the distance M apart, an ensemble of random profiles of a given sinuosity was generated. Different ensembles were generated by varying the distance M and keeping the length L of the chain constant. Expected values of the pertinent observables were found by averaging them over individual states of the ensembles; these values showed a good correspondence with those obtained from observations in nature. In particular, it turned out that the distribution of deviation angles (angle of direction change per unit length) is the same in the chain model as in natural rivers. The same is true for the power spectra of chain sample profiles and river meanders. These facts testify to the essential correctness of the statistical assumptions made.

Similar analog models as that based on a necklace chain have been made with a bent flexible wire (Gedzelman 1974) and with (falling) paper ribbons (Bejan 1982b) with essentially corresponding results.

In conclusion, it may be stated that some "hybrid" stochastic models have also been tried for the explanation of the meander phenomenon. In such models, a random component is superposed upon a deterministic pattern, leading to a further type of "expected" paths. Thus, Ferguson (1976) has assumed a quasi-random variability in valley floor topography and Davies and Tinker (1984) random disturbances in the secondary flow circulation.

4.7.5 Experimental Investigations

The difficulty of giving a satisfactory theoretical description of meander formation has prompted investigators to try empirical methods. This, it may be expected, will at least produce an indication of the factors involved in meander formation if not a true explanation of the origin of the former in terms of basic physical principles. Such experimental studies have been performed, for instance, by Tiffany and Nelson (1939), Fedorov (1954), Leopold and Wolman (1957), Joglekar (1959), Shen and Komura (1968), Schumm and Khan (1972), and Nakagawa (1983).

Of these investigations, that by Tiffany and Nelson (1939) is particularly noteworthy. Accordingly, it is possible to reproduce in flume experiments the meander development of alluvial rivers. The progressive changes in location of the channels as obtained by Tiffany and Nelson is shown in Fig. 68. It was also noted that the rate of introduction of suspended material into the flowing water materially affected the development pattern of the meanders.

4.7.6 Junctions and Braids

4.7.6.1 General Remarks

River patterns in plan do not only comprise meanders, but also braiding and confluences. The problems involved have been studied experimentally and theoretically; a review has been given, e.g., by Chitale (1973), Miall (1977), and Bettess and White (1983).

4.7.6.2 Observational Thresholds

Inasmuch as a river may meander or form braids, studies have been made to identify thresholds between the two possible planar patterns.

Thus, an analysis of natural channels led Leopold and Wolman (1957) to postulate that there is a dividing line between conditions producing braided and

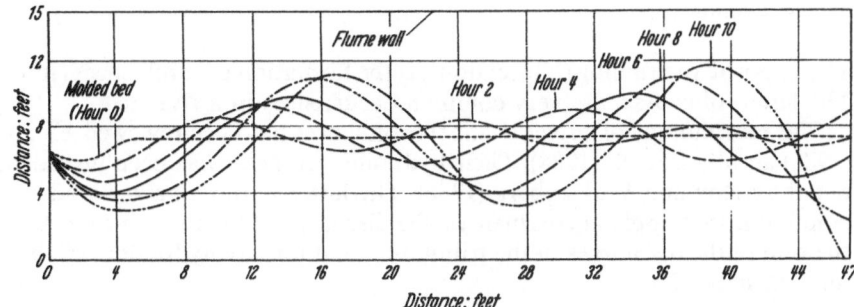

Fig. 68. Progressive changes of a channel in an experiment. (After Tiffany and Nelson 1939)

meandering channels. The dividing line between braids and meanders can be described by the equation

$$S = 0.06Q^{-0.44}, \tag{4.7.28}$$

where S is the slope of the river and Q the bankful discharge in cubic feet per second. The factors influencing the erosion of river banks were studied by Wolman (1959) in the field. Similar studies were undertaken by Fujita and Muramoto (1982a) and by Ashmore (1988).

As noted earlier, empirical relationships do not really provide an explanation of why rivers meander or form braids; they are, at best delineate some of the variables that are important.

4.7.6.3 Mechanism of Braiding

In order to arrive at an actual mechanical explanation of braiding, theoretical considerations have to be introduced.

Thus, Parker (1976) has made an analysis of braiding perturbations in a model alluvial river. This is based on a study of the momentum and mass balance equations and on the theory of small perturbations. The braiding is initiated by patterns of submerged bars on the river bed. A criterion of braiding is arrived at in this fashion which agrees well with observations.

In contrast to Parker's hypothesis, Rundle (1985) claims that bar formations are secondary, and that the primary constructional elements in braid formation are caused by orifice flow entering the flood sheet from a drowned channel.

Computer simulations of braided stream networks will be discussed in Sect. 5.5.2.3.

In all these studies, thresholds between braiding and meandering have been obtained. This leads immediately to response phenomena upon a change of some hydraulic parameters by external processes. Thus, lateral migration of a channel (Knighton 1972), fan deposits (Ori 1982), or an abrupt addition of sandy bed load (Smith and Smith 1984) may change the river pattern.

4.7.6.4 Confluences

The opposite to braiding (bifurcations) is the formation of confluences (junctions). The latter, of course, are very common occurrences in a river net.

The morphology of a river channel confluences has been reviewed, e.g., by Best (1986) and by Bomek (1988). Their hydraulic geometry has been investigated in detail by Roy and Woldenberg (1986). The latter authors have stated the basic conditions that apply at confluences: the discharge of the receiving stream equals the sum of the discharges of the tributaries, and for any hydraulic variable G the fundamental relation is

$$G_0^x = G_1^x + G_2^x, \tag{4.7.29}$$

where the subscript o denotes the receiving stream and 1,2 the tributaries. The exponent x is a characteristic of the junction and may, in fact, be a constant characteristic of the whole drainage network. This model describes adequately the average changes of hydraulic geometry variables (such as width) at a confluence.

4.7.7 Geomorphological Effects of River Motion in Plains

As was noted in Sect. 1.4.4, some field geomorphologists contend that the lateral action of rivers is a most effective agent in the development of drainage basins. In the case of alluvial plains, this lateral action manifests itself primarily in meander formation: meanders swing back and forth across the plain. If a net removal of material (i.e., a net erosion) from the alluvial plain is connected with the meander action, then it stands to reason that terraces will be the result (Mark 1985). At the same time, a characteristic imbrication of cobbles takes place (Kauffman and Ritter 1981).

The above reasoning has been made the basis of a theory of the development of stepped erosion surfaces (Geyl 1960). In this theory it is reasoned that, ordinarily, no net removal takes place of material from an alluvial plain: the river dumps debris here and there and also erodes from here and there. In order to have net erosion, endogenic effects must be involved: if a gradual uplift of the area occurs, then a river can maintain its general course only if it removes material from the region. Stepped erosion surfaces are the result (Pounder 1980).

4.8 Valley Formation

4.8.1 Requirements of a Physical Theory

Our next task is to investigate the formation of valleys. By "valleys" we mean river vallyes which are primarily caused by flowing water. Some of the problems that are involved in this connection have been discussed in general terms, e.g., Hol (1957).

The course of rivers at the bottom of a valley is generally very sinuous. This appears to be a characteristic feature of rivers which is evident not only in the meanders (as discussed in Sect. 4.7) that may be observed in an alluvial plain, but is a completely ubiquitous phenomenon. Thus, it is almost impossible to find a natural flow channel which is straight for any appreciable distance. It is this observation which, above all, calls for an explanation.

We have given earlier (Sect. 4.7) reasons for the sinuous course of graded river reaches, but it is doubtful whether the same explanation obtains under non-equilibrium conditions such as may occur in mountain streams. It is possible that the causes for the sinuosity are different in this case. This problem will be discussed first, in Sect. 4.8.2.

The reasons advanced for the explanation of meanders, however, may be applicable to those effects upon the course of rivers which have been ascribed to the rotation of the Earth. It appears that the Coriolis force is able to induce helicoidal currents in a river and thus affect the meanders. This question will be dealt with in Sect. 4.8.3.

4.8.2 Mountain Valleys

4.8.2.1 General Remarks

The most difficult problem in the development of valleys is to account for the shapes of mountain valleys. The latter are in a transient state which cannot be regarded as even remotely equilibriated. The flow of mountain streams is swift and generally of the shooting (supercritical) variety (cf. Sect. 4.2.1).

The *orientation* of mountain valleys appears to be essentially predesigned by endogenic (tectonic) processes. The evidence for this contention and the consequences to be drawn from it have been discussed at length in the writer's book on geodynamics (Scheidegger 1982; see p 28 ff therein).

4.8.2.2 Sinuosity of Mountain Valleys

As with streams meandering in flood plains, it is again the sinuosity of the river courses in mountain areas which is a most striking feature. Interestingly enough, there seem to be no mathematical attempts in the literature to explain this feature. It is not possible to account for the bends in mountain streams in terms of cross-currents of the type that have been advocated as the cause of alluvial meanders. We have shown in Sect. 4.7.3 that there are theoretical reasons which indicate that helicoidal currents cannot exist in shooting flow. Under such conditions, the arguments presented in Sect. 4.3.7.5 on shooting flow around corners must be advocated: cross-waves will become established which have a high erosive power. Thus, one might argue that the sinuosity of mountain streams is initiated by tributary gulleys eroding the sides. Once kinks have been started in this fashion, the cross-waves originating in the corners will tend to intensify the latter owing to their erosive action. There are no reasons why this action should be regular in any way; this explains the absence of any regularity in the sequence of the bends in mountain streams.

4.8.2.3 Erosion-Deposition Effects

A further complication occurs in mountain valleys because, in addition to eroding the valley sides, a mountain river generally also deposits a lot of debris on the valley floor. Thus, the latter is rising while the river eats its way into the mountain side. An analytical theory of this process has been proposed by Gerber (1959). Accordingly, if the slope of the mountain side is constant, the mountain river action may be envisaged as shown in Fig. 69. During a particular time interval Δt, the valley floor is raised by the amount Δh and the cross-sectional area ΔM is eroded from the mountain side. The ratio

$$\lambda = \frac{\Delta M}{\Delta h} \tag{4.8.1}$$

is indicative of the *erosive power* of the stream. If the mountain side has a constant slope of angle α, then the result of the process is a lowering of the slope; beneath the debris, the slope will be parallel to the original slope. This can be demonstrated as follows:

The (cross-sectional) area eroded is

$$\Delta M = y\Delta x = \lambda\Delta h. \tag{4.8.2}$$

But

$$y = x \tan \alpha - h = \lambda\frac{\Delta h}{\Delta x}. \tag{4.8.3}$$

Hence, in the limit

$$\lambda\frac{dh}{dx} - x \tan \alpha + h = 0. \tag{4.8.4}$$

This is a differential equation whose solution is

$$h = x \tan \alpha - \lambda \tan \alpha, \tag{4.8.5}$$

which is what was to be demonstrated.

Gerber also investigated the influence of a kink in the original mountain side upon the slope developing beneath the debris (cf. Fig. 70). Introducing the origin of the coordinate system (h, x) at that point on the lower slope (i.e., on the slope developing beneath the debris) which is exactly below the kink, we have from an

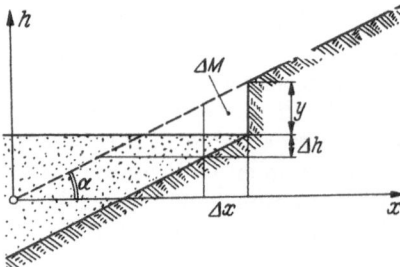

Fig. 69. Geometry of the development of a mountain valley. (After Gerber 1959)

inspection of Fig. 70:

$$\Delta M_2 = \lambda \Delta h \tag{4.8.6}$$

and

$$\Delta M_2 = \Delta x(a_1 + x \tan \alpha_2 - h). \tag{4.8.7}$$

Hence one obtains the differential equation

$$\lambda \frac{dh}{dx} - a_1 - x \tan \alpha_2 + h = 0. \tag{4.8.8}$$

The solution of this equation for the boundary condition $h(x = 0) = 0$ is (note that $\tan \alpha_1 = a_1/\lambda$; cf. Eq. 4.8.5):

$$h(x) = a_1 \left(1 - \frac{\tan \alpha_2}{\tan \alpha_1} \right) \left(1 - \exp\left[-\frac{\tan \alpha_1}{a_1}x \right] \right) + x \tan \alpha_2. \tag{4.8.9}$$

This shows that the valley slope developing beneath the debris is an exponential curve: the kink disappears and is replaced by a rounded corner.

The physical model leading to Eq. (4.8.1) is probably somewhat of an over-simplification. Perhaps, the model could be improved upon by allowing for the fact that the valley becomes *wider* as it fills up. Then, instead of the ratio between ΔM and Δh, the ratio between ΔM and $x\Delta h$ should remain constant. Thus let us set

$$\lambda = \frac{\Delta M}{x\Delta h}. \tag{4.8.10}$$

This changes the differential Eq. (4.8.4) to

$$\lambda x \frac{dh}{dx} - x \tan \alpha + h = 0, \tag{4.8.11}$$

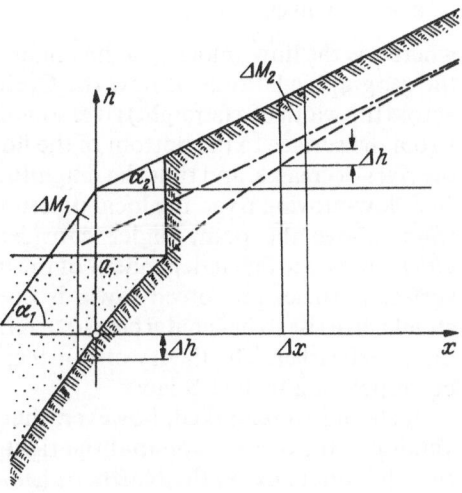

Fig. 70. Effect of a kink in the valley slope upon the development of a mountain valley. (After Gerber 1959)

whose solution is

$$h = \frac{\tan \alpha}{\lambda + 1} x. \tag{4.8.12}$$

This shows that the slope beneath the debris is now no longer parallel to the original mountain side, but has a declivity which is smaller.

The above theories do not exactly explain the evolution of mountain valleys, but at least they provide some indications regarding the phenomena that might be expected. Gerber (1959) gives some examples of geomorphological features of Alpine valleys which can be explained in terms of these theories.

A further peculiar feature is the development of "inner" gorges in a mountain valley (prototype: Colorado canyon). This occurs when a relatively competent layer is encountered during the downcutting by the river due to continued base level fall, the latter most likely being induced by tectonic uplift (Kelsey 1988).

The theories above refer to *lateral* erosion-deposition effects in Alpine valleys. Regarding the *longitudinal* patterns, the reader is referred to Sect. 4.6.5.2 of this book. Similarly, the above theories assume a slow and steady valley evolution. As of recently, the significance of single catastrophic events has been appreciated. Of such events, single huge floods have been particularly effective in shaping valleys. This problem complex has already been dealt with in Sect. 3.4.4.3 of this book.

4.8.3 Influence of the Earth's Rotation

It may be expected that the Earth's rotation has an effect upon the course of rivers. Accordingly, it has been proposed ["Baer's (1860) law"] that on the Northern hemisphere, the Coriolis force would cause rivers to have the tendency to erode preferentially their right bank and on the Southern hemisphere their left bank.

Einstein (1926) has given an explanation of "Baer's law" which is based on the same considerations as his explanation of the tendency of rivers to meander. Accordingly, the Coriolis acceleration a_c in a river is given by

$$a_c = 2v\omega \sin \varphi, \tag{4.8.13}$$

where v is the flow velocity, ω the angular velocity of rotation of the Earth, and φ the geographic latitude. If now the Coriolis force is balanced by a pressure drop across the width of a (straight) river, this can only be achieved for a given velocity v. From the surface to the bottom of the flowing water, however, the flow velocity of the river decreases, and thus the magnitude of a_c will decrease towards the bottom. In a slow-moving river, the local pressure is only conditioned by the height of the water above the point under consideration, and hence any lateral pressure *differentials* must be independent of the depth below the stream surface. Thus, the surface particles are forced towards one side more than the bottom ones. This should give rise to helicoidal cross-currents just as in the case of flow in a river bend discussed earlier. The theory of Einstein thus appears to account for phenomena corresponding to Baer's law.

It should be remarked, however, that the very existence of Baer's law, i.e., the validity of the observation that the right hand bank of rivers is eroded more than the left hand bank on the Northern hemisphere, has been questioned by Schmidt

(1926), who based his criticism not upon actual observations in nature, but upon a series of experiments in which a stream of water was established upon a rotating sand bed. Subsequently, the propriety of Schmidt's experiments was questioned by Exner (1926) because the bed slope in Schmidt's experiments was much greater than what would correspond to nature. Thus, Exner undertook to repeat Schmidt's experiments with some modifications. He took a basin filled with a mixture of sand and clay of 100 cm diameter which he rotated anti-clockwise once in 7 s. In it was carved the channel of a river with a bed slope of 0.01. The flow velocity was chosen as $v = 50$ cm/s. This yields a Coriolis acceleration in Exner's experiments equal to about 90 cm/s^2, which is about 36 times larger than the Coriolis acceleration of, say, the Danube near Vienna. If anything, it would thus appear that the effect of the Coriolis force is larger in Exner's experiments than in nature.

It thus turns out that Exner's experiments show a behavior of rivers which would correspond to Baer's law. Exner proceeded to calculate the difference of erosive power at the two banks of the river, but these estimates do not seem to be based upon the helicoidal cross-currents postulated by Einstein. For an establishment of the latter it would appear that the ratio of depth to width in the river must be of sufficient magnitude; however, no attention was paid by the experiments to this ratio. It is thus not incomprehensible that different results were obtained by different experiments. Nevertheless, the fact that *some* results corresponding to Baer's law were obtained appears to speak for the validity of the latter.

References

Abduraupov, R.R.: Izv. Akad Nauk Uzbek. SSR, Ser. Tekh. Nauk 1958 (5), 67 (1958)
Abrahams, A.D.: Earth Surf. Proc. 7, 469 (1982)
Adams, J.: J. Geol. 85, 209 (1977)
Alexander, D.: Progr. Phys. Geogr. 3(4), 544 (1979)
Allen, J.R.L.: Current ripples. Amsterdam: North Holland (1968)
Allen, J.R.L.: Geogr. Ann. 51A(1–2), 61 (1969a)
Allen, J.R.L.: Proc. Geol. Assoc. 80(1), 1 (1969b)
Allen, J.R.L.: Physical processes of sedimentation. London: Unwin (1970)
Allen, J.R.L.: J. Fluid Mech. 49(1), 49 (1971)
Allen, J.R.L.: Sedimentology 20(2), 189 (1973)
Allen, J.R.L.: Earth Surf. Proc. 1(4), 361 (1976a)
Allen, J.R.L.: Sediment. Geol. 16(4), 255 (1976b)
Allen, J.R.L.: Sediment. Geol. 20(3), 165 (1978)
Allen, J.R.L.: Sedimentary structures. Amsterdam.: Elsevier (1984)
Alonso, C.V.: J. Hydraul. Div. ASCE 107(HY6), 733 (1981)
Amin, M.I., and P.J. Murphy: J. Hydraul. Div. ASCE 107 (HY8), 961 (1981)
Amorocho, J. and W.E. Hart: J. Hydrol. 3, 106 (1965)
Ananyan, A.K.: Dvizhenie zhidkosti na povoroto vodovoda. Erevan: Izdat. Akad. Armyans. SSR (1957)
Anderson, A.G.: Proc. Third Midwest. Conf. Fluid Mechanics, Univ. Minnesota, p. 379 (1953)
Anderson, M.G. and S. Howes: Streamflow modelling. Gainsville: Florida Univ. (1984)
Andrews, E.D.: Bull. Geol. Soc. Am. 94(10), 1225 (1983)
Andrews, E.D.: Bull. Geol. Soc. Am. 95(3), 371 (1984)
Anketell, J.M., J. Cegla and S. Dzulynski: Geol. Romana 8, 41 (1969)
Ashida, K. and K. Sawai: Bull. Disas. Prev. Res. Inst. Kyoto Univ. 26(3), 145 (1976)
Ashida, K., T. Takahashi and T. Sawada: Bull. Disas. Prev. Res. Inst. Kyoto Univ. 26(3), 119 (1976)

Ashida, K., S. Egashira and T. Kanayashiki: Bull. Disas. Prev. Res. Inst., Kyoto Univ. 31(3), 171 (1981)
Ashmore, P.E.: Earth Surface Proc. Landf. 13(8), 677 (1988)
Azia, N.M. and S.N. Prasad: J. Hydraul. Eng. ASCE 111(HY10), 1327 (1985)
Baer, K.E. v.: Bull. Acad. Imp. Sci. (St. Petersburg) 2(2), 218, 353 (1860)
Bagnold, R.A.: U.S. Geol. Surv. Prof. Pap. 282-E, 135 (1960)
Bagnold, R.A.: U.S. Geol. Surv. Prof. Pap. 422-I, 1 (1966)
Bagnold, R.A.: Water Resour. Res. 13(2), 303 (1977)
Baker, V.R.: Bull. Geol. Soc. Am. 100(8), 1157 (1988)
Bardsley, W.E.: J. Hydrol. 52(11–2), 165 (1981)
Barenblatt, G.I.: Prikl. Mat. Mekh. 17, 261 (1953)
Barenblatt, G.I.: Vest. Mosk. Un-ta No. 8, 53 (1955)
Barfield, B.J., E.T. Smerdon and E.A. Hiller: Water Resour. Res. 5(1), 291 (1969)
Barnes, H.H.: Roughness characteristics of natural channels. U.S. Geol. Surv. Water Supply Paper No.
     1849; Washington: U.S. Government Printing Office (1967)
Basset, A.B.: Treatise on hydrodynamics. Cambridge (1888). See Vol. 2, Ch. 4
Bathurst, J.C., C.R. Thorne and R.D. Hey: Nature 269, 504 (1977)
Bathurst, J.C., G.J.L. Leeks and M.D. Newson: Proc. IAHR Symp. Delft, ed. Wessels: Balkema, p. 137
     (1986)
Bechteler, W. (ed.): Transport of suspended solids in open channels. Proc. of Euromech. Conf.
     Dordrecht: Balkema (1986)
Begin, Z.B., D.F. Meyer and S.A. Schumm: Earth Surf. Proc. 6(1), 49 (1981)
Bejan, A.: Geophys. Res. Lett. 9(8), 831 (1982a)
Bejan, A.: Phys. Fluids 25(5), 741 (1982b)
Beltaos, S.: J. Hydraul. Div. ASCE 108 (HY4), 591 (1982)
Benedict, P.: Trans. Am. Geoph. Union 38, 897 (1957)
Bertschler, M.: Wechselwirkung von idealisierten Rauhigkeitselementen mit ihren eigenen Ab-
     lösungswirbeln. Zürich: Inst. Hydromechanik ETH (1972)
Bessrebrennikov, N.K.: Dokl. Akad. Nauk Belorussk. SSR 2(1), 30 (1958)
Best, J.L.: Progr. Phys. Geogr. 10(2), 157 (1986)
Best, J.L. and A. Brayshaw: J. Geol. Soc. Lond. 142, 747 (1985)
Bettess, R. and W.R. White: Proc. Inst. Civil Eng. (2)75, 525 (1983)
Bhomik, N.G. and M. Demissie: J. Hydraul. Div ASCE 108 (HY10), 1227 (1982)
Blench, T.: Regime behaviour of canals and rivers. London: Butterworth's Scientific Publications (1957)
Blench, T.: Mobile-bed fluviology. 2nd edn. Edmonton: Alberta Univ. Press (1969)
Blondeaux, P. and G. Seminara: J. Fluid Mech. 157, 449 (1985)
Bogardi, J.: Acta Tech. Acad. Sci. Hung. 24(3–4), 417 (1959)
Bohlen, W.F.: In Estuarine processes ed. M. Wiley 2, 109 New York: Academic Press (1976)
Bomer, M.B.: Bull. Assoc. Geol. Fr. 65(1), 3 (1988)
Boussinesq, J.M.: Mém. Acad. Sci. Paris 23, Ser. 2(1), Supplément 24, 1 (1877)
Boussinesq, J.M.: Théorie analytique de la chaleur. Paris (1903)
Braden, G.E.: Proc. Oklahoma Acad. Sci. 39, 115 (1959)
Bradley, W.C.: Bull. Geol. Soc. Am. 81, 61 (1970)
Braun, D.D.: Earth Surf. Proc. 8(3) 223 (1983)
Brayshaw, A.C.: Bull. Geol. Soc. Am. 81, 61 (1970)
Brayshaw, A.C.: Mem. Can. Soc. Petrol. Geol. 10, 77 (1984)
Brayshaw, A.C.: Bull. Geol. Soc. Am. 96(2), 218 (1985)
Bretting, A.E.: Acta Polytech. Scand. Ci 1 (245/1958) (1958)
Bridge, J.S.: Earth Surf. Proc. Landforms 6(2), 187 (1981)
Bridge, J.S. and J. Jarvis: Earth Surf. Proc. 2(4), 281 (1977)
Brooks, N.H.: Trans Am. Soc. Civ. Eng. 123, 526 (1958)
Brown, R.J.: Sediment transport in rivers, a Bibliography-2 Vols. Springfield VA: U.S. Tech. Inf. Serf.
     (1979)
Brush, L.M. and M.G. Wolman: Bull. Geol. Soc. Am. 71, 59 (1960)
Bruun, P.: Bull. Geol. Soc. Am. 77, 959 (1966)
Bryan, R.B. and I.A. Campbell: Z. Geomorph. Suppl. 58, 121 (1986)
Buch, E.: Tellus 34(3), 307 (1982)
Burkham, D.E. and D.R. Dawdy: J. Hydraul. Div. ASCE 102 (HY10), 1479 (1976)
Burnett, A.W. and S.A. Schumm: Science 222, 49 (1983)

Butler, P.R.: Bull. Geol. Soc. Am. 88, 1072 (1977)

Callander, R.A.: J. Fluid Mech. 36(3), 465 (1969)

Carling, P.A.: Earth Surf. Proc. Landforms 8(1), 1 (1983)

Carragher, M.J., M. Klein and J.R. Petch: Earth Surf. Proc. 8, 177 (1983)

Carter, A.C.: Critical tractive forces which start movement of sediment in a channel. U.S. Bureau of Reclamation, Hyd. Lab. Rept. HYD-296 (1950)

Chang, F.F.M.: J. Hydraul. Div. ASCE 96 (HY2), 417 (1970)

Chang, H.H.: J. Hydraul. Div. ASCE 105 (HY6), 691 (1979a)

Chang, H.H.: J. Hydrol. 41(3–4), 303 (1979b)

Chang, H.H.: J. Hydrol. 75(1–4), 311 (1984a)

Chang, H.H.: J. Hydraul. Div. ASCE 110 (10), 1398 (1984b)

Chang, T.P. and G.H. Toebes: Water Resour. Res. 6, 557 (1970)

Chang, Y.L.: Trans. Am. Soc. Civ. Eng. 104, 1246 (1939)

Chauvin, J.L.: Bull. Centre Rech. Essais Chatton. No. 1, 64 (1962)

Cheong, H.F. and J.W. Shen: J. Hydraul. Div. ASCE 102 (HY7), 1035 (1976)

Cherkauer, D.S.: Bull. Geol. Soc. Am. 83, 353 (1972)

Cherkauer, D.S.: Water Resour. Res. 9(6), 1613 (1973)

Chien, N.: Proc. Am. Soc. Civ. Eng. 80(565), 1 (1954)

Chitale, S.V.: J. Hydrol. 19(4), 285 (1973)

Chiu, C.L. and K.C. Chen: J. Eng. Mech. Div. ASCE 95 (EM5), 1215 (1969)

Chiu, C.L. and D.E. Hsiung: J. Hydraul. Div. ASCE 107 (HY7), 879 (1981)

Chow, V.T.: Open-channel hydraulics. New York: Mc Graw-Hill (1959)

Christofoletti, A.: Bol. Geogr. Rio de Janeiro 34 (249), 58 (1976a)

Christofoletti, A.: Bol. Geogr. Teoretica Rio Claro 6(11/12), 67 (1976b)

Christofoletti, A.: Geomorfologia Univ. São Paulo 51, 1 (1977)

Ciet, P. and G.S. Tazioli: Geol. Appl. e. Idogeol. (Bari) 11(1), 39 (1976)

Cogley, J.G.: Area 5(1), 33 (1973)

Coleman, N.L.: Water Resour. Res. 6(3), 801 (1970)

Conover, W.J. and N.C. Matalas: U.S. Geol. Surv. Prof. Pap. 575-B. B60 (1967)

Corbel, J.: Z. Geomorph. 3(1), 1 (1959)

Corey, A.T.: Influence of shape on the fall velocity of sand grains. M.S. Thesis, Fort Collins: Colorado A & M College (1949)

Culbertson, J.K. and C.H. Scott: U.S. Geol. Surv. Prof. Pap. 700-B, B 237 (1970)

Daubrée, A.: Études synthéthiques de géologie experimentale, 2 vols. Paris: Dunod (1879)

Davies, T.R. and A.J. Sutherland: Earth Surf. Proc. 5(2), 175 (1980)

Davies, T.R. and A.J. Sutherland: Water Resour. Res. 19(1), 141 (1983)

Davies, T.R. and C.C. Tinker: Bull. Geol. Soc. Am. 95, 505 (1984)

Delleur, J.W. and D.S. McManus: Proc. 6th Midwest Conf. Fluid Mechanics, Austin, Texas, p. 81 (1959)

Desloges, J.R. and J.S. Gardner: Can. J. Earth Sci. 21(9), 1050 (1984)

Dietrich, W.E.: Water Resour. Res. 18(6), 1615 (1982)

Dietrich, W.E., J.D. Smith and T. Dunne: J. Geol. 87, 305 (1979)

Dingman, S.L.: Fluvial hydrology. San Francisco: Freeman (1984)

Dionne, J.C. and C. Laverdière: Can. J. Earth Sci. 9(5), 528 (1972)

Dracos, T.A. and B. Glenne: J. Hydraul. Div. ASCE 93 (HY6), 79 (1967)

Du Boys, P.: Ann. Ponts et Chauss. (5), 18, 141 (1879)

Du Toit, C.G. and J.F.A. Sleath: J. Fluid Mech. 112, 71 (1981)

Duran, L.G.: Rev. Acad. Colomb. Cienc. 12(46), 219 (1964)

Dzulynski, S. and F. Simpson: Geol. Rom. 5, 197 (1966a)

Dzulynski, S. and F. Simpson: Rocz. Polsk. Tow. Geolog. 36(3), 285 (1966b)

Eagleson, P.S., R.G. Dean and L.A. Peralta: The mechanics of the motion of discrete spherical bottom sediment particles due to shoaling waves.—Cambridge MA: M.I.T. Hydrodyn. Lab. Tech. Rep. No. 26 (1957)

Egiazaroff, I.V.: J. Hydraul. Div. ASCE 91 (HY4), 225 (1965)

Egiazaroff, I.F.: Proc. IAHR, 12th Congress A(9), 1 (1967)

Einstein, A.: Naturwissenschaften 14, 223 (1926)

Einstein, H.A.: The bed-load function for sediment transportation in open channel flows. U.S. Dept. Agric. Soil Cons. Serv. Tech. Bull. (1026), 71 pp (1950)

Einstein, H.A.: J. Hydraul. Div. ASCE 94(HY5), 1197 (1968)

Einstein, H.A. and N. Chien: Trans. Am. Soc. Civ. Eng. 121, Pap. 2805, 440 (1956)
Einstein, H.A. and H. Li: Trans. Am. Geoph. Un. 39, 1085 (1958)
Einstein, H.A. and G. Polya: Mitt. Versuchsanst. Wasserbau, E.T.H. Zürich: Verlag Rascher (1937)
Engelund, F.: J. Fluid Mech. 42(2), 225 (1970)
Engelund, F.: J. Fluid Mech. 72(1), 145 (1975)
Engelund, F. and J. Fredsoe: Rep. Danish Ctr. Appl. Math. Mech. TU Denmark 6, 1 (1970)
Engelund, F. and J. Fredsoe: Ser. Pap. Inst. Hydrodyn. TU Denmark 4, 2 (1974)
Engelund, F. and J. Fredsoe: Nord. Hydrol. 7, 293 (1976)
Engelund, F. and O. Skovgaard: J. Fluid Mech. 57(2), 289 (1973)
Ertel, H.: Monatsber. Dt. Akad. Wiss. 5, 69 (1963)
Ertel, H.: Acta Hydrophys. 8(3), 141 (1964)
Ertel, H.: Monatsber. Dt. Akad. Wiss. 7, 552 (1965)
Exner, F.M.: Geogr. Ann. 9, 173 (1926)
Exner, F.M.: Ergebn. Kosm. Physik 1, 431 (1931)
Falcon, M.: Ann. Rev. Fluid Mech. 16, 179 (1984)
Fedorov, N.N.: Trudy Gos. Gidrolog. In-ta 44, 14 (1954)
Ferguson, R.I.: Area 5(1), 38 (1973a)
Ferguson, R.I.: Water Resour. Res. 9(4), 1079 (1973b)
Ferguson, R.I.: Earth Surf. Proc. 11, 337 (1976)
Ferguson, R.I.: Progr. Phys. Geogr. 10, 1 (1986)
Flaxman, E.M.: J. Hydraul. Div. ASCE 98 (HY 12) 2073 (1972)
Folk, R.L.: Sedimentology 23(5), 649 (1976)
Francis, J.R.O.: Engineer. 203 (5280), 519 (1957)
Fredsoe, J.: J. Fluid Mech. 102, 431 (1981)
Fredsoe, J. and F. Engelund: Ser. Pap. Inst. Hydrodyn. TU Denmark 8, 2 (1975)
Friedlander, S.K.: J. Am. Inst. Chem. Eng. 3, 381 (1957)
Fujita, Y. and Y. Muramoto: Bull. Disas. Prev. Res. Inst., Kyoto Univ. 32(1), 49 (1982a)
Fujita, Y. and Y. Muramoto: Bull. Disas. Prev. Res. Inst., Kyoto Univ. 32(2), 115 (1982b)
Fujita, Y. and Y. Muramoto: Bull. Disas. Prev. Res. Inst., Kyoto Univ. 32(1), 1 (1982c)
Furbish, D.J.: Geology 16, 752 (1988)
Galay, V.J.: Water Resour. Res. 19(5), 1057 (1983)
Gani, J. and P. Todorovic: Stochastic Hydrol. Hydraul. 1, 209 (1987)
Garde, R.J. and K.G. Ranga Raju: Mechanics of sediment transportation and alluvial stream problems.
    Bombay: Wiley Eastern Ltd. (1977)
Gardiner, V.: In: Perspectives in geomorphology, ed. Sharma, H.S.: New Delhi: Concept Pub. 2, 19
    (1981)
Gardner, W.D.: Deep-Sea Res. 32(3), 349 (1985)
Gedzelman, S.D.: Pure Appl. Geoph. 112, 265 (1974)
Gerber, E.: Geogr. Helv. 9(3), 160 (1956)
Gerber, E.: Geogr. Helv. 14, 117 (1959)
Gessler, J.: Mitt. Vers. Anst. Wasserbau & Erdbau, Eidg. Techn. Hochsch. Zürich, 69, 1 (1965)
Gessner, F.B.: J. Fluid Mech. 58(1), 1 (1973)
Geyl, W.F.: J. Geol. 68, 154 (1960)
Gilbert, G.K.: Am. J. Sci. 12, 16 (1876)
Gilbert, G.K.: Report on the geology of the Henry Mountains, Washington: U.S. Gov. Printing Office
    (1877)
Goldstein, S.: Modern developments in fluid dynamics, Vol. 2, Oxford: Clarendon Press (1938)
Goossens, D.: Catena 14, 73 (1987)
Gorycki, M.A.: Bull. Geol. Soc. Am. 84, 175 (1973)
Graf, W.: J. Geol. 87, 533 (1979)
Graf, W.: Hydraulics of sediment transport. Fort Collins: Water Resour. Res. 10(4), 870 (1984)
Gregory, K.J.: Search 7(3), 99 (1976a)
Gregory, K.J.: Earth Surface Proc. 1, 273 (1976b)
Gregory, K.J. (ed.): River channel changes. New York: Wiley (1977)
Gregory, K.J.: Progr. Phys. Geogr. 9(3), 414 (1985)
Gregory, K.J. and C. Park: Water Resour. Res. 10(4), 870 (1974)
Griffith, J.C.: J. Geol. 69, 487 (1961)
Griffiths, G.A.: Nature 282, 61 (1979)

Griffiths, G.A. and A.J. Sutherland: J. Hydraul. Div. ASCE 103 (HY11), 1279 (1977)
Gunn, J.: Z. Geomorph. 26(4), 505 (1982)
Hallermeier, R.J.: Continental Shelf Res. 1(2), 159 (1982)
Han, Q.W. and M.M. He: Kexue Tongbao 25(1–2), 89 (1980a)
Han, Q.W. and M.M. He: Sci. Sin. 23(8), 1006 (1980b)
Hand, B.M.: J. Sediment. Petrol. 39(4), 1302 (1969)
Happel, J. and H. Brenner: Low Reynolds number hydrodynamics, Englewood Cliffs: Prentice-Hall (1965)
Happel, J. and R. Pfeiffer: J. Am. Inst. Chem. Eng. 6, 129 (1960)
Haque, M.I. and K. Mahmood: J. Hydraul. Eng. ASCE 109(4), 590 (1983)
Hardisty, J.: J. Sediment. Petrol. 53(3), 1007 (1983)
Hayashi, T.: J. Hydraul. Div. ASCE 96 (HY2), 357 (1970)
Heede, B.H.: Z. Geomorph. 25(1), 17 (1981)
Heggen, R.J.: In: Proc. Symp. Drainage. Basin Sediment Delivery, Albuquerque NM (IAHS Pub. 159), 323 (1986)
Herczynski, R. and I. Pienkowska: Ann. Rev. Fluid Mech. 12, 237 (1980)
Herschel, C.: J. Assoc. Engin. Soc. 18, 363 (1896)
Hey, R.D.: J. Hydraul. Div. ASCE 112(8), 671 (1986)
Hickin, E.J.: Am. J. Sci. 267(8), 999 (1968)
Hickin, E.J.: Can. J. Earth Sci. 15(11), 1833 (1978)
Hinch, E.J. and L.G. Leal: J. Fluid Mech. 76(1), 187 (1976)
Hirsch, P.J. and A.D. Abrahams: J. Sediment. Petrol. 51(3), 757 (1981)
Hjulström, F.: Bull. Geol. Inst. Uppsala 25, 221 (1935)
Hol, J.B.L.: Rev. Quest. Sci. 128, 195 (1957)
Hooke, R.L.B.: J. Geol. 83(5), 543 (1975)
Hormann, K.: Z. Geomorph. 9(4), 437 (1965)
Hung, C.S. and H.W. Shen: Proc. Statist. Hydrol., Tucson (Arizona Univ.) 262 (1971)
Ibad-Zade, Yu. A.: Gidrotekh. Stroitel. 1952 (12) (1952)
Ikeda, H.: Environm. Res. Ctr. Paps. Tsukuba 2, 1 (1983)
Ikeda, H. and F. Iseya: Trans. Jpn. Geoph. Un. 2(2), 231 (1980)
Ikeda, H. and F. Iseya: Internat. Geom. ed. Gardiner, NY: Wiley, 561 (1986)
Iseya, F.: Environmental Research Center Ibaraki Paps. 5, 1 (1984)
Iseya, F. and H. Ikeda: Geogr. Ann. 69A(1), 15 (1987)
Jackson, P.S.: J. Fluid Mech. 111, 15 (1981)
Jackson, R.G.: Bull. Geol. Soc. Am. 86, 1523 (1975)
Jarocki, W.: Ruch rumowieska w riekach. Gdynia: Wydawctwo Morskie (1957)
Joglekar, D.V.: J. Inst. Eng. India 39(7), Part I, 709 (1959)
John, P.T. and V.K. Goyal: Indian J. Technol. 13(2), 69 (1975)
Jones, C.M.: Geology, 5, 567 (1977)
Jopling, A.V.: J. Geoph. Res. 69(16), 3403 (1964)
Karman, T.: Nachr. Ges. Wiss. Göttingen, Math.-phys. Kl. 1930, 58 (1930)
Karolyi, Z.: Publ. Assoc. Int. Hydrol. Scient. 43, 286 (1958)
Kauffman, M.E. and D.F. Ritter: Geology 9, 299 (1981)
Keller, E.A.: Bull. Geol. Soc. Am. 82, 753 (1971)
Kelsey, H.M.: Catena 15, 433 (1988)
Kennedy, J.F.: Annu. Rev. Fluid Mech. 1, 147 (1969)
Kerssens, P.M.J., A. Prins and C. Can Rijn: J. Hydraul. Div. ASCE 105 (HY5), 461 (1979)
Keulegan, G.H.: J. Res. Natl. Bur. Standards 21, 707 (1938)
Kikkawa, H., S. Ikeda and A. Kitagawa: J. Hydraul. Div. ASCE, 102 (HY9), 1327 (1976)
Kitanidis, P.K. and J.F. Kennedy: J. Fluid Mech. 144, 217 (1984)
Klein, M.: Earth Surf. Proc. 6, 589 (1981a)
Klein, M.: Area 13, 47 (1981b)
Klein, M.: Z. Geomorph. 26(4), 491 (1982)
Knapp, R.T.: Trans. Am. Soc. Civ. Eng. 116, 296 (1951)
Knighton, A.D.: Bull. Geol. Soc. Am. 83(12), 3813 (1972)
Knighton, A.D.: Catena 2, 263 (1975)
Knighton, A.D.: Earth Surf. Proc. 6(6), 581 (1981)
Knighton, A.D.: Catena 9, 25 (1982)

Knighton, A.D.: J. Hydrol. 73, 1 (1984)
Koechlin, R.: Mécanisme de l'eau et principes généraux pour l'établissement d'usines hydroélectriques. Paris (1924)
Kolar, V.: Rozpravy Ceskoslov. Akad. Ved., Rada Technic. Ved. 66, 105 (1956)
Kolmogrov, A.N.: Izv. Akad. Nauk, SSSR, Ser. Fiz. 6(1–2) (1942)
Komar, P.D.: J. Geol. 88, 327 (1980)
Komar, P.D. and E.E. Reimers: J. Geol. 86, 193 (1978)
Kondratyev, N.E.: Trudy Gos. Gidrol. In-ta No. 116, 3 (1964)
Krumbein, W.C.: J. Geol. 49, 482 (1941)
Krumbein, W.C. and A.R. Orme: Bull. Geol. Soc. Am. 83, 3369 (1972)
Krumbein, W.C. and F.W. Tisdel: Am. J. Sci. 238, 296 (1940)
Kuenen, P.H.: Leids. Geol. Med. 20, 131 (1955)
Kuenen, P.H.: J. Geol. 64, 336 (1956)
Kuenen, P.H.: Proc. K. Ned. Akad. Wet. B 61, 47 (1958)
Kuenen, P.H.: Am. J. Sci. 257, 172 (1959)
Lacey, G.: Minutes of Proceedings, Inst. Civ. Eng. London 237, 421 (1933)
Lamb, H.: Hydrodynamics. New York: Dover (1945)
Langbein, W.B.: J. Hydraul. Div. ASCE 90 (HY2), 301 (1964a)
Langbein, W.B.: U.S. Geol. Surv. Prof. Pap. 501-B, B 119 (1964b)
Langbein, W.B. and L.B. Leopold: U.S. Geol. Surv. Prof. Pap. 422-H, H1 (1966)
Langbein, W.B. and L.B. Leopold: U.S. Geol. Surv. Prof. Pap. 422-L, L1 (1968)
Laursen, E.M.: J. Hydr. Div. Am. Soc. Civ. Eng. 84, Pap. 1530 (1958)
Lee, B.K. and H.E. Jobson: Open-File Rept., U.S. Geol. Surv., Bay-St. Louis MS. No. 75-358 (1975)
Lee, B.K. and H.E. Jobson: U.S. Geol. Surv. Prof. Pap. 1040, 1 (1977)
Lee, S.H. and L.G. Leal: J. Fluid Mech. 98(1), 193 (1980)
Leeder, M.R.: Earth Surf. Proc. 4(3), 229 (1979)
Leeder, M.R.: J. Geol. Soc. Lond. 137(4), 423 (1980)
Leeder, M.R. and J. Alexander: Sedimentology 34(2), 217 (1987)
Lehmann, E.J.: Sediment transport in rivers; Rept. for 1964–1975 Springfield VA: US Nat. Tech. Inf. Service (1975)
Leighly, J.: Geogr. Rev. 24, 453 (1934)
Leliavsky, S.: An introduction to fluvial hydraulics, London: Constable (1955)
Leopold, L.B.: Bull. Geol. Soc. Am. 84, 1845 (1973)
Leopold, L.B. and W.B. Langbein: Sci. Am. 214(6), 60 (1966)
Leopold, L.B. and J.P. Miller: U.S. Geol. Surv. Prof. Pap. 2822-A, 1 (1956)
Leopold, L.B. and M.G. Wolman: U.S. Geol. Surv. Prof. Pap. 282-B (1957)
Limerinos, J.T.: U.S. Geol. Surv. Prof. Pap. 650-D, D 215 (1969)
Lisle, T.: Bull. Geol. Soc. Am. Pt. I 90(7), 616 (1979)
Liu, H.K.: Proc. Am. Soc. Civ. Eng. 83, Pap. 1197 (1957)
Liu, H.K.: Trans. Am. Geoph. Un. 39, 939 (1958)
Liu, H.K. and S.Y. Hwang: Proc. Am. Soc. Civ. Eng. 85 (HY11), 65 (1959)
Lokhtin, V.: Sur le mécanisme du lit fluvial. St. Petersburg (1897)
Loughran, R.J.: Catena 3, 45 (1976)
Louis, H.: Z. Geomorph. 20(3), 257 (1976)
Maddock, T.: U.S. Geol. Surv. Prof. Pap. 622-A, A1 (1969)
Manning, R.: Trans. Inst. Civ. Eng. Ireland 20, 161 (1890)
Mark, D.J.: Ontario Geogr. 25, 41 (1985)
Martinec, J.: Bull. Internat. Assoc. Scient. Hydrol. 75, 243 (1967)
McColl, P. (ed): Proc. Symposium on Scale Effects in Modelling Sediment Transport Phenomena. Toronto: Nat. Water Res. Inst. (1986)
Meade, R.H.: Environ. Geol. Water Sci. 7(4), 215 (1984)
Meinzer, O.E. (ed.): Hydrology. New York: McGraw-Hill (1942)
Meland, J. and J.O. Norrman: Geogr. Ann. A51(3), 127 (1969)
Mercer, A.G. and M.I. Haque: J. Hydraul. Div. ASCE 99 (HY3) 441 (1973)
Mersi, N. and W.H. Graf: Ann. Geoph. 3(4), 473 (1985)
Meyer-Peter, E. and R. Müller: Trans. 2nd Meeting, Int. Assoc. Hydraul. Struct. Res., Stockholm, Appendix 2, 1 (1948)
Meyer-Peter, E., H. Favre and H. Einstein: Schweiz. Bauztg. 103(13), 147 (1934)

Miall, A.D.: Earth Sci. Rev. 13(1), (1977)
Middleton, G.V.: J. Geol. 84, 405 (1976)
Middleton, G.V. and J.B. Southard: Mechanics of sediment movement. Providence: SEPM Short
    Course (1984)
Miller, T.K. and L.J. Onesti: Bull. Geol. Soc. Am. 88, 85 (1977)
Miller, T.K. and L.J. Onesti: Bull. Geol. Soc. Am. Pt. 1, 90, 301 (1979)
Mitchell, R.: Cah. Géol. 44, 440 (1957)
Morisawa, M.: Rivers: form and process. London: Longmans (1985)
Mosley, M.P.: J. Geol. 84, 535 (1976)
Mulhearn, P.J.: Phys. Fluids 19(6), 796 (1976)
Müller, A., A. Gyr and T. Dracos: J. Hydraul. Res. 9(3), 373 (1971)
Murota, A.: Technol. Rep. Osaka Univ. 10, 85 (1960)
Murphy, P.J. and H. Hooshiari: J. Hydraul. Res. 9(3), 373 (1982)
Murray, S.P.: J. Geoph. Res. 75(9), 1647 (1970)
Nakagawa, H. and I. Nezu: J. Hydraul. Eng. ASCE 110(2), 173 (1984)
Nakagawa, H. and T. Tsujimoto: Bull. Disas. Prev. Res. Inst., Kyoto Univ. 31(2), 115 (1981)
Nakagawa, H., T. Tsujimoto and S. Nakano: Bull. Disas. Prev. Res. Inst., Kyoto Univ. 32(1), 1 (1982)
Nakagawa, T.: Sedimentology 30(1), 117 (1983)
Naot, D. and W. Rodi: J. Hydraul. Div. ASCE 108 (HY8), 948 (1982)
Nasser, M.S.: J. Eng. Univ. Riyadh 4(2), 113 (1978)
Nazaryan, A.G., N.S. Pokhsraryan and M.I. Ter-Astabatsatryan: Izv. Akad. Nauk. Armyan. SSR. Ser.
    Tekh. Nauk 12 (5), 60 (1959)
Nikuradse, J.: Forsch.-h. Ver. Deut. Ing. No. 356 (1932)
Nordin, C.F. and J.H. Algert: J. Hydraul. Div., ASCE 92 (HY5), 95 (1966)
Norgaard, R.B.: Water Resour. Res. 4, 647 (1968)
Nottvedt, A. and R.D. Kreisa: Geology 15(4), 357 (1987)
O'Brien, M.P.: Trans. Am. Geoph. Un. 14, 487 (1933)
O'Brien, M.P. and R.D. Rindlaub: Trans. Am. Geoph. Un. 15, 593 (1934)
Odgaard, A.J.: J. Hydraul. Div. ASCE 108 (HY11), 1268 (1982)
Odgaard, A.J.: J. Hydraul. Div. ASCE 112 (HY12), 117 (1986)
Ori, G.G.: Sediment. Geol. 31, 231 (1982)
Orme, A.R. and R.C. Bailey: Yearbk. Assoc. Pac. Coast Geogr. 33, 56 (1971)
Oseen, C.W.: Hydrodynamik. Leipzig: Akademische Verl-ges. (1927) see p. 132
Ovsepyan, V.M.: Sb. Nauchn. Trud. Erevansk. Politekhn. In-ta No.9, 81 (1955)
Ozer, A. and 4 others: Z. Geomorph. Suppl. 49, 87 (1984)
Parker, G.: J. Hydraul. Div. ASCE 101 (HY2), 211 (1975a)
Parker, G.: Water Resour. Res. 11(4), 551 (1975b)
Parker, G.: J. Fluid Mech. 76(3), 457 (1976)
Parker, G.: Self-formed straight rivers with stable banks and mobile bed in non-cohesive alluvium.
    Edmonton: Dept. Civil Eng. (1977)
Parker, G. and A.G. Anderson: J. Hydraul. Div. ASCE 103 (HY9) 1077 (1977)
Parker, G. and E.D. Andrews: Water Resour. Res. 21 (9), 1409 (1985)
Parker, G., K. Sawai and S. Ikeda: J. Fluid Mech. 115, 303 (1982)
Petit, F.: Z. Geomorph. Suppl. 49, 95 (1984)
Petit, F.: Catena 14, 453 (1987)
Philbrick, S.S.: Bull. Geol, Soc. Am. 85, 91 (1974)
Pickup, G.: J. Hydrol 30(4), 365 (1976)
Pokhsraryan, M.S.: Izv. Akad. Nauk Armyansk. SSR 10(6), 85 (1957)
Pokhsraryan, M.S.: Izv. Akad. Nauk Armyansk. SSR 11(6), 31 (1958)
Potter, P.E.: J. Geol. 86, 423 (1978)
Potter, P.E. and F.J. Pettijohn: Atlas and glossary of primary sedimentary structures. Berlin Göttingen
    Heidelberg: Springer (1963)
Pounder, E.J.: Polytech. Geogr. Soc. Mag. 8, 24 (1980)
Powers, M.C.: J. Sediment. Petrol. 32, 117 (1953)
Prandtl, L.: Trans. 2nd Int. Congr. Appl. Mech. Zürich, p. 62 (1926)
Pratt, C.J.: J. Hydraul. Div. ASCE 99(HY1), 121 (1973)
Pratt, C.J. and K.U.H. Smith: J. Hydraul. Div. ASCE 98(HY5), 859 (1955)
Prestegaard, K.L.: Water Resour. Res. 19(2), 472 (1983)

Prus-Chacinski, T.M.: New Scientist 4(79), 16 (1958)
Rakoczi, L.: Geogr. Ann. A 69(1), 29 (1987a)
Rakoczi, L.: Proc. AIRH-Congress Lausanne (1987b)
Razumikin, N.V.: Vestn. Leningr. Un-ta No. 24, 93 (1960)
Rehbinder, G.: Int. J. Rock Mech. Min. Sci. 14, 229 (1977a)
Rehbinder, G.: J. Appl. Math. Phys. 28 (1977b)
Rendon-Herrero, O.: Water Resour. Res. 14(5), 889 (1978)
Reynolds, O.: Philos. Trans. R. Soc. Lond. A 186, 123 (1894)
Ribberink, J.S.: Comm. Hydraul. & Geotech. Eng., Delft Univ. 87, 2, 1 (1987)
Richards, K.: Progr. Phys. Geogr. 1(1), 65 (1977)
Richards, K.: Earth Surf. Proc. 3, 345 (1978)
Richards, K.: J. Fluid Mech. 99(3), 597 (1980)
Richards, K.: Progr. Phys. Geogr. 10(3), 401 (1986)
Richards, K.: Progr. Phys. Geogr. 11(3), 432 (1987a)
Richards, K.: Inst. Br. Geogr. Spec. Publ. 18, 1 (1987b)
Richards, K.: Inst. Br. Geogr. Spec. Publ. 18, 348 (1987c)
Richards, K. (ed.): River channels: environment and process. London: Inst. Brit. Geogr. Spec. Pub. 18
    (1987d)
Richards, K.: Progr. Phys. Geogr. 12(3), 435 (1988)
Richards, K. and R.I. Ferguson: Proc. 1st Conf. Internat. Geomorphol., ed. V. Gardiner; New York:
    Wiley 1, 541 (1987)
Richardson, P.D.: J. Sediment. Petrol. 38(4), 965 (1968)
Riley, S.J.: Bull. Int. Assoc. Hydrol. Sci. 21(4), 545 (1976)
Romanenko, B.E.: Trudy Akad. Morsk. Flota (4), 158 (1956)
Rosin, P.O. and E. Rammler: Kolloid-Z. 67, 16 (1934)
Rouse, H.: Trans. Am. Soc. Civ. Eng 102, 534 (1937)
Rouse, H.: Engineering hydraulics. New York, Wiley (1950)
Roy, A.G. and M.J. Woldenberg: J. Geol. 94, 402 (1986)
Rozovskiy, I.L.: Dvizhenie vody na povoroto otkrytogo rusla. Kiev: Izd. Akad. Nauk Ukr. SSR (1957)
Rundle, A.: Z. Geomorph. Suppl. 55, 1 (1985)
Samaga, B.R., K.G.R. Raju and R.J. Garde: J. Hydraul. Eng. ASCE 112(11), 1103 (1986)
Sawada, T., K. Ashida and T. Takashi: Z. Geomorph. Suppl. 46, 55 (1983)
Sawai, K.: Bull. Disas. Prev. Res. Inst. Kyoto Univ. 37, 19 (1987)
Scheidegger, A.E.: Physical aspects of natural catastrophes. Amsterdam: Elsevier (1975)
Scheidegger, A.E.: Principles of geodynamics. 3rd edn. Berlin Heidelberg New York: Springer (1982)
Scheidegger, A.E. and P.E. Potter: Sedimentology 8, 39 (1967)
Schelling, H.v.: Trans. Am. Geoph. Un. 32, 222 (1951)
Schelling, H.v.: General Electric Report, No. 64 GL92, Schenectady, N.Y. (1964)
Schick, A.P. and D. Magid: Catena 5, 237 (1978)
Schmidt, M.: Festschr. Zentr. Anst. Met. Geodyn. Vienna (1926)
Schmidt, M.: Gerinnehydraulik. Wiesbaden: Bauverlag (1957)
Schoklitsch, A.: Handbuch des Wasserbaus. Vienna: Springer (1950)
Schumm, S.A.: J. Hydraul. Div. ASCE 95(HY1), 255 (1969)
Schumm, S.A. and H.R. Khan: Bull. Geol. Soc. Am. 83, 1755 (1972)
Schumm, S.A., D.W. Bean and M.D. Harvey: Earth Surf. Proc. 7, 17 (1982)
Shahjahan, M.: Bull. Int. Assoc. Sci. Hydrol. 15(3), 13 (1970)
Shapiro, K.H. SH: Trudy Vsesoyuz. In-ta Gidromekh. i Melior. 28, 171 (1958)
Sharma, T.C. and W.T. Dickson: J. Hydraul. Div. ASCE 105 (HY5), 555 (1979)
Shen, H.: Modeling of rivers. New York: Wiley (1979)
Shen, H.: J. Hydrol 68(1–4), 333 (1984)
Shen, H. and C.S. Hung: J. Hydraul. Eng. ASCE 109(3), 368 (1983)
Shen, H. and S. Komura: Proc. Am. Soc. Civ. Eng. 94, J. Hydr. Div. HY4, 997 (1968)
Shen, H.W. and B.C. Yen: J. Hydrol 68(1–4). 333 (1984)
Shepherd, R.G.: Science 178, 409 (1972)
Shepherd, R.G.: J. Geol 93, 377 (1985)
Sherman, L.K.: Eng. News- Rec. 108, 501 (1932)
Shields, A.: Mitt. Preuss. Vers.-Anst. Wasserbau u. Schiffbau, Berlin 26,1 (1936)
Shifrin, K.S.: The kinetics of precipitation. Trudy Obshch. Geofiz. Obs. No. 31 (1951)

Shifrin, K.S.: Izv. Akad. Nauk SSSR, Ser. Geofiz. 1958, 280 (1958)
Shteynman, B.S.: Meteorol. i. Gidrol. No. 11, 46 (1969)
Shulits, S.: Trans. Am. Geoph. Un. 22, 622 (1941)
Shulits, S. and R. Hill: Bedload formulas. State College: Penn. State Univ. (1968)
Slingerland, R.: Geology 9(10), 491 (1981)
Smart, G.M.: J. Hydraul. Eng. ASCE 110(3), 267 (1984)
Smith, J.D.: J. Geoph. Res. 75(30), 5928 (1970)
Smith, K.V.H. and T.J. Yates: Proc. Inst. Civil Engrs. 59(2), 215 (1975)
Smith, N.D. and D.G. Smith: Geology 12(2), 78 (1984)
Smith, T.: J. Geol. 82, 98 (1974)
Snow, R.S. and S.L. Slingerland: J. Geol. 95, 15 (1987)
Sobey, I.J. and P.G. Drazin: J. Fluid Mech. 171, 263 (1986)
Sölch, J.: Peterm. Geogr. Mitt. Erg.-h. 219/220, 139 (1912)
Soni, J.P.: Water Resour. Res. 17(1), 33 (1981)
Soni, J.P., R.J. Garde and K.G.R. Raju: J. Hydraul. Div. ASCE 106(HY1), 117 (1980)
Southard, J.B. and J.R. Dingler: Sedimentology 16, 251 (1971)
Southard, J.B., L.A. Boguchwal and R.D. Romea: Earth Surf. Proc. 5(1), 17 (1980)
Stall, J.B., R. Rupani and P.K. Kandaswamy: J. Hydr. Div. Am. Soc. Civ. Eng. 84, Pap. 1531 (1958)
Statham, I.: Earth surface sediment transport. Oxford: Clarendon Press (1977)
Stein, R.A.: J. Geoph. Res. 70, 1831 (1965)
Stephenson, D. and M.E. Meadows: Kinematic hydrology and modelling. Amsterdam: Elsevier (1986)
Sternberg, H.U.: Z. Bauw. 25, 483 (1875)
Sumer, B.M. and A. Müller (eds.): Mechanics in sediment transport. Proc. Euromech., 156. Istanbul. Dordrecht: Balkema (1983)
Sundborg, A.: Geogr, Ann. 38, 125 (1956a)
Sundborg, A.: Geogr. Ann. 38, 174 (1956b)
Sundborg, A.: C.R. Ass. Toronto, Assoc. Hydrol. Sci., U.G.G.I., 1, 249 (1957)
Sutherland, A.J.: J. Geoph. Res. 72, 6183 (1967)
Tanner, W.F.: J. Geoph. Res. 65, 993 (1960)
Tanner, W.F.: Earth Surf. Proc. 2(4), 417 (1977)
Tanner, W.F: J. Geol. Educ. 10(4), 116 (1982)
Taylor, E.H.: Trans. Am. Geoph. Un. 20, 631 (1939)
Tchen, C.M.: Mean value and correlation problems connected with the motion of small particles suspended in a turbulent fluid. Diss., Tech. Hoogesch. Delft. The Hague: M. Nijhoff (1947)
Thakur, T.R. and A.E. Scheidegger: J. Hydrol. 12, 25 (1970)
Thomas, R.M.: Nature 277, 281 (1979)
Thomas, W.A. and A.L. Prasuhn: J. Hydraul. Div. ASCE 103 (HY8), 851 (1977)
Thomson, J.: Proc. Roy. Soc. 25, 5 (1876)
Thorne, C.R. and R.D. Hey: Nature 280, 226 (1979)
Thorne, C.R. and 5 others: Nature 315, 746 (1985)
Thornes, J.: Z. Geomorph. Suppl. 36, 233 (1980)
Tiffany, J.B. and G.A. Nelson: Trans. Am. Geoph. Un. 20, 644 (1939)
Todorovic, P.: Water Resour. Res. 11(6), 919 (1975)
Tooby, P.F., G.L. Wick and J.D. Isaacs: J. Geoph. Res. 82(15), 2096 (1977)
Trofimov, A.M. and V.M. Moskovkin: Earth Surf. Proc. 8(4), 383 (1983)
Vanoni, V.A.: Trans. Am. Geoph. Un. 22, 608 (1941)
Vanoni, V.A. and N.H. Brooks: Laboratory studies of the roughness and suspended load of alluvial streams. Pasadena: Report, Sedimentation Laboratory. California Institute of Technology, (1957)
Vanoni, V.A. and G.N. Nomicos: Trans. Am. Soc. Civ. Eng. 125, pt. 1, 1140 (1960)
Velikanov, M.A.: Izv. Akad. Nauk, SSSR, Otd. Tekh. Nauk 1944 (3), (1944)
Velikanov, M.A.: Dvizhenie nanosov. Moscow: Rechizdat (1948)
Velikanov, M.A.: Izv. Akad. Nauk, SSSR (Otd. Tekh. Nauk) (1951)
Velikanov, M.A.: Vest. Mosk. Un-ta No. 12 (1954)
Velikanov, M.A.: Dinamika Ruslovykh Protokov, 2 Vols. Moscow: Gos. Izd. Tekh.-Teoret. Lit. (1955)
Velikanov, M.A.: Izv. Akad. Nauk, SSSR, Ser. Geofiz. 1957(1), 71 (1957)
Velikanov, M.A.: Ruslovoy protsess. Moscow: Gos. Iz-vo Fiz.-Mat. Lit. (1958)
Velikanov, M.A. and N.A. Mikhailov: Izv. Akad. Nauk, SSSR, Ser. Geogr. Geofiz. 1950, No. 5 (1950)
Walling, D.E.: Progr. Phys. Geogr. 10(1), 69 (1986)

Walling, D.E.: Progr. Phys. Geogr. 11(4), 590 (1987)
Walling, D.E. and P.W. Moorehead: Geograf. Ann. A 69(1), 47 (1987)
Werner, P.T.: Trans. Am. Geoph. Un. 32, 898 (1951)
West, E.A.: The equilibrium of natural streams. Norwich: Geoabstracts Pub. (1978)
White, W.R. (ed.): International Conference on River Regime. New York: Wiley (1988)
White, W.R., H. Milli and A.D. Crabbe: Proc. Inst. Civil. Eng. 1975 (2), 265 (1975)
Whiting, P.J. and 4 others: Geology 16, 105 (1988)
Whittaker, J.G. and T.R.H. Davies: in Proc. Exeter Symp. IAHS Pub. 137, 99 (1982)
Whitten, E.H.T.: Bull. Geol. Soc. Am. 75, 453 (1964)
Wilcock, D.N.: Bull. Geol. Soc. Am. 82, 2159 (1971)
Willets, B. and C.G. Murray: J. Fluid Mech. 105, 487 (1981)
Williams, G.P.: J. Hydrol. 88(1–2), 147 (1986)
Williams, J.R. and H.D. Berndt: J. Hydraul. Div. ASCE 98 (HY12), 2087 (1972)
Williams, P.E. and P.H. Kemp: J. Hydraul. Div. ASCE 98 (HY6), 1057 (1972)
Willis, J.C.: J. Hydraul. Div. ASCE 105 (HY7), 801 (1979)
Wittmann, H. and P. Böss: Wasser- and Geschiebebewegung in gekrümmten Flussstrecken. Berlin:
    Springer (1938)
Wolman, M.G.: Am. J. Sci. 257, 204 (1959)
Wolman, M.G.: Geogr. Ann. A 69(1), 5 (1987)
Wolman, M.G. and J.P. Miller: J. Geol. 68, 54 (1960)
Wooding, R.A., E.F. Bradley and J.K. Marshall: Boundary layer meteorology (CSIRO Australia) 5(3),
    285 (1973)
Wundt, W.: Experientia 5(8), 301 (1949)
Yalin, M.S.: Mechanics of sediment transport, 2nd edn. Oxford: Pergamon (1977)
Yalin, S.: Bautechnik 36(3), 96 (1959)
Yalin, S.: Dtsch. Wasserwirtsch. 50, 244 (1960)
Yang, C.D.: J. Hydrol 40(1–2), 123 (1979)
Yang, C.T.: Water Resour. Res. 7(6), 1567 (1971)
Yih, C.S.: J. Fluid Mech. 130, 109 (1983)
Young, G.W. and S.H. Davis: J. Fluid Mech. 176, 1 (1987)
Young, R.W.: Z. Geomorph. Suppl. 55, 81 (1985)
Yu, B. and M.G. Wolman: Water Resour. Res. 23(3), 501 (1987)
Zeller, J.: Schweiz. Bauztg. 83, No. 42 (1965)
Znamenskaya, N.S.: Trudy Gos. Gidrol. In-ta, No. 136, 45 (1966)
Zollinger, F.: Die Vorgänge in einen Geschiebeablagerungsplatz. Zürich: Diss. ETH (1983)
Zuchiewicz, W.: Rocz. Polsk. Tow. Geol. 50(3/4), 311 (1980)

5 System Theory of Landscape Evolution

5.1 Introduction

The phenomena discussed thus far refer only to single features present on the Earth's surface. It is now time to consider the combined effect of all the agents discussed heretofore, and this leads directly into the theory of *systems*.

However, before treating landscapes from the point of view of system theory in its full generality, we shall look at the fundamental principles governing landscape development from a phenomenological standpoint, inasmuch as these form the basis for further considerations.

Subsequently, it is possible to introduce general system theory and to view the fundamental principles of landscape development as expressions of basic system dynamics.

Finally, specific systems, such as drainage basins, will be discussed.

5.2 Fundamental Principles of Landscape Evolution

5.2.1 General Remarks

It has been known for some time that the evolution of landscapes is subject to certain regularities. Some of the arguments bearing out this fact have already been mentioned in this book in connection with the discussion of the morphology of landscapes (cf. Sect. 1.5).

Thus, already Hutton (1788) mentioned regularities of landscape development in his theory of the Earth. Further early attempts were made by advocating landscape development as cyclic, a view which is no longer tenable today (cf. Sect. 1.5.1.2 of this book). The principles of landscape development that are considered as fundamental at present have been summarized by the author (Scheidegger 1987). Accordingly, these are the "Principle of Antagonism", the "Principle of Instability", the "Catena Principle", the "Selection Principle", and the "Principle of Tectonic Predesign". We shall discuss these principles in detail below, if they have not been presented earlier in this book.

5.2.2 Principle of Antagonism

The most fundamental principle of landscape development is the "Principle of Antagonism", which implies that there is an antagonistic action between endogenic and exogenic processes. Its implications have been exposed at length in this book in connection with the morphology of landscapes; viz. in Sect. 1.5.2, to which the reader is referred for further details.

5.2.3 Instability Principle

The idea of antagonistic processes roughly balancing each other in landscape development has to be modified by the observation that this "balance" is quite often unstable. We have already met a case of this type in connection with the theory of ripple and dune formation on river beds (cf. Sect. 4.6.2.4 of this book): if any accidental deviation from uniformity tends to reinforce itself, the dynamic state is necessarily unstable. Phenomenologically, this condition expresses itself not only in the formation of ripples and pools in rivers, but also by the fact that erosion cirques tend to grow, karst holes to increase, valleys to become stepped, etc. Such tendencies are all expressions of the existence of a general "Instability Principle" in geomorphology. An extensive documentation of the operation of this principle has been given by Scheidegger (1983).

The mechanical reason for the operation of the instability principle lies in the frequent existence of a positive feedback mechanism: any accidental nonuniformity frequently increases in proportion to the amount of deviation that has already been reached. Mathematically, this condition corresponds to a Taylor (1950) instability: the dynamic equations of the system are unknown, but the (presumably) nonlinear differential equations can be linearized for states close to the dynamic equilibrium state. Then, the effect of small changes in some variable upon the others can often be specified, at least with regard to the sign. For any parameter x(t) describing a particular landscape feature, the solution of the linearized differential equation is of the form

$$x(t) = \text{const. } \exp(\lambda t). \tag{5.2.1}$$

If λ is positive, the state is unstable: the deviation x(t) grows exponentially, and the value of λ is a measure of the intensity of the instability. Naturally, the Taylor analysis applies only to the *initiation* of an instability; in the long run, a saturation stage may be reached beyond which the instabilities can no longer grow.

5.2.4 Catena Principle

Another fundamental principle of landscape evolution is the "Catena Principle": it states that all landscapes can be considered as a collection of "catenas'", each catena consisting of an *eluvial* region at the top (of flat topography with a lip), a *colluvial* region in the middle (of steep topography with large mass flow rates) and an *alluvial* zone at the bottom (again of flat topography).

Catenas had originally been observed in connection with soil surveys: it has been noted that certain definite sequences (chains, Lat. "catenae") of soil types recurred again and again on slopes; these soil types are connected with specific locations on the slopes (Milne 1935, 1947). In geomorphology, the sequences refer to "chains" of flat-steep-flat elements that follow each other regularly. The catena principle in geomorphology was recognized by Gerber (1986) and Scheidegger (1986). Examples of catenas can be found in nature, e.g., in the structuring of Alpine creeks, of valley steps, and even of scree slopes, amongst many others (Scheidegger 1986; Schmidt 1987; Opp 1985).

The theoretical explanation of the fundamental catena scheme lies in similar considerations as in connection with the instability principle: if one assumes that the erosion rate increases with the topographic gradient, the catena principle is the consequence: the steeper the slope, the faster it recedes; the flat regions above and below tend to remain. In mathematical terms, one arrives at the same models as those that had been applied to slope recession. In fact, Fig. 39 in Sect. 3.5.3.3 of this book also represents the evolution of a catena: the top remains flat, a steep section follows, and the bottom is again flat.

5.2.5 Selection Principle

During the evolution of a landscape by the operation of denudational processes it is seen that a certain directedness occurs: thus, it had been noted by Gerber (1969, 1986) that those landscape forms are "selected" by the natural erosional and weathering processes which are geostatically the most stable ones. Thus, the natural forms and configurations in a landscape are primarily those that are most stable under their own weight. This fact has been ascribed to the action of a general "Selection Principle".

Phenomenological confirmations of the selection principle are seen in the existence of the various types of "structural" features mentioned in Sect. 1.5.2.2, such as "towers", "bastions", etc.

A mechanical explanation for the operation of the selection principle has been given by Scheidegger and Kohlbeck (1985): Regarding towers, a comparison can be made with the form of the Eiffel Tower in Paris, which was constructed as a statically most stable form: the mean vertical pressure (weight per cross-section) is constant at every level if the vertical profile is given by an exponential curve. Other types of forms were investigated in the cited paper of Scheidegger and Kohlbeck (1985) by finite element calculations.

5.2.6 Principle of Tectonic Predesign

It is clear that many geomorphic features are principally controlled by deep-seated (tectonic) processes: this is obvious in the structure of mountain chains, volcanoes, island arcs, etc. However, an element of tectonic or endogenic predesign may also be inherent in many landscape features that are manifestly of exogenic origin. In fact, erosion and other exogenic processes tend to act in a fashion which is concordant with the lithospheric stresses. Naturally, the resulting features are not *directly* caused by the stress field, but the exogenic processes act *preferentially* in conformity with these stresses: either in their shear direction or in a principal stress direction if there is a free surface ("Principle of Tectonic Predesign": Summerfield 1987, 1988).

5.3 General System Theory

5.3.1 Introduction

We are now in a position to view the development of landscapes in terms of general system theory.

Historically, system theory has been applied implicitly to landscape development for a long time. Thus, even Gilbert (1877) already used the word "system" when speaking about "dynamic equilibrium" in drainage basins. Similarly, Davis (1899) implied the existence of "systems" in connection with his cycle theory of landscape evolution. However, the theory of systems as it is understood today (Bertalanffy 1932, 1950) was probably introduced into geomorphology for the first time by Strahler (1950, 1952). Subsequently Chorley (1962) published the first general review of the applications of system theory to geomorphology. A recent critical exposition of the subject matter has been given by Huggett (1985). Furthermore, a brief review of the dynamic aspects of geomorphic systems has been given by the writer (Scheidegger 1988) in connection with a symposium on the relation of system theory with natural hazards (Scheidegger and Haigh 1988).

In this section, we shall first define a geomorphological system, then discuss the equilibrium conditions (which include the response to processes that change some of the parameters, i.e., process-response ideas), and finally analyze the possibilities that lead to a loss of stability.

5.3.2 Concept of a Geomorphic System

5.3.2.1 Definition of a System

A discussion of the theory of landscape systems has to start with a general definition of the concept of a "system".

As noted, this concept was created by Bertalanffy (1932, 1950); a definition of "systems" which is particularly useful in connection with intended geomorphological applications has been given by Haigh (1985). Accordingly, "a system is a wholeness that is created by the integration of a structured set of component parts whose structural and functional interrelationships create an entireness which was not implied by those components in disaggregation".

Thus, in order to speak of a system, one needs (Harvey 1969)

1. a set of elements identified with some variable attribute of objects,
2. a set of relationships between attributes of objects, and
3. a set of relationships between those attributes of objects and the environment.

An integrative modification of the systems concept was achieved by the "synergetic" approach taken by Haken (1983). In "synergetics", the interaction is investigated of quite different systems in physics, chemistry, biology, and other fields which exhibit the phenomena of instability and spontaneous pattern formation (Haken and Weimer 1988). Evidently, this view leads to a truly interdisciplinary approach in the investigation of many natural phenomena, of which landscape evolution is but one.

5.3.2.2 Landscape Systems

If the definition of a "system" given above is to be applied to landscapes, it will be necessary to specify the "objects" and the "attributes" mentioned.

Evidently, the "objects" are landforms (e.g., slopes, rivers, coasts, etc.), the "attributes" are quantifiable properties of these landforms (e.g., drainage density, slope angle, meander curvature, etc.) and the "relationships" consist of exchanges of energy and mass between the landforms so that casual links are formed between the "attributes" (Carson and Kirkby 1972). In this connection, a distinction is generally made between the "system proper" and the "environment". However, this distinction is in fact due to an arbitrary decision made by the observer; a natural delimitation occurs only in systems that are completely "closed" within themselves; i.e., which do not have any interaction whatsoever with any "outside". Most landscape systems are "open" in the sense that there is an interaction with the "outside", i.e., an input and output of mass and energy.

Thus, the model fo a landscape is that of an *open* system through which mass and energy flows or "cascades" from an "input" to an "output" (Ahnert 1981; Terjung 1982; Newman and Terjung, 1990). The input occurs through solar irradiation and by tectonic activity; it cascades through the atmosphere, hydrosphere, lithosphere, and biosphere, and ends by planetary emission as "output" into space. This view is, in effect, nothing but a sophisticated restatement of the "Principle of Antagonism" mentioned earlier.

The idea of a flow through an open system has been further elaborated by the "pansystem" concept introduced by Wu (1981) based on the accident Chinese Ying-Yang philosophy, according to which one looks upon the interaction of the endogenic and exogenic affects as the panenvironment and on the geomorphic patterns as functions and properties therein (Ai and Gu 1984; Ai et al. 1986).

5.3.2.3 Biological Models of Geomorphic Systems

Inasmuch as a landscape can be considered as an open system through which there is a flow of mass and energy, an analogy suggests itself with biological processes (for a review of this matter see Higgins 1983). Such an analogy is already inherent in the old Davisian cycle theory (cf. Sect. 1.5.1.2 of this book), now untenable, according to which landscape evolution runs from a stage of youth through maturity to old age, in analogy with the life cycle of an animal or human. Incidentally, the cycle theory (cf. Thornes 1983) has been extended to a polycyclical interpretation of the evolution of the entire crust of the Earth (Huang and Chen 1985).

The Davisian cycle corresponds, in fact, to the evolution of a closed system, in which energy and mass input occurs only at the very beginning. In reality, however, a landscape is much more like a dynamic, self-regulating open system in which energy and mass input and dissipation occur continuously and concurrently. In this instance, the analogy between a landscape and life processes is seen to be incomplete: at best it applies to landscapes with living begins which either remain in a temporary state of dynamic equilibrium (Hack 1960) or are subject to more or less rapid changes, the latter being principally represented by "growth". In biology, growth occurs commonly "allometrically" (Huxley 1924, 1932), implying that the ratio of relative growth rate of a part to that of the whole organism remains constant throughout the growth process. This implies further that, if two parts of the system are each related allometrically to the whole, then they are related to each other by a power function (Mosley and Parker 1972). If, in addition, the growth rate is proportional to the size which the system has already attained, then the growth is exponential.

The allometric growth concept has been applied to geomorphology primarily by Woldenberg (1966) and by Bull (1975). Since many variables in geomorphic systems are indeed related to each other by power and by exponential functions, this feature has been taken as a confirmation of the validity of biological analogs. However, Scheidegger (1967a) has shown that this feature can also be explained by purely random fluctuations and by Taylor (1950)-type of feedback loops.

Biological analogs of geomorphic systems are, therefore, at best only very superficial.

5.3.2.4 Hierarchical Systems

We have already mentioned (in Sect. 1.5.3.2) that many landscape features show a hierarchical structure. In fact, we have seen that hierarchies can be formalized by the introduction of the concept of a "holon": a holon is a component with regard to a larger-order system, but a wholeness that integrates the operation of its own lower-level components.

In the context of system theory, it is the dynamics of each component (holon) which creates stresses in the hierarchy. In this instance, each component (holon) acts as an individual attractor within the system, i.e., as a relatively stable region in the phase space representing the state of the entire system. Then the competition between the attractors determines the evolution of the system. This competition is influenced by accidental fluctuations of the energy and mass supply and by the past history of the system. The evolution of the hierarchical structure, thus, is a balancing act for all levels of the system as a whole (Haigh 1987).

Inasmuch as the hierarchical structure represents an integrated whole, changes are always discontinuous. If the stresses resulting from the evolution of the components (holons) cannot be accommodated, the system must either decompose to the level of the highest stable subwhole (Simon 1962) or it must restructure the hierarchy (hierarchical jumps; Platt 1970). In effect, a "threshold" is attained, as this will be described in Sect. 5.3.4 on the loss of stability in systems.

5.3.3 Equilibrium Theory

5.3.3.1 Definition

Inasmuch as landscapes can be represented as open systems, their aspect is the result of the interaction of a variety of processes. Of interest are particularly the conditions under which landscapes remain more or less unchanged for some time; these conditions are termed "equilibrium conditions".

The equilibrium conditions are expressed by relationships between specific geomorphological parameters. They are exemplified, for instance, by the conditions for a river to be "at grade" (cf. Sect. 4.6.4.2 of this book). The stability of the equilibrium is determined by the behavior of the parameters if one of them is changed, be it by external influences or by accidental fluctuations (cf. Sect. 5.2.3): if the effect of a small change is such that the deviation from equilibrium is reduced, the equilibrium is stable.

The history of the dynamic equilibrium concept in landscape evolution has been reviewed by Bremer (1984). Originally, it was applied to specific features, such as slopes (Ahnert 1967; Leopold 1970), before it was used in the general system context (Christofoletti 1981; Schumm 1985a).The equilibrium conditions had originally been based on energy and mass balance considerations (Vondran 1979, 1985), but entropy analogs are probably more adequate.

5.3.3.2 Analogy with Phenomenological Thermodynamics

Such entropy analogs can be set up entirely heuristically. Thus, Leopold and Langbein (1962) have noticed that an analogy could be set up between landscape variables and thermodynamic variables. In two dimensions (Cartesian coordinates x,y) the thermodynamic field is described by the temperature $T(x,y)$ and the quantity of heat Q. The landscape field is described by the height $h(x,y)$ of the land above some base line and the mass M. Leopold and Langbein (1962) then introduced heuristically the following analogs

$$T \leftrightarrow h, \tag{5.3.1}$$

$$dQ \leftrightarrow dM. \tag{5.3.2}$$

Based on the above, it is possible to define corresponding entropies S in thermodynamics and landscape evolution:

$$dS = dQ/T \leftrightarrow dM/h. \tag{5.3.3}$$

Similar correspondences between thermodynamic and landscape quantities can be established for other variables. Thus, the quantity of heat introduced in a given substance is given by (Scheidegger, 1967a)

$$dQ = \gamma \, dT, \tag{5.3.4}$$

with γ being a heat capacity coefficient. The analog of this in a landscape is

$$dM = \gamma \, dh, \tag{5.3.5}$$

where γ is now an analog of the heat capacity coefficient.

Our task is now to extend the above correspondences to energy terms. For a regular thermodynamic system, the first principle of thermodynamics states

$$U_2 - U_1 = Q + W \tag{5.3.6}$$

or, in differentials

$$dU = dQ + dW, \tag{5.3.7}$$

where U is the internal energy, Q is the quantity of heat introduced from outside and W the work performed externally on the system. In landscape evolution, one would like to have, therefore, a similar relation, viz.

$$U_2 - U_1 = M + W \tag{5.3.8}$$

or, in differentials

$$dU = dM + dW, \tag{5.3.9}$$

where U now signifies some potential, M the mass that was introduced and W some "fictitious work" whose physical meaning has yet to be defined.

For an ideal gas, W is

$$W = -\int_v p\,dV. \tag{5.3.10}$$

Here, V is the geometric domain in which the variables vary, and p the pressure. Because of the ideal gas law, the latter can be expressed as follows

$$p = \frac{RT}{V} = \text{const}\frac{T}{V}. \tag{5.3.11}$$

The last relation yields a means of setting up an analogy to "pressure" in landscapes. In the latter, V corresponds to the area A under consideration, T is the height h (see above) so that one has

$$p_{landscape} = \text{const}\frac{h}{A}, \tag{5.3.12}$$

at least in the equilibrium case.

If we are essentially interested in an "average" geographical cross-section across a landscape, we have only one space coordinate (x); denoting the total length of the section by L, we have (denoting the constant by α)

$$p_{landscape} = \text{const}\frac{h}{L} = \frac{h}{L}\alpha. \tag{5.3.13}$$

The analog of work is then

$$W = -\int p\,dV = -\int \alpha(h/L)\,dL. \tag{5.3.14}$$

The above analogies allow one to transfer the thermodynamic equilibrium conditions to corresponding equilibrium conditions in landscape development. In thermodynamics, one has as condition for the establishment of a steady state that the entropy production rate σ must be a minimum (Prigogine 1945):

$$\sigma = -\sum \frac{\text{grad}\,T}{T^2}J = \text{minimum}. \tag{5.3.15}$$

Here, J is the heat flux per unit time. Because of the analogy with landscape evolution, this immediately gives a condition for the establishment of a steady state in a landscape also. For a one-dimensional system one has:

$$\sigma = -\int \frac{T'}{T^2} J \, dx = -\int \frac{h'}{h^2} J \, dx = \text{minimum}, \tag{5.3.16}$$

where J is the mass flux per unit time.

The last formula may be applied to a river profile; the latter, indeed, can be regarded as a geomorphological profile. For a steady state, one must have (Scheidegger and Langbein 1966a)

$$\delta\sigma = -\delta\int \frac{h'}{h^2} J \, dx = 0, \tag{5.3.17}$$

where J, being proportional to mass flow, represents sediment transport. Choosing intuitively $J = J(h')$ leads to the following Euler-Lagrange equation for the minimization:

$$h''\left[2\frac{J'(h')}{h^2} + h'\frac{J''(h')}{h^2} \right] - 2\frac{J'(h')}{h^3} h'^2 = 0. \tag{5.3.18}$$

Choosing specifically

$$J = ch' \tag{5.3.19}$$

one obtains

$$\frac{h''}{h'} - \frac{h'}{h} = 0, \tag{5.3.20}$$

which yields when solved

$$h = c_1 e^{-c_2 x} \tag{5.3.21}$$

with c_1 and c_2 as constants of integration.

This formula is analogous to that obtained by a deterministic model.

The above theory may be modified by choosing

$$J = (-qh')^a, \tag{5.3.22}$$

where q is the discharge volume per unit river width and a a constant. One may further assume that q is proportional to x (distance from source in the river basin), for the drainage area is roughly proportional to x^2, the river width proportional to x, hence q to x. This yields

$$J \sim (xh')^a. \tag{5.3.23}$$

Then, the solution of the corresponding Euler-Lagrange equation for the minimization of the entropy production rate [see Eq. (5.3.17)] yields

$$h = \left[c\frac{a-1}{a+1} \ln kx \right]^{(a-1)/(a+1)}, \tag{5.3.24}$$

where c and k are again constants of integration. Thus

$$h' = ch^{2/(1+a)}/x. \tag{5.3.25}$$

A regression analysis on 153–221 streams yields (Scheidegger and Langbein 1966a)

a = 1.5.

This is an excellent confirmation of the theory. Other applications of the theory of minimum rate of entropy production in fluvial equilibrium, with similar success, have been reported by Yang and Song (1979). In spite of this, the basic premises of the theory have been questioned by Davy and Davies (1979), but more on philosophical than on factual grounds.

5.3.3.3 Analogy with Statistical Thermodynamics

It is well known that there are two approaches to thermodynamics: one starting from phenomenology, as used in the last chapter, the other from statistical mechanics.

The idea of using statistical mechanics for representing complicated processes is an old one: it was probably applied for the first time in connection with the kinetic theory of gases (Boltzmann 1897), in which the physical behavior of a complicated mechanical system is described by parameters like temperature, pressure, etc., which play the role of "average" quantities referring to stochastically fluctuating variables.

In connection with landscape evolution, the conceptual problems have been discussed by Scheidegger (1961, 1964) and Scheidegger and Langbein (1966b). As in the kinetic theory of gases, one is basically dealing with a transfer process: in the thermodynamic case, the transferred quantity is energy over the various degrees of freedom, in the landscape, the transferred quantity is mass over the various locations given by the coordinates x, y. In both cases, one is dealing with fluctuating systems. In both cases, one can set up canonical equations of motion, Hamiltonian functions etc., except that in the landscape case the "constant of the motion" to which the Hamiltonian function refers is mass rather than energy.

In fact, the evolution of a landscape is (like that of a gas volume) a mechanically completely determined process. However, it contains so many variables that it is not possible to follow each one individually. It is possible to evade this logistic difficulty by considering, instead of the one system (landscape) actually under consideration, a whole *ensemble* of similar systems that must be regarded as identical within the limits of what one knows about them. All quantities in which one might be interested are then expectation values which are obtained by taking statistical averages over the ensemble.

Thus, in order to apply statistical mechanics properly, the following steps have to be observed: (1) one must specify the ensemble; (2) one must specify the a priori statistics in the ensemble, and (3) one must specify the microdynamic transport law.

Simplifications can be achieved in the above scheme if the validity of the ergodic principle can be assumed: the latter is commonly taken as meaning that time averages can be replaced by ensemble averages in large-scale statistical statements. In a strict sense, this principle does not imply that time averages can be substituted *generally* by ensemble averages: this substitution can be made only if there is some indication that the space sample is, in fact, a good representation of a well-defined statistical ensemble (the problem has been reviewed by Abrahams 1972a, 1973 and by Paine 1985). In specific cases, tests should be made for the (suspected) "randomness" in nature (Mann 1970; Matheron 1970).

With regard to the ensemble, each possible state of the system (landscape) will have a certain probability w. The entropy S of that state is then defined by the well-known equation of Boltzmann (1897)

$$S = k \log w, \tag{5.3.26}$$

where k is some constant that has a very specific significance in gas theory, but not in landscape theory.

If the system under consideration is a large, closed system which consists of a linear combination of fluctuating component systems, then the entropy S of the actual state of the large system can be expressed easily in terms of the probabilities w_i of the actual states of the component systems. It is

$$S = - \sum w_i \log w_i, \tag{5.3.27}$$

In the continuous case, the sum is replaced by an integral. In the equilibrium case, if there exists a constant of the motion, the latter quantity is canonically distributed in the component systems.

The last equation can immediately be applied to calculate the "entropy" of a landscape relief. In cartography, it is common practice to use logarithms to base 2 ("logarithmus dualis", ld). The entropy of a map section is then defined as (Sukhov 1967; Zdenkovic 1977, 1985; Ai 1987)

$$S = - \sum_{i=1}^{n} f_{irel} \, ld f_{irel}, \tag{5.3.28}$$

where n is the number of altitude intervals chosen and f_{irel} the relative frequency of landscape "points" (represented by the nodes of an arbitrary square grid laid on the map) falling into the height interval i. It is easy to show that the landscape entropy is independent of the scale of the map, but a normalization has to be made for a standard number of height intervals, usually 100. It turns out that statistically calculated landscape entropies are highest for plains areas, as compared with those for hills or high mountains.

The statistical definition of landscape entropy given above has been compared with the phenomenological one given in the last section by Lechthaler and Scheidegger (1989). It was shown that the two definitions are entirely equivalent within their respective contexts: they differ solely in that the phenomenological analogy refers to an open isobaric system, whilst the statistical definition refers to a closed equipartioned system.

The above considerations lead directly to the introduction of stochastic models into geomorphology (Scheidegger 1970b; Cheng and Hodge 1976; Whitten 1977): the landscape system is considered as fluctuating according to certain laws (which have to be specified); then the expectation values for specific variables are calculated as time averages (assuming, e.g., Markov processes; cf. Khromchenko 1968; Skala 1977) or as ensemble averages.

It may be useful at this juncture to also mention a procedure of setting up a stochastic theory of geomorphic features which is quite different from that sketched above. Thus, a fluctuating geomorphic system may be considered in which a property, x, exists at many points i so that the values x_i of the property at these points can be regarded as independent. Let us further assume that the fluctuations of the property x_i are given by the same Gaussian probability function p

$$p(x_i) = \text{const} \exp(-x_i^2/\alpha^2),$$ (5.3.29)

which, for the sake of simplicity, has been chosen such that the average

$$\bar{x}_i = 0.$$ (5.3.30)

Then, the joint probability distribution $P(x_1, x_2 \ldots x_n)$ is given by

$$P = \text{const} \exp\left\{\frac{-1}{\alpha^2}[x_1^2 + x_2^2 + \cdots x_i^2 + \cdots x_n^2]\right\}.$$ (5.3.31)

This probability has evidently a maximum if the sum σ^2 of the "variances" x_i^2

$$\sigma^2 = x_1^2 + x_2^2 + \cdots x_n^2$$ (5.3.32)

has a minimum. This has been called the "principle of minimum variance" (Langbein 1964). The result of the minimization is trivial, viz.

$$x_i^2 = 0,$$ (5.3.33)

unless constraints exist between the x_i. In the latter case, the minimization can be effected by using Lagrange multipliers. An example of this procedure has been presented in Sect. 4.7.4.2 of this book on meander theory, where also the drawbacks of this type of approach have been mentioned. Calculating the most probable configuration of the system (implicit in the minimization of σ^2) and taking the value of an observable for it is not in conformity with the general principles of statistical mechanics; the accepted procedure is to calculate expectation values of an observable by averaging all its values over the ensemble rather than by taking its value for the most probable state in the ensemble.

Notions similar to the principle of minimum variance have been introduced by Harvey (1969), Schumm (1985), Trofimov, and Moskovkin (1985) in the form of a "principle of optimality". However, it has been noted (Scheidegger 1970b) that this procedure again does not lead to the expectation values of variables as calculated in statistical mechanics (where not the most probable state in the ensemble, but the average over the ensemble is relevant), and this leads to results that are too regular. Therefore, Knighton (1977) has attempted to deduce the minimum variance principle without explicit recourse to statistical mechanics, by simply assuming that the variables pass through a series of states in which the tendency for the change is given by some a priori probabilistic assumption. This, however, is again not in concert with the usual procedures in physics.

5.3.3.4 Analogy with Stationary Stochastic Processes

Instead of focusing one's attention on the whole of a stochastically fluctuating system consisting of many elements, one can consider the behavior of the individual elements: the latter, then, are subject to stochastic fluctuations regarding specific "random" variables ("observables") attached to them. In fact, a specific case of this type has been encountered in connection with the statistical theory of turbulence (Sect. 2.2.2.3).

Let the random variable be denoted by X; its values $X(t_i)$ at times t_i ($i = 1...n$) form a "time series". In the case where X is measured (almost) continuously, it may be preferable to consider t as a continuous variable, so that our observable $X(t)$ will be a continuously fluctuating random variable.

One can then introduce the cumulative random variable

$$x(t) = \int_{const}^{t} X(t)\,dt. \tag{5.3.34}$$

If the fluctuations of the random variable are many, independent, and additive, one says that "Brownian" conditions obtain, from the analogous case occurring in Brownian motion. Under such conditions, the distribution of X is Gaussian around some mean, and for the cumulative variable x the following relation holds:

$$\overline{(x - \bar{x})^2} = Dt, \tag{5.3.35}$$

where D is some constant.

The above scheme can be modified by assuming that the influences are not independent, but are correlated. Then one can introduce the correlation function C(s) for the random variable X:

$$C(s) = \frac{1}{\overline{X^2}} \int_0^\infty X(t)X(t + s)\,dt. \tag{5.3.36}$$

The correlation time is then defined as follows:

$$L_t = \int_0^\infty C(s)\,ds \tag{5.3.37}$$

assuming that the integral converges. Then one has the following results:

1. The original random variable X is still Gaussian (ultimate independence of the "influences", if one waits long enough).

2. For the cumulative variable x one has the following relations:

$$\overline{x^2(t)} \text{ proportional to } t \text{ for } t \gg L_t \tag{5.3.38}$$

$$\overline{x^2(t)} \text{ proportional to } t^2 \text{ for } t \ll L_t. \tag{5.3.39}$$

Thus, the behavior of the "cumulative" dispersion is bracketed between proportionality to t^2 (short times) and proportionality to t (long times).

If the integral defining L_t [Eq. (5.3.37)] diverges, it is not possible to carry out the scheme outlined in the last section. In this case, the formalism can be salvaged, as outlined by Mandelbrot (1965), by assuming that the time series is entirely self-similar, for large and small time intervals. Then, one has for the cumulative variable x

$$\bar{x}^2 \text{ proportional to } t^{2H}, \text{ with } \tfrac{1}{2} \leq H \leq 1 \tag{5.3.40}$$

for large and small (i.e., all) times t. The quantitiy H has been called the Hurst parameter (cf. Sect. 2.2.2.3), after Hurst (1951), who observed empirically such behavior in floods. Newman and Turcotte (1990) have quite generally claimed that erosional landscapes are scale-invariant and fractal. This may indeed be the case for moderate temporal and spatial scales; however, if the argument is carried over extreme scale ranges, the fluctuations become so severe that the resulting structures are neither differentiable nor integrable ("ultraviolet catastrophe"). Furthermore, such models are entirely empirical and no physical reason is given for the supposed self-similarity.

Of the models discussed above for the representation of time series, the simplest is evidently that corresponding to Brownian motion with an appropriately chosen correlation time L_t (the so-called Gauss-Markov model; for a review see Lloyd 1974). This yields for the cumulative variable a dispersion proportional to t^{2H} with H between 1/2 and 1. This is just, in fact, what is observed in nature. It can also be shown that, in the nonequilibrium case, it leads to a diffusion degradation type at evolution (Culling 1986; cf. Sect. 5.3.4.3 of this book). In order to obtain a self-similar (fractal) landscape evolution, the assumption of the operation of a nonlinear dynamic process in the Fourier transform of the landscape system has been useful (Newman and Turcotte 1990).

5.3.4 Nonequilibrium Theory

5.3.4.1 General Remarks

The systems theory presented thus far deals only with equilibrium conditions. However, what is of interest is evidently the case when a landscape system is not in dynamic equilibrium. In that case, the system is not stable and changes must occur: these can either be slow in the direction of establishing equilibrium or rapid in the direction of a violent complete rearrangement of the parameters.

5.3.4.2 The Process-Response Concept

We consider first small deviations of a landscape system from dynamic equilibrium ensuing in a slow change of the system towards reestablishment of the equilibrium. The original deviation may have been caused by some geomorphic process that affects one or more of the pertinent variables; the system then "responds" by an adjustment of the remaining parameters so that equilibrium is reestablished: this is the process-response theory. A reestablishment of the equilibrium in this fashion presupposes that the latter is and has been a "stable" equilibrium.

The process-response idea has been applied widely to many cases. The first specific case in which it was used seems to have been that of beaches (Krumbein 1963); a general formulation of it was given by Whitten (1964). Modern reviews have been given by Terjung (1982) and by Chorley et al. (1984). We have already mentioned the process-response idea in this book in connection with the discussion of the principle of antagonism (Sect. 1.5.2); specific examples have been given in the sections on the regime theory for rivers (4.6.4.3), and in the sections on meanders (4.7.2.3, 4.7.4).

The classic domain for the application of the process-response idea is slope theory. There are innumerable publications on this aspect of the process-response concept; let us only mention here publications by Leopold (1970), Williams (1970), Kirkby (1971), Carson and Kirkby (1972), Young (1972), Moon (1975), Vondran (1977), Mosley and O'Loughlin (1980), Loyda (1980), Armstrong (1980), Mosley (1981), Selby (1983), Twidale and Milnes (1983), Ahnert (1984), and Burt (1984, 1986, 1987, 1988). In fact, many of the slope models discussed in Chapter 3 of this book could be viewed in the process-response context: thus, the slope retreat due to the undercutting by a river is the "response" of the slope to the "process" of river erosion (Sect. 3.5.4.2). Similarly, climatic (Sect. 3.5.4.5) and endogenic (Sect. 3.5.5.2) "processes" entice a "response" in the slope configuration.

Regarding the response of slopes and landscapes to climate, it may be noted that this idea is at the root of the climatogenic model of relief generations of Büdel (1977, 1982) mentioned in Section 1.5.1.3. A critical assessment of Büdel's model has been given by Stäblein (1984), by Kiewietdejonge (1984) and by Hövermann (1985). Other aspects of the response of landscapes to climatic processes were discussed by Chorley et al. (1984), and by Zuchiewicz (1987).

Other specific applications of the idea of landscape "response" to external "processes" are in connection with tectonic (Loyda 1980; Bremer 1985) and anthropogenic (Woehlke 1969; Gregory 1976a; Ichim and Randoane 1986; Ives 1986) influences.

In all the specific applications mentioned above, the "response" to the external "processes" has essentially been calculated or estimated on a mechanical or heuristic basis. Thus it is known, for instance, which combinations of parameter values make a river to be at grade, so that the effect of the change of one parameter on the others can be calculated deterministically. The same is true for the effect of climatic or tectonic processes (cf. the slope models in Sect. 3.5.5 of this book).

A more systemic approach to the process-response phenomenon has been taken in connection with the discussion of entire drainage basins. Thus, Yonechi (1980, 1984, 1985) considered relations between landforms as of a transmitter-receiver type, as of an actor-reactor type or as independently evolving. Most landscape relations can be fitted into this scheme.

On a somewhat more restricted basis, Ahnert (1976, 1977) investigated three-dimensional landscape processes, and Patton and Schumm (1981), as well as Farrenkopf (1987), stream processes in drainage basins.

The principal aspect of the process-response theory is that the response is assumed to be continuous. This is the case only if no thresholds of instability are reached. Any deviation from equilibrium (due to an external "process") is therefore supposed to be reduced by the response of the system which will adjust its parameters until equilibrium is reestablished.

5.3.4.3 Dissipative Systems

We come now to a discussion of systems in which any possible stability is absolutely lost.

In the first instance, we consider systems in which changes occur slowly and always in the same direction. Because such changes are then generally connected with mass or energy dissipation, one speaks in this instance of "dissipative systems".

The classic case of such systems in physics is found in stochastically fluctuatirg systems: under nonequilibrium conditions, a fluctuating quantity, assuming linear regression of the fluctuations and microscopic reversibility, is subject to a diffusivity equation with a symmetric diffusivity tensor.

Thus, the analogy of landscape systems with open isobaric systems in the equilibrium case (see Sect. 5.3.3.2) suggests its extension to nonequilibrium cases.

In thermodynamics the temperature is subject to a diffusivity equation (D is some constant):

$$\frac{\partial T}{\partial t} = D\left[\frac{\partial^2 T}{\partial x^2} + \frac{\partial^2 T}{\partial y^2}\right]. \tag{5.3.40}$$

In view of the analogy between height and temperature [cf. Eq. (5.3.1)], one also postulates a diffusivity equation for the height h in a landscape:

$$\frac{\partial h}{\partial t} = D\left[\frac{\partial^2 h}{\partial x^2} + \frac{\partial^2 h}{\partial y^2}\right]. \tag{5.3.41}$$

For the diffusivity equation, one has the following well-known solutions (in one dimension)

$$h = \frac{1}{(4\pi Dt)^{1/2}}\exp\left(-\frac{x^2}{4Dt}\right), \tag{5.3.42}$$

$$h = \frac{1}{2} + \frac{1}{2}\operatorname{erf}\left(\frac{x}{(4Dt)^{1/2}}\right). \tag{5.3.43}$$

This is, in effect, the same pattern as had been discussed in Sect. 3.4.4.6. The formulas represent a decay of initial forms: the first of an idealized mountain range (Fig. 25), the second of an idealized slope bank (Fig. 26). It is thus seen that, based upon entirely general principles, one must expect a decay of initial irregularities in a landscape. Obviously, we have an "end" to the decay process when h = const. everywhere. This corresponds to the equipartition theorem in thermodynamics. It also corresponds to a state of maximum entropy in a closed system (Lechthaler and Scheidegger, 1989). The diffusivity equation is thus also seen to embody the tendency of a (closed) system towards attaining a state with maximum entropy. This tendency, incidentally, has been advocated by Haigh et al. (1988) for the successful prediction of the behavior of complex landslide systems, and by Andrews and Hanks (1985; also Hanks and Andrews 1989) for the dating of such geomorphic features as scarps.

Extensions of the diffusivity equation of landscape (slope) evolution to three dimensions have been proposed by Hirano (1976) and by Ohmori (1983, 1984), modifications for varying erodibility (e.g., lithology) of the slope material have been considered by Tödten (1976) and by Pierce and Colman (1986), generally with very suggestive results.

5.3.4.4 Instability Conditions

Dissipative systems can reach conditions in which the stochastic fluctuations lead to an intrinsic instability: this is at the root of the "instability principle" of geomorphology (cf. Sect. 5.2.3). The intrinsic stability of a state can be tested by the Taylor (1950) analysis; the specifics of that procedure have been presented in Sect. 5.2.3.

A modern version of stability analysis has been discussed by Slingerland (1981), who investigated feedback loops in geology. Specific instances have been discussed particularly with regard to erosional phenomena, usually in connection with the geomorphological "instability principle" mentioned above. Thus, Scheidegger (1983) analyzed the growth of an erosion cirque; Patton and Schumm (1975), Finkl (1982), and Bowyer-Bower and Bryan (1986) investigated the initiation of gullies and rills, and Slingerland and Snow (1988) tested the stability of rejuvenated fluvial systems.

5.3.4.5 Thresholds

A landscape or geomorphic system may be stable or slowly adjusting to changing conditions for a wide range of parameters. However, when certain limits of the latter are exceeded, the regular behavior of the system may become abrupt and discontinuous: such critical combinations of the parameters are called thresholds.

The concept of threshold, in relation to geomorphology, has been discussed by Schumm (1979), Fairbridge (1980), and Bremer (1984).

The discontinuous, abrupt behavior of a geomorphic system can be the result of the external parameters reaching a critical combination, or it may be the result of an intrinsic instability being reached.

Thresholds may be the result of the junction in space of different types of transport mechanisms; these junctions, then, represent spatial singularities (Louis 1976), such as waterfalls, abrupt edges of erosion cirques, etc.

More interesting are temporal discontinuities: if several quasi-stable states exist in a system, flips between them can occur. These flips may be induced by small changes in the external parameters or they may arise spontaneously in the course of normal stochastic fluctuations. In fact, there are two possibilities for a system to attain instability in consequence of small changes of certain parameters (Scheidegger 1989).

The first possibility occurs when the region of dynamic (quasi-) stability of the system is described by a "surface" (a space of dimension m < n) in parameter space (of dimension n). This "surface" may contain bifurcations, cusps, or folds. Thus, for a given value of a certain "control" parameter, several combinations of the other parameters may produce a stable state, between which flips are possible. Whitney (1955) has shown that smooth mapping of a "surface" onto a "plane" can contain only cusps and folds as singularities; this was later developed into "catastrophe theory" by Thom (1972; see also Zeeman 1976 or Arnold 1984, for a review). Thus, when in a process-response system the external parameters change by some process, the response must lie on the dynamic stability surface. If, then, one of the mentioned cusps or folds is reached, a sudden flip to another branch of the stability surface *must* occur: this is the catastrophe threshold.

The second possibility of loss of stability occurs if one control parameter possesses one or several critical "bifurcation" values at which the structure of the system changes abruptly. Small further changes of the control parameter lead to further bifurcations which are arranged hierarchically (Haigh 1987). The corresponding states may be run through until a stationary state of complete "chaos" is reached (Eckmann 1981, Kadanoff 1983; Brun 1986; Harrison and Biswas 1986). This type of behavior is characteristic of nonlinear dissipative systems. The determination of the "roads to chaos" is a major achievement of system theory.

In geomorphology, both types of instability are of importance. The first type leads to the recognition of the threshold conditions that lead to flips between several possible regimes in a region (e.g., between beach and cliff coast; between shooting and streaming flow, etc). The second type gives the conditions under which regular features yield to chaotic ones (cataclysmic periods in the Earth's history: Huggett 1988; chronic landslides in the Himalaya: Haigh 1988; etc.).

5.3.4.6 Rare Events

As is implicit in the foregoing discussion, rare events can occur during the temporal evolution of a geomorphic system. In effect, the incidence of rare events can become a near-certainty if long enough time spans are considered. In this instance, such events can have a significant effect on the comportment of the system (Gretener 1967).

The occurrence of rare events has been studied, in the first place, as a purely statistical problem, without any reference to their specific causes. The latter, in fact, can lie in intrinsic instabilities of the system which are reached only at infrequent intervals, or they may lie outside the system proper that is under consideration.

Turning first to the purely statistical aspects of the rare events occurrence problem, we note that the common way is to consider the time intervals t_i between subsequent events in the system under consideration. One thus arrives at a stochastic series of numbers t_i which represent the return periods of the events. Then, one can empirically construct a probability function F(t) giving the probability that a return period has not ended up to the time t.

An alternative way is to count the number n of rare events that have occurred in the system in a unit time interval. This leads directly to a frequency distribution function f(n) regarding unit time.

The probability functions in stationary stochastic processes are often considered as Gaussian (cf. Sect. 5.3.3.4), which leads to "Brownian" conditions. However, in the case of rare events, a Poisson distribution is more adequate (cf. e.g., Mises 1931), for which the probability p(j, t) of j events occurring in the time t is given by

$$p(j, t) = \frac{(kt)^j}{j!} \exp(-kt). \qquad (5.3.44)$$

The recurrence time probability function is then given by

$$F(t) = p(1, t) = kt \exp(-kt) \qquad (5.3.45)$$

and the frequency distribution function by

$$f(n) = p(n, 1) = \frac{k^n}{n!} \exp(-k). \tag{5.3.46}$$

Both the mean and the variance of the number of events in unit time are equal to k.

The Poisson distribution is characteristic for rare events that are *independent*. If the events are mutually correlated, the probability of occurrence of the following event is conditional to the occurrence of the past event. One arrives, thus, at a chain of conditional probabilities, the simplest of which is represented by a Markov chain (cf. Lloyd 1974, for a review).

Actual analyses of rare events, however, do not always seem to fit any of the above simple models. The most celebrated case in this connection is that of the Hurst (1951) phenomenon (cf. Sect. 5.3.3.4), which implies that there is a correlation which would have to be self-similar for all time intervals considered. Inasmuch as this is a difficult assumption to justify theoretically, Klemes (1976) has sought other explanations of the Hurst phenomenon, notably intrinsically nonstationary processes.

Analyses of the above type have been particularly made with hydrological (floods: Scheidegger 1975, Baker et al. 1988; mudflows etc.: Ohmori and Hirano 1988) and with climatological (e.g., droughts: Gupta and Duckstein 1975) disasters. Based on past experience, procedures of risk estimation have been developed (Torelli 1975; Baker 1982; Lutz 1987; Schumm 1988; Mann and Hunter 1988). In this it must be assumed, of course, that the probability of the occurrence of specific rare events does not change with time, an assumption whose validity is by no means assured. Thus, in order to predict the occurrence of rare events, studies have also been made of the processes that actually *cause* them.

In the first place, such processes can be inherent in the system if the latter, during its evolution, proceeds towards an intrinsic instability (a catastrophe in the sense of Thom 1972) or on a road to chaos (cf. Sect. 5.3.4.4 of this book). Models of this type have been investigated by Graf (1979), particularly with regard to hydrological rare events. Graf (1979) has shown that such models have only limited usefulness in the mentioned context, except in special cases. As one such special case, Schumm (1976) mentioned the episodic evolution of stream gradients which alternate between brief periods of instability and long periods of stability.

What remains, then, is the view that a specific rare event occurs in a geomorphic system as the result of a specific cause (Baker 1988). Thus, for instance, the impact of the Tungus meteorite on June 30, 1908 (Zotkin and Chikulin 1966), caused atmospheric and geomorphic effects (Schröder 1972; Brown and Hughes 1977). Similarly, occasional catastrophic floods (Finkl 1982, 1983; Baker 1983) can cause major geomorphic effects such as the channeled scablands of the State of Washington, U.S.A. (Baker and Nummedahl 1978). In all these cases, specific causes have to be invoked (such as the breaking of an ice dam holding back a lake in the scabland case) for the observed effects. For risk estimates, the likelihood of the arising of such causes has to be assessed.

5.4 System Theory and Drainage Basins

5.4.1 General Remarks

System theory has been applied in geomorphology principally to the study of drainage basins and their evolution. In this, the study is directed towards the stream patterns as well as the complementary ridge patterns.

The aspects that are of importance are either topological or metric.

We shall discuss these aspects in detail.

5.4.2 Topological Aspects

5.4.2.1 Empirical Relations

Before making an exposition of the topology of drainage systems, we have to present some empirical relationships regarding this subject.

The most celebrated of these relationships is Horton's (1945) law of stream numbers. Accordingly, if we denote by n_i the number of segments ("links") of order i (cf. Section 1.4.6.2 in a given network, then the numbers n_1, $n_2 \cdots$ form (on the average) a geometric sequence:

$$n_{i+1} = \alpha n_i. \tag{5.4.1}$$

The inverse of the coefficient α, i.e. $1/\alpha$, is called the bifurcation ratio; it is often denoted by R_b. For the rivers in the U.S.A. one finds approximately (Leopold et al. 1964)

$$1/\alpha = R_b = 3.5. \tag{5.4.2}$$

Horton's law was stated above in terms of Strahler segments. However, one can show that, in a system in which the law of stream numbers is satisfied in the Strahler sense, it is also satisfied in the Horton sense and vice versa. We have (the superscript H stands for Horton, the superscript S for Strahler; Scheidegger 1968a)

$$n_i^H = n_i^S - n_{i+1}^S, \tag{5.4.3}$$

Since exactly n_{i+1}^S of the n_i^S Strahler segments have been relabeled to obtain Horton rivers. Thus

$$\frac{n_i^H}{n_{i+1}^H} = \frac{n_i^S - n_{i+1}^S}{n_{i+1}^S - n_{i+2}^S}, \tag{5.4.4}$$

but

$$n_{i+1}^S = \alpha n_i^S \tag{5.4.5}$$

because one is *assuming* the validity of Horton's law. Thus

$$\frac{n_i^H}{n_{i+1}^H} = \frac{n_i^S - \alpha n_i^S}{\alpha n_i^S - \alpha^2 n_i^S} = \frac{n_i^S(1-\alpha)}{\alpha n_i^S - \alpha^2 n_i^S} = \frac{n_i^S(1-\alpha)}{\alpha n_i^S(1-\alpha)} = \frac{1}{\alpha} = \frac{n_i^S}{n_{i+1}^S}, \tag{5.4.6}$$

which leads to the same bifurcation ratio for Horton and Strahler streams. Thus it is seen that it does not matter whether one uses Horton rivers or Strahler segments to state Horton's law of stream numbers.

Based on Horton's law of stream numbers, one can introduce the notion of a Horton net: this is simply a stream net in which Horton's law of stream numbers is numerically satisfied.

In order to test whether a given river net is a Horton net, one performs a "Horton analysis": One plots the logarithm of the observed number of Strahler segments in each order in a drainage basin *versus* the Strahler order; if a straight line is obtained, the net is a Horton net. It turns out that most river nets are indeed Horton nets (Leopold et al. 1969).

Reviews and discussions of the above procedures have not only been given in the cited book of Leopold et al. (1964), but also in publications by Christofoletti (1973), Jarvis (1978), and Zavoianu (1978).

5.4.2.2 Theoretical Explanations of the Law of Stream Numbers

The empirical law mentioned above regarding the characteristics of drainage networks has evidently a very fundamental significance. If it is possible to give a rational explanation for the existence of this law, the crucial phenomena governing the evolution of networks have evidently been explained. Hence much effort has been directed towards obtaining a rational explanation of this law.

The law mentioned above is statistical in nature: it makes a statement about the "average" behavior of a network. Because of this, and because drainage networks are evidently extremely complicated systems, it stands to reason that the methods of statistical mechanics touched upon in Sect. 5.3.3.3 will have to be used. Accordingly, one deals with a "system"; in our case with a river net. In this "system", there are "observables": stream numbers, bifurcation ratios, etc. The system is so complex that it cannot be described in detail; hence, instead of one system, one considers all possible systems (an "ensemble") which, within the limits of one's ignorance, one must consider as equivalent. The expectation value of any observable is simply its average over the ensemble.

Based upon the above general remarks, a variety of probabilistic models for the explanation of the characteristics of river nets have been set up, which we shall discuss below. Incidentally, reviews of the subject matter have been given by Scheidegger (1970b, c), Tokunaga (1972), Werrity (1972), Werner (1973), Dacey (1976), Smart (1978), Verma and Bhattacharya (1978), Cliff et al. (1979), and Dunne (1980).

Accordingly, Horton (1945) himself has, in fact, already proposed a physical explanation for his law of stream numbers: he assumed that a river net is the result of a regular, cyclic growth process; internally, elements of the same structure are being built up, i.e., new parts of the drainage net are created with always the same bifurcation ratio. The result is evidently a "structurally" Hortonian net (Fig. 71a) in which any N-th order stream receives only tributaries of order $N - 1$. Each of these successive generations of rivers forms a cycle; a cycle is complete when the Strahler order increases by 1.

This picture has been formalized by Scheidegger (1966) based on the idea of branching processes, as well as by Woldenberg (1966), Mosley and Parker (1972),

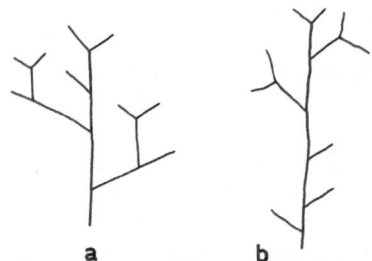

Fig. 71 a, b. Two river nets with nine first-order streams, three second-order streams, and one third-order stream, so that the bifurcation ratio is constant and equal to 3. However, **a** is structurally Hortonian, **b** is not. (After Scheidegger 1968b)

and Faulkner (1974), based on the idea of allometric growth. Assuming that each cycle is geometrically similar to the previous one immediately yields Horton's law of stream numbers. "Allometric" means that the growth rate of the system is proportional to the size of the system. The outcome of this hypothesis is that stream nets should all be completely regular, i.e., structurally Hortonian with constant bifurcation ratio R_b. The network then consists of entirely self-similar cycles. In such regular networks, there is a unique connection between the various characterizations of stream orders, at least for stream links that complete the cycles. Accordingly, Woldenberg (1966) introduced an "absolute" stream order x

$$x = R_b^{N-1},\tag{5.4.7}$$

where N is the Strahler order and R_b the bifurcation ratio. Obviously

$$x = M = I/2 = R_b^{N-1},\tag{5.4.8}$$

where M is the magnitude and I the associated integer to the consistent order (cf. Sect. 1.4.6.2). The growth model has been generalized somewhat by allowing for order-dependent branching probabilities (Van Pelt and Verwer 1983, 1986; Van Pelt et al. 1989) and asymmetries (Smart and Wallis 1971), such as the cis and trans links mentioned in Sect. 1.4.6.3.

Any deviations from complete structural regularity in a river net are, in the growth model, explained away as spurious ("adventitious" streams), a procedure which is evidently not satisfactory, inasmuch as Ranalli and Scheidegger (1968) have shown, by a careful analysis of the Wabash River basin, that "adventitious streams" form a characteristic part of drainage basins.

Thus, different types of stochastic models have been proposed. Such models are based on the assumption that the topology of a network is completely random, i.e., that, for a given number of first-order streams, all possible configurations are equally likely in the ensemble.

The idea of using topologically random networks appears to have been introduced for the first time by Shreve (1966, 1967). Unfortunately, however, Shreve did not then proceed to calculate expectation values of observables, but instead only looked at the most probable configurations and calculated the values of observables for them. This is not generally a satisfactory procedure (see Sect. 5.3.3.3). It was Scheidegger (1967b) who first utilized random networks to calculate actual expectation values of such observables as stream numbers, bifurcation ratios, etc.

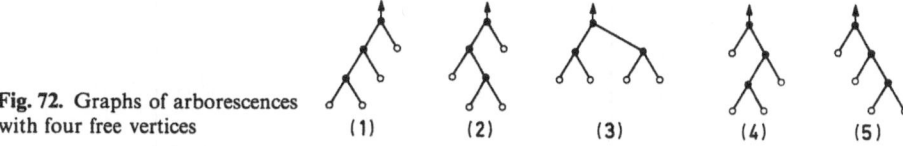

Fig. 72. Graphs of arborescences with four free vertices

(1) (2) (3) (4) (5)

In order to do this, a river network is represented by a special type of topological graph, called a bifurcating arborescence with a root and a given number of free ends. The number N of topologically distinct such graphs with n free ends (first-order streams) was first given by Cayley (1859). It is

$$N = \frac{1}{2n-1}\binom{2n-1}{n-1} = \frac{1}{2n-1}\binom{2n-1}{n}. \tag{5.4.9}$$

As an example, the five graphs with four first-order streams ("free vertices") are shown in Fig. 72.

It is evident that the number N rapidly becomes very large as the number n of first-order streams increases. It is therefore not easily possible to enumerate the ensembles in question for the calculation of expectation values of observables. However, it is possible to random-generate (Monte Carlo technique) a representative sample of the ensembles required on a computer and to calculate the necessary expectation values therefrom. In order to do this, a representation of the graphs in question that is suitable for a computer is required. This can be done by means of "Lukasiewicz words" as outlined in Sect. 1.4.6.2. Then, by random generating words of the above type, river nets can be computer-generated.

The above technique has been used by Liao and Scheidegger (1968) to sample-generate the ensembles in question up to 500 ends and to calculate expectation values for stream numbers in various Strahler orders. The results are displayed in Table 5. It is seen that Horton's (1945) law of stream numbers is fairly well satisfied, at least if a river net of a given order is fully developed. It is not surprising that discrepancies occur when a new order just starts to appear, inasmuch as the sampling procedure is then felt. Although Horton's law is fairly well satisfied, there are systematic deviations inasmuch as R_b decreases somewhat with stream order. This is also observed in nature.

The Monte Carlo results given above were somewhat improved upon by Smart (1968) by using an implicit exact formula to make the required calculations. However, the computing effort in this connection is large and Smart had to stop at 80 free ends.

It may be noted that, in further developments of the random graph model, Tokunaga (1972) has proposed a more efficient algorithm for calculating expectation values than that given above, Sharp (1971) has shown that random graph patterns can also be produced by a suitably chosen (Fibonacci) branching process, and Abrahams (1975a) has taken environmental controls into account.

Reviewing the stochastic models proposed for the explanation of the law of stream numbers, one might well ask oneself whether models other than the two discussed are also possible. This question can be answered as follows.

Essentially, what one has for river nets is a "growth" process that leads to the final configuration which is then observed in nature. The "influences" that affect

Table 5. Results for random graph model of river nets. (After Liao and Scheidegger 1968)

Number of first-order streams	Number of segments, in Strahler order							Bifurcation Ratio R_b					
	1	2	3	4	5	6	7	$i=1$	2	3	4	5	6
3	3.00	1.00						3.00					
4	4.00	1.40	0.40					2.86	3.50				
5	5.00	1.50	0.50					3.33	3.00				
6	6.00	1.62	0.60					3.70	2.72				
7	7.00	1.86	0.70					3.77	2.66				
8	8.00	2.24	0.88	0.002				3.57	2.53	378.00			
9	9.00	2.43	0.94	0.014				3.70	2.57	67.43			
10	10.00	2.61	0.96	0.02				3.83	2.72	48.00			
12	12.00	3.18	1.07	0.08				3.78	2.98	12.88			
14	14.00	3.64	1.16	0.16				3.85	3.13	7.09			
16	16.00	4.13	1.25	0.25				3.88	3.29	5.03			
18	18.00	4.68	1.35	0.34				3.85	3.45	3.97			
20	20.00	5.11	1.46	0.43				3.91	3.50	3.39			
30	30.00	7.62	2.06	0.79	0.004			3.94	3.70	2.61	202.36		
80	80.00	20.20	5.10	1.50	0.46	0.10		3.96	3.96	3.40	3.26		
200	200.00	50.06	12.75	3.31	1.09	0.10		3.99	3.93	3.85	3.04	10.67	
500	500.00	129.78	31.39	8.00	2.06	0.80	0.01	4.01	3.98	3.92	3.88	2.58	120.00

the final configuration are manifold. What are now the possibilities for the structure of the final configuration? Clearly, either there are substructures ("cells" or "cycles") or there are none. In principle, the probability distribution in each cell could be anything. However, if one assumes many fluctuating influences that are (at least asymptotically) independent and additive, then the central limit theorem of probability theory ascertains that the system will have a Gaussian distribution; a special case thereof is a constant distribution. Modifications are possible only by introducing autocorrelation.

An inspection of the river net models given above shows that the two basic possibilities of constructing models have indeed been used up. Possible modifications are only obtained if one assumes that the probability distribution is Gaussian rather than constant in each (or the only) cycle, or the possibility of a change of the mean (autocorrelation) from cycle to cycle.

Thus it is clear that the two types of models discussed above for the explanation of Horton's law of stream numbers are essentially the only ones that are possible, except for minor modifications.

5.4.2.3 Change with Time

The random graph theory allows one to obtain a model for the *evolution* of a river net (Scheidegger 1970a). One simply assumes that a network grows by a given rate of capture of first-order streams. Then, the following questions may be asked:

First, given the rate of capture $\dot{n}_1(t)$ of first-order streams, what is the evolution of the structure of the network, i.e., what are the expectation values $n_i(t)$ in terms of $n_1(t)$, $\dot{n}_1(t)$?

Second, if a particular model for $\dot{n}_1(t)$ is adopted, e.g.,

$$\dot{n}_1 = kn_1 \tag{5.4.10}$$

what is *then* the (expected) evolution of the structure of the network?

Table 5 immediately answers, in tabular form, the first of our questions. The first column corresponds to the "size" of the network, given by $\int \dot{n}_1(t)dt$. One can, therefore, immediately read off the increase in the expectation values n_i for a given increase in n_1. The maximum Strahler order is also determined from that table.

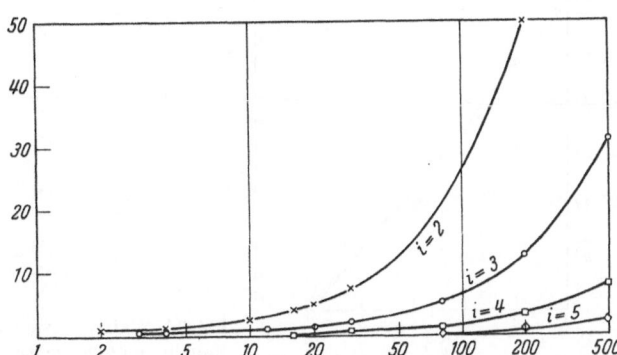

Fig. 73. n_i as a function of n_1. (After Scheidegger 1970a)

Turning to the second question, the assumption of (5.4.10) leads, upon integration, to

$$n_1(t) = e^{kt}. \tag{5.4.11}$$

The most visualizable way to describe the result is simply by plotting the values of Table 5 on semilogarithmic paper (logarithmic with regard to n_1). Since the growth in time of n_1 is given by an exponential function, equal distances on the (logarithmic) axis will simply present growths of $n_2 \cdots n_i$ in equal time intervals. This has been done in Fig. 73.

The above argument assumes the absence of any external (e.g., environmental) agents. In the presence of such agents, mainly of anthropogenic origin (Gregory 1976b; Ovenden and Gregory 1980; Howard and Kerby 1983) adjustments occur rapidly in conformity with the process-response principle (Sect. 5.3.4.2).

5.4.2.4 Tests of Models with Nature

The theories given above purporting to explain Horton's law of stream numbers give correct results, i.e., the correct bifurcation ratio. However, before accepting any of these theories, one must test the validity of the basic hypotheses in nature. This alone will permit one to decide which is the truly correct explanation of Horton's law of stream numbers.

In order to test the basic assumptions of the various theories given above, Ranalli and Scheidegger (1968) made an investigation of the Wabash River system, choosing a root point near Terre Haute, Indiana. In order to test the law of stream numbers, a Horton analysis, plotting the stream numbers n_N against order N, was made, as reproduced here in Fig. 74. The highest-order river obtained was of order 6. The Horton analysis yielded a bifurcation ratio $R_b = 4.088$, a little higher than common for U.S. river basins. An inspection of Fig. 74. (shows) that Horton's law of stream numbers in well satisfied.

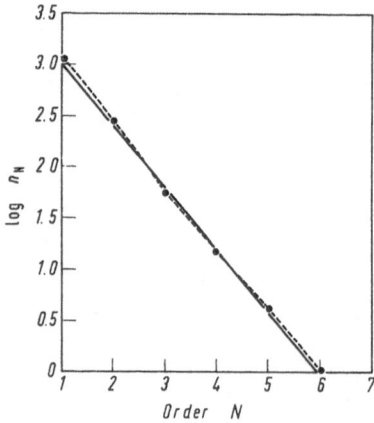

Fig. 74. Wabash river system, Horton diagram. (After Ranalli and Scheidegger 1968)

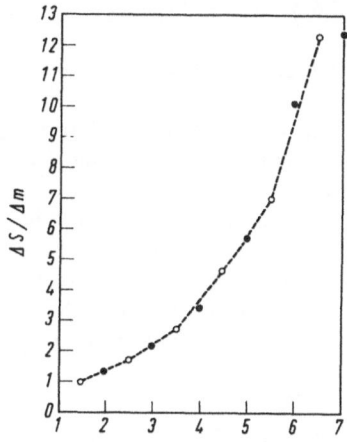

Fig. 75. Wabash river system, number of junctions per constant change in consistent order vs. consistent order. (After Ranalli and Scheidegger (1968)

The next problem was to investigate the structure of the net. For a structurally regular (cyclic) Horton net, Eq. (5.4.8) has to be satisfied, which may also be written

$$m = 1 + (N - 1)\log_2 R_b, \tag{5.4.12}$$

where m is the consistent and N the Strahler order. This can be further transformed to

$$\Delta S = [(R_b - 1)/\log_2 R_b]\Delta m, \tag{5.4.13}$$

where ΔS is the number of junctions required to increase the consistent order by Δm. In a regular net, $\Delta S/\Delta m$ is constant. The result of plotting this relation is shown in Fig. 75, which clearly indicates that $\Delta S/\Delta m$ is not constant so that the Wabash system clearly turns out to be not cyclic.

Finally, a test for the randomness of interconnections in subbasins with four ends was made. The five possibilities denoted by (1) to (5) in Fig. 72 occur with the following relative frequencies: (1) 18.92%, (2) 21.62%, (3) 21.62%, (4) 18.92%, (5) 18.92%. This clearly indicates topological randomness, as assumed in the random graph theory.

Thus, the investigation of the Wabash River net by Ranalli and Scheidegger shows that the random topology explanation of Horton's law of stream numbers is correct in at least that case. On the other hand, the cyclic model does not seem to be applicable, inasmuch as river nets, in nature, do not seem to be structurally Hortonian: adventitious streams appear to be part and parcel of natural networks and cannot be ignored.

Similar tests, with corresponding results, have been made by Werner (1971), Coffman et al. (1972), Jarvis and Werrity (1975), Dacey and Krumbein (1976), and Shimano and Matsukura (1978). Specifically, river nets in Britain (Mosley 1972), Greece (Mariolakos et al. 1976, 1981), and Austria (Drexler 1979) have been investigated in the light of the statistical theory, partly taking account of anthropogenic and lithological factors.

5.4.3 Metric Aspects

5.4.3.1 Empirical Relations

The metric aspects of river nets (cf. Sect. 1.4.6.4) refer to the lengths of river segments (links), to the drainage areas and to the junction angles. For these characteristics, empirical relationships have been found.

Turning first to the link lengths, we note that Horton (1945) has established an empirical relation that is analogous to his law of stream numbers. Horton's law of stream lengths can be stated as follows:

The lengths L_i of (Strahler) segments of order i in a given drainage basin form, on the average, a geometric sequence, so that

$$L_{i+1} = \beta L_i. \tag{5.4.14}$$

The factor β is called the length ratio; it turns out (Strahler 1964) that in nature, the length ratio is on the order of 2.1 to 2.9. Again, it does not really matter whether Horton lengths L^H or Strahler lengths L^S are used; if the length law is satisfied in terms of Strahler segments, then it is also satisfied in terms of Horton rivers (Scheidegger 1968c). One has evidently in approximation (this is only an approximation, because Horton renumbers the longest, not the average streams):

$$L_i^H \cong \sum_{j=1}^{j=i} L_j^S. \tag{5.4.15}$$

Then, assuming the length law to be valid for Strahler segments, one obtains:

$$\frac{L_i^H}{L_{i+1}^H} = \frac{L_1^S + L_2^S + \cdots + L_i^S}{L_1^S + L_2^S + \cdots + L_{i+1}^S} = \frac{L_1^S(1 + \beta + \cdots + \beta^{i-1})}{L_1^S(1 + \beta + \cdots + \beta^i)}$$

$$= \frac{(\beta^i - 1)/(\beta - 1)}{(\beta^{i+1} - 1)/(\beta - 1)} = \frac{\beta^i - 1}{\beta^{i+1} - 1} \sim \frac{1}{\beta} \quad \text{for} \quad i > 1. \tag{5.4.16}$$

The validity of the length law can again be tested by making a corresponding "Horton analysis": one plots the logarithm of the (average) length of (Strahler) segments in a river net versus the order; the result is generally more or less a straight line, implying the validity of Horton's (1945) law of stream lengths. This pattern has been confirmed in many cases (e.g., Leopold et al. 1964; Walsh 1972). In absolute terms (i.e., for the conditioning of absolute lengths, not length ratios) there are, of course, external controls (e.g., mean annual discharge: Morisawa 1967; or network size: Jarvis 1976; Ebisemiju 1979, 1985). The variations of link lengths have been found to be distributed according to log-normal, exponential, and gamma distributions (James and Krumbein 1969). There seem to be some small differences regarding links that end in a tributary and those that end in a main stream (Abrahams 1980a) and in cases where "left" and "right" symmetry is not completely maintained (Russell 1976). On the whole, however, "Horton's law of stream lengths" has been well established.

A logical extension of the relationships discussed heretofore is obtained if one considers drainage areas. This extension leads to the following law: the areas A_i drained by streams of Strahler order i in a given drainage basin form, on the average, a geometric sequence, so that

$$A_{i+1} = \gamma A_i. \tag{5.4.17}$$

The factor γ is called the area ratio. For its value, Leopold et al. (1964) found, in the United States, approximately 4.8.

With regard to the law of areas, it is clear that it is immaterial whether it is stated with regard to Strahler or Horton orders, because, for whole drainage subbasins, these two concepts are identical.

Again, the law can be tested by making a corresponding "Horton analysis": the logarithm of the drained area is plotted against the order. The result is generally a straight line, confirming the law of drainage areas. In general, "Horton's law of drainage areas" has been well established (cf. e.g., Leopold et al. 1964; Christofoletti and Arana 1976).

In absolute terms, there are, of course, again relations between the absolute (as against relative) size of drainage areas and external conditions. Thus, the "diameter" of a (partial) drainage area has been correlated with the order of the highest-order stream it contains and the latter used for interregional comparisons (Kennedy 1978; Jarvis and Sham 1981).

A further observation has been that the stream length L (length of the talweg of the longest continuous downstream water course) and the corresponding drainage area may be related (Hack's law 1957; Mueller 1972; Mosley and Parker 1973):

$$L = CA^B. \tag{5.4.18}$$

The exponent B lies between 0.5 and 0.7 (Sakaguchi 1969); the factor C depends on the units used: if L is in miles, A in square miles, Hack (1957) found $C = 1.4$.

An additional parameter indicative of the metric pattern of a drainage basin is the drainage density (cf. Sect. 1.4.6.4). Several types of correlations have been sought between this indicator and basin area (Gardiner et al. 1977), precipitation (Bandara 1974), and bed load (Bauer 1980), usually with inconclusive results.

Finally, the angles formed by the tributaries at junctions are also metric quantities. Again, Horton (1945) has given a pertinent empirical relationship

$$\cos Z = S_c/g_g, \tag{5.4.19}$$

where Z is the angle of junction (measured in a horizontal plane), S_c the gradient of the main stream, and S_g the gradient of the tributary.

5.4.3.2 Theoretical Explanations of the Metric Relations

First, we turn to the possibilities of giving a rational explanation to Horton's law of stream *lengths*. One attempts of course, to use analogous models as before, but now one has to introduce at some stage or other a *metric* assumption; previously, one was dealing with topology alone.

Thus, the explanation will be based on the previous possibilities: cycle theory and random graph theory, *plus* some metric assumption.

The cycle theory, with the allometric growth assumption implying geometrical self-similarity between cycles, leads immediately to Horton's law of stream lengths. This can be formalized (Woldenberg 1966; Scheidegger 1968a) as follows. Denoting the stream length of order i by L_i, self-similarity implies

$$\frac{L_{i+j+k}}{L_{i+k}} = \frac{L_{i+j}}{L_i} \quad \text{for all } i, k. \tag{5.4.20}$$

Setting $j = k = 1$ yields

$$\frac{L_{i+2}}{L_{i+1}} = \frac{L_{i+1}}{L_i} = \text{const} = \beta \tag{5.4.21}$$

and hence

$$\frac{L_{i+k+1}}{L_{i+k}} = \frac{L_{i+1}}{L_i} = \beta \tag{5.4.22}$$

for all i, which implies a geometric sequence (Horton's law) for L_i. Thus, if the allometric growth idea is correct, Horton's law of stream lengths is the automatic outcome. Furthermore, it turns out that the following relationship should hold between the length ratio β and the bifurcation ratio R_b (Scheidegger 1966)

$$\beta = R_b \tag{5.4.23}$$

However, this relationship is not borne out in nature. Furthermore, cyclic models for the explanation of the law of stream lengths can be invoked only if river nets in nature are truly cyclic. Since this does not seem to be the case, a different explanation for the law of stream lengths must be sought.

The random graph theory employed in connection with the law of stream numbers can be applied to the law of stream lengths provided a metric assumption is introduced. Evidently, the simplest such assumption is to assume that each link length in the graphs is constant; for simplicity's sake this link length may be taken as equal to 1.

With this assumption, it is then possible to calculate the expectation values for lengths of Strahler segments in the various orders in network ensembles with a given number of first-order streams. This is not an easy task, and is best approached again by a Monte Carlo technique of sampling the ensembles in question and calculating the expectation values on such samples. The problem was solved by Liao and Scheidegger (1969), the result, up to networks with 1000 first-order streams, is shown in Table 6.

It is clear that Horton's law of stream lengths is approximately satisfied, but only in the headwater region. In this lowest orders, the stream length ratio starts from an asymptotic value of two which had already been found by Shreve (1969) for "infinite" networks, increases somewhat, and then decreases to values well below 1.

Further attempts at explaining stream length relations assume that a drainage network grows until it reaches a "maximum" extension (Abrahams 1972b). In this connection, direct influences were invoked (Abrahams 1975b; Abrahams and Miller 1982).

Turning next to drainage *areas*, we note that the methodology for finding rational explanations of the pertinent laws is exactly the same as with the law of stream lengths (Scheidegger 1968c), one has the cycle theory and the random graph theory, into which a metric assumption has to be introduced.

Table 6. Theoretical stream lengths and drainage areas in Horton nets of various orders. (After Liao and Scheidegger 1969)

Order	Av. length	Ratio	Av. drainage area	Ratio
Number of free vertices 20				
1	1.00		1.00	
		2.04		5.09
2	2.04		5.09	
		2.38		4.80
3	4.85		24.45	
		0.60		1.60
4	2.92		39.00	
Number of free vertices 500				
1	1.00		1.00	
		1.99		4.99
2	1.99		4.99	
		2.00		4.19
3	4.00		20.97	
		2.03		4.19
4	8.14		87.97	
		2.24		4.21
5	18.29		370.03	
		0.96		2.67
6	17.58		987.63	
		0.06		1.01
7	1.00		999.00	
Number of free vertices 1000				
1	1.00		1.00	
		1.99		4.98
2	1.99		4.98	
		2.00		4.20
3	4.00		21.00	
		2.03		4.12
4	8.13		86.60	
		2.04		4.00
5	16.57		346.35	
		2.08		4.29
6	34.98		1486.47	
		0.54		1.34
7	19.11		1999.00	

Again, the self-similarity between cycles yields immediately the required law. However, one must note that, in complete geometrical similarity, one has to have

$$\gamma^{\frac{1}{2}} = \beta, \qquad (5.4.24)$$

where, as before, γ is the area ratio and β the length ratio. The above relation is not borne out in nature. Smart and Surkan (1967) found

$$\gamma^n = \beta \qquad \text{with } n > \tfrac{1}{2}. \qquad (5.4.25)$$

This is again an indication of the inadequacy of the cycle theory.

In order to extend the random graph theory to the law of drainage areas, one needs again a metric assumption. The simplest such assumption is again that each link (of length L) drains an area proportional to L^2. One does not now have the simple relationship (5.4.24) above, but must figure out the actual relatioship from the ensemble of graphs in question. The problem was again solved by Liao and Scheidegger (1969) by using a Monte Carlo technique; the results thereof show that the law of drainage areas comes approximately out of the theory. However, as with the length law, it is seen that the area law is also only satisfied in the headwater region of a large, finite, network. The drainage area starts around 5 and then decreases towards 1.

Finally, trying to find theoretical explanations for the empirical relations concerning link *angles*, we note that the most successful attempts have been based on some extremum principles (Howard 1971; Woldenberg and Horsfield 1984; Roy 1983, 1985). In these, the angles are thought to adjust themselves in such a fashion that the energy expenditure becomes a minimum. On the other hand, Pieri (1984) has set up a completely deterministic model which is based on the conditions obtaining for graded rivers.

5.4.3.3 Evolution of Link Lengths with Time

As with the topological properties of river nets, studies of the evolution in time have also been made with regard to metric properties, notably with regard to link lengths.

Thus, James and Krumbein (1969) proposed that an initial constant probability of bifurcation would lead to an exponential density of link lengths; then subsequent "adjustments" would lead to a decrease in number and hence to an increase in mean length of the interior links. It is these unspecified adjustments, then, which would condition the evolution of the link lengths with time. As a cause of the adjustments, Blyth and Rodda (1973) have advocated direct physical agents.

Inasmuch as no consistent theory can be built on unspecified, ad hoc adjustments, Dunkerley (1977) has postulated that the evolution is produced by and inherent in the branching mechanism itself. The latter would be subject to the rule that a given catchment area must contain a certain length of drainage channels for the disposal of runoff water. In this fashion, given a constant drainage density (which depends on the prevalent lithological and climatic conditions), the channel length per unit area is also a constant. As the network grows by random branching processes, it was seen (by stochastic simulation) that the mean link length increases slowly with the number of branching generations. Thus, the "adjustments" of James and Krumbein (1969) are explicitly demonstrated to be inherent in the branching process itself.

5.4.3.4 Tests of Models with Nature

As with the theoretical models purporting to explain the topological river net laws, the models purporting to explain the metric river net laws must also be compared with nature.

We have already shown with regard to the topological laws that the cycle theory does not conform to the facts of nature. Hence it cannot be expected to be adequate with regard to stream lengths either, and one must concentrate on the random graph theory. This was done by Ghosh and Scheidegger (1970), who found that the metric assumptions made above are evidently too simple. There are two ways by which one can change the metric assumption: first, by assuming that the mean link lengths in all orders are the same, but that they are drawn from a somewhat lopsided population. Thus Shreve (1967) assumed that the link lengths are subject to a log-normal distribution, but he also took different distributions for "exterior" and "interior" links. A more consistent application of this idea leads to the second possibility, which is to presuppose that the link lengths depend on Strahler order. This was consistently carried through by Ghosh and Scheidegger (1970), who analyzed a series of networks on this basis and found that there is a definite increase of link length with Strahler order. This seems to be a natural law. This analysis was carried out by comparing the observed lengths of Strahler segments in natural networks with the theoretical values given in Table 6. Then, a matching of the observed and theoretical results can only be obtained by increasing the link lengths with order, rather than assuming the latter to be constant and equal to 1. The results show that the link length always has a tendency to increase geometrically with the Strahler order. This leads to the conclusion that there really exists a *law of link lengths* similar to the laws of streams numbers etc. The link length ratios, according to Ghosh and Scheidegger 1970), vary between 1.04 and 2.34.

Finally, the theories purporting to explain the law of drainage areas must be compared with nature. Ghosh and Scheidegger (1970) analyzed the drainage areas of a series of river networks and compared the observed values with the theoretical ones given in Table 6. It was found that the areas drained by each link are indeed constant, independently of the Strahler order of the link in question. This is somewhat startling, inasmuch as the length of the links does increase with Strahler order, contrary to the drained areas. This can only be explained by assuming that the sinuosity of the rivers increases with Strahler order.

Thus, all in all, the random graph theory appears to give a correct and adequate explanation of Hortons (1945) laws: networks simply evolve topologically random. Thus, of all the possibilities by which N first-order streams can be connected, all theoretical possibilities are equally likely, at least in areas where relatively little geological control is present.

5.4.4 Applications to Related Subjects

5.4.4.1 General Remarks

We have noted on several occasions that the system theory developed specifically for river nets can be extended to related subjects.

Thus, the various topological levels (river orders or magnitudes) have been related to other hydraulic variables, such as valley slopes and river parameters.

Furthermore, ridge patterns are obviously complementary to the drainage patterns. Hence, drainage system theory can be applied to the latter.

We shall discuss these possibilities below.

5.4.4.2 Drainage Systems and Hydraulic Variables

Systemic relations have been set up primarily between drainage systems and link slopes (Flint 1976). Observations in nature seem to indicate that the skewness of the link slope distribution moves from left to right as the average stream gradient decreases downstream; this is explained by the decrease of the tendency of streams to move downward with a change in available relief.

A further observation of Flint (1976) notes that interior link slopes vary with river magnitude (cf. definition in Sect. 1.4.6.2) according to a (decreasing) power function. This is taken by Flint as an a priori fact; however, one can understand this characteristic as an analog to the circumstance that the river gradient decreases with distance from the source (cf. Sect. 4.6.5.2). Then a very realistic mathematical model of link slope distribution can be set up.

Furthermore, Leopold and Miller (1956) suggested that generalized multi-variate relations hold between stream order and the hydraulic geometry variables of Leopold and Maddock (1953; see Sect. 1.4.2 of this book). These relations, however, were tested by Miller and Onesti (1977) with equivocal results. Hence a search for possible causes thereof does not seem to be warranted.

5.4.4.3 Ridge Patterns

Finally, we note that ridge patterns are simply complementary to drainage patterns. Hence, similar methods can be applied in the two cases.

The most extensive studies of the subject matter have been published by Mark (1979, 1981). In the first of these, a substantial ridge anisotropy was found in some landscapes which cannot be explained in terms of stream topology alone. However, if certain extremum principles ("minimum spanning tree") are introduced, the general patterns can be successfully explained.

A corresponding study published by Abrahams (1980b) focuses on the "divide angles", i.e., the angles formed by the two divides delineating a tributary basin joining a main stream basin. In areas of uniform environment, the link lengths are inversely related to these divide angles. This seems to be the expression of the condition that the topological as well as the link length properties are controlled by the way in which the partial channel networks fit together in space.

5.5 Simulations of Landscapes

5.5.1 General Remarks

We have already used simulations of landscape systems in order to investigate their general features: thus, we have random-generated arborescences to explain the various laws of Horton (cf. Sect. 5.4.2.2) etc.

Now we want to apply such simulations to the explanation of more specific features, such as valley sectors, braids, network growth, etc.

We note that, quite generally, simulations can be made, on the one hand, one a computer, based on certain assumptions regarding every step of an iteration procedure. On the other hand, simulatios can be obtained from physical analogs such as scale model experiments (Anderson 1988).

We shall discuss these possibilities in turn.

5.5.2 The Stochastic Simulation of Landscapes

5.5.2.1 The Idea

We turn first to the computer simulation of landscapes, in which stochastic iteration procedures are set up which produce statistical ensembles which allow one to calculate expectation values of general observables. Such schemes have been applied above to the explanations of Horton's empirical laws.

Such schemes can be further exploited by applying them to specific cases. The writer has reviewed these possibilities in earlier editions of this book (Scheidegger 1970c; a newer review has been published by Beaumont (1979).

5.5.2.2 Random Model of a Stream Network

We consider first specific generation models of stream networks. The best known of these is a model of Leopold and Langbein (1962). Beginning with a series of points 1

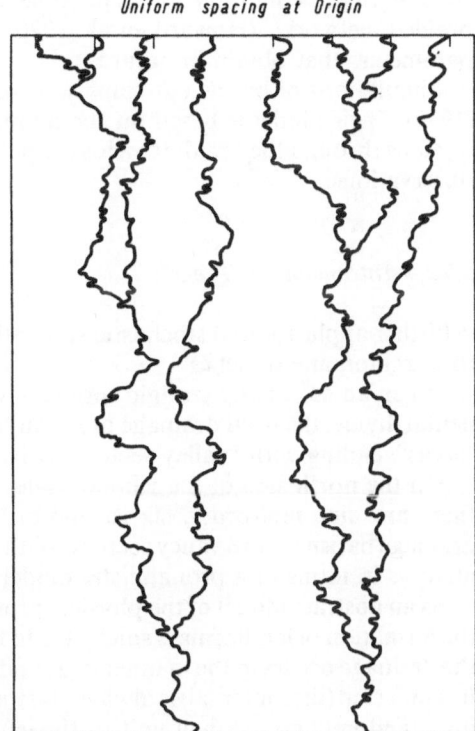

Fig. 76. Random-walk model of stream network. (After Leopold and Langbein 1962)

inch apart on a sheet of paper, these authors constructed random walks by restricting the motion to steps to the right, left, and forward (with equal probability). The several random walks were continued until junctions, representing confluences of rivers, were obtained. The random walk of the joined rivers was then continued; the pattern obtained by Leopold and Langbein (1962) is reproduced here in Fig. 76. Evidently, the pattern is very reminiscent of a natural stream network. One can draw Horton diagrams for the observed patterns; the latter bear out the validity of Horton's laws and therefore of the modeling procedure.

Similar drainage patterns were also produced on a computer by Schenck (1963), Smart et al. (1967), Seginer (1969), Valencia and Schaake (1973), and Morandi et al. (1976).

5.5.2.3 Braided Streams

Features related to "normal" river nets are systems of braided streams. Stochastic simulations have also been attempted for the latter.

The actual mechanism of braiding has already been discussed on a deterministic basis (cf. Sect. 4.7.6.3). At this point, we wish to add some remarks regarding braids that are based on system dynamics.

Thus, one of the principal aspects of braided streams seems to be that similarity is preserved in streams with the same average number of channels, but of different sizes. With this condition, it was possible to make numerical simulation models of braided networks (Howard et al. 1970) which produced relations of system parameters that obtain in nature.

Simulations of braided streams were also proposed by Krumbein and Orme (1972). Their model is based on the number of channels in equally spaced cross sections through the braid. It yields outputs in fair topological agreement with the observations.

5.5.2.4 Intramontane Trench

A further application of stochastic simulation has been made (Scheidegger 1967c) to intramontane trenches.

When an essentially straight, primary valley runs roughly parallel to a continental divide, the total drainage area can be subdivided into first, second ... order "areas", ending with "valley sectors" (Fig. 77). [Gerber's (1944) classification.]

On the north side of the Rhone Valley in the Canton of Valais, Switzerland, there are nine first-order, six second-order, two third-order, two fourth-order drainage basisn and 18 valley sectors. With these numbers, it is possible to make an analysis in terms of a probabilistic model (Scheidegger 1967c).

As an abstract model of the physical processes that are involved we assume that the formation of landforms is solely due to fluvial erosion. We further suppose that the drainage occurs in the manner of a random walk. We see thus that the drainage in a unit step (time interval) is always "forward" (toward the main valley), but that it may randomly go one-half unit to the left or to the right. Thus, it is possible to

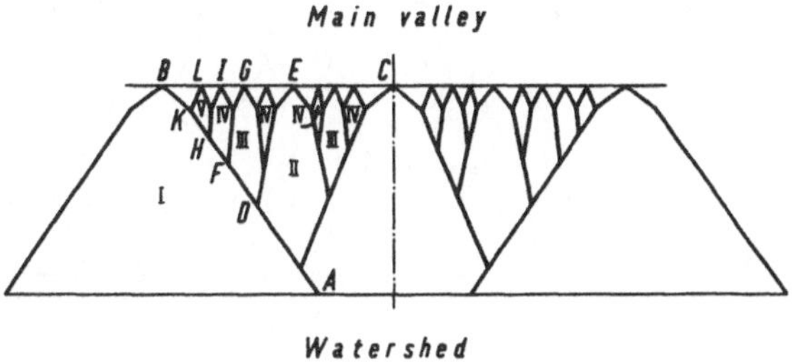

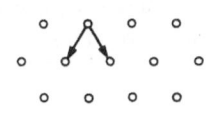

Fig. 77. Gerber's (1944) scheme for the ordering of drainage areas. The *smallest triangles* touching the main valley are the "valley sectors"

model the drained strip by a grid of points; the point are arranged in rows where each subsequent row is displaced sideways by one-half of the lattice distance with regard to the former row. For the drainage of each grid point there are two possibilities: to the nearest downward point to the left or to the right. The choice as to which of the two possibilities actually takes place is completely random. The grid, the "down" direction, and the possible choices (these may take place at each point) are illustrated in Fig. 78.

The simplest way for a deduction of the drainage patterns induced by the stochastic process described above is by simulating the latter directly on a computer (Monte Carlo method). Accordingly, random numbers were generated on a computer. The computer was instructed to print "1" when the number was larger than the mean, "0" when it was smaller. Twenty rows of 50 numbers each were printed, subsequent rows were staggered by one character so as to make the correspondence with the grid of Fig. 78 evident.

Once the random numbers are printed out, it is a simple matter to construct the drainage net. We start at the top of the sheet, then "0" represents drainage to the left (seen from below) "1" to the right. The drainage net that is obtained is shown in Fig. 79. The next step is to draw the boundaries between the drainage areas generated by the random numbers. If this is done, the result is as shown in Fig. 80.

One can now classify the various drainage areas according to order, in the same manner as this was done by Gerber. In this fashion, the valley sectors became more divided up than is inherent in Gerber's scheme. In order to be in conformity with

Fig. 78. The model grid showing the possibilities for each drainage point. (After Scheidegger (1967c)

Fig. 79. Random-generated drainage net. (After Scheidegger 1967c)

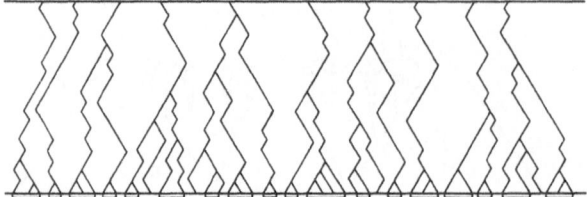

Fig. 80. Random-generated drainage basins. (After Scheidegger 1967c)

Gerber's procedure, one must count into one valley sector all areas containing only single narrow channels; this convention of designating valley sectors has been indicated in Fig. 80 by brackets underneath the drained strip. Counting the various drainage areas gives the result that there are ten basins of first, seven of second, three of third, two of fourth order, and 18 valley sectors. These numbers are close to those observed in the Rhone valley; the results thus strongly suggest that the model employed is correct.

The above prodecure has been extended by Kondoh et al. (1987) to metric properties, simply by focusing their attention to the length and areal properties of Fig. 79. They showed that the model also automatically explains such observational features as Hack's (1957) law (see Sect. 5.4.3.1 of this book) which represents a self-affine structure of river nets but obviates the self-similarity inherent in fractals.

5.5.2.5 Stochastic Simulation of Time Changes

Computer simulations of time changes of river nets have already been mentioned in connection with the general discussion of such time changes (Sect. 5.4.3.3). In addition to the work of Dunkerley (1977) mentioned there, it may be noted that stochastic simulations referring more specifically to the evolution of channel networks have also been reported by Nogami (1982), and to that of river terraces by Boll et al. (1988).

Furthermore, it may be noted that stochastic simulations have been used in connection with the problem of *initiation* of networks in the first place. A notable attempt along these lines was made by Kashiwaya (1978). The model is based on a stochastic transition process from sheet flow to streaky flow which then causes rills to form. This transition is represented by a stochastic differential equation.

5.5.3 The Physical Simulation of Landscapes

5.5.3.1 The Idea

It is evident that models of landscape systems can be made not only by numerical simulation, but also by the study of scale models.

In all scale model experiments, scaling relations have to be applied which are based on a dimensional analysis of the parameters involved.

Then, scale models can be constructed in a laboratory. In some studies, "scale" models have been made on a 1:1 scale, i.e., by the use of experimental full-size landscape plots.

We shall review these possibilities.

5.5.3.2 Dimensional Analysis

In landscape systems, it is possible to define a series of dimensionless numbers. It is possible, in fact, to deduce some semi-empirical relationships from such numbers. The dimensionless numbers must be the same in the models and in the original landscapes. One of the most notable attempts at finding dimensionless numbers is due to Strahler (1958) and involves mainly the drainage density D.

The variables which must be assumed to affect the drainage density are: (1) The runoff intensity Q, measured as the volume rate of flow per unit area. This quantity Q is the difference between the precipitation per unit of area and time and the infiltration per unit of area and time. It thus has the dimension of a velocity. (2) The erosion proportionality factor k_e as defined by Horton. It is the quotient of the mass rate of erosion per unit area and the eroding force per unit area. It thus has the dimension of an inverse velocity. The definition of the "erosional force" is somewhat vague, but it must be understood that, for various landscapes, it ought to be described in an analogous fashion so that the factor k_e will be indicative of the resistance to erosion that is offered by that landscape. (3) The relief H. It is measured as the total difference of height between the summit in the area and the mouth of a river of a given, pre-selected order (see Sect. 1.5.3). The relief has the dimension of a length. (4) The drainage density obviously must also be affected by the density of the fluid ρ (dimension ML^{-3}), the viscosity η of the fluid (dimension $ML^{-1}T^{-1}$) and the gravity acceleration g (dimension LT^{-2}).

Because these are *all* the variables that might interact with each other, there must exist a relationship of the form

$$f(D, Q, k_e, H, \rho, \eta, g) = 0. \tag{5.5.1}$$

The usual way to attack this is by assuming that the relation (5.5.1) can be written as follows

$$D^{\alpha_1} Q^{\alpha_2} k_e^{\alpha_3} H^{\alpha_4} \rho^{\alpha_5} \eta^{\alpha_6} g^{\alpha_7} = \text{const}, \tag{5.5.2}$$

with the right hand side dimensionless. Since there are three dimensions (L, M, T), one obtains three linear equations for the seven alphas, which means that there are four linearly independent solutions. Each linearly independent solution represents a dimensionless number; the latter can be multiplied together with arbitrary exponents to yield a relationship of the required form.

It is thus evident that there must exist four independent dimensionless numbers in the present case. The latter may be chosen as follows

1. $R_u = HD,$ (5.5.3)

which is called the ruggedness number;

2. $H_0 = Qk_e,$ (5.5.4)

which is called the Horton number;

3. $R_e = \dfrac{HQ\rho}{\eta}$,   (5.5.5)

which is the well-known Reynolds number; and

4. $F_r = \dfrac{Q^2}{Hg}$,   (5.5.6)

which is the Froude number.

The drainage density equation must therefore be of the following form:

$$D = \frac{1}{H} f(H_0, R_e, F_r).$$ (5.5.7)

The four dimensionless products introduced above establish the criteria for scaling erosional patterns. As usual, scaling does not "explain" what is happening, but it does at least yield a means of determining the course of natural events by providing the means for making model experiments.

An interesting remark which may be made in connection with what was said above is that Horton (1945) noted that the ruggedness number R_u cannot assume entirely arbitrary values. If ϑ denote the average slope (tangent of slope angle) in the drainage basin under consideration, it turns out that for natural conditions

$$G_e = \frac{R_u}{\vartheta} = \frac{HD}{\vartheta} \sim \frac{1}{2}.$$ (5.5.8)

Here G_e is a dimensionless number which has often been called the *geometry number*. The factor 1/2 denotes an order of magnitude only; the actual values of G_e vary from about 0.38 to 1.00.

The relative constancy of the geometry number in natural basins is significant. It indicates, for instance, that if in a given drainage basin the drainage density be increased through removal of the plant cover, the slope must also become steeper; this will produce the appearance of badlands.

5.5.3.3 Scale Models

Base on the scaling relations deduced above for drainage basins, some notable laboratory studies of characteristic landscape patterns have been made.

We wish to mention here only a few of these studies. Thus Bryan et al. (1978) have investigated the factors controlling the runoff and piping patterns in badlands, Best and Reid (1984) have analyzed the confluence angles in basins in dependence of confluence separation, and Phillips and Schumm (1987) designed a study to investigate the effect of slope on the evolution of dendritic drainage patterns. Schumm et al. (1987) have also published a monograph on experimental fluvial geomorphology, to which the reader is referred for further details.

5.3.3.4 Experimental Field Studies

The ultimate in "scale" model experimentation are 1:1 scale models: this, of course, means experimental field studies in especially constructed or chosen plots.

Inasmuch as the subject matter of this book is *theoretical* geomorphology, we only mention some of these studies as an aside.

Thus, we note that Flint (1973) has investigated the headward growth of channel networks, Cannon (1976) has studied the explicit effect of certain parameters in a drainage basin, Le and Davies (1979) have studied stream braiding experimentally, and Abrahams (1980c) the effect of ground slope on channel link density. For further details, the reader is again referred to the monograph of Schumm et al. (1987) already mentioned earlier.

References

Abrahams, A.D.: Bull. Geol. Soc. Am. 83, 1523 (1972a)
Abrahams, A.D.: J. Geol. 80(6), 730 (1972b)
Abrahams, A.D.: Bull. Geol. Soc. Am. 84, 353 (1973)
Abrahams, A.D.: Bull. Geol. Soc. Am., 86, 1459 (1975a)
Abrahams, A.D.: Geology 3, 307 (1975b)
Abrahams, A.D.: J. Geol. 88, 681 (1980a)
Abrahams, A.D.: Geogr. Anal. 12(2), 157 (1980b)
Abrahams, A.D.: Ann. Assoc. Am. Geogr. 70(1), 80 (1980c)
Abrahams, A.D. and A.J. Miller: Water Resour. Res. 18(4), 1126 (1982)
Ahnert, F.: In: L'évolution des versants, Proc. Symp. Int. Geomorph. ed. P. Macar, Liège: Université de Liège, 23 (1967)
Ahnert, F.: Z. Geomorph. Suppl. 25, 29 (1976)
Ahnert, F.: Earth Surface Proc. 2, 191 (1977)
Ahnert, F.: Z. Geomorph. Suppl. 39, 1 (1981)
Ahnert, F.: Am. J. Sci. 284(9), 1035 (1984)
Ai, N.S.: Acta Conservat. Soli et Aquae Sin. 1(2), 1 (1987)
Ai, N.S. and H.Y. Gu: Chongqing Jiaotong Inst. 2(9), 72 (1984)
Ai, N.S., H.Y. Gu and X.M. Wu: Explor. Nature 4(2), 125 (1986)
Anderson, M.G. (ed.): Modelling geomorphological systems. New York: Wiley (1988)
Andrews, D.J. and T.C. Hanks: J. Geophys. Res. 94(B1), 10193 (1985)
Armstrong, A.C.: Catena 7, 327 (1980)
Arnold, V.I.: Catastrophe theory. Berlin Heidelberg New York Tokyo: Springer 1984
Baker, V.R.: In: Scientific basis of water resource management. Washington: National Acad. Press. 109 (1982)
Baker, V.R.: In: Background to paleohydrology ed. K.J. Gregory. p. 453, New York: Wiley (1983)
Baker, V.R.: Z. Geomorph. Suppl. 67, 25 (1988)
Baker, V.R. and D. Nummedahl: The channeled scabland (field guide). Washington: NASA (1978)
Baker, V.R., R.C. Kochel and P.C. Patton: Flood geomorphology. New York: Wiley (1988)
Bandara, C.M.M.: J. Hydrol. 21. 187 (1974)
Bauer, B.: Z. Geomorph. 24(3), 261 (1980)
Beaumont, C.: Progr. Phys. Geogr. 3(3), 363 (1979)
Bertalanffy, L.v.: Theoretische Biologie. Berlin: Borntraeger (1932)
Bertalanffy, L.v.: Science 111, 23 (1950)
Best, J.L. and I. Reid: J. Hydraul. Eng. ASCE 110(11), 1588 (1984)
Blyth, K. and J.C. Rodda: Water Resour. Res. 9(5), 1454 (1973)
Boll, J. and 3 others: Z. Geomorph. 32(1), 31 (1988)
Boltzmann, L.: Vorlesungen über Gastheorie. Leipzig: Borth (1897)
Bowyer-Bower, T.A.S. and R. Bryan: Z. Geomorph. Suppl. 60, 161 (1986)

Bremer, H.: Z. Geomorph. Suppl. 50, 11 (1984)
Bremer, H.: Z. Geomorph. Suppl. 54, 11 (1985)
Brown, J.C. and D.W. Hughes: Nature 268 (5620), 512 (1977)
Brun, E.: Ordnungs-Hierarchien. Zürich: Neujahrsbl. Natf. Ges. (1986)
Bryan, R.B., A. Yair and W.K. Hodges: Z. Geomorph. Suppl. 29, 151 (1978)
Büdel, J.: Klima-Geomorphologie. Berlin: Borntraeger (1977)
Büdel, J.: Climatic geomorphology. Princeton: Univ. Press (1982)
Bull, W.B.: Geol. Soc. Am. 86, 1489 (1975)
Burt, T.P.: Progr. Phys. Geogr. 8(4), 570 (1984)
Burt, T.P.: Progr. Phys. Geogr. 10(4), 547 (1986)
Burt, T.P.: Progr. Phys. Geogr. 11(4), 598 (1987)
Burt, T.P.: Progr. Phys. Geogr. 12(4), 583 (1988)
Cannon, P.J.: Oklahoma Geol. Notes 36(1), 3 (1976)
Carson, M.A. and M.J. Kirkby: Hillslope form and process. Cambridge: University Press (1972)
Cayley, A.: Philos. Mag. 18, 374 (1859)
Cheng, R.T. and D.S. Hodge: J. Int. Assoc. Math. Geol. 8(1), 43 (1976)
Chorley, R.J.: U.S. Geol. Surv. Prof. Pap. 500B, 1 (1962)
Chorley, R.J., S.A. Schumm and D.E. Sudgen: Geomorphology. London: Methuen (1984)
Christofoletti, A.: Bol. Geogr. Teoret. Rio Claro 3(6), 4 (1973)
Christofoletti, A.: Rev. Geograf. Norte Grande 8, 69 (1981)
Christofoletti, A. and J. Arana: Bol. Paulista Geogr. 52, 5 (1976)
Cliff, A.D., P. Haggett and J.K. Ord: In: Application of graph theory, ed. R.J. Wilson and L.W. Beineke, p. 293. London: Academic Press (1979)
Coffman, D.M., E.A. Keller and W.N. Melhorn: Water Resour. Res. 8(6), 1497 (1972)
Culling, W.E.H.: Trans. Jpn. Geoph. Un. 7(4), 221 (1986)
Dacey, M.F.: In: Random processes in geology, ed. D.F. Merriam, p. 16. Berlin: Springer (1976)
Dacey, M.F. and W.C. Krumbein: J. Geol. 84, 153 (1976)
Davis, W.M.: Geogr. J. 14, 481 (1899)
Davy, B.W. and T.R.H. Davies: Water Resour. Res. 15(1), 103 (1979)
Drexler, O.: Einfluss von Petrographie und Tektonik auf die Gestaltung des Talnetzes im oberen Rissbachgebiet (Karwendelgebirge, Tirol), München: Geobuch Verlag (1979)
Dunkerley, D.L.: J. Geol. 85, 459 (1977)
Dunne, T.: Progr. Phys. Geogr. 4(2), 211 (1980)
Ebisemiju, F.S.: Geogr. Ann. A61(1–2), 103 (1979)
Ebisemiju, F.S.: Catena 12, 261 (1985)
Eckmann, J.P.: Rev. Mod. Phys. 53(4), 643 (1981)
Fairbridge, R.W.: In: Thresholds in geomorphology, ed. D. Coates and J. Vitek, 43 London: Allen & Unwin (1980)
Farrenkopf, D.: Z. Geomorph. Suppl. 66, 73 (1987)
Faulkner, H.: Z. Geomorph. Suppl. 21, 76 (1974)
Finkl, C.W.: Z. Geomorph. 26(2), 137 (1982)
Finkl, C.W.: Notic. Geomorfol. 21(41), 111 (1983)
Flint, J.J.: Bull. Geol. Soc. Am. 84, 1087 (1973)
Flint, J.J.: Water Resour. Res. 12(4), 645 (1976)
Gardiner, V., K.J. Gregory and D.E. Walling: Area 9(2), 117 (1977)
Gerber, E.: Morphologische Untersuchungen im Rhonetal zwischen Oberwald und Martigny. Zürich: Diss. Geogr. Inst. ETH (1944)
Gerber, E.: Z. Geomorph. Suppl. 8, 94 (1969)
Gerber, E.: Mitt. Aarg. Naturforsch. Ges. 31, 75 (1986)
Ghosh, A.K. and A.E. Scheidegger: Water Resour. Res. 6(1) 336 (1970)
Gilbert, G.K.: Report on the geology of the Henry Mountains. Washington: U.S. Geol. Surv. (1877)
Graf, W.L.: In: Proc. 10th Geomorph. Symp., ed. D.D. Rhodes, G.P. Williams, London: Allen and Unwin (1979)
Gregory, K.J.: Geogr. Polonica 34, 155 (1976a)
Gregory, K.J.: Geogr. J. 142(2), 237 (1976b)
Gretener, P.E.: Bull. Am. Assoc. Petrol. Geol. 51(11), 2197 (1967)
Gupta, V.K. and L. Duckstein: Water Resour. Res. 11(2), 221 (1975)
Hack, J.T.: U.S. Geol. Surv. Prof. Pap. 294B, B1 (1957)

Hack, J.T.: Am. J. Sci. 258A, 80 (1960)
Haigh, M.J.: Geoforum 16(2), 191 (1985)
Haigh, M.J.: Catena Suppl. 10, 181 (1987)
Haigh, M.J.: Z. Geomorph. Suppl. 67, 79 (1988)
Haigh, M.J., J.S. Rawat and S.K. Bartarya: Current Sci. 57(18) 1000 (1988)
Haken, H.: Synergetics: an introduction. 3rd edn. Berlin Heidelberg New York: Springer (1983)
Haken, H. and W. Weimer: Z. Geomorph. Suppl. 67, 103 (1988)
Hanks, I.C. and D.J. Andrews: J. Geoph. Res. 94 (B1), 565 (1989)
Harrison, R.G. and D.J. Biswas: Nature 321, 394 (1986)
Harvey, D.: Explanation in geography. London: Arnold (1969)
Higgins, C.G.: In: Revolution in the earth sciences, ed. S.J. Boardman. Dubuque Ia.: Kendall/Hung,
    p. 181 (1983)
Hirano, M.: Z. Geomorph. Suppl. 25, 50 (1976)
Hövermann, J.: Z. Geomorph. Suppl. 56, 143 (1985)
Horton, R.E.: Bull. Geol. Soc. Am. 56, 275 (1945)
Howard, A.D.: Water Resour. Res. 7(4), 863 (1971)
Howard, A.D. and G. Kerby: Bull. Geol. Soc. Am. 94(6), 739 (1983)
Howard, A.D., M.E. Keech and C.L. Vincent: Water Resour. Res. 6(6), 1674 (1970)
Huang, J.Q. and B.W. Chen: Explor. Nature 4(2), 53 (1985)
Huggett, R.J.: Earth surface systems. Berlin Heidelberg New York Tokyo: Springer (1985)
Huggett, R.J.: Earth Surf. Proc. and Landforms 13, 45 (1988)
Hurst, H.E.: Trans. Am. Soc. Civ. Eng. 116, 770 (1951)
Hurst, H.E.: Proc. Inst. Civil. Engrs. Part I, 519 (1956)
Hutton, J.: Trans R. Soc. Edingburgh 1, 209 (1788)
Huxley, J.S.: Nature 114, 895 (1924)
Huxley, J.S.: Problems of relative growth. London: Methuen (1932)
Ichim, I. and M. Radoane: Efectele barajeolor in dinamica reloiefului. Bucuresti: Ed. Acad. Rep. Soc.
    Romania (1986)
Ives, J.D.: Progr. Phys. Geogr. 10(3), 437 (1986)
James, W.R. and W.C. Krumbein: J. Geol. 77(5), 544 (1969)
Jarvis, R.: Water Resour. Res. 12(6), 1215 (1976)
Jarvis, R.: Progr. Phys. Geogr. 1, 271 (1978)
Jarvis, R. and C.H. Sham: Water Resour. Res. 17(4), 1019 (1981)
Jarvis, R. and A. Werrity: Water Resour. Res. 11(2), 309 (1975)
Kadanoff, L.P.: Physics Today 1983(12), 46 (1983)
Kashiwaya, K.: Bull. Disas. Res. Inst. Kyoto Univ. 28(3–4), 69 (1978)
Kennedy, B.: Earth Surf. Proc. 3(3), 219 (1978)
Khromchenko, A.I.: J. Int. Assoc. Math. Geol. 1968, 123 (1968)
Kiewietdejonge, C.J.: Progr. Phys. Geogr. 8(2), 218, (1984)
Kirkby, M.J.: Spec. Publ. Inst. Br. Geogr. 3, 15 (1971)
Klemes, V.: Catastrophist Geol. 1(1), 43 (1976)
Knighton, A.D.: Bull. Geol. Soc. Am. 88, 364 (1977)
Kondoh, H., M. Matsushita and Y. Fukuda: J. Phys. Soc. Jpn 56(6), 1913 (1987)
Krumbein, W.C.: Annu. Bull. Beach Erosion Board 17, 1 (1963)
Krumbein, W.C. and A.R. Orme: Bull. Geol. Soc. Am. 83, 3369 (1972)
Langbein, W.B.: J. Hydr. Div. ASCE HY2, 1964, 301 (1964)
Le, B.H. and T.R.H. Davies: Bull. Geol. Soc. Am. 90(I, 12), 1094 (1979)
Lechthaler, M. and A.E. Scheidegger: Z. Geomorph. 33(3), 361 (1989)
Leopold, L.B.: Z. Geomorph. Suppl. 9, 57 (1970)
Leopold, L.B. and W.B. Langbein: U.S. Geol. Surv. Prof. Pap. 500A, 1 (1962)
Leopold, L.B. and T. Maddock: U.S. Geol. Surv. Prof. Pap. 252, 1 (1953)
Leopold, L.B. and J.P. Miller: U.S. Geol. Surv. Prof. Pap. 282A, 1 (1956)
Leopold, L.B., M.C. Wolman and J.P. Miller: Fluvial processes in geomorphology. San Francisco:
    Freeman & Co (1964)
Liao, K.H. and A.E. Scheidegger: Bull. Int. Assoc. Sci. Hydrol. 13(1), 5 (1968)
Liao, K.H. and A.E. Scheidegger: Water Resour. Res. 5, 744 (1969)
Lloyd, E.H.: J. Hydrol. 22, 1 (1974)
Louis, H.: Z. Geomorph. Suppl. 20(3), 257 (1976)

Loyda, L.: Sb. Cesk. Geogr. Spol. 1980 (1.85), 29 (1980)
Lutz, T.M.: Geology 15(12), 1115 (1987)
Mandelbrot, B.: C.R. Acad. Paris 260, 3274 (1965)
Mann, C.J.: Bull. Geol. Soc. Am. 81, 95 (1970)
Mann, C.J. and R.L. Hunter: Z. Geomorph. Suppl. 67, 39 (1988)
Mariolakos, I.D., S.P. Lekkas and D.J. Papanikolaou: Arb. Geogr. Inst. Univ. Salzburg 6, 231 (1976)
Mariolakos, I.D. and 3 others: Ann. Geol. Pays Hellen. 30(2), 515 (1981)
Mark, D.M.: Bull. Am. Geol. Soc. 90, 164 (1979)
Mark, D.M.: Geology 9(8), 370 (1981)
Matheron, G.: Rev. Inst. Int. Statistique 38(1), 1 (1970)
Miller, T.K. and L.J. Onesti: Bull. Geol. Soc. Am. 88, 85 (1977)
Milne, G.: Soil Res. 4, 183 (1935)
Milne, G.: J. Ecol. 27, 192 (1947)
Mises, R.V.: Wahrscheinlichkeitsrechnung. Leipzig: Deuticke (1931)
Moon, B.P.: S. Afr. Geogr. J. 57(2), 111 (1975)
Morandi, M.C., A. Del Grosso and B. Limoncelli: Geol. Appl. Idrogeol. Bari 11(1), 7 (1976)
Morisawa, M.: Proc. Int. Hydrol. Symp. Fort Collins 1, 173 (1967)
Mosley, M.P.: East Midland Geogr. 5, 235 (1972)
Mosley, M.P.: Progr. Phys. Geogr. 5(1), 114 (1981)
Mosley, M.P. and C. O'Loughlin: Progr. Phys. Geogr. 4(1), 97 (1980)
Mosley, M.P. and R.S. Parker: Bull. Geol. Soc. Am. 83, 3669 (1972)
Mosley, M.P. and R.S. Parker: Bull. Geol. Soc. Am. 84, 3123 (1973)
Mueller, J.E.: Bull. Geol. Soc. Am. 83, 3471 (1972)
Newmann, W.I., and D.L. Turcotte: Geoph. J. Int. 100, 433 (1990)
Nogami, M.: Geogr. Rev. Jpn 55(7), 490 (1982)
Ohmori, H.: Trans Jpn. Geomorph. Un. 4(1), 107 (1983)
Ohmori, H.: Bull. Dept. Geogr. Univ. Tokyo 16, 5 (1984)
Ohmori, H. and M. Hirano: Z. Geomorph. Suppl. 67, 44 (1988)
Opp, C.: Peterm. Geogr. Mitt. 1985(1), 25 (1985)
Ovenden, J.C. and K.J. Gregory: Earth Surf. Proc. 5, 47 (1980)
Paine, A.D.M.: Progr. Phys. Geogr. 9(1), 1 (1985)
Patton, P.C. and S.A. Schumm: Geology 3, 88 (1975)
Patton, P.C. and S.A. Schumm: Quat. Res. 15, 24 (1981)
Phillips, L.F. and S.A. Schumm: Geology 15, 813 (1987)
Pierce, K.L. and S.M. Colman: Bull. Geol. Soc. Am. 97, 869 (1986)
Pieri, D.C.: J. Geophys. Res. 89 (B8), 6878 (1984)
Platt, J.: Gen. Syst. 15, 49 (1970)
Prigogine, I.: Bull. Cl. Sci. Acad. R. Belge 31, 600 (1945)
Ranalli, G. and A.E. Scheidegger: Bull. Int. Assoc. Sci. Hydrol. 13(2), 142 (1968)
Randara, C.M.M.: J. Hydrol. 21, 187 (1974)
Roy, A.G.: Geogr. Anal. 15(2), 87 (1983)
Roy, A.G.: In: Models in geomorphology, Binghamton Symposia, ed. M.J. Woldenberg 14, 269, London: Allen & Unwin (1985)
Russell, J.R.: S. Afr. Geogr. J. 58(1), 25 (1976)
Sakaguchi, Y.: Dept. Geogr. Univ. Tokyo 1, 67 (1969)
Scheidegger, A.E.: Can. J. Phys. 39, 1573 (1961)
Scheidegger, A.E.: Bull. Int. Assoc. Sci. Hydrol. 9(1), 12 (1964)
Scheidegger, A.E.: Water Resour. Res. 2, 199 (1966)
Scheidegger, A.E.: Water Resour. Res. 3(1), 103 (1967a)
Scheidegger, A.E.: Proc. Internat. Assoc. Sci. Hydrol. Berne Meeting. Vol. "Hydrological Aspects of the Utilization of Water" p. 415 (1967b)
Scheidegger, A.E.: Bull. Int. Assoc. Sci. Hydrol. 12(1), 15 (1967c)
Scheidegger, A.E.: Bull. Int. Assoc. Sci. Hydrol. 12(4), 57 (1967d)
Scheidegger, A.E.: Water Resour. Res. 4, 167 (1968a)
Scheidegger, A.E.: Water Resour. Res. 4, 655 (1968b)
Scheidegger, A.E.: Water Resour. Res. 4, 1015 (1968c)
Scheidegger, A.E.: Bull. Int. Assoc. Sci. Hydrol. 15(1), 109 (1970a)
Scheidegger, A.E.: Water Resour. Res. 6(3), 750 (1970b)

Scheidegger, A.E.: Theoretical geomorphology, 2nd edn. Berlin Heidelberg New York: Springer (1970c)
Scheidegger, A.E.: Physical aspects of natural catastrophes. Amsterdam: Elsevier (1975)
Scheidegger, A.E.: Principles of geodynamics, 3rd, edn. Berlin Heidelberg New York: Springer (1982)
Scheidegger, A.E.: Z. Geomorph. 27(1), 1 (1983)
Scheidegger, A.E.: Z. Geomorph. 29(2), 223 (1985)
Scheidegger, A.E.: Z. Geomorph. 30(3), 257 (1986)
Scheidegger, A.E.: Catena Suppl. 10, 199 (1987)
Scheidegger, A.E.: Z. Geomorph. Suppl. 67, 5 (1988)
Scheidegger, A.E.: Landschaftsformen und Naturgewalten. Zürich: Neujahrsbl. Natf. Ges. (1989)
Scheidegger, A.E. and N.S. Ai: Tectonophysics 126, 285 (1986)
Scheidegger, A.E. and M.J. Haigh (eds.): Dynamic systems approach to natural hazards. Stuttgart/Berlin: Borntraeger (Z. Geomorph. Suppl. Bd. 67) (1988)
Scheidegger, A.E. and F. Kohlbeck: Proc. Int. Symp. Erosion, Debris Flow and Disaster Prev., Tsukuba, p. 285 (1985)
Scheidegger, A.E. and W.B. Langbein: Bull. Int. Assoc. Sci. Hydrol. 11(3), 43 (1966a)
Scheidegger, A.E. and W.B. Langbein: U.S. Geol. Surv. Prof. Pap. 500C, 1 (1966b)
Schenck, H.: J. Geoph. Res. 68, 5739 (1963)
Schmidt, K.H.: Z. Geomorph. Suppl. 66, 23 (1987)
Schröder, W.: Z. Geoph. 38, 179 (1972)
Schumm, S.A.: Bull. Geol. Soc. Am. 67, 597 (1956)
Schumm, S.A.: Episodic erosion: a modification of the geomorphic cycle. Fort Collins: Colorado State Univ. (1976)
Schumm, S.A.: Trans. Inst. Br. Geogr. 4(4), 485 (1979)
Schumm, S.A.: Trans. Jpn. Geomorph. Un. 6(1), 1 (1985)
Schumm, S.A.: Z. Geomorph. Suppl. 67, 17 (1988)
Schumm, S.A., M.P. Mosley and W.E. Weaver: Experimental fluvial geomorphology. New York: Wiley 1987
Seginer, I.: Water Resour. Res. 5, 591 (1969)
Selby, M.J.: Hillslope materials and processes. Oxford: University Press (1983)
Sharp, W.E.: Water Resour. Res. 7(6), 1548 (1971)
Shimano, Y. and Y. Matsukura: Annu. Rep. Inst. Geosci. Univ. Tsukuba 1978 (4), 41 (1978)
Shreve, R.L.: J. Geol. 74, 17 (1966)
Shreve, R.L.: J. Geol. 75(2), 178 (1967)
Shreve, R.L.: J. Geol. 77(4), 397 (1969)
Simon, H.A.: Proc. Am. Philos. Soc. 106(6), 467 (1962)
Skala, W.: Math. Geol. 9(5), 519 (1977)
Slingerland, R.: Geology 9, 491 (1981)
Slingerland, R.L. and R.S. Snow: Z. Geomorph. Suppl. 67, 93 (1988)
Smart, J.S.: Bull. Int. Assoc. Sci. Hydrol. 13(4), 61 (1968)
Smart, J.S.: Earth Surf. Proc. 3(2), 129 (1978)
Smart, J.S. and A.J. Surkan: Water Resour. Res. 3, 963 (1967)
Smart, J.S. and J.R. Wallis: Water Resour. Res. 7(5), 1346 (1971)
Smart, J.S., A.J. Surkan and J.P. Considine: General Assembly, Int. Assoc. Sci. Hydrol., Bern, Vol. "on river morphology", p. 88 (1967)
Stäblein, G.: Z. Geomorph. Suppl. 50, 137 (1984)
Strahler, A.N.: Am. J. Sci. 248, 673 (1950)
Strahler, A.N.: Bull. Geol. Soc. Am. 63, 923 (1952)
Strahler, A.N.: Bull. Geol. Soc. Am. 69, 270 (1958)
Strahler, A.N.: Quantitative geomorphology. In: Handbook of applied hydrology, ed. V.T. Chow, see p. 4–46 (1964)
Sukhov, V.I.: Geodez. i Aerofotosemka 1967 (4), 11 (1967)
Summerfield, M.A.: Progr. Phys. Geogr. 11, 384 (1987)
Summerfield, M.A.: Progr. Phys. Geogr. 12, 389 (1988)
Taylor, G.I.: Proc. R. Soc. A201, 192 (1950)
Terjung, W.H.: Process-response systems in physical geography. Bonn: Dümmler (1982)
Thom, R.: Stabilité structurelle et morphogénèse. Reading MA: Benjamin (1972)
Thornes, J.B.: Geography 68(3), 225 (1983)

Tödten, H.: Ein Analogmodell für den Feststofftransport bei der Hangerosion. Aachen: Diss. T.U. (1976)

Tokunaga, E.: Geogr. Rept. Tokyo Metrop. Un. 1972 (6/7), 39 (1972)

Torelli, L.: Ann. Geofis. (Roma) 28(2–3), 271 (1975)

Trofimov, A.M. and V.M. Moskovkin: Z. Geomorph. 29(3), 257 (1985)

Twidale, C.R. and A.R. Milnes: Z. Geomorph. 27(3), 343 (1983)

Valencia, D. and J.C. Schaake: Water Resour. Res. 9(3), 580 (1973)

Van Pelt, J. and R.W.H. Verwer: Bull. Math. Biol. 45(2), 269 (1983)

Van Pelt, J. and R.W.H. Verwer: Bull. Math. Biol. 48(2), 197 (1986)

Van Pelt, J., M.J. Woldenberg and R.W.H. Verwer: J. Geol. 97, 281 (1989)

Verma, V.K. and G. Bhattacharya: Nat. Geogr. J. India 24(3–4), 62 (1978)

Vondran, G.: Z. Geomorph. Suppl. 28, 124 (1977)

Vondran, G.: Augsburger Geogr. H. 1, 1 (1979)

Vondran, G.: Regensburger Geogr. Schr. 19, 93 (1985)

Walsh, F.: Water Resour. Res. 8(1), 141 (1972)

Werner, C.: Proc. Assoc. Am. Geogr. 3, 181 (1971)

Werner, C.: Proc. Assoc. Am. Geogr. 5, 287 (1973)

Werrity, A.: In: Spatial Analysis in gemorphology, ed. R.J. Chorley, p. 167. London: Methuen (1972)

Whitney, H.: Ann. Math. 62, 374 (1955)

Whitten, E.H.T.: Bull. Geol. Soc. Am. 75, 455 (1964)

Whitten, E.H.T.: J. Geol. 85, 321 (1977)

Williams, F.J.: Z. Geomorph. Suppl. 9, 67 (1970)

Woehlke, W.: Geogr. Rdsch. 1969 (8), 298 (1969)

Woldenberg, M.J.: Bull. Geol. Soc. Am. 77, 431 (1966)

Woldenberg, M.J. and K. Horsfield: J. Theor. Biol. 104, 301 (1984)

Wu, X.M.: J. Huazong Inst. Technol. Engl. Ed. 3(1), 15 (1981)

Yang, C.T. and C.C.S. Song.: J. Hydraul. Div. ASCE 105(HY7), 769 (1979)

Yonechi, F.: Bull. Yamagata Univ. Nat. Sci. 10(1), 143 (1980)

Yonechi, F.: Sci. Rep. Tohoku Univ. Geogr. 34(2), 82 (1984)

Yonechi, F.: Sci. Rep. Tohoku Univ. Geogr. 35(2), 86 (1985)

Young, A.: Slopes. London: Oliver and Boyd (1972)

Zavoianu, I.: Morfometria bazinelor hidrografice. Bucuresti: Ed. Acad. Rep. Soc. Rom. (1978)

Zdenkovic, M.: Informat. Yugosl. 9(1–4), 19 (1977)

Zdenkovic, M.: Entropija prikaza reljefa na topografskim kartama. Zagreb: Diss. (1985)

Zeeman, E.C.: Sci. Am. 234, 65 (1976)

Zotkin, I.T. and M.A. Chikulin: Dokl. Akad. Nauk. SSSR 167(1), 59 (1966)

Zuchiewicz, W.: Stud. Geom. Carpatho-Balc. 21, 183 (1987)

6 Theory of Aquatic Effects

6.1 General Remarks

We have seen in the descriptive section (1.6) on aquatic morphology that aquatic effects occur on land as well as on coasts and in the deep sea.

The interaction of water with the landscape occurs in the first place through reduction of the rocks. The theory of the latter has already been described in connection with slope evolution (cf. Sect. 3.2); the explanations given there are also basic for the understanding of aquatic effects.

Next, movements in large bodies of water are of prime importance in the present context. Thus, the mechanisms of waves, currents, tides, and turbidity currents will have to be described (Sect. 6.2)

It will then be possible to proceed to a detailed description of the various aquatic features that may be of interest. We will begin with the theory of aquatic land morphology (Sect. 6.3), then continue with the theory of the evolution of coasts (Sect. 6.4) and that of the dynamics of river mouths (Sect. 6.5). Finally, the present chapter will be concluded with a theoretical analysis of submarine geomorphology.

6.2 Movements in Large Bodies of Water

6.2.1 Principles

As noted above, the first task in understanding marine geomorphology is to develop an understanding of the movements of large bodies of water.

In this connection, the prime phenomenon which comes to mind is that of *surface waves*. The theory of water waves is a very complex subject and much information exists in the literature. Those features of water waves which are of most importance in context with the subject matter of the present book will be discussed in Sect. 6.2.2.

We shall next proceed to a discussion of the phenomenon of *turbidity currents*, which is of great importance with regard to the shaping of the submarine bottom relief of the Earth. Turbidity currents will be discussed at length in Sect. 6.2.3.

Finally, we shall give a brief description of the phenomena of *tides* (in Sect. 6.2.4) and of *ocean currents* (in Sect. 6.2.5) as far as they are of importance in connection with theoretical geomorphology.

6.2.2 Waves

6.2.2.1 General Remarks

The surface of any large body of water is generally agitated. The agitation manifests itself basically in an up-and-down movement of the surface; it may start at one region and travel across large distances. Because of these characteristics, one speaks of *waves*.

In connection with geomorphology, the main problem is the accounting for the changes in wave action due to the presence of a bottom relief. Hence, it is the theory of shoaling waves (Sect. 6.2.2.2) and of the wave refraction and diffraction on obstacles (Sect. 6.2.2.3) which is of greatest interest. A limit case of water waves occurs in deep ocean water (Sect. 6.2.2.4). In enclosed basins, internal waves are of interest (Sect. 6.2.3.5). Finally, the generation of waves also merits some attention.

Comprehensive treatises on waves have been written by Stoker (1957), Schuleikin (1959), Tricker (1965), and Le Mehaute (1976), who discussed all aspects of the problem. Other, brief expositions of the wave problem have been given, e.g., by Roll (1957) and by Pierson (1961).

6.2.2.2 Theory of Shoaling Waves

The study of shoaling water waves goes back to the last century (cf. Stokes 1847) and has since become a classical subject of textbooks (cf. literature cited above).

In connection with geomorphological questions, a series of aspects of shoaling waves are particularly important: (a) the problem of wave velocity in dependence on water depth, which is generally treated by linearizing the basic hydrodynamical differential equations; (b) the problem of the generation of breaking waves (white water), for which some form of nonlinear theory is required and (c) the bahavior of large or long "solitary" waves on shoals. To this, some remarks on solution attempts of the basic problem based on (d) general system consideration, on (e) numerical techniques, on (f) scale model experiments and (g) on field studies will be added.

a) *Wave velocity.* Beginning with the problem of wave velocity, we note that one can describe fluid motion in two dimensions (in x, z-space; upward; u, w are the velocity components; cf. Fig. 81) by a velocity potential φ so that

$$u = -\frac{\partial \varphi}{\partial x} \tag{6.2.1}$$

$$w = -\frac{\partial \varphi}{\partial z}. \tag{6.2.2}$$

The velocity potential satisfies a Laplace equation (continuity equation)

$$\frac{\partial^2 \varphi}{\partial x^2} + \frac{\partial^2 \varphi}{\partial z^2} = 0 \tag{6.2.3}$$

if the motion is assumed to be irrotational. The equations of motion reduce to

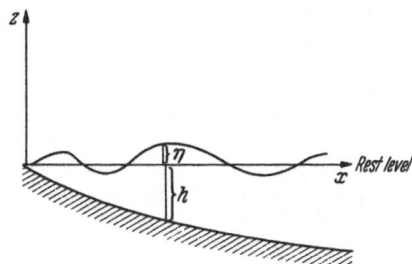

Fig. 81. Geometry of water waves

Bernoulli's equation if the external forces (gravity) have a potential:

$$\frac{p}{\rho} = \frac{\partial\varphi}{\partial t} - gz - \frac{1}{2}\left[\left(\frac{\partial\varphi}{\partial x}\right)^2 + \left(\frac{\varphi}{\partial z}\right)^2\right]. \tag{6.2.4}$$

Here p is the fluid pressure, ρ the density, and g the gravity acceleration. The equations of motion enter only into the boundary conditions; the latter are

1. $\dfrac{d\varphi}{\partial n} = 0$ for $z = -h(x)$ $\hspace{2cm}$ (6.2.5)

if $z = -h(x)$ represents the equation of the bottom of the body of water (cf. Fig. 81) and n denotes the normal to the bottom surface;

2. $\eta = \dfrac{1}{g}\dfrac{\partial\varphi}{\partial t} - \dfrac{1}{2g}\left[\left(\dfrac{\partial\varphi}{\partial x}\right)^2 + \left(\dfrac{\partial\varphi}{\partial z}\right)^2\right]$ for $z = \eta$, $\hspace{1cm}$ (6.2.6)

where $z = \eta(x, t)$ represents the equation of the surface of the water, and we assume that $p = 0$ thereon; finally, expressing the condition that

$$\frac{d}{dt}(z = \eta) = 0$$

yields:

3. $-\dfrac{\partial\varphi}{\partial x}\dfrac{\partial\eta}{\partial x} + \dfrac{\partial\varphi}{\partial z} = -\dfrac{\partial\eta}{\partial t}$ for $z = \eta$. $\hspace{1.5cm}$ (6.2.7)

To this, one will add initial conditions of the form $\eta = \eta_0$ for $t = 0$.

The above system of equations is nonlinear and has never been fully solved. There are two types of approximate solutions that have been obtained. The first is due to Stokes (1847), who assumed that the amplitudes are small. The second assumes that the water is shallow; in this it is a nonlinear theory.

We shall first discuss the theory assuming that the amplitude and therewith the velocity potential φ are so small that squares of φ and of its derivatives can be neglected. Then, the system of equations becomes linear; it may be written as follows

$$\frac{\partial^2\varphi}{\partial x^2} + \frac{\partial^2\varphi}{\partial z^2} = 0. \tag{6.2.8}$$

$$\frac{p}{\rho} = \frac{\partial \varphi}{\partial t} - gz, \tag{6.2.9}$$

$$\frac{\partial \varphi}{\partial n} = \quad \text{for } z = -h, \tag{6.2.10}$$

$$\eta = \frac{1}{g} \frac{\partial \varphi}{\partial t} \quad \text{for } z = 0, \tag{6.2.11}$$

$$\frac{\partial \eta}{\partial t} = -\frac{\partial \varphi}{\partial z} \quad \text{for } z = 0. \tag{6.2.12}$$

Eliminating η from (6.2.11) and (6.2.12) yields

$$\frac{1}{g} \frac{\partial^2 \varphi}{\partial t^2} + \frac{\partial \varphi}{\partial z} = 0 \quad \text{for } z = 0. \tag{6.2.13}$$

One may now seek solutions of the above system which represent progressive waves. The best known of these solutions is (its correctness may be verified directly by substituting it into the above system):

$$\varphi = -a \cosh m(z + h) \cos(mx \pm \sigma t + \delta) \tag{6.2.14}$$

with m and σ satisfying

$$\sigma^2 = gm \tanh mh. \tag{6.2.15}$$

The frequency ν and the wave length λ of the waves are

$$\nu = \sigma/(2\pi), \tag{6.2.16a}$$

$$\lambda = 2\pi/m. \tag{6.2.16b}$$

Furthermore, a and δ are arbitrary constants. The phase velocity c of the waves is (Stokes formula):

$$c = \lambda \nu = \frac{\sigma}{m} = \frac{\sigma \lambda}{2\pi} = \sqrt{\frac{g\lambda}{2\pi} \tanh \frac{2\pi h}{\lambda}}. \tag{6.2.17}$$

If $\lambda \ll 2h$, one has $\tanh(2\pi h/\lambda) \approx 1$, and hence one obtains the following formula for waves in deep water:

$$c = \sqrt{\frac{g\lambda}{2\pi}}. \tag{6.2.18}$$

Conversely, if $\lambda \gg h$, one has $\tanh(2\pi h/\lambda) \approx (2\pi h/\lambda)$, and hence one has for waves in shallow water:

$$c = \sqrt{gh}. \tag{6.2.19}$$

The displacements Δx, Δy corresponding to the above solution can also be calculated (simply by integrating the velocities over time). One obtains

$$\Delta x = \frac{am}{\sigma} \cosh m(z + h) \cos(mx + \sigma t + \delta), \tag{6.2.20a}$$

$$\Delta z = \frac{am}{\sigma} \sinh m(z + h) \sin (mx + \sigma t + \delta), \qquad (6.2.20b)$$

which shows that the wave motion is elliptical. Of particular interest is that the tangential motion parallel to the bottom does not vanish for $z = -h$. One also sees that the amplitude A (half the height from crest to trough) is connected with the constant a as follows:

$$A = \frac{am}{\sigma} \sinh mh.$$

There are instances where the above theory of "infinitesimal" waves is not entirely satisfactory. In this case, a second approximation of the theory may be considered. A particularly important result of the second approximation is that there is a net mass transport in traveling waves. The velocity U of net mass transport at depth h in a wave of height H (crest to trough) is

$$U = \frac{1}{2} \left(\pi \frac{H}{\lambda} \right)^2 c \frac{\cosh \dfrac{4\pi(h - z)}{\lambda}}{\left(\sinh \dfrac{2\pi h}{\lambda} \right)^2}, \qquad (6.2.21)$$

where the wave velocity c is cs given by (6.2.17).

Incidentally, the "second" approximation of the Stokes theory has been extended to a third linear approximation by Tsuchiya and Yasuda (1981) using a perturbation method. The reaction of the net mass transport in the wave mentioned above upon a movable bed was investigated by MacPherson (1980). An approximation model which is based on a kinematic wave model instead of on the hydrodynamic equations has been proposed by Miller (1984).

Inasmuch as the surf contains waves of many frequencies, the characteristics of the spectrum can be studied in the linear approximation (Collins 1972; Schwind and Reid 1972; Huang et al. 1983).

b) *Breaking waves*. Next, we shall discuss some of the results of the *nonlinear shallow water theory*. The procedures of the shallow water theory (Lamb 1954; Defant 1957) are based upon the assumption that the vertical accelerations can be neglected with regard to the horizontal ones. Consequently, the pressure at any point in the body of water equals the static pressure.

$$p = g\rho(\eta - z), \qquad (6.2.22)$$

where the pressure at the surface of the water is taken as zero and the other symbols have the same meaning as before (cf. Fig. 81). The (Eulerian) equation of motion $(du/dt = -(1/\rho)\partial p/\partial x)$ therefore yields

$$\frac{\partial u}{\partial t} + u \frac{\partial u}{\partial x} = -g \frac{\partial \eta}{\partial x}. \qquad (6.2.23)$$

The continuity equation yields

$$-\frac{\partial \eta}{\partial t} = \frac{\partial}{\partial x} [u(\eta + h)] \qquad (6.2.24)$$

since, in Fig. 81, $(\eta + h)$ can be regarded as a "linear density" σ; the continuity equation then is simply

$$-\partial\sigma/\partial t = \text{div}(\sigma u).$$

The two first-order differential Eqs. (6.2.23) and (6.2.24) are the differential equations of the shallow water theory; they are sufficient to determine the two functions η and u.

If we now linearize the shallow water theory by assuming that u and η are so small that square terms can be neglected, we end up with

$$\frac{\partial u}{\partial t} = -g\frac{\partial \eta}{\partial x}, \tag{6.2.25}$$

$$\frac{\partial}{\partial x}(uh) = -\frac{\partial \eta}{\partial t}. \tag{6.2.26}$$

Assuming that the depth h is constant, we find the wave equation

$$\frac{\partial^2 \eta}{\partial x^2} - \frac{1}{gh}\frac{\partial^2 \eta}{\partial t^2} = 0, \tag{6.2.27}$$

which has solutions representing waves progressing with the speed c given by

$$c = \sqrt{gh}. \tag{6.2.28}$$

This is the same result as was obtained from the "small amplitude theory" if it was applied to shallow water.

The nonlinearized shallow water theory has solutions corresponding to shock fronts, as is usually the case with nonlinear hyperbolic partial differential equations. A corresponding situation occurs in gas dynamics; in water wave theory the shock fronts correspond to breakers. The analogy of the nonlinear water wave theory with the equations of gas dynamics is complete. Introducing a ficticious density $\bar{\rho}$ (See e.g., Stocker 1957)

$$\bar{\rho} = \rho(\eta + h) \tag{6.2.29}$$

and a ficticious pressure $\bar{p}$

$$\bar{p} = \int\limits_{-h}^{\eta} p\,dy \tag{6.2.30}$$

the hydrostatic pressure law yields

$$\bar{p} = \frac{g}{2\rho}\bar{\rho}^2 \tag{6.2.31}$$

and the equations of motion become

$$\bar{\rho}\left(\frac{\partial u}{\partial t} + u\frac{\partial u}{\partial x}\right) = -\frac{\partial \bar{p}}{\partial x} + g\bar{\rho}\frac{\partial h}{\partial x}, \tag{6.2.32}$$

$$\frac{\partial}{\partial x}(\bar{\rho}u) = -\frac{\partial \bar{\rho}}{\partial t}. \tag{6.2.33}$$

These are the exact analogs to the equations of gas dynamics. The "velocity of sound" is given by

$$c = \sqrt{\frac{\partial \bar{p}}{\partial \rho}} = \sqrt{\frac{g\bar{\rho}}{\rho}} = \sqrt{g(\eta + h)}.$$ (6.2.34)

The basic equations (6.2.23) and (6.2.24) can be reformulated in terms of this speed

$$\frac{\partial u}{\partial t} + u\frac{\partial u}{\partial x} + 2c\frac{\partial c}{\partial x} - \frac{\partial H}{\partial x} = 0$$ (6.2.35)

$$2\frac{\partial c}{\partial t} + 2u\frac{\partial c}{\partial x} + c\frac{\partial u}{\partial x} = 0$$

with

$$H = gh.$$ (6.2.36)

It can be shown that c is indeed the propagation velocity of a disturbance. For infinitesimal disturbances, this can be verified directly. For constant bed slope ($\partial H/\partial x = m$) the system of equations can be treated by the method of characteristics. Along the lines

$$C_1: \frac{dx}{dt} = u + c$$

$$C_2: \frac{dx}{dt} = u - c$$ (6.2.37)

the change of the following quantities

$$u + 2c - mt = k_1 \quad \text{on} \quad C_1$$
$$u - 2c - mt = k_2 \quad \text{on} \quad C_2$$ (6.2.38)

is nil. Of course, on different characteristic lines, the respective constants (k_1 and k_2) are different.

Breaking of waves occurs when the characteristics of one family intersect each other.

Let us consider a case where the bottom is horizontal (h = const). We attempt to find a solution for the characteristic equations where, at x = 0:

$$\eta(0, t) = A \sin \omega t$$ (6.2.39)

and where we assume that the C_1-family of characteristics are straight lines. Then, assuming that these lines issue from the t-axis at $t = \tau$, we have

$$\frac{dx}{dt} = u(\tau) + c(\tau).$$ (6.2.40)

On any characteristic of the family C_2 we have then (note that m = 0):

$$u(\tau) - 2c(\tau) = u_0 - 2c_0.$$ (6.2.41)

Hence, the slope of any straight characteristic of the family C_1 is given by

$$\frac{dx}{dt} = \frac{1}{2}[3u(\tau) - u_0] + c_0,$$ (6.2.42)

or

$$\frac{dx}{dt} = 3c(\tau) - 2c_0 + u_0. \tag{6.2.43}$$

The values with the index 0 refer to the area where there is no disturbance, i.e.,

$$c_0 = \sqrt{gh}. \tag{6.2.44}$$

We have then for the C_1 family:

$$x = (t - \tau)[3c(\tau) - 2c_0 + u_0]. \tag{6.2.45}$$

As noted above, breaking occurs when the characteristics of one, say the C_1 family, intersect each other. This occurs on the points forming an envelope to the family. The latter is obtained by differentiation of the C_1-equation (6.2.45) with regard to the parameter (τ). The result is

$$x_{env} = \frac{[3c(\tau) - 2c_0 + u_0]^2}{3c'(\tau)}$$

$$t_{env} = y + \frac{3c(\tau) - 2c_0 + u_0}{3c'(\tau)}. \tag{6.2.46}$$

Corresponding to our assumption for η, we have

$$c = \sqrt{g(h + A \sin \omega\tau)}. \tag{6.2.47}$$

Taking the first characteristic $(\tau = 0)$, we obtain the first breaking point

$$x_b = \frac{2c_0(c_0 + u_0)^2}{3gA\omega}$$

$$t_b = \frac{2c_0(c_0 + u_0)}{3gA\omega}. \tag{6.2.48}$$

This shows that for higher frequencies, the breaking point occurs ealier. Hence, the response of the breakers to a beach slope can be calculated (Munk 1969; Divoky et al. 1970; Elgar and Guza 1986); the maximum breaker height (Sunamura 1985) to be expected can be estimated.

However, in spite of 100 years of studies, some questions of the problem remain still unanswered (Cokelet 1977). Bubbles may have an influence (Thorpe and Humphries 1980; Melville and Rapp 1985). On a system basis, Tsuchiya and Tsutsui (1982) have sought the breaking mechanism in terms of an imbalance of the partition of wave energies.

c) *Solitary waves.* The breaking point relation can be improved upon by using solitary wave theory. This has been done by Munk (1949) and by Sverdrup and Munk (1946). These authors quote the following relationship between the wave height η, the depth of breaking h, and the cross-sectional area Q of the breaking wave;

$$Q = 4h^2 \sqrt{\frac{\eta}{3h}}. \tag{6.2.49}$$

In fact, it is not at all clear whether a long, solitary wave remains stable over a sloping bottom (Berryman 1976; Yeh 1985). Osborne et al. (1982) thus considered such a solitary wave as a localized linear packet of waves and treated it by Fourier analysis applying Stokes wave theory to each component.

d) *System dynamics.* We have already mentioned [Sect. (b) above] some system dynamic energy considerations in connection with the wave-breaking problem.

In addition, it may be noted that some form of response theory (based on transmission line theory) has been used to model the resonance at a coast to the incoming surf (Miles 1972; Bowen and Guza 1978). Furthermore, Bampi and Morro (1979) have tried to apply general variational principles.

However, these procedures also cannot answer the open question in wave theory mentioned in (b) above.

e) *Numerical simulations.* Because of the difficulty in dealing with the shoaling wave problem analytically, attempts have been made to solve the fundamental nonlinear hydrodynamic differential equations by various numerical techniques.

Thus, Hibberd and Peregrine (1979) studied the behavior of a uniform bore on a beach by numerical methods. Mattioli (1981) has approximated the three-dimensional basic equations by a set of two-dimensional equations which he solved by numerical techniques. Elgar et al. (1984) applied such methods to wave group statistics and Richards and Taylor (1981) calculated the effect on the sea bottom.

Unfortunately, such numerical solutions only provide for single individual cases. The general behavior of the solutions in dependence of a change of the characteristic parameters does not become evident.

f) *Scale models.* The next possibility of studying the shoaling wave problem is evidently by scale model experimentation.

The general requirements for the dynamic similarity between model and nature and the limits of possible similarity conditions (usually based on Froude's number) have been discussed by Huber (1976), who also gave a number of nomographs exhibiting pertinent relations. Nakamura (1976) studied the transformation of wave frequency during straight run-up of waves on a beach, Guza and Chapman (1979) investigated the same problem for oblique wave incidence. Other studies concerned the drift velocity at the wave-breaking point (Wang et al. 1982) and the horizontal accelerations in shoaling gravity waves (Elgar et al. 1988).

The limitations of scale models lie quite generally in the fact that the scaling relations are always based on some approximate wave theory (Collins 1976; Huber 1976). This, in fact, limits the results derived from scale model studies in the same fashion as the analytical solutions are limited by the approximations of the equations in question.

g) *Field studies.* The difficulties with analytical, numerical, and scale-model studies of the shoaling-wave problem are evidently avoided if field studies (1:1 models!) are undertaken. The one trouble with such field studies is, of course, that they concern only specific cases and do not "explain" anything.

Nevertheless, Guza and coworkers (Huntley et al. 1977; Thornton and Guza 1983; Elgar and Guza 1985a, b; Guza and Thornton 1985) have attempted to find some universal forms for wave run-up spectra. Such spectrum studies have also been made by Le Mehaute and Wang (1982) and by Mase (1988). On the other hand, Arhan and Plaisted (1981) and Sunamura (1982a) have made explicit amplitude studies. The limitations of all such attempts have been outlined above.

6.2.2.3 Wave Refraction and Diffraction

Inasmuch as the water wave velocity in shallow water depends on the depth of the basin, shoals cause refraction phenomena. In addition, obstacles such as islands cause wave diffraction. This is easily explained by reference to the linearized water wave theory.

From Stokes' (1947) analysis of water waves, summarized in Sect. 6.2.2.2, it is seen that the velocity of water waves depends on the depth of the water. The Stokesian formula for the wave velocity c can be written as follows [cf. (6.2.17)]

$$c^2 = \frac{g\lambda}{2\pi} \tanh \frac{2\pi h}{\lambda}, \tag{6.2.50}$$

where h is the water depth, λ the wave length of the waves, and g the gravity acceleration. If ocean swell is striking a coast, it stands to reason that its period T stays invariant. However, the period, the wave velocity c, and the wave length are related by the following formula

$$c = \frac{\lambda}{T}, \tag{6.2.51}$$

which is a general relationship valid for waves. It is therefore possible to eliminate the wave length λ from the two Eqs. (6.2.50) and (6.2.51), which yields a relationship expressing c as a function of h for any given period T.

If the bottom contours of an underwater region be known, it is possible to construct a corresponding chart, for each chosen period T, showing the wave velocity at each point. The areal variation of the wave velocity will cause incoming ocean waves of the corresponding period to be refracted in a pattern which is predictable. Starting with parallel wave crests in the area where the depth is great enough so as not to affect the wave velocity, one can construct the subsequent positions of the crests by successive approximations. O'Brien and Mason (1940) coined the name "refraction diagram" for the resulting chart.

From the refraction patterns it becomes quite obvious that the *wave crest tend to assume the shape of the depth contours*. This has been enunciated as a *law* by Munk and Traylor (1947)

Wave refraction becomes significant only if the wave velocity is materially different from that for "deep" water, i.e., if the hyperbolic tangent in (6.2.50) is materially different from 1. This will be the case if, e.g.,

$$\frac{h}{\lambda} \lessgtr \frac{1}{2}. \tag{6.2.52}$$

The wave length λ for a given period T can be calculated approximately (using the deep water formula) as follows:

$$c^2 = \frac{\lambda^2}{T^2} = \frac{g\lambda}{2\pi};$$ (6.2.53)

thence

$$\lambda = \frac{gT^2}{2\pi} \approx 160T^2,$$ (6.2.54)

where the last value is for the c.g.s.-system. Thus, the critical depth at which wave refraction becomes important, is from (6.2.52):

$$h = \frac{1}{4}\frac{gT^2}{\lambda} \approx 80T^2 cm.$$ (6.2.55)

Ocean swell has seldom a period of more than 14 s, hence the depth at which it becomes refracted is approximately

$$h = 160 \, meters.$$ (6.2.56)

For a short period wind wave (T = 7 s), the corresponding value is

$$h = 40 \, meters.$$ (6.2.57)

The corresponding variation of wave height in refracted waves has been calculated by Iwagaki et al. (1977) by a numerical integration. The numerical treatment has been extended by Liu and coworkers (Tsay and Liu 1982; Liu et al. 1985) to (weakly) nonlinear approximations. Mathiesen (1987) has investigated refraction not caused by a variation of bottom topography, but by "external" flow structures, such as current whirls.

Related to the problem of refraction is that of diffraction: here, the wave nature of the water motion comes into play on obstacles whose dimensions are of the order of the wave length. According to Huyghens' principle, each point of a wave field can be regarded as the source of an elemental wave; the resulting wave field, then, is obtained by a superposition of all the elemental waves. On this basis, Miles (1973) calculated the wave diffraction caused by a transverse discontinuity in depth, and Mattioli (1979) the wave diffraction by islands.

Wave diffraction leads to the scattering of wave energy, for instance, by estuaries (Momoi 1976; Hsieh and Buchwald 1984) and by islands (Foda 1988). Wave energy loss is also produced by the viscosity of the water (Bampi and Morro 1981) and by fields of foreign bodies on the surface, such as sea weed (Ertel 1968a).

6.2.2.4 Waves in the Open Ocean

The wave propagation in the deep ocean is, in approximation, described by Stokes' (1847) deep-water formula [cf. Eq. (6.2.18)].

However, in the open ocean, the individual wave trains are not independent. In this instance, Hasselmann (1966) gave a theory of wave interaction based on statistical mchanics, which, however, leads to very complicated mathematical formalism. On a more direct basis, Longuet-Higgins and Stewart (1960) and Keyon et al. (1983) calculated the interaction between two wave trains which propagate parallel to each other and the interaction between short waves and long waves. Wave trains, however, can interact not only with other wave trains, but also with currents (Smith 1976). In the latter case, giant waves may be the result. The net effect of such interactions is that the ocean surface becomes a random wave system (Chakrabarti and Cooley 1977; Mysak 1978; Longuet-Higgins 1980; Glazman 1986), best treated by the theory of stochastic processes. Measurements (partly from satellites) have generally confirmed the random nature of the ocean surface, for which spectral analyses can be given (Sellars 1975; Rufenach and Alpers 1978; Osborne et al. 1983).

For geomorphological purposes, the wave patterns in the open ocean are of limited importance, inasmuch as it is the interaction of the water with the *solid* surface of the Earth that is of interest.

6.2.2.5 Waves in Enclosed Basins

Somewhat peculiar conditions obtain in enclosed basins: the latter include lakes and, possibly, nearly enclosed bays and lagoons on an ocean coast. The main features are (a) internal oscillations (seiches) and (b) surges caused by external events.

Turning first to internal oscillations (seiches), we note that any basin has a series of eigenfrequencies for surface gravity waves. In shallow water, the wave velocity for the latter is given by Eq. (6.2.19). In a basin of linear dimension L, the wave length λ of an internal seiche must be $2L/n$, with n an integer, so that the two ends of the length L can be on a wave node. For waves of period T one has the general relation

$$\lambda = cT \tag{6.2.58}$$

so that one obtains the following relation for the period T_n of a seiche with n nodes:

$$T_n = \frac{1}{n} \frac{2L}{\sqrt{gh}}. \tag{6.2.59}$$

Refinements of this simple formula for basins of non-constant depth or irregular form have been given by Ertel (1962a). Other refinements have been reviewed in the writer's book on catastrophes (Scheideger 1975), as well as applications to many actual lakes. In addition, it may be noted that internal spiral waves in circular basins have been investigated by Mei (1973), coastal trapped waves have been studied by Suginohara (1981), and nonlinear waves (breakers, whitecaps) by Thorpe et al. (1987). Further applications of the seiche theory to lakes and enclosed seas are due to Rao et al. (1977: Lake Ontario), Wübber and Krauss (1979: Baltic Sea) and Mysak et al. (1983: Lake of Lugano).

With regard to surges, it may be said that these are mainly due to material falling suddenly into a basin: this material can be soil (rock) due to land slide or snow (ice) due to an avalanche. Studies of the problem have been mainly made by civil engineers with a view to the protection of shore installations (Volkart 1974; Huber 1979, 1980, 1982, 1987). For geomorphology, it is the behavior of surges on sloping boundaries (Hay 1983) which is of interest. Since such events occur only rarely, their real importance in the context of this book is limited.

6.2.2.6 Generation of Waves

The *generation* of waves is commonly ascribed to the action of wind. This involves resonance and instability phenomena. A theory of this has been proposed and a review of earlier work has been given by Barnett (1968) according to which the wave spectrum caused by a wind fields can be predicted.

Wave generation occurs by both resonance and instability mechanisms. The actual generation can be ascribed to a transfer of momentum from wind to (preexisting) waves which grow in consequence thereof (Valenzuela and Wright 1976; Naeser 1979; Mitsuyasu 1985; Davies 1986). Such a mechanism has, in fact, been reproduced in model experiments by Mitsuyasu and Honda (1982) and Mitsuyasu and Kusaba (1984). Other theoretical models assume wind pressure to act vertically on a segment of the free surface of a ponded fluid (Stoker 1957; Fangmeir 1970), forcing the latter to oscillate. An alternative view is that the wind causes boundary layer shear (cf. Sect. 2.4.3.3) to act upon the water surface, dragging the latter into the form of waves (Amorocho and De Vries 1980; Cavaleri and Zecchetto 1987)

Quite generally, the wind-caused distribution of wave heights on the open ocean has been postulated (Longuet-Higgins 1952) to be a Rayleigh distribution

$$P = \exp(-a_0^2/\bar{a}^2), \tag{6.2.60}$$

where P is the probability that the amplitude a of any given wave in a wind-disturbed ocean patch exceeds the value a_0; $\bar{a}$ is the root-mean-square amplitude of the waves in the patch. The formula is based on linear theory and may therefore overpredict somewhat the maximum wave heights (Forristall 1978; Longuet-Higgins 1980).

6.2.3 Turbidity Currents

6.2.3.1 General Remarks

As noted in Sect. 1.6.4.3, turbidity currents have been advocated as the explanation of some features of submarine geomorphology. It is therefore necessary to investigate the dynamics of such turbidity currents.

First of all, turbidity currents have been regarded as a case of stratified flow of fluids of different density and viscosity (Harleman 1961; Benjamin 1968). If such flow is laminar, however, one runs into difficulty with the investigations mentioned in Sect. 2.2.3.3 according to which the interface of two fluids of different densities superimposed upon each other and moving at different velocities cannot possibly be stable under any circumstances. Hydrodynamics, therefore, requires that no stratified flows are possible for any length of time.

Upon further investigation it becomes obvious that turbidity currents cannot possibly represent laminar flow. Such currents consists of a suspension of various kinds of particles in water and it has been remarked earlier (in Sect. 4.4.4) that the lifting force keeping such particles in suspension must of necessity be due to prevailing turbulence. Measurements of the Reynolds number in turbidity currents have subsequently yielded the result that this number is of the order of 100 000, which is much higher than the critical Reynolds number at which turbulence is commonly thought to set in. Thus, turbidity currents may safely be assumed to be highly turbulent, and it turns out that, physically, turbidity currents are "slugs" of fairly confined turbulent water mixed with debris which move downslope for a long distance until they come to rest.

If it is assumed that each turbidity flow is triggered by some external cause (earthquake, collapse of a pile of loose material, etc.) which is of no interest for the intrinsic mechanics of a turbidity current, there remain three basic problems to be investigated: (a) Why does the turbulence remain confined in a steady-slug? (b) how does the turbulent slug move? and (c) how does the turbulence decay after the slug has come to rest?

6.2.3.2 Confinement of Turbulence

Turning to the *first problem*, we note that a solution has been proposed by Einstein (1941), who ventured the opinion that the macroscopic rheological state of a sediment-laden stream is similar to that in a plastic substance. There are some indications that this might be so, particularly if the sediment is flocculated. The turbulence, thus, might be somehow confined inside the current, and the latter would move like a plastic body through the surrounding water. Naturally, Einstein's plasticity assumption is purely heuristic. It does not explain why the turbulence does not become dissipated into the surrounding fluid, but simply assumes that this is a fact and tries to give a semitheoretical description of what happens in consequence of this assumption.

Subsequent studies have yielded the result that the critical region for the confinement of turbulence in a slug is its head. It turns out (Allen 1971) that the head always becomes transversely fingered-out (with clefts and lobes); only then is a steady state possible. The problem can also be tackled from the standpoint of statistical mechanics: the conditions for the development of a steady state are that the production rate of entropy be a minimum; applying this to the energy "flux" between the various size eddies in the turbulent fluid yields an explanation in principles as to why the phenomenon of turbidity currents is mechanically possible (Tomkoria and Scheidegger 1967).

6.2.3.3 Movement of Turbulent Slugs

Turning now to the *second problem*, viz. the question as to how the turbulent slug moves downhill, we note that most investigators in some fashion use a formula analogous to the Chézy relation (4.2.19) in rivers. Thus, one would expect for a turbidity current, assuming that it acts like a river (Kuenen 1951)

$$v = \text{const} \sqrt{RS\rho_e} \tag{6.2.61}$$

where v is the velocity of the slug, R is a "hydraulic radius" (the thickness of the current), S the (sine of the) slope, and ρ_e the effective density (i.e. the difference between the density of the current and that of the surrounding water). A somewhat more elaborate relationship is due to Blanchet and Villate (1954), who take into account that there is friction not only at the bottom of turbidity current (against the bed) but also at the top (against the still water; in a river no friction against the atmosphere is assumed). Further refinements can be made by directly applying boundary layer theory to the turbulent slug (Bata 1959), or by applying general balance equations (Lüthi 1981; Siegenthaler and Buhler 1985). Furthermore, as a turbidity current moves over regions of varying topography, adjustments of thickness, velocity, particle concentration, etc. (flow transformations) occur (Kuenen 1971; Komar 1973b; Lüthi 1980; Fisher 1983).

The above remarks relate to the main part of the turbidity current. Its head, however, behaves differently. First, its thickness h_h is double that of the steady-state tail

$$h_h = 2R. \tag{6.2.62}$$

Second, for the velocity v_h of the head, one has empirically from tank experiments (Keulegan 1958; Middleton 1966)

$$v_h = K \sqrt{\frac{\rho_e}{\rho} gh_h}, \tag{6.2.63}$$

where K is some constant varying from 0.4 to 0.7, ρ_e is the density difference between current and surrounding water, ρ the density of the fluid, and g the gravity acceleration. The last formula shows that the velocity of the head is largely independent of the slope S. Incidentally, the velocity of the head may cause the latter to move upslope on a counterslope; the run-up may be 1.53 times the flow thickness (Muck and Underwood 1990)

6.2.3.4 Decay of Turbulence

Finally, the *third problem* concerns the mechanics of decay of the turbulence in a turbidity current. It stands to reason that this decay occurs at the end of the flow process, when owing to the lack of further resplenishment, the turbulence decays according to the well-known laws presented in Sect. 2.2.2.3 of this book.

6.2.4 Tides

6.2.4.1 Introductory Remarks

Another phenomenon that may have a great effect upon coastal and submarine geomorphology is that of the tides.

Tides are due to the attraction of the Sun and the Moon upon the water masses in the oceans, and manifest themselves in a periodic rise and fall of the sea level with a period of about 12 h or so. These fluctuations set up tidal currents, particularly in narrows. These questions will be discussed below.

6.2.4.2 Tides in the Open Ocean

The tides in the open ocean are, in effect, forced oceanic waves on a large scale (Bartels 1957; Defant 1957; Webb 1974; Wunsch 1975; Philander 1978; Anderson et al. 1979). Measurements in the open ocean (Schott 1942; Kuo and Jachens 1977; Won et al. 1978; Won and Miller 1979) have shown that the amplitude of the tidal fluctuations of the sea level reach only about a meter or so. However, in enclosed areas such as bays and estuaries, the tidal range may be very great (up to 15 m in the Bay of Fundy on the Canadian East Coast).

6.2.4.3 Tidal Currents

The tidal variation of the sea level causes currents to flow which are especially noticeable in shallows between different bodies of water. The particular conditions obtaining in river estuaries will be discussed in Sect. 6.5; here, we will give a general discussion of the tidal current phenomenon.

Thus, currents are induced on a shelf slope front (Barbee et al. 1975; Beardsley et al. 1977; Battisti and Clarke 1982; Davies et al. 1985; Ou and Maas 1986, 1988). Particularly strong currents are induced in channels, such as the Bristol Channel (up to 4.5 m/s surface speed; see Kuenen 1950a) or the English Channel (Provost 1981). Around obstacles, vorticity is generated (Pingree and Maddock 1979), which may lead to unstable eddies (Black and Gay 1987).

The structure of a tidal current has been measured and analyzed by Bowden et al. (1959). These authors measured the flow velocity at various heights in Red Wharf Bay in North Wales where the water was 16 m deep. The result is shown in Fig. 82. Bowden et al. then assumed that the profile of the velocity v near the

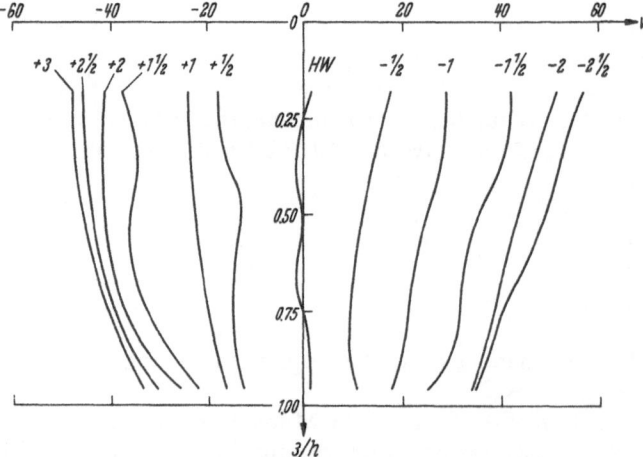

Fig. 82. Velocity as a function of relative depth in a tidal current, from $2\frac{1}{2}$ h before to 3 h after high water. (After Bowden et al. 1959)

bottom is given by the usual logarithmic law of Karman [cf. Eq. (4.2.30)], *viz*:

$$v = \frac{1}{k} \sqrt{\frac{\sigma_m}{\rho}} \operatorname{lognat} \frac{z}{z_0},$$ (6.2.64)

where z denotes the height, k the Karman constant (= 0.4), and z_0 the bottom roughness. Furthermore, ρ is as usual the density, and σ_m the bottom shearing stress. By a least-squares solution using near-bottom data, σ_m and z_0 can be determined for a tidal cycle. The variation of bottom stress which corresponds to the velocity measurements given in Fig. 82. is shown in Fig. 83.

Like all currents, tidal currents are effective in the transport and diffusion of bottom sediment (Seymour 1980; Zimmerman 1981). In this connection, it may be interesting to calculate the size of bottom particles which, according to the drag theory (see Sect. 4.4.3.2) will just be dislodged by the tidal bottom stress as estimated by Bowden et al. Referring to Eq. (4.4.30), we note that the bottom stress σ is related to the diameter d of the particles which are just being dislodged by

$$\sigma = Ad,$$ (6.2.65)

where A is a constant. From Fig. 57, we take the value of this constant A as equal to:

$$A = 166$$ (6.2.66)

if d is given in millimeters and σ in units of g/m^2 (0.1 $dynes/cm^2$). Taking Bowden's value for σ as approximately 5 $dynes/cm^2$ (50 g/m^2) yields

$$d = 0.3 \text{ mm}.$$ (6.2.67)

Thus, it may be seen that a tidal stress of some 5 $dynes/cm^2$ is able to start particles moving of about $\frac{1}{4}$ mm in diameter. In stronger tidal currents, the bottom effect will be correspondingly larger.

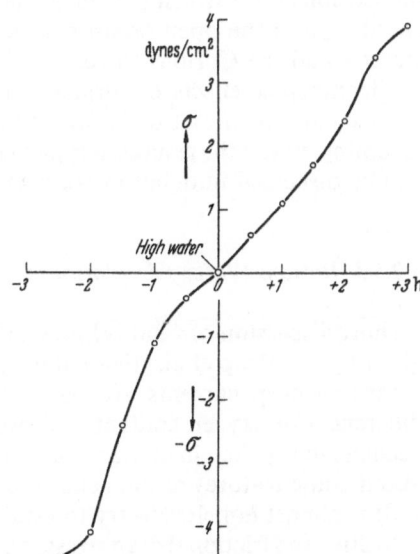

Fig. 83. Variation of bottom stress in a tidal current between $2\frac{1}{2}$ h before and 3 h after high water: drawn from data supplied by Bowden et al. (1959)

It cannot be assumed that the tidal current follows the logarithmic law of velocity distribution at all heights; in fact the velocity profiles shown in Fig. 82 speak against this. This is due to the circumstance that, in general, one has to deal with an unsteady state.

6.2.5 Ocean Currents

6.2.5.1 Introductory Remarks

We turn our attention now to effects in the open ocean. Here, it is ocean currents which are of interest, whose existence is well known from observations of mariners. The currents observed by mariners are essentially surface currents; of course, these cannot exist alone, inasmuch as there has to be a counterflow of water at depth to rectify the mass balance. Thus, the circulation in the world ocean forms a single system (Broecker and Li 1970; Pond and Bryan 1976; Kilworth 1983; Woods 1985).

6.2.5.2 Surface Currents

Surface currents are features that affect large parts of the world. Well known is the Gulf Stream which runs from the Gulf of Mexico across the Atlantic to Norway and Spitsbergen. The surface velocity of such currents is of the order of 9 km/h (Stommel 1965).

The dynamical theory of such currents is well established; it has been reviewed, for instance, by Sverdrup et al. (1946) and Dietrich and Kalle (1947).

The driving force of the surface currents is the wind (Langmuir 1938; Pedlosky 1979; Katz and Garzoli 1982; McCreary and Lukas 1986; Katz 1987). The surface wind stress [Eq. (2.4.14)] sets the water in motion; subsequently, it is deflected by the Coriolis force which provides a lateral acceleration. In fact, the entire surface circulation in the open ocean can be explained in terms of the prevailing wind stresses and the Coriolis force.

Geomorphic effects of surface currents occur mainly in coastal areas. Here, surface currents are also set up by the shore-parallel velocity component of the incoming waves: the result is a typical near-shore circulation system. This problem will be discussed in detail in connection with the theory of coasts (Sect. 6.4.2.2).

6.2.5.3 Bottom Water Circulation

A short discussion of what is known about the bottom water circulation has been given by Sverdrup et al. (1946) and by Sverdrup (1957). Accordingly, conclusions regarding deep currents are based upon the temperature distribution, taking differences of oxygen content and other variable constituents into account. It is needless to say that, in this fashion, not much more than perhaps a qualitative idea about some features of the general deep water circulation can be obtained.

It is almost hopeless to try to establish a quantitative picture of the prevailing velocities and frictional drag stresses at the ocean bottom. It is uncertain, therefore, whether these stresses are large enough to cause much transport of material. Until rather recently, it had been thought that the currents are not large enough, but photographs of scour marks (Heezen et al. 1959) at depths of 6 km seem to indicate that deep water currents may cause at least some effects upon the bottom of the oceans (cf. also Sect. 6.6.5). Calculations by Wüst (1957) indicate that values of bottom current velocities up to 25 cm/s do not seem to be unreasonable. Direct measurements (Swallow and Worthington 1957; Laird and Ryan 1969; Rabinowitz and Eittreim 1974) have also indicated velocities of the same order of magnitude. Doebler (1967) even found peak velocities of 40 cm/s in a deep ocean current near Bermuda. In such currents, there are temporal ("benthic storms": Gardner and Sullivan 1981) and seasonal (Dickson et al. 1982) velocity variations. The bottom velocities in the oceans have also varied throughout geological history in response to the climatic changes mentioned in Sect. 2.5 (Ledbetter 1981; Brass et al. 1982; Duplessy and Shackleton 1985). Bottom velocities of the orders mentioned above can have a geomorphological significance (Heezen 1966).

6.3 Aquatic Effects on Land

6.3.1 General Remarks

Our next task is a discussion of the mechanical aspects of aquatic effects on land. The corresponding descriptive aspects have been given in Sect. 1.6.2. The theoretical sections following here will be keyed to the chapter mentioned.

6.3.2 Theoretical Limnology

6.3.2.1 Introduction

The problems arising in lakes differ from those in oceans because the water bodies involved are smaller by an order of magnitude. Thus, except for some specific instances, the difference to oceans is just a matter of scale: this concerns particularly the shore dynamics, the generation and shoaling of waves, etc. These problems will therefore be relegated to the chapter on coasts.

The subjects considered here are specific limnological questions: these concern the origination of lakes, the hydrodynamics in lakes, and the retention effects.

6.3.2.2 Origination of Lakes

The theoretical possibilities for the origination of lakes have been enumerated in the chapter on physiography (Sect. 1.6.2.2). There is little theory that can be added to this enumeration, except for some specific cases.

Thus, the groundwater-fed lakes (Sect. 1.6.6.2-8) communicate with the water table. The inflow is governed by the groundwater flow in accordance with the theory of flow through porous media (Sect. 2.2.4.3). On a large scale, the lake acts as a sink (in consequence of evaporation) or as an "injection well" (in consequence of water inflow and precipitation into the lake). The solution for the pressure drawdown (or rise) around a singularity point is well known (Theis 1935): the pressure is an exponential-integral function of the radial distance from the singularity (see also Scheidegger 1974, p. 110).

Features genetically related to groundwater lakes are African *dambos*: in this case, the groundwater level falls occasionally below the bottom of the "lakes" so that the latter become seasonal features (Acres et al. 1985).

6.3.2.3 Circulation in Lakes

The circulation in lakes is different from that in the open sea on account of the closedness of the lake basin. Specific features are the eigenoscillations ("seiches") which have already been discussed in Sect. 6.2.2.5.

The general hydrodynamic conditions obtaining in lakes have been summarized by Hutter and Trösch (1975). Accordingly, the fundamental equations account for the conservation of mass, the conservation of momentum, and the balance of energy (cf. also Hankanson 1981). In addition, one has dynamic and kinematic boundary conditions as well as a friction condition at the lake bottom.

In a lake, a closed circulation springs up (Taus and Gerber 1977) which can be calculated by a numerical integration of the fundamental equations (Lake Erie: Gendney and Lick 1972). An important feature is the upwelling part of this circulation (Csanady 1977), which may be intermittent.

6.3.2.4 The Retention Problem

A problem particular to lakes is their response to (temporary) imbalances between inflow and outflow: lakes act as hydrological buffers. In fact, Q_i is the inflow, Q_e the outflow, mass balance requires (Hutchinson 1957):

$$Q_i - Q_e = a \frac{dh}{dt},$$ \hfill (6.3.1)

where h is the water height above some base, a the area of the lake and t time.

Equation (6.3.1) contains a feedback feature, because, whilst $Q_i(t)$ may be an externally given (stochastic) function of time t, Q_e is a function of h. A solution of the retention equation can therefore only be obtained if the function $Q_e(h)$ has been (empirically or theoretically) determined. Some cases have been considered by Ertel (1962b). Inasmuch as the function $Q_e(h)$ can be influenced by the design of the outlet(s) of an artificial lake, it is possible to optimize the retention capabilities of water reservoirs for the expected (stochastic) behavior of $Q_i(t)$. (Phatarfod 1976; Spreafico 1977; Vischer 1977; Pegram 1978). However, this is a problem of engineering rather than of geomorphology and will not be further discussed here.

6.3.3 Theory of Solution and Deposition Effects

6.3.3.1 Introduction

We consider next the solution and deposition effects of water on land.

Solution effects concern mainly *karst* features. In the latter the chief agent is the dissolution of limestone by water. The chemistry of this process has been outlined in the chapter (Sect. 3.2.2.3) on weathering. Therefore, we shall be concerned here only with the effect of such dissolution: i.e., with the genesis of karst features and caves.

Deposition effects have been described morphologically in Sect. 1.6.2.4. It is our task to review here the little that is known about their kinetics.

6.3.3.2 Surface Karst

The flutes and rills referred to as "karren" structures in Sect. 1.6.2.3 are obviously solution features. General reviews of the dynamics of a karst system have been given by Sweeting and Pfeffer (1976) and by Ford (1985).

Karst features are easily explained in terms of the instability principle of geomorphic forms (Sect. 5.2.3): after random catching of runoff (Kemmerly 1982) accidental channels form, which reinforce themselves in the flow dynamics (Gunn 1981) due to a positive feedback mechanism. In this fashion, the karst rills are formed.

An experimental confirmation of karst genesis in the above fashion has been obtained by Dzulynski et al. (1988).

6.3.3.3 The Genesis of Caves

The most spectacular karst features are caves. Their genesis is intimately tied up with the hydrological conditions prevalent in an area. In an essentially permeable ground, one distinguishes between two regions: the region *above* the water table (vadose zone) and the region *below* the water table (phreatic zone). The water table may be curved, since a steady state will develop between the supply of atmospheric water and the outflow into the nearest stream. An example of this is shown at the top of Fig. 84.

There are several theories regarding the development of caves. Reviews of these theories have been given by Ford and Ewers (1978) and by Waltham (1981). They can be classified as to "phreatic", "vadose", and "water-table" theories, corresponding to the groundwater zone in which the cave is assumed to be created. All theories start with minute channels in the limestone rock, which are supposed to be due to original heterogeneities; the problem is then how these microchannels are enlarged into caves.

The phreatic theory was proposed by Grund (1910) and supported by Weller (1927) and Davies (1960). Accordingly, a cave forms through leaching in the phreatic zone, inasmuch as flow velocities are highest in the biggest channels, and becomes dry when the water table sinks owing to the entrenchment of nearby streams.

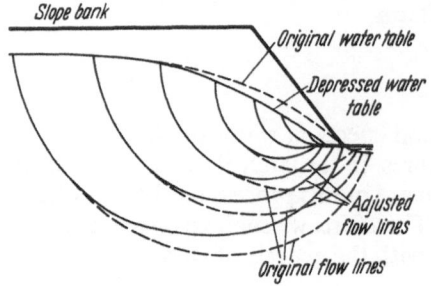

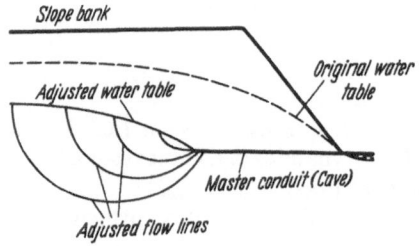

Fig. 84. Stages of the development of a cave. (After Rhoades and Sinacori 1941)

The vadose theory is due to Matson (1909). It assumes that the leaching occurs mostly in channels through which the atmospheric water flows to the groundwater table.

Finally, Swinnerton (1932) assumed that cave formation occurs exactly at the water table because there the flow velocities are supposed to be highest. A modification of this idea, due to Rhoades and Sinacori (1941) is shown in Fig. 84.

It may be that all the possibilities of cave formation are operative in nature. Bretz (1942), in field studies, observed both vadose and phreatic features in natural cave systems.

A cave system cannot grow indefinitely: eventually the cavities become unstable and collapse. However, in an elastic medium, the stability of a cavity is independent of its size. Thus, the maximum shearing stress in a pure isotropic compression (overburden stress) far away from a spherical cavity is (Lame 1852):

$$\sigma_{\text{Shear max}} = \tfrac{3}{2} p, \tag{6.3.2}$$

which is independent of the radius of the sphere. Thus, if the shearing strength of the material is greater than the above maximum shear, a cavity that is stable at all could grow indefinitely. Since this is obviously not the case, it must be assumed that the material must show some type of inelastic rheological behavior (Serata and Gloyna 1960; Rehbinder 1984), whose exact nature, however, is not entirely clear.

6.3.3.4 Sink Holes

Sink holes are the result of the collapse of an underground cave. Generally, it is assumed that the subsidence due to a collapsing underground cavity makes itself felt along the shear planes rising from the rim of the hole into the medium above. In a granular medium, the angle α of these planes with the horizontal is

$$\alpha = \tfrac{1}{2}\phi + 45°, \tag{6.3.3}$$

where $\tan \phi$ is the coefficient of internal friction in the granular medium.

An alternative model of the subsidence above a cavity is by regarding the latter as the result of a stochastic motion process of the individual grains. Then it turns out that the vertical subsidence distance w of a layer at height z above the cavity obeys a diffusivity equation (Litwiniszyn 1963):

$$\frac{\partial w}{\partial z} = K \left(\frac{\partial^2 w}{\partial x^2} + \frac{\partial^2 w}{\partial y^2} \right) \tag{6.3.4}$$

Thus, a collapsing cavity "diffuses" to the surface; the constant K must be estimated from field observations.

Both models account for the appearance of sink holes on the surface. The phenomenon has also been duplicated by scale model experiments (Balwierz and Dzulynski 1976).

6.3.3.5 Theory of Deposition Effects

The deposition effects on land are the opposite of the solution effects. The chemical reactions, mainly involving water, Ca, and CO_2, are the same as in the latter instance (Dreybrodt 1982) but operating in the opposite direction.

The deposition is caused by a negative feedback between the solubility of the precipitate and the deposition. A typical case is the genesis of spring tufa terracettes (Scheidegger 1983): once a hump has been accidentally formed in a spring flow, the rate of loss of CO_2 is larger on it than in the surrounding region, which leads to an increased rate of deposition on the hump (cf. Instability Principle, Sect. 5.2.3).

Similar considerations explain the origination of stalagmites and stalactites.

Unfortunately, no further theoretical studies of the processes involved seem to be available.

6.4 Theoretical Coastal Morphology

6.4.1 General Remarks

Coasts are the most striking aquatic features on the Earth's surface that meet the eye of a casual observer. Whereas it is clear that the large-scale trend of coast lines is determined by endogenic movements of the Earth, the finer features, such as the configuration of beaches etc., are due to exogenic influences and are thus properly dealt with in a treatise in geomorphology.

Monographs on the dynamics of coasts have been published by King (1959), Zenkovich (1967), and Ippen (1966). An older book, almost a classic, was authored by Johnson (1919). Recent reviews of the subject have been published by McCann (1981) and by Viles (1986).

The background for a study of the dynamics of coasts is supplied by an investigation of the *nearshore circulation system* (Sect. 6.4.2). It will then be possible to discuss the dynamics of beaches (Sect. 6.4.3) and of other characteristics of *gently dipping* shore lines (Sect. 6.4.4). Then, we shall analyze the dynamics of *steep coasts* such as cliffs and shore terraces (Sect. 6.4.5). Finally, we shall discuss the possible explanations of large-scale features of shore lines (Sect. 6.4.6).

An inspection of the theories which will be presented below shows that many individual traits of coast lines have been satisfactorily explained. However, most of these explanations are rather qualitative. In several instances, conditions for the equilibrium of a shore line have been devised; the process of the *evolution* of the various coastal features, assuming that there is no equilibrium, is frequently only very poorly understood.

6.4.2 The Nearshore Circulation System

6.4.2.1 Description

Near a coastline there exists generally a characteristic circulation system consisting of longshore, rip (normal to the coast) and upwelling currents. These are generated by the hydraulic action of the incoming and outgoing waves and tides and/or also by the action of the wind.

The nearshore circulation system is responsible for the sediment transport near and at the coast.

Specific local observations of the patterns of such nearshore circulation systems are legion. Thus, we mention here only the review by Winant (1979) and the book by Csanady (1982).

6.4.2.2 Hydraulic Generation

The breaking waves, as they roll into the shore, give rise to currents near the shore.

If the waves break into the shore not absolutely orthogonally, but at an angle $\frac{1}{2}\pi - \alpha$, then it can be shown that a longshore current will be set up. A simple theory of how this may occur has been proposed by Putnam (1949; see also Putnam et al. 1949).

Thus, let us consider waves striking the shore in a pattern as sketched in Fig. 55. We denote by Q the cross-sectional area of a breaking wave crest, by λ its wave length and by c its velocity. Then the average momentum M per unit surface area is (ρ is the density of water)

$$M = \frac{\rho Q c}{\lambda} \tag{6.4.1}$$

and hence the mean flux of momentum dF_M into the volume ABCDE (cf. Fig. 85) is (u denoting some velocity constant)

$$dF_M = u \frac{c \rho Q}{\lambda} \cos \alpha \, dx. \tag{6.4.2}$$

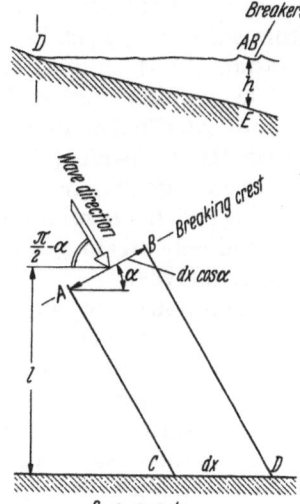

Fig. 85. Geometry of waves breaking at an angle: top view in section; bottom, view in plan. (After Putnam et al. 1949)

Consequently, the mean flux parallel to the shore line is

$$dF_{M\parallel} = u \frac{c\rho Q}{\lambda} \sin\alpha \cos\alpha dx. \tag{6.4.3}$$

After the water has been slowed down to form the longshore current with velocity v, the flux of momentum *after* the slowdown $dF'_{M\parallel}$ is

$$dF'_{M\parallel} = u \frac{v\rho Q}{\lambda} \cos\alpha\, dx. \tag{6.4.4}$$

The difference of $dF'_{M\parallel}$ and $dF_{M\parallel}$ represents the net flux of momentum, or the equivalent force, applied by the breakers upon the water mass in the surf zone. The latter is balanced by the frictional force acting on the volume ABCDE in the longshore current which we assume, in conformity with the drag theory, as equal to (with k equal to some constant and *l* denoting the distance from the shore to the breaker line):

$$dF_{friction} = ku\rho v^2 l dx. \tag{6.4.5}$$

Writing down the force balance equation yields an equation for v

$$\frac{c\rho Q}{\lambda} \sin\alpha \cos\alpha\, dx - \frac{v\rho Q}{\lambda} \cos\alpha dx - k\rho v^2 l dx = 0. \tag{6.4.6}$$

Solving for v yields

$$v = \frac{a}{2}\left\{ -1 + \sqrt{1 + 4\frac{c\sin\alpha}{a}} \right\} \tag{6.4.7}$$

with

$$a = \frac{Q\cos\alpha}{kl\lambda}. \tag{6.4.8}$$

In the above equations, one can replace the value of Q by a relationship obtained from wave theory [cf. Eq. (6.2.49)]. Then, the formula (6.4.8) contains only one constant (k) which must be adjusted *a posteriori*.

The simple theory of Putnam et al. (1949) has been made somewhat more sophisticated, but on the same physical basis (momentum theory) by Galvin (1967), Hogg (1971), Noda (1974) Ryrie (1983), and Dolata and Rosenthal (1984). Computer simulations have been attempted by Wu et al. (1985) and Lakhan and Jopling (1987). Leontiev (1988) has shown that longshore currents are also set up by randomly breaking waves. The state of the art has been reviewed recently by Sherman (1988).

In addition to longshore currents, another type of currents may be caused by the incoming waves. These are *rip currents*, which have been described, for instance, by Shepard (1948), Shepard and Inman (1950), Bowen and Inman (1969), and Bowen (1969). Such rip currents are seaward-moving water masses which are restricted to relatively narrow lanes. They carry some of the backflow which has to compensate for the onshore water movement caused by the swell (cf. Fig. 86).

In effect, it may be argued that rip currents are not independent facets of the nearshore circulation system, but are intimately tied up with wave refraction and longshore currents. Thus, Bruun (1963) assumes that, as a wave breaks at an angle towards the shore, it contributes water to the surf zone which, because of mass conservation, creates alternate longshore and rip currents behind and across a system of bars (Fig. 86). Introducing a Chézy type of friction for the longshore current, Galvin (1967) deduced from Bruun's (1963) theory the following formula for the average velocity of the induced longshore current:

$$v = C_f [H_{bs}^{3/2} \beta (\sin 2\alpha)/T]^{1/2}, \qquad (6.4.9)$$

where C_f is a Chézy type of friction factor, β the beach slope and the other symbols have the same meaning as in the equation before. Thus, introducing mass conservation and a rip current system, a new theory of the origination of longshore currents is in fact arrived at.

Again, more sophisticated theories of the rip current generation mechanism have been proposed by Bowen (1969), Tam (1973), Leblond and Tang (1974), and

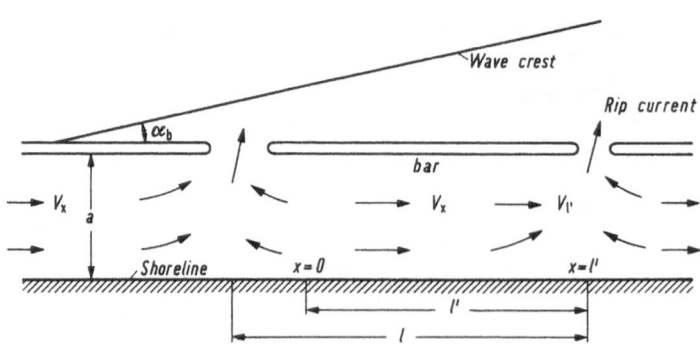

Fig. 86. System of rip and longshore currents. (After Bruun 1963)

Dalrymple and Lozano (1978). However, for geomorphological purposes, the model of Bruun (1963) is quite sufficient.

A corollary to rip currents are coastal *upwelling* currents. They seem to be mainly induced by temperature and density variations with depth (Clarke 1979; Heaps 1980; Tomczak 1981; Suginohara 1982), although their existence has also been attributed to the longshore circulation (Ertel 1964), particularly to the action of the Coriolis force (Hsueh and O'Brien 1971). They are, however, of biological rather than of geomorphological significance.

Finally, tidal flows and tidal currents are of obvious geomorphological significance. Their mechanical aspects have already been discussed in Sect. 6.2.4.3.

6.4.2.3 Wind-Driven Circulation

The nearshore circulation may also be driven by wind. In this, the latter may simply be the shorepart of the general wind driven oceanic circulation (cf. Sect. 6.2.5.2) or may be purely locally generated.

The wind stress (cf. Sect. 2.4.3.3) that generates waves (cf. Sect. 6.2.6.6) also causes currents to form. Again, local studies to bear this out are legion. Thus, amongst many others, a comprehensive study of wind-driven circulation was undertaken on the coast of California (review by Beardsley and Lentz 1987).

In geomorphology, wind-driven circulation is important in connection with smaller coastal features. We discuss here the example of nearly enclosed bays. In such bays, a special type of circulation may become established. Thus, one might reason that an onshore wind will, in addition to causing waves, also create an appreciable onshore flow of surface water which must be compensated by a seaward counter-current at the bottom. This type of circulation might be expected to be particularly pronounced in relatively small bodies of water such as a bay; it is called "vertical" circulation. Measurements of vertical currents have been made by Ryzhkov (1959) on natural coasts.

The above ideas have been tested in model experiments by King (1959) and by Nakano (1955). A schematical view of Nakano's experiment is shown in Fig. 87. The length and the width of the basin were 48 and 15 cm, respectively, and its greatest depth was 2.5 cm. It was filled with water to a depth of 1.5 cm at the center and mounted into a corresponding opening of a wind tunnel. Nakano investigated the circulation by inserting into the water about 12 g of coal grains ($\rho = 1.42$ g/cm^3) of a diameter of $1 - 2$ mm. The wind velocity in the tunnel was arranged to be about 7 m/s. The result of the experiment showed that a circulation as indicated in Fig. 87 became indeed established by which the coal grains were transported little by little towards the side from which the wind blew until a certain equilibrium condition was reached.

Fig. 87. Sketch of Nakano's experiment showing the establishment of vertical currents. (After Nakano 1955)

6.4.2.4 Sediment Entrainment

The nearshore circulation system is mainly responsible for the sediment transport around a coast. General reviews (partly in monographic form) of the problem have been given by Tanner (1973, 1983), Hails and Carr (1975), Jolliffe (1978), Fuhrböter (1979), Greenwood and Davis (1984), and Allen (1985).

The nearshore circulation system gives rise to littoral drift of sedimentary material. Attempts to calculate theoretically the amount of littoral drift from the wave pattern have been reported by Shvartsman and Makarova (1966), but many assumptions regarding carrying capacity of certain types of currents etc. have to be made. Inman and Frautschy (1966) state that the sediment transport rate along most oceanic beaches is from about 45 900 to 764 500 m³ of sand per year, the average being near 153 000 m³ of sand per year.

The basic entrainment conditions for currents generally have been discussed in connection with tidal currents (Sect. 6.2.4.3) to which the reader is referred.

6.4.3 Theory of Beaches

6.4.3.1 Introduction

We now turn our attention to the theory of friable material coasts. The most common manifestation of such coasts is in the form of beaches: therefore, this whole section (6.4.3) is devoted to the theory of beaches, with other forms of friable material coasts being relegated to the next section.

6.4.3.2 Flow Regime on a Beach

Our first interest concerns the *flow regime on a beach*. The theory of water waves presented in Sect. 6.2.2 can be applied to a discussion of the behavior of waves ascending a sloping beach. Most of the results that we are to use are based upon one of the two possible approximations (cf. Sect. 6.2.2), the complete problem of wave action being intractable.

Studies of specific flow phenomena on beaches have been made by Suhayda (1974), who investigated standing waves, by Mahoney and Pritchard (1980), who analyzed the reflection of waves from beaches, and by Guza and Davis (1974) and Evans (1988), who studied the mechanisms for the generation of edge waves. The nonlinear wave-breaking phenomenon on beaches was studied further by Guza and Thornton (1981), Peregrine (1983), and Fu (1987). In this connection, Carrier and Greenspan (1958) and Greenspan (1958) showed that it is, in fact, not always necessary that waves break on a shore. Quite generally, analytical solutions of the partial differential equations of beach flow are difficult to obtain, so that much effort has been applied to developing computational (numerical) techniques (Keller et al. 1960; Colonell and Goldsmith 1972; Alliney 1981; Malone and Kuo 1981). Direct observations of the beach flow regime in nature have been reported by Huntley (1976), Huntley et al. (1977), and Suhayda and Pettigrew (1977).

For our further purposes, the theory as given in Sect. 6.2.2 will generally be adequate; for further refinements, the reader is referred to the papers cited above or to the pertinent monographs mentioned in Sect. 6.2.2.

An interesting modification of the wave theory has been made by allowing for the fact that the beach surface may be *permeable* to water. Essentially, this produces a dissipation of energy whilst the wave is ascending the beach. The problem has been studied by Putnam (1949), by Reid and Kajiura (1957; actually only for deep water), by Hunt (1959) and by Nago and Maeno (1984) from a theoretical standpoint, and by Grantham (1953), Saville (1956) and Savage (1958) from an experimental standpoint. From these investigations it appears that out picture of the wave dissipation process which occurs on a beach is still quite incomplete. Nevertheless, Savage (1958) presented graphs relating the run-up to wave steepness, slope roughness, and slope permeability, which he claims to be applicable to prototype conditions.

6.4.3.3 Particle Size Distribution on a Beach

a) *Introduction.* The particles making up a beach show some degree of sorting. Thus, we turn now to the problem of finding the corresponding sorting mechanism.

This mechanism depends on the beach zone that is involved. In fact, there are several dynamic zones on a beach which are illustrated in Fig. 88. In the various zones, different sorting mechanisms may apply. Furthermore, in the region where the nearshore circulation is important (in addition to the shoaling waves), dispersion by currents may occur.

It has been noted earlier (in Sect. 1.6.3.2) that there are two types of beaches: sandy beaches and shingle beaches. There do not seem to be any gradations from one type to the other, although there are "mixed" (sand and shingle) beaches (McLean and Kirk 1969; Kirk 1980). Therefore, the particle-sorting mechanisms are probably different in the two cases.

b) *Sandy beaches.* Most of the studies of sorting of grains on sandy beaches are concerned with the wave zone. Thus, we shall discuss first the sorting of the grains due to shoaling waves, and examine the other dynamic zones later.

The mechanism that is presumably responsible for the sorting of grains by wave action on a beach in the zone of shoaling waves has been discussed by Kolp (1958), in qualitative terms. Accordingly, the sorting is due to the differential motion of

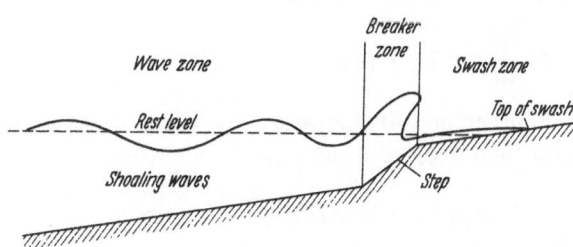

Fig. 88. Schematic representation of the dynamic zones on a beach

individual grains, depending on their size, as they are being dragged along at the bottom by the orbital motion of the water caused by the waves.

A discussion of the phenomenon under consideration, which can be regarded somewhat more justifiably as a "theory", was given long ago by Cornaglia (1891). Accordingly, the crest velocity in a wave is greater than the trough velocity in shallow water. Hence the force moving a sand grain will tend to move the grain in the direction in which the crests are moving, a distance proportional to the difference between crest and trough velocities. The difference between crest and trough velocities becomes less as the water depth increases, and at the same time the influence of gravity becomes predominant, tending to pull the particle seaward. Somewhere on the beach (null point), all the forces that act on one particular grain size are equal; seaward to this point the particles are drawn seaward, landward to it, they are dragged landward. The null point is the only place where a particle can remain in an equilibrium condition.

Cornaglia's "theory" was further tested by Ippen and Eagleson (1955), who from a series of very careful experiments, arrived at the following dimensionless relation for the null points

$$\left(\frac{H}{h}\right)^2\left(\frac{\lambda}{H}\right)\left(\frac{c}{w}\right) = 11.6, \tag{6.4.10}$$

where H is the wave height, λ the wave length, c the wave velocity, h the still water depth, and w the settling velocity of the grains in question. The use of Eq. (6.4.10) allows one to construct theoretical null line maps for a given type of swell on a beach.

Eagleson et al. (1957) and Eagleson and Dean (1959) expanded their empirical approach by going back to the equation of motion for bottom sediments (4.5.19) in a river. However, they now replace the friction term (for the notation see Sect. 4.5.3.2)

$$(M_s - M_f)\varepsilon\, g \cos\alpha$$

by

$$C_{Rx}\frac{\pi d^2}{8}(v_f - v_s)^2, \tag{6.4.11}$$

where C_{Rx} is a new coefficient. This, in fact, implies that the bottom resistance is not that which one would expect from the law of sliding friction, but allegedly that corresponding to rolling friction. In fact, the coefficient C_{Rx} is difficult to measure, and it is therefore usually incorporated into C_D (now denoted by C_D'), the two terms having the same form. Thus

$$M_s\frac{dv}{dt} = C_D'\frac{\pi}{8}d^2\rho(v_f - v_s)^2 - (M_s - M_f)\,g \sin\alpha. \tag{6.4.12}$$

One can now average the above equation over one wave cycle. The acceleration term then drops out, and one is left with

$$C_D'\frac{\pi}{8}d^2\rho\overline{(v_f - v_s)^2} = (M_s - M_f)\,g \sin\alpha. \tag{6.4.13}$$

Eagleson et al. made experiments to check this equation. Using a special assumption for the shoaling wave form, Eagleson et al. (1957) showed that the average in (6.4.13) can also be written

$$C_D'\frac{\pi}{8}d^2\rho(\bar{v}_f - \bar{v}_s)^2 = (M_s - M_f)g\sin\alpha. \tag{6.4.14}$$

The reason that this can be done lies in the fact that, if the wave motion is quasi-oscillatory, the sediment motion is also quasi-oscillatory, without any phase shift. This immediately allows one to form the averages as indicated above.

The last Eq. (6.4.14) was also investigated experimentally by Eagleson et al. This was done by plotting the "coefficient" C_D', as found experimentally by measuring all remaining quantities in Eq. (6.4.14) against a Reynolds number

$$R_e = (\bar{v}_f - \bar{v}_s)d/v, \tag{6.4.15}$$

where v is as usual the kinematic viscosity. The result obtained is that shown in Fig. 89. In that figure, the result is then compared with an empirical curve obtained by Carty (1957; solid line in Fig. 89) giving the resistance coefficient (corresponding to C_D' above) for a sphere rolling down an inclined slope in a liquid. According to Eagleson et al., the agreement is excellent which, in turn, would establish the fact that sand particles on a beach are actually being rolled by the wave action. The curve around which the measured points scatter, is roughly of the form

$$C_D' = \text{const}/R_e. \tag{6.4.16}$$

Now, it has been found empirically in shoaling waves that the following relationship holds approximately

$$\bar{v}_f/\bar{v}_t = \text{const}\,(d/T^{1.43})^{\frac{1}{4}}, \tag{6.4.17}$$

where T is the period of the waves and v_t is the fictitious bottom velocity calculated

Theory of Beaches

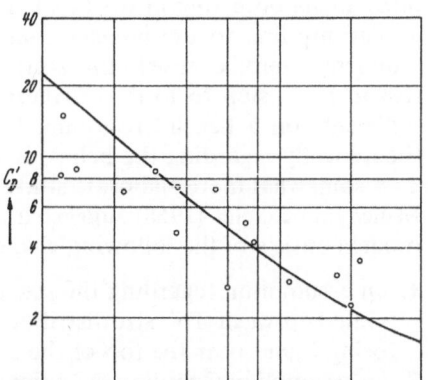

Fig. 89. Plot of resistance coefficient versus Reynolds number. (After Eagleson et al. 1957)

from wave motion theory according to a formula by Stokes [cf. (6.2.21)]

$$\bar{v}_t = \frac{1}{2}\pi^2 \frac{H^2}{\lambda^2} \frac{\text{const}}{\left(\sinh\frac{2\pi h}{L}\right)^2}.$$

(6.4.18)

Here, H is the height of the wave, h its depth, and λ its wavelength. Taking all the equations together yields

$$d^{7/4}\sin\alpha = \text{const } \bar{v}_t\, T^{-0.36}.$$

(6.4.19)

This shows that a particular size d of particles will collect at that point of the beach where the corresponding wave period prevails. This gives an explanation for the empirically observed sorting of particles on a beach.

A somewhat different approach to the sorting problem has been taken by Ertel and Hartke (1971) and by Zimmerman (1973), who did not consider variations in the resistance of the grains, but variations in the bottom stress caused by the topography as of prime importance.

The particle sorting theory has been refined by Rusnak (1957) to include an analysis of the orientation of sand grains under conditions of unidirectional flow. He found that the most stable position which an elliptical particle acquires in depositional transport is one with its long axis lying parallel to the direction of the fluid motion.

Measurements of the size distribution and the motion of beach sand have been reported, e.g., by Blau (1956), Inman and Chamberlain (1959), Ingle (1966), and Miller and Zeigler (1958). The last paper is of particular importance in that it contains a direct evaluation of Cornaglia's and Eagleson's theories. Null lines are constructed and compared with the actual sand distribution on a natural beach. The agreement with the natural cases was quite good. Further studies to test the sorting theory were made later by Cacchione and Southard (1974), by Komar and Miller (1974), and by Miller and Komar (1979) with similar results.

As was noted earlier, the above discussion referred to the sorting of grains in the wave zone only: most investigations reported in the literature are concerned with the wave zone only. The paper by Miller and Zeigler cited above forms a notable exception in this regard, as it contains at least a qualitative discussion of the phenomena occurring in the breaker zone and in the swash zone.

Turning first to the *breaker zone*, we note that the inshore velocity of the sediment is zero because in this zone the wave is met by the backwash. This is the reason for a step to form. Furthermore, it is in this zone where the coarsest sediments on a beach are found. Unfortunately, not much more can be said theoretically regarding the behavior of grains in the breaker zone.

A somewhat more elaborate study has been made of the *swash zone*, for which Miller and Zeigler (1958) suggested a rather elaborate theoretical model. Their model consists of the following elements:

1. an assumption regarding the size distribution left just at the end of any individual upswash. This size distribution is taken as graded with the finest particles being found near the top of the swash;
2. an assumption regarding the minimum velocity required to drag a given particle downslope by the backwash and

3. an assumption regarding the velocity distribution during the backwash.

Miller and Zeigler (1958) use empirical relationships for (2) and (3) above. It is then possible to find the "theoretical" size gradation that should be left by a given backwash. Miller and Zeigler (1958) claim reasonable agreement with observations. On a *purely* empirical basis, Sunamura (1984a) has set up correlations between onshore-offshore sediment transport rates in the swash zone and grain size, wave period, etc. These correlations could also be used to explain the observed sorting.

c) *Shingle beaches.* So far, we have dealt with the sorting of grains on *sandy* beaches. However, not all beaches consist of sand, some consist of *shingle*. By the term "shingle" we mean, as noted earlier, pebbles of about a centimeter or more in diameter.

It is an interesting observation that there are no beaches intermediate between shingle beaches and sandy beaches; in other words, the size distribution function of beach detritus is bimodal. This is a fact which requires an explanation in terms of physical principles.

The contrition of the original material to pebbles presumably takes place by the Rayleigh process which also applies to river pebbles; in the latter connection it was discussed in Sect. 3.2.3.2. The action of the waves upon the shingle is very similar to the action of the rotating can in Rayleigh's experiments; the contrition thus corresponds to the first set of experiments in which steel nuts were used as an abrasive material (cf. Sect. 3.2.3.2). Corresponding experiments with actual beach pebbles have been made by Bigelow (1984). Thus, the standard shape of shingle particles may be assumed to be rather oblong, as usual in corraded particles (cf. Sect. 3.2.3.2). Very large pebbles may be spherical (Kuznetsov and Kuznetsov 1959).

As long as the shingle particles remain fairly large, the Rayleigh process is the only process causing size diminution. However, it has been suggested (Bluck 1969) that, once the shingle particles become smaller than a certain critical size, they also begin to break up while being thrown around by the surf. This would greatly accelerate the contrition and the particles, once they are below the critical size, will rapidly be reduced to sand. Hence shingle particles found on beaches will always be above a certain critical size, depending on the action of the surf; any particles below that size will be ground to sand almost immediately. This explains the bimodal size distribution of beach material.

Apart from the above abrasion processes, shingle particles, like sand particles, are also subject to a sorting mechanism (Gleason and Hardcastle 1973; Ozer 1978) regarding size as well as regarding orientation (Williams and Gulbrandsen 1977).

d) *Dispersion by currents.* Finally, it should be mentioned that particle size sorting can also be caused by longsore currents.

The fact that materials are dispersed by currents has mainly been established by observation (Komar and Inman 1970; Komar 1977). The process has been modeled mathematically by Ostendorf (1982). Essentially, it is governed by a dispersivity equation. The matter is mainly of interest in connection with the spread of pollutants over beaches.

6.4.3.4 Beach Cusps and Ripples

After having discussed the sorting of particles on a beach, we now turn to effects of a somewhat larger scale, such as beach cusps and ripples.

Beach cusps are small, crescentic features open to the sea whose linear dimensions are from 1 to 10 m. Genetically, there exist two types of cusps (Inman and Guza 1982): those formed in the surf zone by the hydrodynamic system, and those formed on the beach face by the swash and backwash.

The cusps formed in the surf zone may again have two different mechanical origins. First, their existence was ascribed long aso (Branner 1900) to an interference phenomenon between two sets of waves (cf. Fig. 90). This idea was further specificed by Guza and Inman (1975), Guza and Bowen (1981), and Dalrymple and Lanan (1976). On certain beaches, the interference structure can actually be seen as a rhomboidal lattice (Stauffer et al. 1976). Accordingly it is mainly edge and other reflective waves that are responsible for cusp formation. Second, contrary to this view, Russell and McIntyre (1965) and also Schwartz (1972b) ascribe cusp formation to cellular flow patterns connected with the nearshore circulation system.

The cusps formed in the swash zone are directly related to the breaking waves and to the "sheet floods" set up by their retreat (Gorycki 1973; Dubois 1978). The process has been carefully studied on typical natural and laboratory beaches by Takeda and coworkers (Takeda and Sunamura 1983; Takeda et al. 1986), who deduced a set of empirical dimensionally homogeneous relationships for such cusps. The most important of these is the relation connecting the spacing S_c of beach cusps to the swash length S_1

$$S_c = 1.5 S_1. \tag{6.4.20}$$

Turning now to *ripple marks*, we note that we have met such features already in the case of river beds. One can, therefore, regard the origin of beach ripples as due to essentially the same causes as river bed ripples.

However, there is one difference: on a beach, the motion of the water is oscillatory, whereas on a river bed it is unidirectional. The analogy with river flow may therefore be best valid for those coastal ripple marks which are caused by tidal currents that remain unidirectional at least for some hours (Mosley 1973; Allen 1980). Furthermore, the "traffic jam" idea (cf. Sect. 4.6.2.4) of ripple formation in

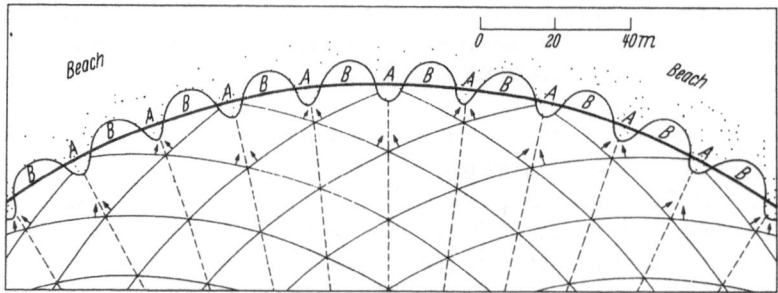

Fig. 90. Diagram illustrating Branner's (1900) theory of beach cusp formation

rivers could possibly still be applied to a beach inasmuch as kinematic interference between particles must be expected in oscillatory as well as in unidirectional flow.

Specific studies of ripple formation in oscillatory flow have been made by Baker (1970) in the field and by Trenhaile (1973) and Sunamura (1981) in a laboratory wave tank. From these studies, empirical correlations have been deduced. Thus, Sunamura (1981) found the following dimensionless relation

$$I = 0.54 - 0.080 \log U, \tag{6.4.21}$$

where I is the ripple symmetry parameter $I = \beta/\lambda$ with β denoting the horizontal distance measured from the ripple crest to the onshore ripple trough, λ the ripple length and U a dimensionless number

$$U = HL^2/h^3, \tag{6.4.22}$$

with H equal to the wave height and L to the wave length at water depth h.

6.4.3.5 Equilibrium Beach

Next, we investigate the conditions under which a beach will remain relatively stable.

Evidently, there are some quite general conditions that have to be satisfied for a beach to be stable. Thus, the gains and losses of sediment (Nagata 1961; Clayton 1980), as well as the influx and dissipation of energy (Pravotorov 1965) have to balance each other, otherwise changes occur. With regard to energy, it is interesting to note that only about 1% of the total wave energy is used in sediment transport (May and Tanner 1975), the rest is dissipated by the breaking waves themselves.

Attempts at ascertaining the equilibrium profile of a beach may be based on theoretical mechanical models. Thus, Wells (1967) tried to obtain direct stability conditions from second-order gravity-wave theory, and Breeding (1981) showed that rhythmic topography is an equilibrium beach form owing to the refraction of water wave packets (hydrons). However, these studies do not arrive at general relationships.

Therefore, there have been a great number of surveys of beaches to ascertain the "standard" form and to develop fundamental nondimensional conditions for an equilibrium beach. In this connection, it has been noted by Tanner (1958) that the concepts of the regime theory of rivers (Sect. 4.6.4) can also be applied to beaches. The most pertinent investigations along these lines have been made by Sunamura (1975), who came up (from an analysis of many natural and experimental data) with the following "regime equations" for equilibrium beaches:

$$\tan \beta = 0.10 \, g^{0.5} \, d^{0.5} \, T/H_b \tag{6.4.23}$$

$$\tan \beta = 0.45 \, (d/H_o)^{0.5} \, (H_o/L_o)^{-0.3}, \tag{6.4.24}$$

where $\tan \beta$ denotes the beach slope, g the gravity acceleration, d the (mean) grain diameter of the beach material, T the wave period, H_b the breaker height, H_o the deep water wave height, and L_o the deep water wave length. From these equations, it is evidently possible to predict the equilibrium beach face slopes quantitatively under given hydrodynamic and sedimentary conditions (Sunamura 1984b).

6.4.3.6 Changes in Beaches

Finally, we are turning toward the changes that may be expected in the course of the life of a beach. These changes may be due to the intrisically prevailing hydrodynamic conditions, or they may be due to external changes occurring in the environment.

Regarding the hydrodynamically produced beach changes, it stands to reason that the latter are linked with the phenomenon of wave refraction. The problem has been examined from this angle in a study by Munk and Traylor (1947). The effect of the swell manifests itself in two ways: first, sediment is transported by *direct action* associated with the swell, and second, *convection* occurs which is associated with the secondary currents created (such as longshore currents) by the swell.

The direct action manifests itself mainly by the bottom drag of the waves. This phenomenon has been studied carefully by Sunamura (1980), who used many data from natural beaches. He arrived at a delimitation condition between erosional and accretionary beaches which is given by the equation

$$H_o/L_o = 18 \, (\tan \beta)^{-0.27} (d/L_o)^{0.67}, \tag{6.4.25}$$

where H_o is the deep water wave height, L_o the deep water wave length, $\tan \beta$ the (initial) beach slope, and d the (mean) beach material diameter.

The direct action of the swell is rather small. The secondary nearshore currents set up by the swell seem to be of much greater importance inasmuch as they may cause large-scale convection. It turns out that it is difficult to find the effect of the currents on the shore by mathematical analysis. Thus, attention has been drawn to model experiments (e.g., Saville 1950); however, difficulties are again encountered, this time because of the impossibility of achieving sedimentological and hydrodynamic similarity at the same time (Griesseier and Vollbrecht 1956). Numerical models (Watanabe 1982; Swain and Houston 1985) are more tractable, but they can never produce general relationships; they can only be made so flexible that they can be re-run for a great number of desired conditions.

Beach changes, as noted, may also be induced by changes occurring in the environment.

The most common environmental changes are represented by seasonally changing weather patterns: these induce seasonal changes in the beaches of an area; observations are plentiful. The beach changes are caused by the meteorologically induced changes in the nearshore hydrodynamic conditions (Owens 1977).

Another type of environmental change is embodied by eustatic variations (cf. Sect. 2.5.2.5). For a rise of sea level, Bruun (1962; see also Bruun and Schwartz 1985) has enounced a simple rule: as water level rises, sediments are eroded from an equilibrium beach and deposited in the nearshore zone; the latter, in turn, is elevated proportionally to the eustatic rise. Out of this, a volumetric relationship predicting beach erosion as a function of rising water level can be abstracted (Dubois 1977).

Instead of by a direct mechanical model, the response of an equilibrium beach can be studied in terms of process-response theory. We have given (in Sect. 6.4.3.5) the "regime equations" for an equilibrium beach. Evidently, if one parameter is changed therein, the others will adjust themselves so that a new equilibrium is reached (Krumbein 1963; Howd and Holman 1987). A most notable result of external changes is the appearance of beach *scarps* (Sherman and Nordstrom 1985; Hughes and Cowell 1987) marking the response point in the beach slope. Finally, in terms of system theory, beach profiles have been modeled by first-order Markov processes in time (Sonu and James 1973).

6.4.4 Theory of Special Features on Shallow Coasts

6.4.4.1 Introduction

In Sect. 1.6.3.2 we have mentioned some features which occur specifically on shallow friable coasts. We shall now discuss the possible mechanical theories for their explanation.

6.4.4.2 Spits and Hooks

First, turning to spits and hooks, we note that their formation (Holmes 1944; Tanner 1960) has been generally linked with the wave refraction phenomenon (cf. Sect. 6.2.2.3).

The spits, when curved, form hooks. The manner in which a hook may be formed has been illustrated by Holmes (1944) and is shown in Fig. 91. As the waves are refracted around the tip of the spit, the attendant longshore currents become weaker and hence material is deposited. This phenomenon was studied by Castanho (1958) experimentally and by Rea and Komar (1975) by computer simulation.

6.4.4.3 Offshore Bars and Barrier Islands

Next, we turn to a discussion of offshore features such as bars and barrier islands. Studies of their genesis have been made for a long time; some of the "benchmark

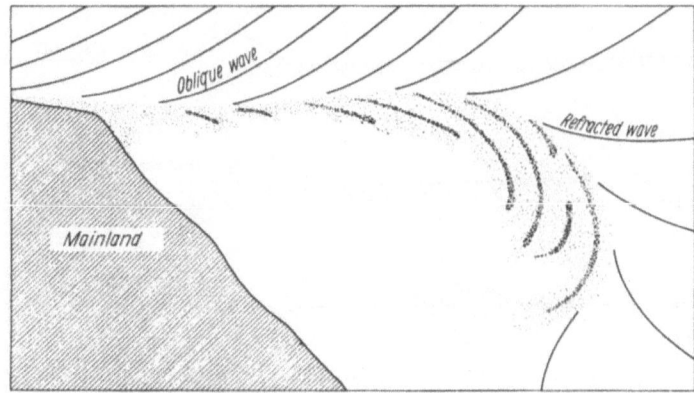

Fig. 91. Formation of a curved hook. (After Holmes 1944)

papers" on the subject (going back to the last century) have been collected and
republished by Schwartz (1972a, 1973).

A review of the literature indicates that there does not seem to be a single cause
for all offshore features (Shepard 1960; Dolan et al. 1980). The causative processes
that are involved include eustatic changes, wave refraction patterns, nearshore
circulation currents, tidal currents, and storm surges.

The cause of the formation of offshore features has most frequently been sought
in eustatic changes, notably in a lowering of the sea level, i.e., in the emergence of
the coast line. A phenomenological theory based upon this hypothesis has been
developed by Johnson (1919). Accordingly, bars are supposed to develop at the
youthful stage of the emergence of the coast line, owing to a process of transfer of
bottom material from the shallow water towards the sea. First, a submerged bar is
formed which later grows owing to the lowering of the sea level.

Although Johnson's theory has met with general acceptance, it has been pointed
out by Zenkovich (1957) that bars are found on *submerging* coasts as well as on
emerging ones. A mechanism different from that suggested by Johnson must be
advocated in coasts of submergence. Zenkovich suggested such a mechanism; it is
illustrated in Fig. 92. In order for this mechanism to be operative, two elements are
required: the existence of a rise above the water on the shore and the existence of an
underwater slope which is steeper than the general slope of the land behind the
ridge. As original rise, Hoyt (1967) assumed the existence of a beach ridge, whereas
Fisher (1968) proposed that it might be a subaerial dune. When the destruction of
the whole complex begins, i.e., when the sea level rises, the rise migrates landward
as the waves pass over its crest, at the same time heightening it through the
deposition of material. This process keeps on being operative as long as there is a
slow rise of sea level (landward migration of offshore features: Otvos 1970;
Rampino and Sanders 1981; Leatherman 1983; Sunamura and Takeda 1984). In
this instance, it may be noted that bar migration may be enhanced by storm surges
(Greenwood and Hale 1982; Greenwood and Sherman 1984; Sallenger et al. 1985),
particularly by the overwash involved (Leatherman and Williams 1983; Kochel
and Dolan 1986).

Further explanations of the origin of barrier features have been based on wave
refraction processes (Niedoroda and Tanner 1970; Heathershaw and Davies 1985;

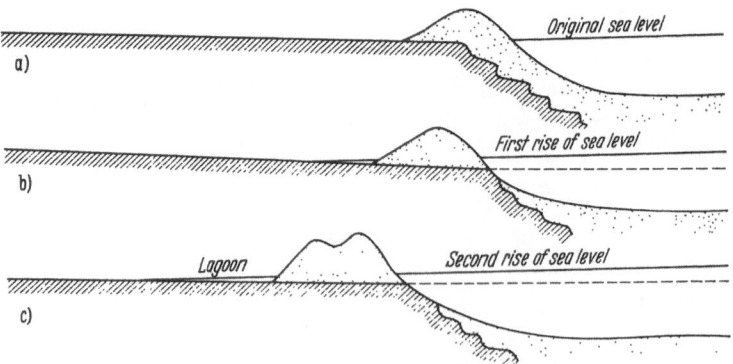

Fig. 92 a–c. Creation of an offshore bar on a submerging coast line. (After Zenkovich 1957)

Boczar-Karakiewicz and Davidson-Arnott 1987), and the longshore currents connected with them. Thus, Gilbert (1885) supposed that these currents would deposit material where the features are. A phenomenological theory of such a process was set up by Ertel (1968b), who based it on some general assumptions regarding continuity and the erosion-deposition mechanisms. In this fashion, he arrived at a linear differential equation of the third order for the developing profile. A variation of the above idea ascribes barrier formation to tidal rather than to wave-caused currents (Mardell and Pingree 1981; Huthnance 1982). This type of process is important in the shaping of the large tidal marshes on the Atlantic coast of the Netherlands and Germany ("Wadden" areas; cf. Dijkema 1987; Heyer et al. 1986).

Finally, it may be noted that the migrating offshore features are responsible for the hummocky cross-bedding often found in shallow water sediments (Allen 1985).

6.4.4.4 Theory of the Depth Distribution in Shallow Bays

Next, we shall investigate the depth distribution in shallow bays. In such bays, a special type of circulation becomes established as shown in Sect. 6.4.2.3, which has been advocated by Nakano (1955) for the explanation of the depth distribution in bays.

Thus, let us consider a bay as schematically shown in Fig. 93. We take an arbitrary point P on the shore line, whose distance from an arbitrary initial point along the shore is denoted by s. Then we denote by W a unit vector in the direction in which the prevailing wind blows and by L the unit vector normal to the shore (pointing seaward). The angle between W and L is denoted by ϑ. Thus, the region near the coast becomes shallower if the wind blows offshore ($\vartheta = 0$) and deeper if the prevailing wind blows onshore ($\vartheta = \pi$). Therefore, we take as the depth effect on

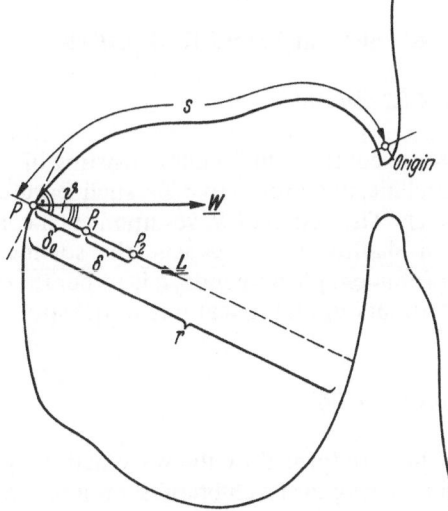

Fig. 93. Schematic view of a bay

the wind a function $\Phi(s)$ which we write as follows

$$\Phi(s) = F(r) \cos \vartheta(s). \tag{6.4.26}$$

Here, r denotes the "fetch" of the wind representing the distance from P to the opposite shore along **L**; F(r) is a monotonously increasing function which may be assumed as proportional to r for $r \leqq R$, R being a limiting value equal to several dozen km, and as constant for $r > R$.

We now consider two points P_1 and P_2 on the normal L such that

$$\overline{PP_1} = \delta_0, \tag{6.4.27}$$

$$\overline{P_1P_2} = \delta \tag{6.4.28}$$

and denote by h the mean depth on $\overline{P_1P_2}$. If then δ_0 and δ are kept constant, h becomes a function of s. Denoting by $\bar{h}$ the average value of h(s) for all s, we can define a depth variation Δ in the following fashion:

$$\Delta(s) = h(s) - \bar{h}. \tag{6.4.29}$$

If the depth structure of a bay is determined by the wind, then h will be small if Φ is large. However, since

$$h(s) + \Delta(s) = \bar{h} = \text{const}, \tag{6.4.30}$$

it follows that Δ must be large if Φ is large, i.e., there must be a positive correlation between $\Delta(s)$ and $\Phi(s)$. Thus, the above theory predicts a parallelism betwen $\Delta(s)$ and $\Phi(s)$; this parallelism is in form of an ordinal correspondence: if one quantity increases, so must the other.

The above discussion indicates that in many shallow bays the vertical currents caused by the prevailing winds are the determining agents of the depth distribution. It must be expected, however, that there are bays where other influences, such as tides and notably wave-generated vortical longshore currents (O'Rourke and Leblond 1972), may be able to obscure the effect of the vertical currents.

6.4.5 Steep and Hard Rock Coasts

6.4.5.1 Introduction

Steep coasts include cliffs consisting of rocky or friable materials; hard rock coasts include, in addition to cliffs, such features as rocky shore platforms and limestone reefs. The action of waves upon such coasts is somewhat different from the action on shallow friable coasts: the sorting and transport of sand is no longer a prominent phenomenon; it is rather the direct effect of the surf and of the nearshore currents upon the features in question which requires an explanation.

6.4.5.2 Cliffs

Cliffs are formed by the wave action against their foot. The phenomenon of surf beating against a cliff represents a process of extreme complexity. An inspection of the mere geometry of wave trains beating against a cliff, however, shows that a considerable amount of energy may be available for work on the cliff. The energy flux F carried by waves of wave length $\lambda = 2\pi/m$ across a strip of unit width in water of depth h can be calculated (according to the theory of waves of small amplitude); it is (Stoker 1957; see p. 50 therein)

$$F = U \frac{A^2 \rho \sigma^2}{2g} \cosh^2 mh \tag{6.4.31}$$

with

$$U = \frac{1}{2} c \left(1 + \frac{2mh}{\sinh 2mh} \right), \tag{6.4.32}$$

where c is the wave velocity (cf. 6.2.17)

$$c = \frac{\sigma}{m} = \sqrt{\frac{g\lambda}{2\pi} \tanh \frac{2\pi h}{\lambda}} \tag{6.4.33}$$

and σ is given by

$$\sigma = \sqrt{gm \tanh mh}. \tag{6.4.34}$$

It is clear that this energy flux is available for the destruction of the cliff if the waves are stopped by it, i.e., if no reflection is taking place.

It is, in fact, possible to give a solution (in the small amplitude theory) for harmonic waves traveling from infinity in infinitely deep water toward a vertical cliff, their energy becoming absorbed there. In order to have the energy absorbed, a mechanism must be introduced to achieve this for which one can make the following suitable theoretical model: Let an external pressure fluctuation be operative between the cliff and the distance a from it:

$$p(x, t) = \begin{cases} P \sin \sigma t & \text{for } |x| \le a \\ 0 & \text{for } |x| > a \end{cases} y = 0, \tag{6.4.35}$$

where x denotes a co-ordinate parallel to the sea and normal to the cliff (the latter is at $x = 0$) (see Fig. 94). Then it can be shown that the solution for the velocity

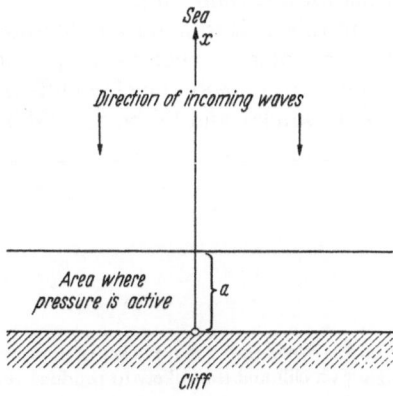

Fig. 94. Waves traveling against a cliff

potential φ in the small amplitude theory which represents waves travelling towards the cliff and being absorbed there, is given by (z represents the vertical coordinate, cf. Stoker 1957, p. 67)

$$\varphi(x, z, t) = -\frac{2P\sigma}{\rho gm} \sin m\, a\, e^{my}[\sin(mx + \sigma t)] + 0(1/r), \qquad (6.4.36)$$

where $0(1/r)$ is a function which behaves like $1/x$ at infinity. The surface $z = \eta$ is then given by (6.2.9).

The amplitude A of the surface elevation η at infinity is given, from (6.2.11) and (6.4.36) by

$$A = \frac{2P}{\eta g} \sin ma. \qquad (6.4.37)$$

It is obvious that P becomes infinite if sin ma is zero. This shows that, in the present model at least, conditions occur where the pressure fluctiations needed to absorb the wave energy must be infinitely great. The absorption mechanism introduced here is only a simple model, but the treatment does show that waves beating against a cliff may exert very great pressures upon the latter.

The amount of energy expended toward the destruction of a surf-beaten cliff depends on how much of the wave energy is reflected. It has been argued by Russell and Macmillan (1952) that the reflection is least (and thus the destruction greatest) if the waves break against a cliff in such a fashion that a pocket of air is trapped between the water and the cliff at impact time (cf. Fig. 95). No mathematical theory of the reflection coefficient can as yet be given. For a wave 3 m high and 46 m long, pressures of 7.25 MPa have been recorded. The maximum pressures ever recorded are about ten times greater (cf. King 1959, p. 289).

Pressures of the above magnitude make it understandable why a cliff can be eroded at its foot, since cavitation (cf. Sect. 3.2.3.4) must occur. Wave tank experiments to investigate the erosive action of waves beating against a rocky coast have been reported by Sanders (1968). If it be assumed that the water is laden with rock debris, the erosive action is even more plausible. The action of the sea against a steep coast may thus be likened to that of a meandering river undercutting a slope. The development of a steep sea coast should therefore be explainable in terms of the same theory as that given in Section 3.5.4.2 for undercutting rivers; at least for a certain time.

In this, it is immaterial whether the cliff consists of the usual rock or else of friable materials, such as the pumice flow cliffs of the Asama mountain in Japan (Matsukura 1988), the clay cliffs on the Great Lakes of North America (Brayan 1970; Quigley and Di Nardo 1980), or the chalk cliffs of Dover (May and Heeps

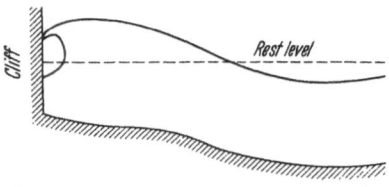

Fig. 95. Wave trapping a pocket of air when breaking against a cliff and thus likely to produce cliff recession

1985). Recession always occurs by erosion at the foot as mentioned above. Incidentally, the process has also been modeled successfully in a wave tank (Sunamura 1976, 1977, 1981) and on a computer (Kirkby 1984).

6.4.5.3 Shore Platforms

Since the wave action is only effective near the still water line, erosion will proceed only above a certain level. Below it, a *shore platform* remains. After the latter has attained a certain breadth, it will retard the incoming waves and hence hinder further cliff recession, unless eustatic changes occur. With regard to such shore platforms, we note that there is a certain amount of controversy concerning their origin. They are due to erosion; some authors ascribe the erosion to chemical effects of the sea water, some to the action of the waves, and others to subaerial erosion near the coast. General reviews of the problem have been given, for instance, by Wentworth (1938), Fairbridge (1952), Gill (1967), and Trenhaile (1980).

In general, the existence of shore platforms, sometimes much above the present shore line, is taken as evidence of the former position of the sea level at that locality. Thus, the shore platforms are regarded as evidence of an uneven change of the relative position of the sea level with regard to the land. However, Popov (1957) has maintained that this does not necessarily have to be so. Popov considered a case where the speed of lowering of sea level is constant and claimed that a series of terraces may be created. Again, it is seen that there is some controversy as to whether or not a series of shore platforms can result from a steady eustatic change. What is needed is evidently a theory of the creation of shore platforms and of their evolution.

Such a theory can be obtained (Scheidegger 1962) by considering the erosion on a shore profile as a form of slope recession. Thus, referring to Fig. 96 as exhibiting the general geometry of a cross-section of a coast, we can start with the fundamental equation of slope development (3.5.23)

$$\frac{\partial y}{\partial t} = -\Phi\sqrt{1 + (\partial y/\partial x)^2} \tag{6.4.38}$$

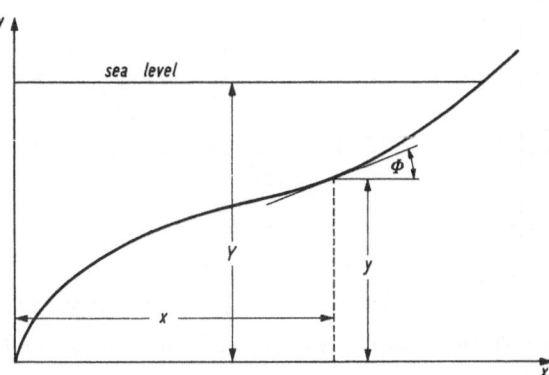

Fig. 96. Geometry of a shore profile. (After Scheidegger 1952)

adding suitable boundary conditions: in order to describe the erosive action on a shore, we assume that it becomes less and less the deeper we go below the surface of the water. We denote the height of the water surface above some base line by $Y(t)$; this may be a function of time. We then take our erosional function Φ proportional to an exponential and the sine of φ:

$$\Phi = a \sin \varphi \exp[-\alpha(Y-y)^2], \tag{6.4.39}$$

where α and a are some constants. This function has indeed the property that the erosion decreases with distance downward from the water level, and it also increases with increasing declivity of the slope. Expressing $\sin \varphi$ in terms of $\partial y/\partial x$ and inserting (6.4.39) into (6.4.38) yields the fundamental equation of shore development:

$$\frac{\partial y}{\partial t} = -a \exp[-\alpha(Y-y)^2] \cdot \frac{\partial y}{\partial x}. \tag{6.4.40}$$

Naturally, this equation is only valid for $y \leq Y$. Above the water line, no erosion occurs.

The basic shore development Eq. (6.4.40) is nonlinear and cannot be integrated in closed form. Thus, numerical methods may be used as in connection with the slope development theory presented in Section 3.5.3.3. The results of the calculations, starting with an initially straight profile, are shown for a constant sea level in Fig. 97.

Similar calculations, for a static sea level, were made by Sunamura (1978), Trenhaile and Layzell (1981), and Trenhaile and Bryne (1986), leading to results very similar to those of Scheidegger (1962). The pattern was also confirmed in wave tank experiments by Sunamura (1975).

As noted, the above results apply only to a static sea level. However, similar calculations can be made for a sinking or rising sea level, assuming

$$Y(t) = Y_0 \pm t_{const} t. \tag{6.4.41}$$

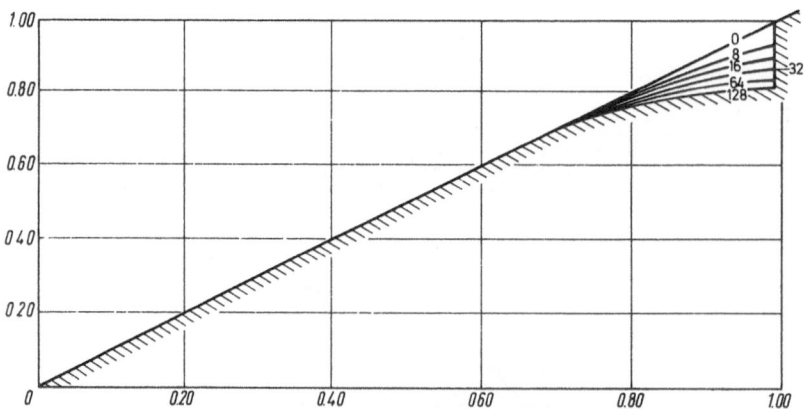

Fig. 97. Development of a marine terrace with constant sea level ($a = 1$, $\alpha = 64.0$). (After Scheidegger 1962)

The method of characteristics allows one to get an asymptotic solution of Eq. (6.4.40) with Y given by (6.4.41), i.e., for a steadily rising or falling sea level. The result is that the horizontal recession R(y) at the level y is asymptotically independent of y (i.e., constant) and given by

$$ R = \frac{a}{\sqrt{\alpha\, t_{const}}} \frac{\sqrt{\pi}}{2} \qquad\qquad (6.4.42) $$

Hence, we see that a eustatic change proceeding at a constant rate does not bear out Popov's contention that a sequence of shore platforms should result: it simply leads to a recession by a constant amount of the slope profile. "Platforms" result only if there is a change in the rate of the eustatic changes. This result was also confirmed by more elaborate calculations by Esin (1966, 1980), who investigated slightly different expressions for the function Φ in Eq. (6.4.38).

Finally, it may be noted that shore platforms can also be caused by primary lithological and tectonic conditions not related to coastal hydrodynamics (Read 1982; Weaire and O'Carroll 1983; Hanks et al. 1984). In this instance, the platforms would be formed like any other rocky plateau not connected with the coastal situation at all.

6.4.5.4 Limestone Reefs

As a final topic of this section, let us mention the formation of limestone reefs by biological agents.

Actually, on the whole, biological agents tend to be destructive rather than constructive (Jones and Goodbody 1984). Such destructive agents include animals burrowing into rocks (Duerden 1902; Gardiner 1930; Ginsburg 1953), algae growing below the rock surface (Nadson 1927), and snails (North 1954) ingesting sand. In addition, it also appears that some fish eat coral, crabs and some gastropods may enter fissures in coastal rock and thereby help disintegrate it. Often the disintegration is caused not directly, but the animals etc. create holes which lay the rock open to accelerated attack by the physicochemical processes described in Section 3.2.2.4.

Contrary to the destructive action of the biological agents mentioned above, coral polyps and calcareous algae build up rock by depositing their skeletons. The limits for the environmental conditions under which corals can exist are rather narrow. The temperature of the water must never fall below 18 °C or rise above 36 °C, with 25 to 30 °C being most favorable. Light is also essential, which restricts the depth at which reef-building corals can grow to about 30 m. Salinity must be between 27 and 40 parts per thousand. Much wind and sand are unfavorable (Guilcher 1958).

According to Darwin (1842), the formation of actual reefs is tied to the existence of shores of submergence: if the sinking rate is just keeping pace with the rate of growth of the corals and calcareous algae, a reef is created. Naturally, it is difficult to give a mechanical theory of this process, other than stating that some type of equilibrium must be established between the biological and eustatic factors. In this, there is some question as to whether it is the windward or leeward side of a barrier

that has the highest biological growth rates (Davies and Marshall 1980;
Braithwaite 1982). Hydrodynamic features (such as upwelling currents containing
many nutrients) undoubtedly also play a major role (Thompson and Golding 1981).
Remnants of coral reefs may give rise to island chains; thus the origin of the Florida
Keys has been ascribed to the activity of corals (Hoffmeister and Multer 1968).

6.4.6 Large-Scale Features on Coasts

6.4.6.1 Introduction

Finally, one may look at coasts from a more macroscopic standpoint. Thus, one
finds ever-recurring features on coasts, such as headlands and bays.

Next, the large-scale (cumulative) changes of coastal features are of interest. For
individual forms, such as beaches, a discussion of changes has already been given in
Section 6.4.3.6; it will now be our task to study effects on a grander scale.

Finally, last but not least, the effect of rare events (such as storm surges) is of
importance.

6.4.6.2 The Headland-Bay Problem

One finds generally that a coast consists typically of a sequence of headlands and
bays, the former representing a steep coast, the latter a shallow one with a
concomitant beach. The question arises, therefore, why in one case a beach
develops, whereas in another it does not. From the discussions given earlier it is
clear that this question is tied up with the nearshore circulation system: if the
sediment supply is higher than what littoral drift can remove, a beach develops,
otherwise it does not. The sequence of head-land bays is therefore somehow tied up
with a similarly sequential (cellular) structure of the near shore circulation system.

Qualitatively, the above contentions have been tested in the field by Dolan et al.
(1974), by Rosen (1975) and by Mardell and Pingree (1981). In the laboratory, the
envisaged hydrodynamic pattern has been simulated by Komar (1971), analytical
studies thereof have been made by Le Blond (1979) and Willmott (1983). The
author mentioned last was able to model parts of the Californian coast rather
realistically, based on an approximation assuming a slowly varying coastline
which leads to a linearized vorticity equation that can be solved by means of a
Green's function technique.

In addition to a purely "exogenic" origin of headlands and bays, an endogenic
("structural") component (principle of tectonic predesign of geomorphic features,
cf. Sect. 5.2.6) cannot be excluded (Hanks and Wallace 1985).

6.4.6.3 Large-Scale Changes on Coasts

Coastline changes are of great importance with regard to human existence,
inasmuch as many large cities have been constructed at the sea shore. Long-term
studies of coastline changes have therefore been made by bodies concerned with
coastal development (Bird 1985).

Qualitatively, the global development of coastal morphodynamics during the last 20 000 years has been described by Ellenberg (1983). On such a scale, the processes have been mainly governed by the climatic evolution and the ensuing eustatic changes.

For a theoretical analysis, it is clear that a "coast" represents a geomorphic "system", and it should be possible to deal with it by the usual means of system dynamics and process-response theory. This approach has been applied successfully to such small-scale features as beaches (cf. Sect. 6.4.3.6), but not many investigations along these lines seem to be available for coasts on a large scale, except for a notable study by Trofimov and Moskovkin (1985): these authors investigated the process-response pattern for a cliff-beach system based on a phenomenological mass balance equation. The analysis was then applied to optimize shore protection measures.

6.4.6.4 Storm Surges

Finally, we come to the discussion of rare events. Here, "rare" is to be understood with regard to the human view; in geological time spans, such events may be sufficiently frequent and regular that they entail regular effects.

The "rare" events of concern in the coastal context are almost exclusively storm surges which are commonly regarded as "events of the century", although they may have average return periods of much less than 100 years.

Because of the human context, the prediction of storm surges in coastal areas is of importance (so that the population can be evacuated in the face of approaching danger). Methods to achieve such predictions have been based on observations of cyclone tracks (El-Sabh and Murty 1989) and on observations of the tides of the solid earth (Zschau and Kumpel 1979). The general frequency of storm events has to be determined from records of past occurrences.

Then, the actual changes of the coastline in the wake of such events have to be analyzed (and predicted). In this, the storm waves cause deposition (Orford 1977) as well as erosion (Kobayashi 1987). The combined effects are caused by the bottom stress induced by the storm waves; this stress has been investigated analytically by Kielmann and Kowalik (1980) and in a wave tank by Sunamura (1983). In the field, the changes have been monitored by Lins (1984). Together, these studies provide a base for potential risk estimates. In the geomorphic context, storm surges are sufficiently frequent so that their effects can be considered together with all the other effects that cause large-scale changes on coasts over long time spans.

6.5 Dynamics of River Mouths

6.5.1 General Remarks

Rivers entering the sea or a large body of water cause a variety of peculiar features on the coast line concerned. In Section 1.6.3.4, we have described the morphology of river mouths, and it will now be our endeavor to investigate the mechanical causes of the physiographic features mentioned there.

The physical background for a dynamical theory of river mouths will be supplied by a discussion of the general hydrodynamic conditions in a river mouth (Sect. 6.5.2). We shall then proceed to a description of the origin of the two most common types of river mouths: estuaries (Sect. 6.5.3) and deltas (Sect. 6.5.4). Finally, we shall conclude our survey with a brief description of the origin of sand bars off estuaries.

It will be found that the general dynamics of the evolution of a river mouth is not very well developed as of yet. There is a series of criteria indicating when a particular river mouth may be expected to be stable, but the problem of greatest interest to the geoscientist, viz. that of elucidating what happens when a river mouth is known to be unstable, has not yet been solved.

The theories to be presented here are contained in a variety of papers, mostly originating in engineering laboratories. Notable recent monographs dealing specifically with river mouths have been written by Samoilov (1956) and by Lauff (1967). Although they are mostly descriptive, some references to theoretical work are given.

6.5.2 General Hydrodynamic Conditions in a River Mouth

6.5.2.1 Basic Equations

Where a river enters the sea, two agents act both at the same time: first, there is the influence of the river and second, there is the influence of the sea. The combination of these two influences has the result that the hydrodynamic conditions in a river mouth become very complicated.

By itself, the river water entering the sea forms a turbulent jet that diffuses in all directions (Hutter and Hofer 1978). However, this fact is more of interest in connection with the dispersion of sediments and pollutants than in geomorphic context. In the latter case, the scouring action as in a river, discussed in Chapter 4, is of greater importance. In this connection, a river jet may become a local density flux (Ikeda 1984), and it may be reckoned that for such a flux the arguments

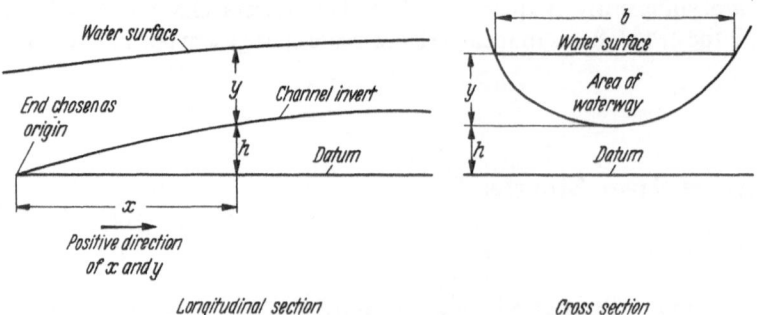

Fig. 98. Geometry of a river channel. (After Scholer and Germanis 1959)

presented for rivers are still valid. Since the dynamics of rivers has already been discussed in Chapter 4, it is primarily the influence of the *sea* which deserves our attention in the present context. This influence is mostly due to the effect of the *tides* upon a river mouth. What one has to consider is the unsteady-state problem of the motion of water in a channel under given time-dependent boundary conditions at its mouth. Combining the non-linear shallow water wave theory [cf. (4.2.12)] with Manning's friction formula (4.2.12), one obtains as equation of motion (see, e.g., Scholer and Germanis 1959)

$$\frac{\partial}{\partial x}(y + h) + \frac{1}{g}\frac{\partial v}{\partial t} + \frac{v}{g}\frac{\partial v}{\partial x} \pm \text{const}\,\frac{v^2}{y^{\frac{4}{3}}} = 0, \tag{6.5.1}$$

where the meaning of most of the symbols is evident from an inspection of Fig. 98. In addition, v denotes the mean flow velocity in the channel. Further to the equation of motion, one requires a continuity condition. The latter can be written as follows:

$$\frac{\partial Q}{\partial x} + \frac{\partial B}{\partial t} = 0, \tag{6.5.2}$$

where Q is the total discharge and B the area of the waterway. The area B has to be expressed as a function of height y and distance x which defines the geometry of the channel.

6.5.2.2 Analytical Simplifications

The task, then, is to integrate Eqs. (6.5.1) and (6.5.2). Needless to say, this problem is very difficult to solve as one has to integrate a system of nonlinear partial differential equations. The available methods have been collected in a book by Dronkers (1964). Accordingly, there are in essence three types of methods available: The harmonic method, in which solutions for the individual harmonic components of the wave spectrum are sought, the method of characteristics, and digital numerical methods.

Because of the difficulty of solving the system of Eq. (6.5.1) and (6.5.2) exactly, a variety of approximation methods for obtaining a solution has been suggested. These range from the application of the linearized shallow water theory [cf. Eq. (6.2.25)] to various less severe simplifications of the exact differential equations. A review of some of the possiblities has been given by Scholer (1958), by Bruun and Gerritsen (1958) and in the book of Dronkers mentioned above. Stoker (1957) also discussed a variety of solutions that have been applied to the design of breakwaters in harbors.

Accordingly, the simplifying assumptions that may be introduced are the following:

1. the average elevation of the water surface is constant along a horizontal channel
2. tide is sinusoidal
3. friction term is linearized
4. differential equation itself is linearized.

According to how many simplifications are introduced, one obtains a more or less satisfactory description of the phenomena.

The basic nonlinear character of the fundamental differential equations (6.5.1) and (6.5.2) has some well-known consequences: a "shock" front may develop which represents a tidal "bore". This feature, of course, gets lost if the equations are linearized. The more exact calculations (particularly the digital ones) yield very acceptable results that seem to check with actually observed phenomena. Dronkers' method was used for the calculation of the tides entering through gaps in the dykes of the Zuiderzee and the predictions are found to be very accurate.

Because of the many variables involved, it is difficult to make general statements regarding the tidal currents that may be expected in any particular estuary. The calculations, preferably by means of electronic computers, have to be carried out separately for each case under consideration.

6.5.2.3 Results

The tides are the principal influences induced by the sea in a river mouth (Zimmerman 1981). However, there are also other ones. Of some importance may be the direct wind drag (Samoilov 1956; Elliott 1982) upon the water. The action of a strong upstream wind can be very similar to that of the tide entering the river mouth (cf. Fig. 99). In other cases, strong secondary currents may be set up which may form rather well-defined solenoidal *convection cells* whose axes are parallel to the wind direction. The Coriolis force also may be of some significance (Sekherzh-Zen'kovich 1959; Labeish 1959).

The hydrodynamic conditions in a river mouth are also affected by the *swell*. However, the latter can be geomorphological importance only if the river mouth is very wide indeed.

For a theoretical explanation of the development of the various types of river mouths, the hydraulics of the river and the effect of the tides are the most important factors. It is probably safe to neglect the other influences, at least in a first approximation.

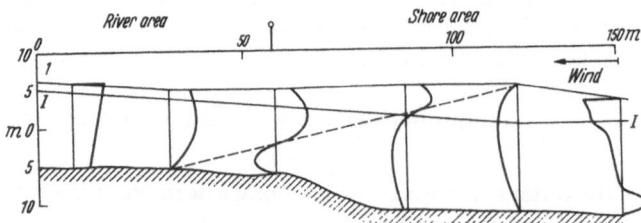

Fig. 99. Velocity distribution (*heavy* lines) and wave caused by wind. I Level without wind. (After Samoilov 1956)

6.5.3 River Estuaries

6.5.3.1 General Remarks

The background supplied in Sect. 6.5.2 makes it clear what would be required to set up an exact theory of the mechanics of the development of a given river estuary: First, the hydrodynamic regime (taking into account the river flow and the tides) at a specified time t has to be calculated. Then, the effect of the resulting currents upon the bottom and the sides of the estuary (using the methods of Chap. 4) has to be established. This yields the *changes* that are taking place at time t by which it is possible to calculate the configuration of the estuary at time t + dt. The procedure can then be repeated over and over again to yield the development of the estuary with time. The above outline of the requirements for an exact theory of the evolution of a tidal estuary indicates that the procedure is rather involved. Reviews of the problem have been published, e.g., by Dyer (1973, 1979) Carter (1979), Officer (1976, 1981), Mehta (1986), and Kreeke (1986).

Because of the difficulties with an exact solution of the problem, the description of the evolution, or at least the establishment of equilibrium conditions, of a river estuary had also been tried by other means. First, there is a *shoal theory* by Bruun and Gerritsen (1958) which is based upon an extremely simplified theoretical model of the hydrodynamic conditions in the estuary, and second, there are two theories aiming at criteria regarding whether a certain estuary is stable or not. The first of these, a *gorge theory*, is also due to Bruun and Gerritsen; it only considers the material balance across the entrance ("gorge") to the estuary. The second is a *regime theory* of Blench (1953), who tried to establish equilibrium criteria in a similar fashion as he had done this for rivers (cf. Sect. 4.6.4). Furthermore, computer solutions have been tried and the geomorphological significance of the various solutions has been investigated.

6.5.3.2 Shoal Theory

Turning first to the *shoal theory*, we assume (with Bruun and Gerritsen 1958) that the whole of the estuary is subject to tidal currents. The change in cross-sectional area of the channel (ΔA) caused by deposits must be proportional to the net amount of material deposited (per unit length) ΔM:

$$\Delta A = c_1 \Delta M \tag{6.5.3}$$

or, integrated (the c_i's are constants)

$$M = \frac{A - c_2}{c_1}. \tag{6.5.4}$$

Using one of the bed load formulas (cf. Sect. 4.5.3.2), we can write for the material q_s transported per unit channel width

$$q_s = c_3 \sigma^{\frac{5}{2}}, \tag{6.5.5}$$

where σ is the shear stress at the bottom. The last formula is a modification of

(4.5.22) with $\sigma_{cr} = 0$ and $m = \frac{5}{2}$. However, we can use the expression (4.2.20) for σ, viz.

$$\sigma = g\rho \, RS, \qquad (6.5.6)$$

where ρ is the density of the water, g the gravity acceleration, R the hydraulic radius and S the slope. We also have [from (4.2.12)]

$$S = c_4 v^2 \qquad (6.5.7)$$

and thus

$$q_s = c_5 v^5. \qquad (6.5.8)$$

Let us assume that the velocity in the estuary due to the tides is (valid for one half-cycle)

$$v = v_{max} (\sin \omega t)^{\frac{1}{2}} \qquad (6.5.9)$$

with $\omega = 2\pi/T$ (T being the tidal period). Then, the total amount of material dislodged during a tidal half-cycle is

$$M = c_5 v_{max}^5 \int_0^{T/2} (\sin \omega t)^{\frac{5}{2}} dt$$

$$= c_5 \frac{1.432}{2\pi} v_{max}^5 T. \qquad (6.5.10)$$

It is convenient now to introduce the *tidal prism* Ω which represents the total amount of water that flows into (and out of, if there is no river discharge) the estuary during one flood or ebb period. One obviously has a relationship of the form

$$v_{max} = c_6 \frac{\Omega}{AT}. \qquad (6.5.11)$$

Thus,

$$M = c_7 \frac{1}{T^4} \left(\frac{\Omega}{A} \right)^5 \qquad (6.5.12)$$

and, with (6.5.4)

$$A = c_1^{\frac{1}{6}} \left[c_7 \frac{1}{T^4} \Omega^5 + \frac{c_2}{c_1} A^5 \right]^{\frac{1}{6}} \approx \text{const } \Omega^{\frac{5}{6}}. \qquad (6.5.13)$$

This is the equilibrium condition over one-half of a tidal cycle according to the shoal theory. As is noted, no modification can be made for the effect of the river. The shoal theory is thus restricted in its applicability to river mouths but does apply to tidal estuaries.

6.5.3.3 Gorge Theory

The *george theory* mentioned above (which is also due to Bruun and Gerritsen 1958), however, lends itself to a modification to take the river discharge into account. As mentioned above, the gorge theory considers only the flow through the

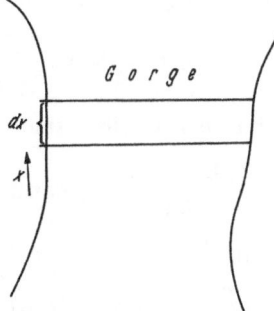

Fig. 100. Geometry of a gorge

entrance of the estuary. Following Bruun and Gerritsen, we investigate a section dx of a gorge (see Fig. 100) and we denote by $Q_s dt$ the total amount of sand passing through that section during the time element dt. The mass erosion on the distance dx is then obviously given by $(\partial Q/\partial x)dt\,dx$. If the estuary is to be stable, then the net erosion over a complete tidal cycle (time T) must be zero everywhere:

$$\int_0^T \frac{\partial Q_s}{\partial x} dt = 0. \tag{6.5.14}$$

We assume that the total transport of material can be represented in the following form

$$Q_s = c_8 v^n, \tag{6.5.15}$$

where n is some empirical exponent. The velocity in the estuary is also expressed in terms of a semi-empirical relationship of the type

$$v = v_{max} \cos^p \omega t, \tag{6.5.16}$$

which can be written, using (6.5.11) as follows

$$v = c_6 \frac{\Omega}{AT} \cos^p \omega t. \tag{6.5.17}$$

The exponent p is purely empirical. Thus, the expression for Q_s becomes

$$Q_s = c_9 \left(\frac{\Omega}{AT} \right)^n \cos^{pn} \omega t. \tag{6.5.18}$$

Stability is achieved if Eqs. (6.5.14) and (6.5.18) can be satisfied. Since only Ω and A are functions of x, this is obviously the case, if

$$\Omega = \text{const } A. \tag{6.5.19}$$

This is the stability condition.

The above theory can be modified to take the river flow into account. Instead of (6.5.15) we must set

$$v = v_{river} + v_{max} \cos^p \omega t = \frac{Q_{river}}{A} + c_6 \frac{\Omega}{AT} \cos^p \omega t, \tag{6.5.20}$$

where Q_{river} is the total river discharge. Hence, (6.5.17) becomes

$$Q_s = c_9 \left(\frac{Q_{river}}{A} + c_6 \frac{\Omega}{AT} \cos^p \omega t) \right)^n.$$
(6.5.21)

Thus, it is obvious that stability is achieved if

$$\frac{TQ_{river} + \Omega \cdot \xi}{AT} = \text{const},$$
(6.5.22)

or if

$$\xi \frac{\Omega}{A} = \text{const } T - \frac{Q_{river} T}{A},$$
(6.5.23)

where ξ is a certain coefficient which originates from the integration with regard to time in the stability Eq. (6.5.14). Relationship (6.5.23) is exact for $n = 1$ and approximate otherwise.

6.5.3.4 Numerical Attempts

Finally, in a discussion of specific theoretical treatments of the tidal inlet problem it should be noted that numerical solutions have been attempted by Festa and Hansen (1976, 1978), Bäumer (1983) and Jenkins (1983).

However, as in all similarly structured situations, numerical solutions simply cannot provide a comprehensive answer to the real problems concerned: numerical solutions always refer only to a single specific case; at best, a number of cases can be "run through" by varying certain parameters; an overview over the whole situation can never be obtained.

6.5.3.5 System Theory

The direct mechanical approaches to the tidal inlet problem obviously do not lead to satisfactory results. Therefore, approaches referring to system theory have been tried.

The first of these was one by Blench (1953), corresponding to the regime theory of rivers: stability conditions are abstracted from a great number of empirical data. The condition of Blench is essentially similar to that of the shoal theory [cf. Eq. (6.5.13)]; another similar equation corresponds to that of the gorge theory (6.5.19); tests were made by Bruun and Gerritsen (1958), by Johns (1967), and by Bruun (1978, 1986) and yielded the result [in (6.5.13)]:

$$\text{const} = 1.56 \times 10^4 \, \text{m}.$$
(6.5.24)

Finally, it may be mentioned that Langbein (1963) has extended his minimum variance principle (cf. Sect. 4.6.4.1) to estuarine geometry, and that Uncles (1982) has applied a random walk model to the dispersion problem in an inlet.

As usual, the existence of "regime" conditions enables one to take the process-response approach to predict system changes caused by changes in the external parameters.

6.5.3.6 Geomorphic Processes in Estuaries

The flow and stability conditions applying to estuaries have morphodynamic effects on the environment. These have been studied in scale models in the laboratory (Thorn 1982) and in the field (see Fairbridge 1980, for a review). Specific field observations are legion and cannot be referenced individually; more interesting are the generalizations that have been obtained from such studies.

In this connection, it is the resulting sedimentary structures that are most notable (De Boer et al. 1988). In particular, currents in estuaries cause sedimentary furrows in cohesive materials (Flood 1981a) and characteristic facies sequences at the riverine-estuarine transition (Ashley and Renwick 1983; cf. proceedings of a conference on sediment transport in estuaries edited by Perillo and Lavelle 1989). Ancient estuarine structures found occasionally in sedimentary areas are known as tidal paleomorphs.

6.5.4 Theory of Delta Formation

6.5.4.1 Introduction

When a river enters a large body of water, the velocity of the flow is retarded and it stands to reason that the sediment-carrying capacity of the water is thereby reduced. Thus, a large amount of material ought to be deposited at a river mouth, the latter is thereby lenghtened and a delta should be the result. The theory of erosion and accumulation by water (see Sect. 3.4.4) should be adequate to explain the observed phenomena qualitatively (Bruun and Gerritsen 1958).

It is possible, however, to refine the above statements somethat. In Sect. 1.6.3.4, we have noted that there are two types of deltas: bird's foot deltas and arcuate deltas. One might expect that there should be a rational explanation for the occurrence of these two types.

General reviews of the subject have been published by Moore (1971), Coleman (1975, 1976) and Kelletat (1984).

Accordingly, theories of delta formation are based either on hydraulic considerations or else on sedimentological precepts. Reports on scale model experiments and on geomorphological effects will conclude the present section.

6.5.4.2 Hydraulic Theories

An explanation of delta formation based on hydraulic principles has been proposed by Bates (1953). Accordingly, the flow of water from the river mouth into the stagnant body of water can be regarded as a turbulent jet of fluid entering still fluid from a nozzle. There are two types of flow that can occur: either one has an *axial* jet or one has a *planar* jet. These two types of jets simply refer to the three-dimensional and to the two-dimensional case, respectively. In their appearance, it turns out that the axial jet becomes slowed down by mixing much more rapidly than the planar jet. Thus a certain fraction of the original velocity (say one-tenth) is reached ten times sooner in the axial jet than in the planar jet. Furthermore, the planar jet always stays much narrower than the axial jet. There is a significant drop of velocity from the center line of the planar jet towards the sides, whereas in the axial jet this effect is small if it be compared with the slowing down of the jet as a function of a distance from the mouth.

If the density of the river water (including the suspension) is about the same as that of the still water which it is entering, then there is every reason to expect the jet to be an axial one. However, if the river water is considerably less dense than the still water (for instance if the latter is highly saline), then the jet will stay at the surface of the still water and one will have the characteristics of a planar jet. Similarly, if the river water is much denser than the still water, it will follow the bottom and one obtains a turbidity current (see Sect. 6.2.3).

Bates (1953) now assumes that the two types of jets give rise to the two types of deltas, respectively: in an axial jet, the decrease of velocity is so rapid that sediments are deposited in an arc around the mouth of the river. Thus, the river constantly blocks its way, forces new channels and builds up a delta by adding new arcs. This, according to Bates, leads to an *arcuate* delta.

Contrariwise, in a planar jet, the decrease of velocity with distance from the river mouth is fairly small, but there is a significant velocity decrease sideways from the jet. Thus, sediments are deposited in lines paralleling the flow of the jet and the result is a *bird's foot* delta.

The above ideas can be further expanded by considering quite generally the prograding of a river *system* into a standing body of water. It has been noted that the deposition of sediments causes the river to become braided (McPherson et al. 1987); the result is a distributary network as it is well known, for instance, from the Amazon Delta (Damuth et al. 1983) or the Guadalupe (San Antonio Bay, Texas) Delta (Morton and Donaldson 1978). The topological properties of such braid have been studied by Smart and Moruzzi (1972) and by Morisawa (1985); they are similar to those of braided stream networks (cf. Sect. 5.5.2.3).

6.5.4.3 Sedimentation Theories

The hydraulic theories of delta formation have been criticized by Crickmay (1955), who maintained that jet diffusion will have a plane (two-dimensional) pattern, no matter whether the density of the inflow is the same as that of the surrounding water or not. This view seems to be, in effect, correct (cf. Axelsson 1967). The *type* of

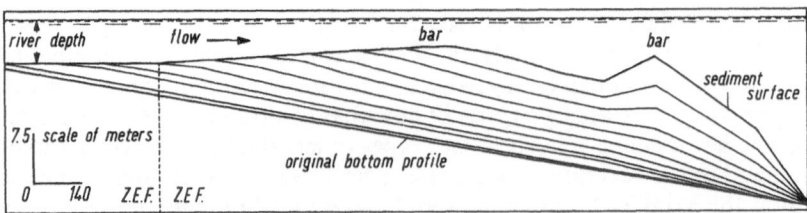

Fig. 101. Stratigraphic cross-section along the jet centerline, showing "foreset" bedding. (After Bonham-Carter and Sutherland 1967)

delta, then, which results would have something to do with the type and quantity of sediment transported by the river.

Assuming the delta formation to be connected with the diffusion of a planar jet, theoretical calculations of the attendant flow patterns and settling of sediments have been made by Bonham-Carter and Sutherland (1967), using an electronic computer. The deposition of sediment is found by calculating the trajectories of nominal sediment particles which are subject to the forward transport by the jet and settling due to gravity. A sample of the results obtained by Bonham-Carter and Sutherland (1967) is shown in Fig. 101. The bar is the result of permitting the water depth to decrease as a function of current velocity. It is seen that a good simulation of a cross-section of a delta is obtained.

On a similar basis, computer models of delta sedimentation have been made by Komar (1973a) and by Chang and Hill (1976). Wang (1984) has tried an analytical solution.

The sedimentation processes in deltas lead to the buildup of unstable mounds: slumps are common in delta regions. The processes involved have been studied qualitatively by Postma (1984) and by Lindsay et al. (1984).

6.5.4.4 Physical Studies

Finally, the problem of delta formation has been studied physically in the laboratory and in the field. In the laboratory, Jopling (1963) was able to produce "deltaic" type structures in scale model flume experiments. In the field, the pertinent processes were studied at the Mississippi River mouth by Wright and Coleman (1974), on the Amazon Cone by Damuth and Kumar (1975), and on the Nile Delta by Murray et al. (1981). The stratigraphic features, notably the slump structures, mentioned earlier were thereby confirmed.

6.5.5 Barred River Mouths

6.5.5.1 Introduction

It is possible that sand accumulations develop off a river mouth which cause the latter to become *barred* (Oertel 1985). This can be caused by the river flow being slowed down when in contact with the sea; it can be caused by the littoral drift, by the tidal flow, or by eustatic changes.

6.5.5.2 River Regime and Bars

Evidently, a river jet entering (relatively) still water is slowed down so that its sediment-carrying capacity decreases. This does not only cause levees to form leading to the development of a birds foot delta as discussed in Sect. 6.5.4.1, but also to the formation of bars in the sea.

A mathematical theory of this process has been developed by Ertel and Kobe (1964). Basically, the relation between suspension capacity and velocity of the river is exploited as the latter is slowed down by dispersion in the sea. Gill (1972) shows that the secondary (helicoidal) currents in the river are slowed down as well, leading to the formation of point bars. A synthesis of the various possible factors has been made by Wright (1977) showing the rationale for the development of bars.

6.5.5.3 Bars and Longshore Drift

Another cause of the development of bars off river mouths has been sought in the behavior of longshore drift.

A study has been reported by Bruun and Gerritsen (1959) bearing upon the bar problem, in which an investigation was made regarding how the sand bypasses the river mouth over the bar. The contention is that the offshore bars are not solid features but exist only owing to a delicate balance between the sand which is deposited and the sand which is removed. This causes sand to "bypass" the river mouth from the updrift to the downdrift side. The general mechanics of the process is illustrated in Fig. 102.

From an analysis of observational data, Bruun and Gerritsen (1959) derived an empirical relationship expressing the conditions when bar-bypassing of the littoral drift can take place by an inlet. Involved is the ratio between the amount of tidal flow through the gorge of the inlet and the magnitude of the littoral drift. If this ratio is below a certain critical value, no bar can develop as the tides will sweep the channel clean.

6.5.5.4 Tides and Bars

Tidal currents act like river currents with regard to sediment transport. Hence, under certain conditions, sedimentation as well as erosion can be caused by them in river mouths.

Unfortunately, the hydraulic complexity of tidal entrances is very great (Ludwick 1974): tides, wind, waves, density gradients, etc. combine to make a theoretical analysis almost impossible. Some simple models have been proposed by Allen (1982); the process-response approach has been tried in specific cases

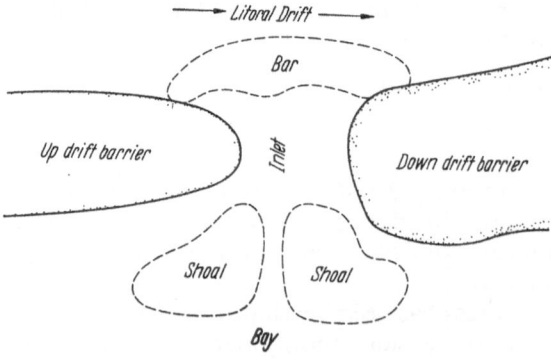

Fig. 102. Bypassing of a river mouth by sand moving across a bar. (After Bruun and Gerritsen 1959)

(Thames: Bowen and Pinless 1977). Unfortunately, all this does not provide an understanding of the mechanics of bar formation in terms of basic physics.

6.5.5.5 Eustatic Changes

Finally, bar formation may be connected with eustatic changes. The general contention is that the whole inlet-bar system simply moves as a whole with the changing sea level over long times without regard to short-term effects (Leatherman 1979).

6.6 Theoretical Submarine Geomorphology

6.6.1 General Remarks

The last section in the present chapter has to deal with submarine geomorphology. The features to be analyzed have been listed in Sect. 1.6.4. Unfortunately, no coherent attempts at an explanation of these features in theoretical-mechanical terms seem ever to have been reported in the literature. Thus although many qualitative arguments can be found scattered over a variety of sources, the substantiation of these arguments by numbers and equations is still mostly lacking. The treatises on the physiography of the sea bottom mentioned in Sect. 1.6.4 contain some information on theoretical submarine geomorphology; other items have been collected from various sources which will be mentioned in their proper context.

The organization of the present section will correspond to that of 1.6.4: we will first deal with shelves and other shallow regions, then proceed to the continental slope regions and finally deal with the deep sea areas.

6.6.2 Shallow Regions

The characteristic morphological features of shelves and other shallow regions have been described in Sect. 1.6.4.2.

Regarding the genesis of such features, it has already been mentioned that many of them can be compared with similar features found on river beds (Gibbs 1977; Butman et al. 1979). The currents which cause the features may be of tidal origin (Pierini 1981; Komen and Riepma 1981) or they may be wind-driven (Hayes and Schumacher 1976; Beardsley and Hart 1978; Petrie 1983; Ou 1984). In the latter case, evaporite deposition may result (Stuart 1973).

In addition, the oscillatory wave motion causes a corresponding oscillatory bottom stress as this has already been discussed in connection with the study of beaches (Davies 1982; Grant and Kaplan 1986; Hammond and Heathershaw 1981).

Furthermore, a specifically oceanic feature is represented by the existence of extremely low-angle flow slides on tidal flats and at the edge of deltas (Wells et al.

1980; Prior et al. 1981b; Syvitski et al. 1988). The mechanical explanation of the occurrence of such slides lies in the Terzaghi relation (see Sect. 2.2.4.2) for fluid-filled porous media which shows that all frictional resistance may, in fact, disappear.

6.6.3 Submarine Slope Areas

6.6.3.1 Introduction

We turn next to the actual submarine slope area: these occur mainly on the continental slope; other instances may be found at the edge of delta and other river deposits.

The principal feature of slopes is that they may become unstable: thus, submarine slides occur. The latter trigger turbidity currents which cause graded deposits to form further out in the ocean.

The continental slopes are often cut by large submarine canyons whose genesis is not at all clear.

We shall now discuss the mechanical aspects of the features mentioned above.

6.6.3.2 Submarine Slides

As noted, submarine slides occur on the continental slope and on the edge of accumulations of sediments. Because of their possible effects on engineering works (such as submarine pipelines or cables) they have been extensively studied. Thus, Nastav et al. (1980) compiled a bibliography, Saxov and Nieuwenhuis (1982) convened a workshop, and Schwarz (1982) published a review on the subject.

In essence, subaqueous slides are governed by the same mechanisms as subaerial landslides, if only the effect of the water pressure in the interstices of the sea floor material is taken into account: there exists (Scheidegger 1982) a principle of correspondence between subaerial and subaqueous phenomena based on the notion of effective pressure introduced by Terzaghi [cf. Eq. (2.2.21)]. The principle of correspondence states that the deformation of a mass in an aqueous environment is the same as that in a subaerial environment if the total pressure acting on the mass in the subaerial environment is replaced by the effective pressure in the aqueous medium.

Based upon this correspondence principle, the submarine slide initiation and the mechanics of sliding can be fully explained. Thus, triggering may be due to an intrinsic instability caused by sediment deposition, to current scour at the slope foot or to earthquakes (Karlsrud and Edgers 1982). Specific instances are legion and have been discussed in the review literature cited above.

6.6.3.3 Turbidity Currents and Graded Deposits

A submarine slide will generally trigger a turbidity current whose mechanics has been discussed in Sect. 6.2.3. The turbidity currents, in turn, cause specific types of deposits to form, as mentioned in Sect. 1.6.4.3. What has to be done is to discuss the mechanics of sedimentation of these deposits.

This problem is a difficult one. Two types of approaches have been used to tackle it: theoretical models and scaled experiments.

Turning first to the theoretical approach, we note that Scheidegger and Potter (1965; also Potter and Scheidegger 1966) have set up a model based on the following assumptions:

(1) A turbidity current represents a slug of turbulent water moving downslope. It contains a size distribution of sediment which is given by the sediment source. (2) When the current comes to rest, it decays according to standard laws. (3) The sediment-carrying capacity of the turbulent water is like that of a river.

The reason why a graded bed is deposited is that a sorting takes place during the decay period of the turbulence: the coarse grains fall out first, then smaller and smaller ones, and lastly the finest ones. The actual calculations for this process are rather tedious and the reader is referred to the original papers for the details. However, the final result can be stated by the following simple formula:

$$H = R(d_{max}^{2p+5} - d_m^{2p+5}), \tag{6.6.1}$$

where R is a constant, d_{max} is the maximum grain size (a diameter) in the bed, d_m is the mean grain size found at the height H above the bottom of the graded bed in question, and p is a parameter connected with the grain size distribution of the source material. Equation (6.6.1) is a relation between grain size and the level H in the bed, i.e., it is a grading relation. One sees that for $-1 > p > -2$ this relation is convex upward, for $p = -2$ it is a straight line, and for $-2.5 < p < -2$ it is concave upward. Thus, the three characteristic grading curves observed in nature (cf. Sect. 1.6.4.3) are shown to be the result of the particle size distribution of the source material.

If d_m is set equal to zero in Eq. (6.6.1) the latter yields a relationship between the total thickness H_{TOT} of the bed and the maximum grain diameter found therein:

$$H_{TOT} = R d_{max}^{2p+5}. \tag{6.6.2}$$

The last relation can be used to explain the correlations between thickness and maximum grain size of naturally observed turbidite *sequences*: the physical model is based on the idea that the source material has always the same p (particle size distribution), but that, depending on the force of the particular turbidity flow in question, only particles below a certain maximum size can be carried along. The "stronger" currents, as is natural, will deposit thicker beds; this is expressed by Eq. (6.6.2). The theory of Scheidegger and Potter thus gives an excellent representation of the facts observed in nature.

Turning now to the experimental efforts for studying the depositional effects of turbidity currents, we note that the classic investigations have been made by Kuenen (1950b, 1951b, 1953, 1965, 1966): turbidite type structures were shown to be the result of turbidity flows. Middleton (1967) has made similar tank-type experiments and claims that the deposition of sediment takes place immediately behind the "head" of the flow. Finally, Riddell (1969) studied the suspension effect of turbidity currents.

Graded deposits are the most common sedimentary features caused by submarine turbidity flows. In addition, hummocky cross-stratification (Nottvedt and Kreisa 1987), ungraded muds (McCave and Jones 1988), and large "floating" clasts (Postma et al. 1988) have on occasion been observed. The study of the orientation of the various deposits can be used for paleocurrent and basin analysis (Potter and Pettijohn 1963; De Cellis et al. 1983).

6.6.3.4 Submarine Canyons

Immediately after the discovery of submarine canyons, geoscientists tried to explain their existence. The similarity of submarine canyons with subaerial valleys at first prompted investigators to ascribe a similar, subaerial origin to the two types of features. It soon became apparent, however, that tremendous changes in sea level would have been required to maintain such an explanation, particularly if one includes the Midocean Canyon in the Atlantic (see Sect. 1.6.4) with the submarine canyons.

Thus, an explanation different from subaerial erosion had to be sought. Some of the proposals that have been advanced border on the fantastic (see reviews by Kuenen 1937; Shepard 1956, 1972; Whitaker 1976); at present it is generally agreed that it is currents which are responsible for the scouring out of at least the deeper submarine canyons. The first thought is that these currents are turbidity currents (Daly 1936; Kuenen 1951a).

There is little doubt that turbidity currents do exist (cf. Sect. 6.2.3) and also that they have some erosive power. The problem, therefore, is to investigate whether a reasonable number of turbidity flows can account for the size and shapes of the submarine canyons as the latter have been observed. Kuenen (1937) makes some estimates in this regard; but they do not seem to be too well founded, either upon theoretical or experimental considerations. In a later paper, Kuenen (1951 b) notes that in any experiments he performed, the "velocity was never sufficient for the initial erosion to exceed the subsequent deposition". This would mean that all the material moving down a canyon would have to come from the collapse of a canyon wall or such like, with no net erosion taking place at all. However, Kuenen also notes that, if the scale of the experiment were enlarged, the reverse might be true.

In fact, it was later contended that oceanic water currents alone could have sufficient erosive power to scour out the canyons (Shepard 1976; Shepard and Marshall 1973). In addition, sedimentation could help to build up the canyon walls (Andrews and Hurley 1978). Thus, multistage models have been proposed (Farre et al. 1983; O'Connell et al. 1987) starting with slides off deltas (Prior et al. 1981a) creating small gullies, followed by headward growth and indentation in the shelf. Then, deep sea processes take over with buildup of canyon walls (levees) by sedimentation and with the action of turbidity flows farther out.

6.6.4 Deep Sea Region

6.6.4.1 Agents Effective in Submarine Geomorphology

The striking similarity between many submarine and subaerial features has prompted some investigators to postulate an identical subaerial origin for the two.

However, it is now beyond doubt that at least the deep water features are caused by *truly submarine processes*. One is therefore faced with the problem of finding agents that can produce submarine erosion, transportation, and redeposition of material.

General reviews of the agents that may be effective in submarine geomorphology have been published by Kuenen (1957, 1959) and by Heezen (1959a, b).

Accordingly, *erosion* in the general sense (i.e., the removal of material from its resting place) may be caused by various types of rare events in addition to the obvious drag of bottom currents.

Submarine *transportation* of material occurs primarily by turbidity currents although the effect of ocean bottom currents is not entirely negligible. Material which is eventually deposited on the sea floor may also arrive there through the air: dust from the Sahara was found far out in the Atlantic. It obviously got there by the action of winds carrying it over large distances.

Finally, the *deposition* of material on the sea floor occurs by the transporting agent, whichever it may be, losing its carrying capacity: turbidity currents coming to a halt, the wind letting the dust settle out, or ocean bottom currents slowing down.

In general, it may be stated that, although subaerial and submarine features are similar in many regards, *erosion* is the decisive factor on land, but *deposition* below the water level.

6.6.4.2 Theory of Submarine Erosion

The circulation in the sea at large has the effect that, in addition to the well-known surface currents, there are bottom currents (cf. Sect. 6.2.5.3) in the deep ocean. These currents are the most obvious cause of submarine erosion. In Sect. 6.2.5.3, their speeds have been quoted to reach 40 cm/s. The bottom drag conditions are the same as in shallower (tidal) currents (cf. Sect. 6.2.4.3) and can therefore also be applied to deep sea erosion. In effect, simulation experiments (Hammond and Collins 1979) and direct observations (Broecker and Bainbridge 1978; Lavelle et al. 1984) have confirmed this view.

The bottom currents may not be constant throughout a year; there are definite seasonal signals (Dickson et al. 1982). Oceanic rises and slopes may cause "shadow" (insulating) effects (Csanady and Shaw 1983).

The erosive action may produce sediment truncation of entire layers that expresses itself in the geological record (Johnson 1972). Smaller features, like furrows, are also caused (Embley et al. 1980; Flood 1983).

In addition to currents, rare events like submarine slides (Morelock 1969) triggered by earthquakes, tsunamis (Kastens and Cita 1981), and turbidity currents cause submarine erosion as well.

6.6.4.3 Theory of Submarine Sediment Transport Features

The sediment transport by ocean bottom currents (see Nowell and Hollisted 1985, for a review) occurs in an analogous fashion as in rivers. Thus, there exists suspended sediment transport (Halper and McGrail 1988) and some sort of bed load transport. The latter entails a reaction on the ocean bottom (Heezen and Hollister 1964).

Of geomorphological interest is particularly the mentioned reaction of the ocean bottom to the drag by currents. Entirely in analogy with river bottom features, ripples (less than 20 cm high, meters long: Flood 1981b; Tucholke 1982), dunes (Fox et al. 1968), and long bottom-material (sand or mud) waves (1 km wavelength: Hall 1979; 6 km wave length in mud: Flood 1988) have been found. The possible theoretical explanations of these features are the same as those for river bed features (see Sect. 4.6), the very long waves in mud might be the result of wavy streamlines in water (Cartwright 1959) which would be similar to those in air behind an obstacle (cf. Sect. 2.4.3.2).

6.6.4.4 Deposition Processes

We have noted (cf. Sect. 1.6.4.4) that abyssal plains are covered by sediments. The general slowness of marine sedimentation, therefore, raises the question as to the origin and mode of deposition of the sediments that smoothen the original hilly relief of up to 300 m on the plains.

The answer seems to lie only partly in the existence of the deep ocean currents. In them, the fallout of sediment occurs as in rivers in slack water regions. In addition, it may also be seasonal (Katz et al. 1981). The fallout is enhanced if two currents meet, obstructing each other: this may be the origin of the Blake Outer Ridge off Cape Hatteras (Bryan 1970).

Agents that are more efficient than deep ocean currents with regard to causing sedimentation are turbidity currents (Heezen 1959a) particularly where the deposits encircle islands as archipelagic aprons. In addition, marine slides have been found to occur even in deep ocean regions (Holler 1988).

References

Acres, B.D. and 5 others: Z. Geomorph. Suppl. 52, 63 (1985)
Allen, J.R.L.: J. Sediment. Petrol. 41(1), 97 (1971)
Allen, J.R.L.: Sediment. Geol. 26(4), 281 (1980)
Allen, J.R.L.: Mar. Geol. 48(1–2), 51 (1982)
Allen, J.R.L.: Geogr. Rev. 78(2), 148 (1988)
Allen, P.A.: Nature 313, 562 (1985)
Alliney, S.: Appl. Math. Model. 5(5), 321 (1981)
Amorocho, J. and J.J. De Vries: J. Geoph. Res. 85(C1), 433 (1980)
Anderson, D.L.T. and 3 others: J. Geoph. Res. 84(C8), 4795 (1979)
Andrews, J.E. and R.J. Hurley: Mar. Geol. 26(3–4), M47 (1978)
Arhan, M. and R.O. Plaisted: Oceanol. Acta 4(2), 107 (1981)
Ashley, G. and W.H. Renwick: Spec. Publ. Int. Assoc. Sediment. 6, 203 (1983)
Axelsson, V.: Geograf. Ann. 49A, 1 (1967)
Baker, R.A. Bull. Geol. Soc. Am. 81, 1589 (1970)
Balwierz, J. and S. Dzulynski: Ann. Soc. Geol. Pol. 46(4), 419 (1976)
Bampi, F. and A. Morro: Nuovo Cim. 2C(3), 352 (1979)
Bampi, F. and A. Morro: Nuovo Cim. 4C(5), 551 (1981)
Barbee, W.B. and 4 others: J. Geoph. Res. 80(15), 1965 (1975)

Barnett, T.P.: J. Geoph. Res. 73, 513 (1968)
Bartels, J.: Encycl. Phys. 48, 734 (1957)
Bata, G.L.: Internat. Assoc. Hydraul. Res. 8th Congr., Montreal, Proc. 2, 12C1 (1959)
Bates, C.C.: Bull. Am. Assoc. Petrol. Geol. 37, 2119 (1953)
Battisti, D.S. and A.J. Clarke: J. Geoph. Res. 87(C9), 7873 (1982)
Bäumer, H.P.: Catena 10, 41 (1983)
Beardsley, R.C. and J. Hart: J. Geoph. Res. 83(C2), 873 (1978)
Beardsley, R.C. and S.J. Lentz: J. Geophys. Res. 92(C2), 1455 (1987)
Beardsley, R.C. and 4 others: J. Geoph. Res. 82(21), 3175 (1977)
Benjamin, T.B.: J. Fluid Mech. 31, 209 (1968)
Berryman, J.G.: Phys. Fluids 19(6), 71 (1976)
Bigelow, G.E.: Earth Surf. Proc. Landforms 9(4), 383 (1984)
Bird, E.C.F.: Z. Geom. Suppl. 57, 1 (1985)
Black, K.P. and S.L. Gay: J. Geoph. Res. 92(C9), 9514 (1987)
Blanchet, C. and H. Villate: Courants de densite. Confer. technique regionale, Tokyo (1954)
Blau, E.: Wasserwirtsch.-Wassertechn. 6, 257 (1956)
Blench, T.: Proc. Minn. Internat. Hydraul. Conv. p. 77 (1953)
Bluck, B.J.: Geol. Mag. 106(1), 1 (1969)
Boczar-Karakiewicz, B. and R.G.D. Davidson-Arnott: Mar. Geol. 77(3–4), 287 (1987)
Bonham-Carter. G.F. and A.J. Sutherland: Trans. Gulf Coast Assoc. Geol. Soc. 17, 326 (1967)
Bowden, K.F., L.A. Fairbairn and P. Hughes: Geoph. J. R. Astron. Soc. 2, 988 (1959)
Bowen, A.J.: J. Geoph. Res. 74, 5467 (1969)
Bowen, A.J. and R.T. Guza: J. Geoph. Res. 83(C4), 1913 (1978)
Bowen, A.J. and D.L. Inman: J. Geoph. Res. 74, 5479 (1969)
Bowen, A.J. and S.J. Pinless: Estuarine Coastal Mar. Sci. 5(2), 197 (1977)
Braithwaite, C.J.R.: Progr. Phys. Geogr. 6(4), 505 (1982)
Branner, C.J.: J. Geol. 8, 481 (1900)
Brass, G.W., J.R. Southam and W.H. Peterson: Nature 296, 620 (1982)
Breeding, J.E.: J. Geol. 89(2), 260 (1981)
Bretz, J.H.: J. Geol. 50, 675 (1942)
Broecker, W.S. and A. Bainbridge: J. Geoph. Res. 83(C4), 1963 (1978)
Broecker, W.S. and Y.H. L: J. Geoph. Res. 75 (18), 3545 (1970)
Bruun, P.: J. Waterways & Harbors Div. ASCE 88, 117 (1962)
Bruun, P.: J. Geoph. Res. 68, 1065 (1963)
Bruun, P.: Stability of tidal inlets. Amsterdam: Elsevier (1978)
Bruun, P.: J. Coastal Res. 2(2), 123 (1986)
Bruun, P. and F. Gerritsen: J. Waterways & Harbors Div. ASCE 84(3), 1644 (1958)
Bruun, P. and F. Gerritsen: J. Waterways & Harbors Div. ASCE 85(4), 75 (1959)
Bruun, P. and M.L. Schwartz: Z. Geomorph. Suppl. 57, 33 (1985)
Bryan, G.M.: J. Geoph. Res. 75(24), 4530 (1970)
Butman, B., N. Noble and D. W. Folger: J. Geoph. Res. 84(C3), 1187 (1979)
Cacchione, D.A. and J.B. Southard: J. Geoph. Res. 79(15), 2237 (1974)
Carrier, G.F. and H.P. Greenspan: J. Fluid Mech. 4, 97 (1958)
Carter, H.H.: Rev. Geoph. Space Phys. 17(7), 1585 (1979)
Cartwright, D.E.: Proc. Roy. Soc. A253 (1959)
Carty, J.J.: Resistance coefficients for spheres on a plane boundary. B. Sc. Thesis, Cambridge: M.I.T.
    (1957)
Castanho, J.P.: Mem. Lab. Nac. Engenharia Civil, Lisboa 139, 1 (1958)
Cavaleri, L. and P.M. Rizzoli: J. Geoph. Res. 86(11), 10961 (1986)
Cavaleri, L. and S. Zecchetto: J. Geoph. Res. 92(C4), 3984 (1987)
Chakrabarti, S.K. and R.P. Cooley: J. Geoph. Res. 82(9), 1363 (1977)
Chang, H.H. and J.C. Hill: J. Hydraul. Div. ASCE 102 (HY10), 1461 (1976)
Clarke, A.J.: J. Geoph. Res. 84(C7), 3743 (1979)
Clayton, K.M.: Progr. Phys. Geogr. 4(4), 471 (1980)
Cokelet, E.D.: Nature 267, 769 (1977)
Coleman, J.M.: Deltaic processes. Baton Rouge: coastal studies Inst. (1975)
Coleman, J.M.: Deltas: Processes of deposition and models for exploration. Champaign IL: Continuing
    Educ. Co. (1976)

Collins, J.I.: J. Geoph. Res. 77(15), 2693 (1972)
Collins, J.I.: Spec. Publ. Soc. Econ. Paleontol. Mineral. 24, 54 (1976)
Collonel, J.M. and V. Goldsmith: In: Quantitative geomorphology; Proc. 2nd Ann. Geomorph.
    Sympos. Binghampton p. 198 (1972)
Cornaglia, P.: Sul regime delle spaggie e sulla regolazione dei porti. Torino (1891)
Crickmay, C.H.: Bull. Am. Assoc. Petrol. Geol. 39, 1 (1955)
Csanady, G.T.: J. Geoph. Res. 82(3), 397 (1977)
Csanady, G.T.: Circulation in the coastal ocean: Dordrecht: Reidel (1982)
Csanady, G.T. and P.T. Shaw: J. Geoph. Res. 88(C12), 7519 (1983)
Dalrymple, R.A. and G.A. Lanan: Bull. Geol. Soc. Am. 87(1), 57 (1976)
Dalrymple, R.A., and C.J. Lozano: J. Geophys. Res. 83(C2), 6063 (1978)
Daly, R.A.: Am. J. Sci. 31, 401 (1936)
Damuth, J.E. and N. Kumar: Bull. Geol. Soc. Am. 86, 863 (1975)
Damuth, J.E. and 7 others: Geology 11(2), 94 (1983)
Darwin, C.: The structure and distribution of coral reefs. London: Smith, Elder (1842)
Davies, A.G.: J. Mar. Res. 40(2), 478 (1982)
Davies, A.G.: Continental Shelf Res. 6(6), 715 (1986)
Davies, A.M., J. Sauvel and J. Evans: Continental Shelf Res. 4(3), 341 (1985)
Davies, P.J. and J.F. Marshall: Nature 287, 37 (1980)
Davies, W.E.: Bull. Natl. Speleol. Soc. 22, 5 (1960)
De Boer, P.L., A. Van Gelder and S.D. Nio (eds.): Tide-influenced sedimentary environments and facies.
    Dordrecht: Reidel (1988)
De Celles, P.G., R.P. Langford and R.K. Schwartz: J. Sediment. Petrol. 53(2), 629 (1983)
Defant, A.: Encycl. Phys. 48, 846 (1957)
Dickson, R.R. and 3 others: Nature 295, 193 (1982)
Dietrich, G. and K. Kalle: Allgemeine Meereskunde. Berlin: Borntraeger (1947)
Dijkema, K.S.: Z. Geomorph. 31(4), 489 (1987)
Divoky, D., B. Le Mehaute and A. Lin: J. Geoph. Res. 75(9), 1681 (1970)
Doebler, H.J.: J. Geoph. Res. 72, 511 (1967)
Dolan, R., B. Hayden and H. Lins: Am: Sci. 68(1), 16 (1980)
Dolan, R., L. Vincent and B. Hayden: Z. Geomorph. 18(1), 1 (1974)
Dolata, L.F. and W. Rosenthal: J. Geoph. Res. 89(C2), 1973 (1984)
Dreybrodt, W.: Earth Planet. Sci. Lett. 58(2), 293 (1982)
Dronkers, J.J.: Tidal computations in rivers and coastal waters. Amsterdam: North Holland (1964)
Dubois, R.N.: J. Geol. 85, 470 (1977)
Dubois, R.N.: Bull. Geol. Soc. Am. 89, 1133 (1978)
Duerden, J.E.: Bull. Am. Mus. Nat. Hist. 16, 326 (1902)
Duplessy, J.C. and N.J. Shackleton: Nature 316, 500 (1985)
Dyer, K.R.: Estuaries: a physical introduction. London: Wiley (1973)
Dyer, K.R. (ed.): Estuarine hydrography and sedimentation, a hand-book. Cambridge: University Press
    (1979)
Dzulynski, S., E. Gil and J. Rudnicki: Z. Geomorph. 32(1), 1 (1988)
Eagleson, P.S. and R.G. Dean: J. Hydraul. Div. ASCE 85(10/1), 53 (1959)
Eagleson, P.S. and R.G. Dean and L.A. Peralta: The mechanics of the motion of discrete spherical
    bottom sediment particles due to shoaling waves. -M.I.T. Hydrodyn. Lab. Tech. Rep. No. 26 (1957)
Einstein, H.A.: Trans. Am. Geoph. Union 22, 597 (1941)
Elgar, S. and R.T. Guza: J. Fluid Mech. 158, 47 (1985a)
Elgar, S. and R.T. Guza: J Fluid Mech. 161, 425 (1985b)
Elgar, S. and R.T. Guza: J. Fluid Mech. 167, 1 (1986)
Elgar, S., R.T. Guza and R.J. Seymour: J. Geoph. Res. 89(C3), 3623 (1984)
Elgar, S., R.T. Guza and M.H. Freilich: J. Geoph. Res. 93(C8), 9261 (1988)
Ellenberg, L.: Geogr. Rundsch. 35(1), 9 (1983)
Elliott, A.J.: Oceanol. Acta 5(1), 7 (1982)
El Sabh, M.I. and T.S. Murty: Natural Hazards 1, 371 (1989)
Embley, R.W. and 4 others: Bull. Geol. Soc. Am. Pt. I, 91, 731 (1980)
Ertel, H.: Monatsber. Dtsch. Akad. Wiss. Berlin 4, 467 (1962a)
Ertel, H.: Monatsber. Dtsch. Akad. Wiss. 4, 707 (1962b)
Ertel, H.: Monatsber. Dtsch. Akad. Wiss. 6, 97 (1964)

Ertel, H.: Gerl. Beitr. Geoph. 77(2), 177 (1968a)
Ertel, H.: Acta Hydrophys. 13, 77 (1968b)
Ertel, H. and D. Hartke: Acta Hydrophys. 15(4), 249 (1971)
Ertel, H. and G. Kobe: Gerl. Beitr. Geoph. 73(2), 127 (1964)
Esin, N.V.: In: Issledovaniye gidrodin. i morfodin. Prots. beregovoy zony morya. Izdat. "Nauka", p. 170 (1966)
Esin, N.V.: Okeanologiya 22(6), 1970 (1980)
Evans, D.V.: J. Fluid Mech. 186, 379 (1988)
Fairbridge, R.W.: Proc. 7th Pac. Sci. Congr. 3 (1952)
Fairbridge, R.W.: In: Chemistry and biochemistry of estuaries; eds. E. Olausson & I. Cato, p. 1, New York: Wiley (1980)
Fangmeier, D.D.: Water Resour. Res. 6(4), 1216 (1970)
Farre, J.A. and 3 others: Spec. Publ. Soc. Econ. Paleont. Min., 33, 25 (1983)
Festa, J.F. and D.V. Hansen: Estuarine Coastal Mar. Sci. 4(3), 309 (1976)
Festa, J.F. and D.V. Hansen: Estuarine Coastal Mar. Sci. 7(4), 347 (1978)
Fisher, J.J.: Bull. Geol. Soc. Am. 79, 1421 (1968)
Fisher, R.V.: Geology 11(5), 273 (1983)
Flood, R.D.: Sedimentology 28, 511 (1981a)
Flood, R.D.: Mar. Geol. 39, M13 (1981b)
Flood, R.D.: Bull. Geol. Soc. Am. 94, 630 (1983)
Flood, R.D.: Deep-Sea Res. 35(6), 973 (1988)
Foda, M.: J. Geoph. Res. 93(C8), 9295 (1988)
Ford, D.C.: Ann. Soc. Geol. Belg. 108, 283 (1985)
Ford, D.C. and R.O. Ewers: Can. J. Earth Sci. 15, 1783 (1978)
Forristall, G.Z.: J. Geoph. Res. 83(C3), 2353 (1978)
Fox, P.J., B.C. Heezen and A.M. Harian: Nature 220, 470 (1968)
Fu, S.: Acta Oceanol. Sin. 6(2), 169 (1987)
Führböther, A. (ed.): Sandbewegung im Küstenraum, Ergebnisse und Ausblick. Boppard: Boldt (1979)
Galvin, C.J.: Rev. Geoph. 5, 287 (1967)
Gardiner, J.S.: Proc. Linn. Soc. Lond. 1930(1), 65 (1930)
Gardner, W.D. and L.G. Sullivan: Science 213, 329 (1981)
Gedney, R.T. and W. Lick: J. Geoph. Res. 77(15), 2714 (1972)
Gibbs, R.J. (ed.): Transport processes in lakes and oceans. New York: Plenum Press (1977)
Gilbert, G.K.: U.S. Geol. Surv. Annu. Rep. 5, 87 (1885)
Gill, D.: Proc. R. Soc. Victoria 80(2), 183 (1967)
Gill, D.: Can. J. Earth Sci. 9, 1382 (1972)
Ginsburg, R.N.: Bull. Mar. Sci. Gulf Caribbean 3, 59 (1953)
Glazman, R.E.: J. Geoph. Res. 91(C5), 6629 (1986)
Gleason, R. and P.J. Hardcastle: Estuarine Coastal Mar. Sci. 1(1), 11 (1973)
Gorycki, M.A.: J. Geol. 81, 109 (1973)
Grant, W.D. and A.M. Kaplan: Annu. Rev. Fluid Mech. 18, 265 (1986)
Grantham, K.N.: Trans. Am. Geoph. Un. 38, 662 (1953)
Greenspan, H.P.: J. Fluid Mech. 4, 330 (1958)
Greenwood, B. and R.A. Davis (eds.): Hydrodynamics and sedimentation in wave-dominated coastal environments. Amsterdam: Elsevier (1984)
Greenwood, B. and P.B. Hale: Can. J. Earth Sci. 19(3), 424 (1982)
Greenwood, B. and D.J. Sherman: Mar. Geol. 60, 31 (1984)
Griesseier, H. and K. Vollbrecht: Wasserwirtsch.-Wassertechn. 6, 247 (1956)
Grund, A.: Penck's Geogr. Abh. 9, 345 (1910)
Guilcher, A.: Coastal and submarine geomorphology. Transl. from French. London: Methuen and Co. (1958)
Gunn, J.: Z. Geomorph. 25(3), 313 (1981)
Guza, R.T.: J. Fluid Mech. 95(1), 199 (1979)
Guza, R.T. and A.J. Bowen: J. Geoph. Res. 86(C5), 4125 (1981)
Guza, R.T. and D.C. Chapman: J. Fluid Mech. 95(1), 199 (1979)
Guza, R.T. and R.E. Davis: J. Geoph. Res. 79(9), 1285 (1974)
Guza, R.T. and D.L. Inman: J. Geoph. Res. 80(21), 2997 (1975)
Guza, R.T. and E.B. Thornton: J. Geoph. Res. 86(C5), 4133 (1981)

Guza, R.T. and E.b. Thornton: J. Geoph. Res. 90(C2), 3161 (1985)
Hails, J. and A. Carr: Nearshore sediment dynamics and sedimentation, an interdisciplinary review. New York: Wiley (1975)
Hakanson, L.: Can. J. Earth Sci. 18, 899 (1981)
Hall, J.K.: Sediment. Geol. 23, 269 (1979)
Halper, F.P. and D.W. McGrail: Continental Shelf Res. 8(1), 23 (1988)
Hammond, F.D.C. and A.D. Heathershaw: Nature 293, 208 (1981)
Hammond, T.M. and M.B. Collins: Sedimentology 26(6), 795 (1979)
Hanks, T.C. and R.E. Wallace: Bull. Seismol. Soc. Am. 75(3), 835 (1985)
Hanks, T.C. and 3 others: J. Geoph. Res. 89(B7), 5771 (1984)
Harleman, D.R.F.: In: Handbook of fluid dynamics, ed. V.L. Streeter, p. 26–2, New York: Mc Graw-Hill (1961)
Hasselmann, K.: Rev. Geoph. 4(1), 1 (1966)
Hay, A.E.: J. Geoph. Res. 88(C1), 751 (1983)
Hayes, S.P. and J.D. Schumacher: J. Geoph. Res. 81(36), 6411 (1976)
Heaps, N.S.: Oceanol. Acta 3(4), 449 (1980)
Heathershaw, A.D. and A.G. Davies: Mar. Geol. 62(3–4), 321 (1985)
Heezen, B.C.: Geoph. J. R. Astron. Soc. 2, 142 (1959a)
Heezen, B.C.: J. Geol. 67, 713 (1959b)
Heezen, B.C. and C. Hollister: Mar. Geol. 1, 141 (1964)
Heezen, B.C., M. Tharp and M. Ewing: The floors of the ocean, I: The North Atlantic. Geol. Soc. Am. Spec. Pap. 65, New York (1959)
Heezen, B.C. et al.: Nature 211, 611 (1966)
Heyer, H., R. Hewer and J. Sundermann: Küste 43, 167 (1986)
Hibberd, S. and D.H. Peregrine: J. Fluid Mech. 95(2), 323 (1979)
Hoffmeister, J.E. and H.G. Multer: Bull. Geol. Soc. Am. 79, 1487 (1968)
Hogg, N.G.: Geoph. Fluid Dyn. 2, 361 (1971)
Holler, P.: Ber. Geol-Paläont. Inst. Univ. Kiel 23, 1 (1988)
Holmes, A.: Principles of physical geology. London: T. Nelson & Sons (1944)
Howd, P.A. and R.A. Holman: Mar. Geol. 78(1–2), 11 (1987)
Hoyt, J.C.: Bull. Geol. Soc. Am. 78, 1125 (1967)
Hsieh, W.W. and V.T. Buchwald: Geoph. Astroph. Fluid Dyn. 28 (3–4), 257 (1984)
Hsueh, Y. and J.J. O'Brien: J. Phys. Oceanogr. 1(3), 180 (1971)
Huang, N.E. and 4 others: J. Geoph. Res. 88(C14), 9579 (1983)
Huber, A.: Mitt. Vers.-Anst. Wasserbau, Hydrol. Glaziol. ETH Zürich 21, 1 (1976)
Huber, A.: E.T.H.—Bulletin 1979, 7 (1979)
Huber, A.: Mitt. Vers.-Anst. Wasserb & c. ETH Zürich 47, 1 (1980)
Huber, A.: Eclogae Geol. Helv. 75(3), 563 (1982)
Huber, A.: Wasser, Energie, Luft 79(11), 309 (1987)
Hughes, M.G. and P.J. Cowell: J. Coastal Res. 3(2), 153 (1987)
Hunt, J.N.: J. Geoph. Res. 64, 437 (1959)
Huntley, D.A.: J. Geoph. Res. 81(36), 6441 (1976)
Huntley, D.A., R.T. Guza and A.J. Bowen: J. Geoph. Res. 82(18), 2577 (1977)
Hutchinson, G.E.: A treatise on limnology. New York: Wiley (1957)
Huthnance, J.M.: Estuarine, Coastal Shelf Sci. 14(1), 79 (1982)
Hutter, K. and K. Hofer: Mitt. Vers.-Anst. Wasserb. & c. ETH Zürich 27, 1 (1978)
Hutter, K. and J. Trösch: Mitt. Vers.-Anst. Wasserb. & c. ETH Zürich 20, 3 (1975)
Ikeda, M.: J. Geoph. Res. 89(C5), 8008 (1984)
Ingle, J.C.: The movement of beach sand. 221 pp. Amsterdam: Elsevier (1966)
Inman, D.L. and T.K. Chamberlain: J. Geoph. Res. 64, 41 (1959)
Inman, D.L. and J.D. Frautschy: Proc. Santa Barbara Coastal Eng. Conf., Coastal Eng. 511 (1966)
Inman, D.L. and R.T. Guza: Mar. Geol. 49(1–2), 133 (1982)
Ippen, A.T. (ed.): Estuarine and coastline hydrodynamics. New York: Mc Graw-Hill (1966)
Ippen, A.T. and P. Eagleson: A study of sediment sorting by waves shoaling on a plane beach. M.I.T. Hydrodynamics Lab. Technol. Rept. No. 18 (1955)
Iwagazi, Y. and 3 others: Bull. Disas. Prev. Res. Inst. Kyoto Univ. 27(2), 73 (1977)
Jenkins, A.d.: Int. J. Numer. Math. Fluids 3(1), 61 (1983)
Johns, B.: Geoph. J. R. Astr. Soc. 13, 377 (1967)

Johnson, D.A.: Bull. Geol. Soc. Am. 83, 3121 (1972)
Johnson, D.W.: Shore processes and shoreline development. New York: Wiley (1919)
Jolliffe, I.: Progr. Phys. Geogr. 2(2), 264 (1978)
Jones, B. and Q.H. Goodbody: Bull. Can. Petrol. Geol. 32(2), 201 (1984)
Jopling, A.V.: Sedimentology 2, 115 (1963)
Karlsrud K. and L. Edgers: In: Marine slides and other mass movements, eds. S. Saxov and J.K. Nieuwenhuis, p. 61, New York: Plenum Press (1982)
Kastens, K.A. and M.B. Cita: Bull. Geol. Soc. Am. 92(11), 845 (1981)
Katz, E.J.: J. Geoph. Res. 92(C2), 1885 (1987)
Katz, E.J. and S. Garzoli: J. Mar. Res. 40 Suppl., 307 (1982)
Katz, E.J. and 5 others: Oceanol. Acta 4(4), 445 (1981)
Keller, H.B., D.A. Levine and G.B. Whitham: J. Fluid Mech. 7, 302 (1960)
Kelletat, D.: Deltaforschung: Verbreitung, Morphologie, Entstehung und Ökologie von Deltas. Darmstadt: Wiss. Buchges. (1984)
Kemmerly, P.R.: Bull. Geol. Soc. Am. 93, 1078 (1982)
Kenyon, K., D. Sheres and R. Bernstein: J. Geoph. Res. 88(C12), 7589 (1983)
Keulegan, G.H.: Repts. U.S. National Bureau of Standard Nos. 5168 (1957) and 5831 (1958)
Kielmann, J. and Z. Kowalik: Ocenaol. Acta 3(1), 51 (1980)
Kilworth, P.D.: Rev. Geoph. Space Phys. 21(1), 1 (1983)
King, C.A.M. Beaches and coasts. London: E. Arnold & Co. (1959)
Kirk, R.M.: Progr. Phys. Geogr. 4(2), 189 (1980)
Kirkby, M.J.: Z. Geomorph. 28(4), 405 (1984)
Kobayashi, N.: J. Waterways., Port Ocean. Eng. ASCE 113(4), 401 (1987)
Kochel, R.G. and R. Dolan: J. Geol. 94, 902 (1986)
Kolp, O.: Peterm. Geogr. Mitt. 102(3), 173 (1958)
Komar, P.D.: Bull. Geol. Soc. Am. 82, 2643 (1971)
Komar, P.D.: Bull. Geol. Soc. Am. 84, 2217 (1973a)
Komar, P.D.: Bull. Geol. Soc. Am. 84, 3329 (1973b)
Komar, P.D.: J. Sediment. Petrol. 47(4), 1444 (1977)
Komar, P.D. and D.L. Inman: J. Geoph. Res. 75(30), 5914 (1970)
Komar, P.D. and M.C. Miller: Proc. 14th Coastal Eng. Conf. Copenhagen (ASCE) p. 756 (1974)
Komen, G.J. and H.W. Riepma: Geoph. Astroph. Fluid Dyn. 18, 93 (1981)
Kreeke, J.v.: Physics of shallow estuaries and bays. Berlin Heidelberg New York Tokyo: Springer (1986)
Krumbein, W.C.: Ann. Bull. Beach Erosion Board 17, 1 (1963)
Kuenen, P.H.: Leidse Geol. Med. 8, 327 (1937)
Kuenen, P.H.: Submarine geology, New York: Wiley (1950a)
Kuenen, P.H.: Rept. 18th Int. Geol. Congr. 8, 44 (1950b)
Kuenen, P.H.: Spec. Publ. Soc. Econ. Min. Paleont. 2, 14 (1951a)
Kuenen, P.H.: Bull. Am. Assoc. Petrol. Geol. 37, 1044 (1953)
Kuenen, P.H.: J. Alberta Soc. Petrol. Geol. 5, 59 (1957)
Kuenen, P.H.: Geol. Mijnbouw 21, 191 (1959)
Kuenen, P.H.: Proc. 17th Symp. Colston Res. Soc. 47 (1965)
Kuenen, P.H.: Sedimentology 7, 267 (1966)
Kuenen, P.H.: Geol. Mijnbouw 50(3), 429 (1971)
Kuo, J.T. and R.C. Jachens: Ann. Geoph. 33(1/2), 73 (1977)
Kuznetsov, A.M. and V.A. Kuznetsov: Izv. Akad. Nauk. SSSR, Ser. Geofiz. 1959(8), 1247 (1959)
Labeish, V.G.: Izv. Akad. Nauk SSSR, Ser. Geofiz. 11, 1714 (1959)
Laird, N.O. and T.V. Ryan: J. Geoph. Res. 74, 5433 (1969)
Lakhan, V.C. and A.V. Jopling: Geogr. Ann. 69A(2), 251 (1987)
Lamb, H.: Hydrodynamics, 6th edn., p. 254, New York: Dover (1954)
Lamé, G.: Leçons sur la théorie mathématique de l'élasticité des corps solides. Paris: Dalmont (1852)
Langbein, W.B.: Bull. Int. Assoc. Sci. Hydrol. 8(3), 84 (1963)
Langmuir, I.: Science 87, 119 (1938)
Lauff, G.H. (ed.): Estuaries. Washington: Am. Assoc. Advnace Sci. (1967)
Lavelle, J.W., H.O. Mofjeld and E.T. Baker: J. Geoph. Res. 89 (C4), 6543 (1984)
Leatherman, S.P.: Sediment. Geol. 33(1-2), M15 (1979)
Leatherman, S.P.: Nature 301, 415 (1983)
Leatherman, S.P. and A.T. Williams: Earth Surf. Proc. Landf. 8, 141 (1983)

Le Blond, P.H.: J. Sediment. Petrol. 49(4), 1093 (1979)
Le Blond, P.H. and C.L. Tang: J. Geoph. Res. 79(6), 811 (1974)
Ledbetter, M.T.: Nature 294, 554 (1981)
Le Mehaute, B.: An introduction to hydrodynamics and water waves. Berlin Heidelberg New York:
    Springer (1976)
Le Mehaute, B. and J.D. Wang: J. Waterways Port, Coastal Div. ASCE 108 (WW1), 33 (1982)
Leontiev, I.O.: Coastal Eng. 12(1), 83 (1988)
Lindsay, J.F., D.B. Prior and J.M. Coleman: Bull. Am. Assoc. Petrol. Geol. 68 (11), 1732 (1984)
Lins, H.F.: Mar. Geol. 62(1-2), 13 (1984)
Litwiniszyn, J.: Rock Mech. Eng. Geol. 1, 186 (1963)
Liu, P.L.F., S.B. Yoon and J.T. Kirkby: J. Fluid Mech. 153, 185 (1985)
Longuet-Higgins, M.S.: J. Mar. Res. 11, 245 (1952)
Longuet-Higgins, M.S.: J. Geoph. Res. 85(C3); 1519 (1980)
Longuet-Higgins, M.S. and R.W. Stewart: J. Fluid Mech. 8, 565 (1960)
Ludwick, L.D.: Bull. Geol. Soc. Am. 85, 717 (1974)
Lüthi, S.: Eclogae Geol. Helv. 73(3), 881 (1980)
Lüthi, S.: Sedimentology 28(1), 97 (1981)
Macpherson, H.: Fluid Mech. 97(4), 721 (1980)
Mahoney, J.J. and W.G. Pritchard: J. Fluid Mech. 101 (4), 809 (1980)
Malone, F. and J.T. Kuo: J. Geoph. Res. 86(C5), 4029 (1981)
Mardell, G.T. and R.D. Pingree: Oceanol. Acta 4(1), 63 (1981)
Mase, H.: Coastal Eng. 12(2), 175 (1988)
Mathiesen, M.: J. Geoph. Res. 92(C4), 3905 (1987)
Matson, G.C.: U.S. Geol. Surv. Water Supply Pap. 223, 42 (1909)
Matsukura, Y.: Z. Geomorph. 32(2), 129 (1988)
Mattioli, F.: Nuovo Cim. 2C(5), 619 (1979)
Mattioli, F.: Nuovo Cim. 4C(6), 621 (1981)
May, V. and C. Heeps: Z. Geomorph. Suppl. 57, 81 (1985)
May, V. and W.F. Tanner: Z. Geomorph. Suppl. 22, 1 (1975)
McCann, S.B.: Progr. Phys. Geogr. 5(2), 286 (1981)
McCave, I.N. and K.P.N. Jones: Nature 333, 250 (1988)
McCreary, J.P. and R. Lukas: J. Geoph. Res. 91(C10), 11691 (1986)
McLean, R.F. and R.M. Kirk: New Zeal. J. Geol. Geoph. 12(1), 138 (1969)
McPherson, J.G., G. Shanmugam and R.J. Moiola: Bull. Geol. Soc. Am. 99, 331 (1987)
Mehta, A.J. (ed. 1984): Estuarine cohesive sediment dynamics; Proc. Workshop Tampa 1984. Berlin
    Heidelberg New York Tokyo: Springer (1986)
Mei, C.C.: Geoph. Res. 78(6), 977 (1973)
Melville, W.K. and R.J. Rapp: Nature 317, 514 (1985)
Middleton, G.V.: Can. J. Earth Sci. 3, 523 (1966)
Middleton, G.V.: Can. J. Earth Sci. 4, 475 (1967)
Miles, J.W.: J. Fluid Mech. 54(1), 63 (1972)
Miles, J.W.: J. Fluid Mech. 57(2), 401 (1973)
Miller, J.E.: U.S. Geol. Surv. Prof. Pap. 1302, 1 (1984)
Miller, M.C. and P.D. Komar: J. Geol. 87(6), 593 (1979)
Miller, R.L. and J.M. Zeigler: J. Geol. 87(66), 417 (1958)
Mitsuyasu, H.: J. Geoph. Res. 90(C2), 3343 (1985)
Mitsuyasu, H. and T. Honda: J. FLuid Mech. 123, 425 (1982)
Mitsuyasu, H. and T. Kusaba: Nat. Disaster Sci. 6(2), 43 (1984)
Momoi, T.: J. Phys. Earth 24(1), 1 (1976)
Moore, G.T.: Bull. Geol. Soc. Am. 82, 2563 (1971)
Morelock, J.: J. Geoph. Res. 74, 465 (1969)
Morisawa, M.: in Models in geomorphology, Binghampton Symp. 14, 239 (1985)
Morton, R.A. and A.C. Donaldson: Bull. Geol. Soc. Am. 89, 1030 (1978)
Mosley, M.P.: J. Sediment. Petrol. 43(3), 795 (1973)
Muck, M.T., and M.B. Underwood: Geology 18, 54 (1990)
Munk, W.H.: Proc. N.Y. Acad. Sci. 51, 376 (1949)
Munk, W.H.: Okeanologiya USSR 9(1), 71 (1969)
Munk, W.H. and M.A. Traylor: J. Geol. 55, 1 (1947)

Murry, S.P., J.M. Coleman and H.H. Roberts: Nature 293, 51 (1981)
Mysak, L.A.: Rev. Geoph. Space Phys. 16(2), 233 (1978)
Mysak, L.A. and 3 others: Nature 306, 46 (1983)
Nadson, G.: C.R. Acad. Sci. (Paris) 184, 896, 1015 (1927)
Naeser, H.: Geoph. Astroph. Fluid Dyn. 13(4), 335 (1979)
Nagata, Y.: Rec. Oceanogr. Wks. Jpn 6(1), 53 (1961)
Nago, M. and S. Maeno: Mem. School of Eng., Okayama Univ. 19 (1) 13 (1984)
Nakamura, S.: Bull. Disas. Prev. Inst. Kyoto Univ. 26(4), 195 (1976)
Nakano, M.: Rec. Oceanogr. Wks. Jpn 2(2), 68 (1955)
Nastav, F. and 3 others: Research elements, study areas, sample locations, tracklines and publications
    of the marine seafloor stability program. Miami: NOAA Env. Res. Lab. (1980)
Niedoroda, A.W. and W.F. Tanner: Mar. Geol. 9, 41 (1970)
Noda, E.K.: J. Geoph. Res. 79(27), 4097 (1974)
North, W.J.: Biol. Bull. 106, 185 (1954)
Nottvedt, A. and R.D. Kreisa: Geology 15(4), 357 (1987)
Nowell, A.R.M. and C.D. Hollister (ed.): Deep ocean sediment transport. Amsterdam: Elsevier (1985)
O'Brien, M.P. and R. Mason: A summary of the theory of oscillatory waves. Tech. Rep. 2, Beach
    Erosion Board (1940)
O'Connell, S., W.B.F. Ryan and W.R. Normark: Marine and Petrol. Geol. 4, 308 (1987)
Oertel, G.F.: Mar. Geol. 63(1–2), 1 (1985)
Officer, C.B.: Physical oceanography of estauries (and Associated Costal Waters) London: Wiley (1976)
Officer, C.B.: Mar. Geol. 40(1–2), 1 (1981)
Orford, J.D.: Earth Surf. Proc. 2(4), 381 (1977)
O'Rourke, J.C. and P.H. Le Blond: J. Geoph. Res. 77(3), 444 (1972)
Osborne, A.R., A. Provenzale and L. Bergamasco: Nuovo Cim. 5C(6), 597 (1982)
Osborne, A.R., A. Provenzale and L. Bergamasco: Nuovo Cim. 36(18), 593 (1983)
Ostendorf, D.W.: J. Geoph. Res. 87(C6), 4241 (1982)
Otvos, E.G.: Bull. Geol. Soc. Am. 81, 241 (1970)
Ou, H.W.: J. Phys. Oceanogr. 14(6), 985 (1984)
Ou, H.W. and L. Maas: Continental Shelf Res. 5(6), 611 (1986)
Ou, H.W. and L.R.M. Maas: Continental Shelf Res. 8(5–7), 729 (1988)
Owens, E.H.: J. Sediment. Petrol. 47(1), 168 (1977)
Ozer, A.: Bull. Soc. Geogr. Liege 14, 117 (1978)
Pedlosky, J.: Geoph. Fluid Dynamics. Berlin Heidelberg New York: Springer (1979)
Pergram, G.G.S.: Geoph. Res. Lett. 5(1), 13 (1978)
Peregrine, D.H.: Annu. Rev. Fluid Mech. 15, 149 (1983)
Perillo, G.M.E. and J.W. Lavelle (ed.): J. Geoph. Res. 94 (C10), 14287 (1989)
Petrie, B.D.: J. Geoph. Res. 88(C14), 9567 (1983)
Phatarfod, R.M.: J. Hydrol. 30(3), 199 (1976)
Philander, S.G.H.: Rev. Geoph. Space Phys. 16(1), 15 (1978)
Pierini, S.: Nuovo Cim. 4C(4), 458 (1981)
Pierson, J.D.: Appl. Mech. Revs. 14, 1 (1961)
Pilkey, O.H.: Spec. Publ. Geol. Soc. 31, 1 (1987)
Pingree, R.D. and L. Maddock: Mar. Geol. 32(3/4), 269 (1979)
Pond, S. and K. Bryan: Rev. Geoph. Spac Phys. 14(2), 243 (1976)
Popov, B.A.: Trudy Okeanogr. Kommiss. Akad. Nauk SSSR 2, 111 (1957)
Postma, G.: Geology 12(1), 27 (1984)
Postma, G., W. Nemec and K.L. Kleinspehn: Sediment. Geol. 58, 47 (1988)
Potter, P.E. and F.J. Pettijohn: Paleocurrents and basin analysis. Berlin Heidelberg New York:
    Springer (1963)
Potter, P.E. and A.E. Scheidegger: Sedimentology 7, 233 (1966)
Pravotorov, I.A.: Okeanologiya 5(3), 473 (1965)
Prior, D.P., W.J. Wisman and W.R. Bryant: Nature 290, 326 (1981a)
Prior, D.B., W.J. Wisman and R. Gilbert: Geo-Mar. Lett. 1, 85 (1981b)
Provost, C.L.: Oceanol. Acta 4(3), 279 (1981)
Putnam, J.A.: Trans. Am. Geoph. Un. 39, 349 (1949)
Putnam, J.A., W.H. Munk and M.A. Traylor: Trans. Am. Geoph. Un. 30, 337 (1949)
Quigley, R.M. and L.R. Di Nardo: Z. Geomorph. Suppl. 34, 39 (1980)

Rabinowitz, P.D. and S.L. Eittreim: J. Geoph. Res. 79(27), 4085 (1974)
Rampino, M.R. and J.E. Sanders: Sedimentology 28(1), 37 (1981)
Rao, S.T., U. Czapski and L. Sedefian: J. Geoph. Res. 82(12), 1725 (1977)
Rea, C.C. and P.D. Komar: J. Sediment. Petrol. 45(4), 866 (1975)
Read, J.F.: Tectonophys. 81(3–4), 195 (1982)
Rehbinder, G.: Rock Mech. Rock Eng. 17, 129, (1984)
Reid, R.O. and K. Kajiura: Trans. Am. Geoph. Un. 38, 662 (1957)
Rhoades, R. and M.N. Sinacori: J. Geol. 49, 785 (1941)
Richards, K.J. and P.A. Taylor: Geoph. J. RAS 65(1), 103 (1981)
Riddell, J.F.: Can. J. Earth Sci. 6, 231 (1969)
Roll, H.U.: Encycl. Phys. 48, 671 (1957)
Rosen, P.S.: Estuarine Res. 2, 77 (1975)
Rufenach, C.L. and W.R. Alpers: J. Geoph. Res. 83(C10), 5011 (1978)
Rusnak, G.A.: J. Geol. 65, 384 (1957)
Russel, R.C.H. and D.H. Macmillan: Water waves and tides. New York: Hutchinson (1952)
Russel, R.J. and W.G. McIntyre: Bull. Geol. Soc. Am. 76, 307 (1965)
Ryrie, S.C.: J. Fluid Mech. 137, 273 (1983)
Ryzhkov, YU. G.: Izv. Akad. Nauk SSSR, Ser. Geofiz. (9), 1432 (1959)
Sallenger, A.H., R.A. Holman and W.A. Birkmeier: Mar. Geol. 64(3–4), 237 (1985)
Samoilov, I.V.: Die Flußmündungen. Transl. from Russian by F. Tutenberg. Gotha: Hermann Haack
    (1956)
Sanders, N.K.: Pap. & Proc. Roy. Soc. Tasman. 102, 11 (1968)
Savage, R.P.: J. Waterways. Div. Am. Soc. Civ. Eng. 84, WW3 (1958)
Saville, T.: Trans. Am. Geoph. Un. 31, 555 (1950)
Saville, T.: J. Waterways. Div. ASCE 82(WW2), 1 (1956)
Saxov, S. and J.K. Nieuwenhuis (eds.): Marine slides and other mass movements. New York: Plenum
    Press (1982)
Scheidegger, A.E.: Geofis. Pura Appl. 52, 69 (1962)
Scheidegger, A.E.: The physics of flow through porous media. 3rd edn. Toronto: Univ. Press (1974)
Scheidegger, A.E.: Physical Aspects of Natural Catastrophes. Amsterdam: Elsevier (1975)
Scheidegger, A.E.: In: Marine slides and other mass movements, eds. S. Saxov and J.K. Nieuwenhuis,
    p. 11, New York: Plenum Press (1982)
Scheidegger, A.E.: Z. Geomorph. 27(1), 1 (1983)
Scheidegger, A.E. and P.E. Potter: Sedimentology 5, 289 (1965)
Scholer, H.A.: J. Inst. Eng. Austr. 1958, 125 (1958)
Scholer, H.A. and E. Germanis: Inst. Eng. Austr., Civ. Eng. Trans. CEI 1, 27 (1959)
Schott, G.: Geographie des Atlantischen Ozeans. Hamburg: Boysen (1942)
Schuleikin, W.: Theorie der Meereswellen (Transl. from Russian by E. Bruns) Berlin: Akademie-Verlag
    (1959)
Schwartz, M.L. (ed.): Spits and bars. Stroudsburg PA: Wiley (1972a)
Schwartz, M.L.: Bull. Geol. Soc. Am. 83, 1115 (1972b)
Schwartz, M.L. (ed.): Barrier islands. Stroudsburg PA: Wiley (1973)
Schwarz, H.U.: Subaqueous slope failures—experiments and modern occurrences. Stuttgart:
    Schweizerbart (1982)
Schwind, J.J.v. and R.O. Reid: J. Geoph. Res. 77(39) 420 (1972)
Sekherzh-Zenkovich, T.YA: Izv. Akad. Nauk SSSR, Ser. Geofiz. 10, 1460 (1959)
Sellars, F.: J. Geoph. Res. 80(3), 398 (1975)
Serata, S. and E.F. Gloyna: J. Geoph. Res. 65, 2979 (1960)
Seymour, R.J.: J. Geoph. Res. 85(C4), 1898 (1980)
Shepard, F.P.: Submarine geology. New York: Harper (1948)
Shepard, F.P.: Proc. 8th Pac. Sci. Congr. 2A, 820 (1956)
Shepard, F.P.: In: Recent sediments northwest Gulf of Mexico, 1951 to 1958, Tulsa: Am. Assoc. Petrol.
    Geol. Spec. Publ. (1960)
Shepard, F.P.: Earth Sci. Rev. 8, 1 (1972)
Shepard, F.P.: J. Geol. 84, 343 (1976)
Shepard, F.P. and D.L. Inman: Trans. Am. Geoph. Un. 31, 196 (1950)
Shepard, F.P. and N.F. Marshall: Bull. Am. Assoc. Petrol. Geol. 57(2), 244 (1973)
Sherman, D.J.: Geogr. Rev. 78(2), 158 (1988)

Sherman, D.J. and K.F. Nordstrom: Z. Geomorph. 29(2), 139 (1985)
Shvartsman, A. YA and A.I. Makarova: Trudy Gos. Gidrol. In-ta 132, 57 (1966)
Siegenthaler, C. and Buhler: Mar. Geol. 64(1-2), 19 (1985)
Smart, J.S. and V.L. Moruzzi: Z. Geomorph. 16(3), 268 (1972)
Smith, R.: J. Fluid Mech. 77(3), 417 (1976)
Sonu, C.J. and W.R. James: J. Geoph. Res. 78(9), 1462 (1973)
Spreafico, M.: Mitt. Vers.-Anst. Wasserb., & c. ETH-Zürich 25, 1 (1977)
Stauffer, M.R., A. Hajna and D.J. Gendzwill: Can. J. Earth Sci. 13(12), 1667 (1976)
Stoker, J.J.: Water Waves. New York: Interscience Publ. Inc. (1957)
Stokes, G.G.: Trans. Camb. Philos. Soc. 8, 441 (1847)
Stommel, H.M.: The Gulf Stream. Berkeley: Univ. Calif. Press (1965)
Stuart, W.D.: Bull. Geol. Soc. Am. 84, 2691 (1973)
Suginohara, N.: J. Phys. Oceanogr. 11(8), 1113 (1981)
Suginohara, N.: J. Phys. Oceangor. 12(3), 272 (1982)
Suhayda, J.N.: J. Geoph. Res. 79(21), 3965 (1974)
Suhayda, J.N. and N.R. Pettigrew: J. Geoph. Res. 82(9), 1419 (1977)
Sunamura, T.: J. Geol. 83, 389 (1974)
Sunamura, T.: Geogr. Rev. Jpn 48(7), 485 (1975)
Sunamura, T.: J. Geol. 84, 427 (1976)
Sunamura, T.: J. Geol. 85, 613 (1977)
Sunamura, T.: Math. Geol. 10(1), 53 (1978)
Sunamura, T.: Annu. Rep. Inst. Geosci. Univ. Tsukuba 6, 51 (1980)
Sunamura, T.: Sci. Rep. Inst. Geosci. Tsukuba A2, 31 (1981)
Sunamura, T.: Ann. Rep. Inst. Geosci. Univ. Tsukuba 8, 53 (1982a)
Sunamura, T.: Earth Surf. Proc. Landf. 7, 333 (1982b)
Sunamura, T.: Trans. Jpn Geomorph. Un. 4(1), 1 (1983)
Sunamura, T.: Coastal Eng. Jpn 27, 207 (1984a)
Sunamura, T.: Bull. Geol. Soc. Am. 95(2), 242 (1984b)
Sunamura, T.: Trans. Jpn Geomorph. Un. 6(4), 361 (1985)
Sunamura, T. and I. Takeda: Mar. Geol. 60, 63 (1984)
Sverdrup, H.U.: Encycl. Phys. 48, 608 (1957)
Sverdrup, H.U. and W.H. Munk: Trans. Am. Geoph. Un. 27, 828 (1946)
Sverdrup, H.U., M.W. Johnson and R.H. Fleming: The oceans, revised ed. Englewood Cliffs N.J.: Prentice-Hall (1946)
Swain, A. and J.R. Houston: Can. J. Civ. Eng. 12(1), 231 (1985)
Swallow, J.C. and V. Worthington: Nature, 179, 1183 (1957)
Sweeting, M. and K.H. Pfeffer: Z. Geomorph. Suppl. 26, 1 (1976)
Swinnerton, A.C.: Bull. Geol. Soc. Am. 43, 663 (1932)
Syvitski, J.P.M. and 3 others: J. Geoph. Res. 93(C6), 6895 (1988)
Takeda, I., and T. Sunamura: Coastal Eng. Jpn 26, 121 (1983)
Takeda, I., T. Terrasaki and T. Sunamura: Annu. Rep. Inst. Geosci. Tsukuba 12, 55 (1986)
Tam, C.K.W.: J. Geoph. Res. 78(12), 1937 (1973)
Tanner, W.F.: Trans. Am. Geoph. Un. 39, 889 (1958)
Tanner, W.F.: Science 132 (3433), 1012 (1960)
Tanner, W.F.: Shore and Beach 41(1), 22 (1973)
Tanner, W.F. (ed.): Near-shore sedimentology. Proc. 6th Sympos. Coastal Sedimentology. Tallahassee: Florida State Univ. (1983)
Taus, K. and U. Gerber: Material and Technik 5(4), 197 (1977)
Theis, C.V.: Trans. Am. Geoph. Un. 16, 519 (1935)
Thompson, R.O.R.Y. and T.J. Golding: J. Geoph. Res. 86(C7), 6517 (1981)
Thorn, M.F.C.: In: Proc. Conf. ICE London 1981; London: Telford for Inst. Civ. Eng. p. 65 (1982)
Thornton, E.B.: J. Geoph. Res. 84(C8), 4931 (1979)
Thornton, E.B. and R.T. Guza: J. Geoph. Res. 88(C10), 5925 (1983)
Thorpe, S.A. and P.N. Humphries: Nature 283, 463 (1980)
Thorpe, S.A., M.B. Belloul and A.J. Hall: Nature 330, 740 (1987)
Tomczak, M.: Oceanol. Acta 4(2), 161 (1981)
Tomkoria, B.N. and A.E. Scheidegger: Can. J. Phys. 45, 3569 (1967)
Trenhaile, A.S.: J. Sediment. Petrol. 43(2), 558 (1973)

Trenhaile, A.S.: Prog. Phys. Geogr. 4(1), 1 (1980)

Trenhaile, A.S. and M.L. Bryne: Geogr. Ann. 68A(1–2), 1 (1986)

Trenhaile, A.S. and M.G.L. Layzell: Trans. Inst. Brit. Geogr. 6, 82 (1981)

Tricker, R.A.R.: Bores, breakers, waves and wakes Amsterdam: Elsevier (1965)

Trofimov, A.M. and V.M. Moskovkin: Z. Geomorph. 29(3), 257 (1985)

Tsay, T.K. and P.L.F. Liu: J. Geoph. Res. 87(C10), 7932 (1982)

Tsuchiya, Y. and S. Tsutsui: Bull. Disas. Prev. Res. Inst. Kyoto Univ. 32(3), 143 (1982)

Tsuchiya, A. and T. Yasuda: Bull. Disas. Prev. Inst. Kyoto Univ. 31(1), 15 (1981)

Tucholke, B.E.: Nature 296, 735 (1982)

Uncles, R.J.: Oceanol. Acta 5(4), 403 (1982)

Valenzuela, G.R. and J.W. Wright: J. Geoph. Res. 81(33), 5795 (1976)

Viles, H.A.: Progr. Phys. Geogr. 10(3), 429 (1986)

Vischer, D.: Wasserwirtschaft 67(7/8), 1 (1977)

Volkart, P.: Wasser-Energiewirtschaft 1974 (7/8), 2 (1974)

Waltham, A.C.: Progr. Phys. Geogr. 5(2), 242 (1981)

Wang, F.C.: J. Geoph. Res. 89(C5), 8054 (1984)

Wang, H., T. Sunamura and P.A. Hwang: Coastal Eng. 6, 121 (1982)

Watanabe, A.: Coastal Eng. Jpn 25, 147 (1982)

Weaire, D. and C.O'Carroll: Nature 302, 240 (1983)

Webb, D.J.: Revs. Geoph. Space Phys. 12(1), 103 (1974)

Weller, J.M.: Kentucky Geol. Surv., Ser. 6, 38, 42 (1927)

Wells, D.R.: J. Geoph. Res. 72, 497 (1967)

Wells, J.T., D. Prior and J.M. Coleman: Geology 8(6), 272 (1980)

Wentworth, C.K.: J. Geomorph. 1, 5 (1938)

Whitaker, J.H. (ed.): Submarine canyons and deep-sea fans: modern and ancient. New York: Wiley (1976)

Williams, A.T. and L.F. Gulbrandsen: Cambria 4(2), 174 (1977)

Willmott, A.J.: Geoph. Astroph. Fluid Dyn. 23(4), 273 (1983)

Winant, C.D.: Rev. Geoph. Space Phys. 17(1), 89 (1979)

Won, I.J., J.T. Kud and R.C. Jachens: J. Geoph. Res. 83(B12), 5947 (1978)

Won, I.J. and L.S. Miller: J. Geoph. Res. 84(B8), 3844 (1979)

Woods, J.D.: Nature 314, 501 (1985)

Wright, L.D.: Bull. Geol. Soc. Am. 88(6), 857 (1977)

Wright, L.D. and J.M. Coleman: J. Geol. 82, 751 (1974)

Wu, C.S., E. Thronton and R.T. Guza: J. Geoph. Res. 90(C3), 4951 (1985)

Wübber, C.S. and W. Krauss: Oceanol. Acta 2(4), 435 (1979)

Wunsch, C.: Rev. Geoph. Space Phys. 13(1), 167 (1975)

Wüst, G.: Wiss. Erg. dtsch. Atlant. Exped. METEOR 6, Pt. 2, 261 (1975)

Yeh, H.H.: Fluid Mech. 152, 479 (1985)

Zenkovich, V.P.: Trudy Inst. Okeanologii Akad. Nauk, SSSR 21, 3 (1957)

Zenkovich, V.P.: Processes of coastal development, 738 pp. Translated from Russian by D.G. Fry. London: Oliver & Boyd (1967)

Zimmerman, J.T.F.: Netherl. J. Sea Res. 6(4), 542 (1973)

Zimmerman, J.T.F.: Nature 290, 549 (1981)

Zschau, J. and H.J. Kümpel: Geoph. Astroph. Fluid Dyn. 13, 245 (1979)

7 Theory of Niveal, Glacial, and Periglacial Features

7.1 Introduction

We are turning next to the discussion of the genesis of the various morphological features that have been caused directly or indirectly by the action of water in its frozen state. The phenomenology of the features in question has been presented in Sect. 1.7.

What is involved, therefore, is the analysis of the entire complex of features caused by snow or by ice above or below the ground. This includes features which are commonly called "niveal", "glacial", or "periglacial". In this connection, it must be recalled that the extent of the snow cover and of glaciation is much less at the present time than during the recent "ice ages" (cf. Sect. 2.5.2). Thus, features of the type under consideration may be found in regions that are not subject to a particularly cold climate today.

The subject is vast, and the writer is not aware of any comprehensive reviews of it. Except for some collections of papers (cf. Vischer 1988), most reviews have been directed specifically either to snow dynamics, to glacier physics, or to periglacial effects.

7.2 Snow Problems

7.2.1 General Remarks

Snow, by itself, is not a particularly important geomorphic agent; it becomes so more effectively after it has been transformed into ice (cf. Sect. 2.3.2.4).

By itself, snow can fill hollows, it may cause pressure ridges to form, it may be blown around by winds and, particularly, it may form snow and slush avalanches which can cause morphodynamic effects upon impact. However, these effects are more important in connection with hazards to human endeavors than in connection with landscape evolution. This whole section on snow problems will therefore be held brief, stressing mainly geomorphological implications; the reader is referred to the writer's book on catastrophes for other details (Scheidegger 1975).

7.2.2 Snow as Material

7.2.2.1 Snow Fields and Firns

Snow accumulates as a precipitate from the atmosphere on the ground. In this, the vegetation present may influence its accumulation pattern (Daly 1984). The distribution of the snow is seasonal at middle latitudes and at low elevation, and it is also determined globally by the prevailing world climatic conditions. At high latitudes and high elevations, the snow may last year-round and then forms permanent "firns". Firns creep slowly; in the Alps their age may reach 200 to 300 years (Haeberli et al. 1988), corresponding to surface creep rates of 4 m/a. Corresponding stresses and the strain rates have been calculated (Haefeli and Sury 1975).

7.2.2.2 Wind-blown Snow

The surface of a snow field, when dry, is much like that of a sandy desert. Hence, the snow crystals may be picked up by the wind like the sand grains in a desert (cf. Sect. 8.2.2.3). The theories presented in that connection, however, are not quite adequate for snow, because in addition to purely mechanical considerations, thermodynamic conditions have to be observed. In fact, threshold speeds for blowing snow vary over such a wide range in nature that it must be considered as a separate parameter (Schmidt 1982).

The problem is of great importance with regard to the prevention of snow drifts on highways etc., but does not have too many geomorphological applications.

7.2.2.3 Geomorphological Effects

As noted, the geomorphological effects of snow as a material by itself are rather limited.

Nevertheless, hollows filled with snow may give cause for lakes to form later. Wind-blown snow may accumulate behind obstacles to precisely fill such hollows.

However, the most notable geomorphic effect originates in snow fields and firns which like ice (see below) also are in a state where they creep (McClung et al. 1984). Thus, a snow-bank push mechanism has been advocated for the formation of some seasonal (annual) moraine ridges (Birnie 1977). The "push" in the snow field may, in fact, be against an ice contact.

7.2.3 Avalanches

7.2.3.1 General Remarks

The most important geomorphological effects of snow are caused by avalanches.

There are many types of avalanches which, however, fall into two broad categories: snow avalanches proper and slush flows in which a mixture of melt water and snow makes for a torrent of a peculiar kind.

Again, the main concern with avalanches is from the angle of human endeavors. The writer has dealt with the latter aspects in his book on catastrophes (Scheidegger 1975); here we shall confine ourselves to geomorphological implications.

7.2.3.2 Snow Avalanches

Beginning with snow avalanches, we note that there are three phases in their evolution: (1) their initiation by the instabilization of a snow accumulation, (2) their progressive motion down a slope, and (3) their runout or impact when the moving masses come to a standstill.

The initiation of an avalanche is tied to the stability of various types of snow covers. Accumulations and semiconsolidated snow slabs behave somewhat differently: in either case, tensile failure within the snow at right angles, shear failure parallel to the (future) motion, and friction at the bottom of the potentially moving mass are of importance (cf. Scheidegger 1975). Instability of a snow cover can be adduced by external influences, such as by gun shots fired into or by animals or skiers crossing over a potentially unstable slope. Recently, system-dynamic probabilistic considerations have also been applied to the avalanche initiation problem (Conway and Abrahamson 1988).

Regarding the downhill motion of the snow, it must be noted, first of all, that there are many possible types of snow avalanches (De Quervain 1973), of which the most important are powdersnow and snowflow avalanches (Scheiwiller and Hutter 1983). In the former, particles are essentially carried in a turbulent suspension in air (turbidity current), in the latter, one is faced with a shear flow in the free snow particles with a free surface (much like the "flow" of any granular material).

Again, because of the human implications of avalanches, the literature is vast. A general summary of the subject is contained in the cited book by Scheidegger (1975); specifically, the dynamics of powder snow avalanches has been comprehensively presented by Scheiwiller (1986; also Scheiwiller et al. 1987) and that of snow flow avalanches by Scheiwiller and Hutter (1982).

Finally, snow avalanches will come to a standstill in the runout region. The runout process depends on topographic parameters for which empirical influence relations have been established (Lied and Bakkehoi 1980; McClung and Lied 1987). It is here where most direct mechanical effects occur; if an obstacle is hit, great destructive forces are released (Voellmy 1955 reviewed by Scheidegger 1975; Laatsch et al. 1981). The impact pressures can reach more than 600 kPa (Fitzharris and Owens 1984).

7.2.3.3 Slush Flows

Niveal processes recognized specifically as such only in very recent times are *slush flows*.

Such flows represent rapid, linear, channeled mass movements of water-saturated snow. They occur mainly during the beginning stages of spring breakup in arctic and high subarctic regions (Washburn 1980; Clark and Seppälä 1988).

As with ordinary snow avalanches, there are three stages in slush flows: (1) the initiation, (2) the flow, and (3) the runout stage.

The initiation stage has been particularly investigated by Onesti (1985). Evidently, it is necessary that there be sufficient snow present and that the latter become water-saturated. Unlike snow avalanches, slush flows do not require steep slopes for their initiation (slope angles of as little as 2° may suffice). However, the conditions necessary to actually start a slush flow process are complex and mainly of a meteorological nature, such as accelerated snow melt etc.

The actual dynamics of slush flows, once the latter have been started, has been studied only cursorily. Slush flow tracks may follow preexisting channels or form of their own down open slopes. Varying or alternating gradients affect the width of the flow whereby a decrease in inclination causes a widening of the track (Hestnes 1985). In this, slush flows resemble water torrents more than mass avalanches.

In a similar fashion, the runout of a slush flow is much like that of a torrent squall hitting flat bottom lands.

7.2.3.4 Geomorphological Effects of Avalanches

The geomorphological effects of avalanches consist in erosional effects, impact effects, and deposition effects.

At the beginning of their run, avalanches pick up material on their way, often cutting huge lanes into a hillside: the ground is torn open, lumps are loosened and pushed up, leaving wide scars (Hope 1988) that can become the centers for the attack by other exogenic agents. Cumulatively, the erosive action of avalanches may contribute substantially to the denudation of a moutain range (e.g., the New Zealand Alps: Ackroyd 1986).

If avalanches find an obstacle in their main motion, impact effects result. The high impact pressures mentioned in Sect. 7.2.3.2 have been advocated as the cause of some "tarns": small scooped-out basins that may contain a lake (Fitzharris and Owens 1984). Impact traces may also be seen on slopes opposite to the actual avalanche slopes (Huber 1982).

A further important geomorphological effect of avalanches consists in the deposition of debris material in the runout region. This material may contribute to the damming up of lakes and generally affect the nature of a landscape (Gardner 1970).

7.3 Ice Problems

7.3.1 General Remarks

Ice problems are important in geomorphology mainly in connection with the behavior of glaciers. The extent of glaciation has been different in different geological epochs (cf. the problem of "ice ages" discussed in Sect. 2.5.2). Even at the present time, some glaciers are growing and others decreasing: this is determined by the prevailing mass balance.

Glaciers are basically huge rivers of ice steadily flowing downhill. The explanation of the flow of glaciers still poses some problems; it appears that there are two models of motion: internal motion and sliding. The glacier motion is most easily discussed in two dimensions i.e., along its bed if the thickness of the ice varies. The analysis of three-dimensional ice motion is much more complicated than the investigation of longitudinal glacier flow. However, its application to the explanation of the motion of ice caps and of cirque glaciers is of great geomorphological significance. Finally, glaciers may reach or ice may form on the sea, giving rise to problems of "sea ice" which have a certain effect in coastal regions.

General studies of the physics of glaciers have been published by Paterson (1969) and, notably, by Hutter (1983), who collected and worked through most of the pertinent physical theories. In the present book, only brief resumes of some of the theories can be given, which seemed most appealing to the present writer. For all further details, the reader is referred to the 510-page monograph by Hutter (1983).

7.3.2 Mass Balance

The mass balance (snow/ice accumulation versus melting/ablation) of a glacier is responsible for the latter's (kinematic) response to seasonal and climatic changes.

There are several aspects of the advance/retreat problem of glacier.

First, there is the purely observational background. Volume changes of glaciers, represented by the position of their tongues, mainly in the Alps, were measured during the last 100 years or so, generally with the result that glaciers have been retreating during that period (Lang and Patzelt 1971; Kasser and Aellen 1976; Holzhauser 1982).

Second, internal correlations for each glacier were sought, for instance between advance/retreat and the internal ice temperature (Haeberli 1975) or between the former and thickness and velocity (Reynaud 1977; Vallon 1977).

The first step was an attempt at finding correlation between glacier fluctuations and external climatic variables, usually by applying multiple correlation analysis (Tollner 1974; Martin 1974, 1978; Müller 1988), unfortunately with the result that these relationships are complex and equivocal (Haeberli et al. 1989).

Finally, the problem has been treated entirely theoretically by Hutter (1983, p. 333 ff.), who considered the local effect of a change in accumulation of glacier mass at the surface or base, of a change of energy input from the atmosphere, and of a change of geothermal heat flux. Needless to say, it is difficult to estimate the rates of such changes in the external parameters and thus all further conclusions are uncertain to this extent. Nevertheless, if the changes in the external parameters are given, it can be shown that kinematic waves (Nye 1963a, b) are started in the glacier which, when arriving at the tongue, effect an advance or retreat thereof.

However, as long as the rate of change of the external parameters (as functions of climate variations) is not known, it seems somewhat futile to attempt accurate calculations.

7.3.3 Longitudinal Movement of Glaciers

7.3.3.1 General Remarks

The study of the movement of glaciers is an involved discipline on which much effort has been expended by various research workers. In connections with theoretical geomorphology, we are interested in glacier flow insofar as it has a bearing upon the landforms that have been affected by it. Unfortunately, when it comes to a study of the interaction of a glacier with its bed and the valley walls, studies available are very rare and incomplete indeed.

Of the various theories of glacier flow available at the present time, one of the variants of Nye's theories, based upon plasticity theory, is the most suitable for our purpose. It will be presented in Sect. 7.3.3.2. Nye's theory explains the internal flow of a glacier only, it does not account for the sliding of a glacier over its bed. Questions concerned with the latter problem will be discussed in Sect. 7.3.3.3, Subsequently, we shall turn our attention to some special problems, viz. to the formation of glacier snouts and crevasses. Finally, some of the geomorphological implications of the above phenomena will be discussed.

All the discussions in these chapters are concerned with the longitudinal movement of glaciers only. If the three-dimensional movement of ice is under consideration, the mathematics involved becomes much more complicated. This will be investigated in Sect. 7.3.4.

7.3.3.2 Longitudinal Creep of Glaciers

The flow of glaciers has been studied by many investigators. In the earliest attempts (Finsterwalder 1907), a heuristic assumption was used for the connection between glacier thickness and flow velocity. Subsequently, the theory of glacier flow was based on a viscous-type flow law, later a plastic flow law was considered more appropriate and, finally, a power flow law was considered as most adequate. The plastic flow law can also be considered as a limiting case of the power flow law; the two laws lead, in many cases, to very similar results. Today, the best-known treatment of the glacier flow problem is probably that of Nye, who first regarded it from the standpoint of plastic flow (Nye 1951) and then from the standpoint of a power flow law (Nye 1955). It turned out (Nye 1957) that the solutions for glacier flow obtained for the general flow law are very similar to those found for the plastic flow law. We shall therefore mainly discuss Nye's earlier theory, where a plastic flow law has been assumed as basic for glacier flow.

The theory applying in this case has been mentioned in Sect. 3.3.2.2. Accordingly, the flow may occur in an *active* or in a *passive* Rankine state. On a uniform slope of slope angle α (see Fig. 103) and in a plane strain state, the equations that have to be satisfied are

$$\frac{\partial \sigma_x}{\partial x} + \frac{\partial \tau_{xy}}{\partial y} - \rho g \sin \alpha = 0, \tag{7.3.1}$$

$$\frac{\partial \tau_{xy}}{\partial x} + \frac{\partial \sigma_y}{\partial y} - \rho g \cos \alpha = 0, \tag{7.3.2}$$

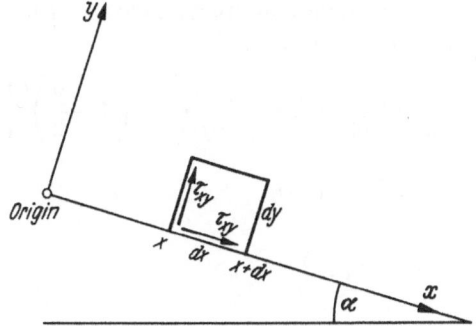

Fig. 103. Geometry of glacier flow

where σ and τ are normal and shear stresses, respectively, ρ is the density of the ice, g the gravitational acceleration and $+x$ is taken parallel to the slope, pointing downhill. To the above equation one must add the yield criterion (k being a constant):

$$(\sigma_x - \sigma_y)^2 + 4\tau_{xy}^2 = 4k^2.$$

If the velocity of flow is sought, one also has to heed the condition of incompressibility and the condition that the principal axes of stress and strain coincide. This will be dealt with below (cf. 7.3.8 and 7.3.9).

Assuming that the boundary conditions imply (k a positive number)

$$\tau_{xy} = -k$$

at the bed, that the velocity at the bed is zero, and that the shear stress vanishes at the surface of the glacier $(y = h)$, the two possible stress solutions to the problem are:

$$\sigma_x = x\left(-\frac{k}{h} + \rho g \sin\alpha\right) + y\rho g \cos\alpha \pm 2k \sqrt{\left\{1 - \left(1 - \frac{y}{h}\right)^2\right\}} + a$$

$$\sigma_y = x\left(-\frac{k}{h} + \rho g \sin\alpha\right) + y\rho g \cos\alpha + a \qquad (7.3.3)$$

$$\tau_{xy} = -k\left(1 - \frac{y}{h}\right).$$

Here, a is a constant which must be set equal to

$$a = -h\rho g \cos\alpha \qquad (7.3.4)$$

in order to have a free surface for $y = h$. Furthermore, for the same reason, the following condition must be imposed upon h

$$h = \frac{k}{\rho g \sin\alpha} = \frac{h_0}{\sin\alpha} \qquad (7.3.5)$$

with

$$h_0 = k/(\rho g) \qquad (7.3.6)$$

for then σ_y and σ_x will also vanish on $y = h$.

Provided the condition (7.3.5) is satisfied, the possible stress solutions for glacier flow are therefore

$$\frac{\sigma_x}{k} = \frac{y-h}{h_0} \cos\alpha \pm 2\sqrt{\left\{1 - \left(1 - \frac{y}{h}\right)^2\right\}}$$

$$\frac{\sigma_y}{k} = \frac{y-h}{h_0} \cos\alpha \qquad\qquad\qquad (7.3.7)$$

$$\frac{\tau_{xy}}{k} = -1 + \frac{y}{h}.$$

The corresponding velocity solutions (u parallel to x, v parallel to y) are obtained from the condition of incompressibility

$$\frac{\partial u}{\partial x} + \frac{\partial v}{\partial y} = 0 \qquad\qquad\qquad (7.3.8)$$

and from the condition that the principal axes of stress and stain rate coincide

$$\frac{\partial v/\partial x + \partial u/\partial y}{\partial u/\partial x - \partial v/\partial y} = \frac{2\tau_{xy}}{\sigma_x - \sigma_y}. \qquad\qquad\qquad (7.3.9)$$

Hence:

$$u = C \pm \frac{Vx}{h} + 2V\sqrt{\left\{1 - \left(1 - \frac{y}{h}\right)^2\right\}} \qquad\qquad (7.3.10)$$

$$v = \mp Vy/h, \qquad\qquad\qquad (7.3.11)$$

where C and V are constants. The two solutions are illustrated in Fig. 104. It should be noted that in both solutions, a vertical velocity component is present. Thus, in order to keep condition (7.3.5) satisfied, material (snow) has to be either added or subtracted from the surface.

The slip lines (trajectories of maximum shear stress) can be calculated for the above cases; in fact, this problem had been solved long ago by Frontard (1922) and Prandil (1923). The result is shown in Fig. 105.

The above calculations are also roughly valid for a slope of varying slope angle. The interpretation of the approximate solutions for this case is that the ice always adopts the type of flow that maintains the critical thickness [Eq. (7.3.5)]. Accordingly, a concave bottom and loss of material would tend to favor the passive, a convex bottom and addition of material would tend to favor the active Rankine state.

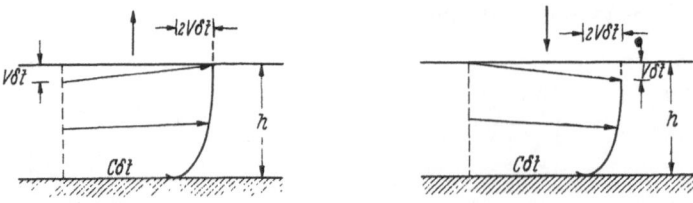

Fig. 104. The two velocity solutions for a glacier. (After Nye 1951)

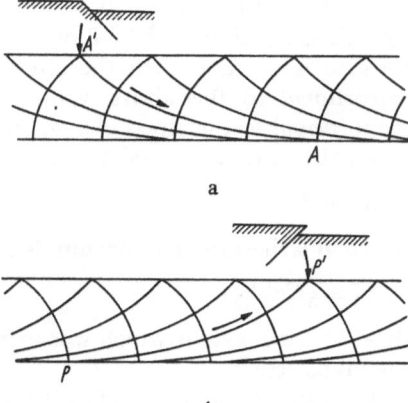

Fig 105 a, b. The slip line fields and fault planes **a** in active flow and **b** in passive flow. (After Nye 1951)

Of particular importance is a discussion of glacial erosion. It appears that the slip lines discussed above have a connection with this. If the slip line is of the type PP′ (Fig. 105, the differential movement at P might tend to suck debris into the ice from the bottom, whereas in the case of a slip line of the type A′A, it would merely roll debris along. Thus, it might be expected that it is passive flow which causes erosion.

The above theory is for a cross-section of an infinitely wide glacier. Nye (1965) has also produced solutions of the basic flow equations for glacier flow in rectangular, elliptic, and parabolic cross-sections. Mackay (1965) has made analog models for the solution of glacier flow problems.

As was noted earlier, the above theory has been extended to the use of flow laws different from that adopted in plasticity theory. However, the results obtained are very similar to those presented above. Details may be found in the cited monograph by Hutter (1983). Since our interest is essentially in the geomorphological effects of glacier flow, improved theories of the latter are of little significance for us.

7.3.3.3 Theory of Longitudinal Sliding of Glaciers

The theory of glacier movement described above assumes as boundary condition that the velocity of the glacier vanishes at the bed. This is almost certainly an oversimplification, as it may be expected that a certain amount of sliding will always take place. The bed of a glacier is very uneven so that there are only two effects that can contribute much to sliding: the first is pressure melting, the second stress concentrations. A theory of glacier sliding based upon these two effects has been proposed by Weertman (1957, 1964).

In order to set up his theory, Weertman had to assume an idealized glacier bed. Such a bed consists of cubic protuberances of edge length L which are centered on a square grid with grid length L′.

From the above model, it is first of all possible to calculate the difference ΔT in melting point temperature in front and behind an obstacle. If τ be the macroscopic shear stress at the glacier bottom, the average normal stress on one side of an obstacle is *equal* to $\tau L'^2/L^2$ if no hydrostatic pressure is present. If a hydrostatic pressure is present, then the stress increase caused by the obstacle must be *proportional* to the above value. Since the change of melting point ΔT is proportional to the stress change, we have the result that the latter will also be proportional to (cf. Fermi 1937)

$$\tau L'^2/L^2.$$

We shall write this relationship as follows

$$\Delta T = \tfrac{1}{3} C \tau L'^2/L^2. \tag{7.3.12}$$

Here, C is a constant which was experimentally found to be equal to 7.4×10^{-9} deg/(dynes/cm^2).

The speed of sliding S_A of a glacier is equal to the volume of ice melted per unit time in front of an obstacle divided by the cross-sectional area of the obstacle. The melting process, however, will be caused by heat flow across each obstacle at which the difference in temperature ΔT exists as calculated above. Thus

$$S_A = \frac{1}{L^2} \Delta T \frac{DL}{H\rho} = \frac{\tau CD}{3H\rho L} \left(\frac{L'}{L}\right)^2, \tag{7.3.13}$$

where D is the coefficient of heat conductivity of the rock, H is the heat of fusion of the ice, and ρ is its density.

If there is a whole spectrum of protuberances present in the glacier bed (L'/L being constant), then it becomes obvious that the larger the obstacle, the less likely it is that sliding by the above mechanism can take place.

Fortunately, it can be shown that the second effect (that of stress concentration) can precisely account for the sliding of glaciers over *large* obstacles. Stress concentrations will again be of the order of $\tau(L'/L)^2$. Using the power law of creep for ice (cf. Sect. 2.3.3.3), the creep rate K due to this stress concentration will be

$$K = \text{const } \tau^n \left(\frac{L'}{L}\right)^{2n}. \tag{7.3.14}$$

Furthermore, it stands to reason that this creep rate is effective over the distance L; hence the sliding velocity S_B due to the stress concentration is (writing the constant somewhat differently)

$$S_B = B(\tfrac{1}{2}\tau L'^2/L^2)^n L. \tag{7.3.15}$$

This vanishes for very small obstacles. Weertman now argues that the sliding will take place with just such a speed which is the same for the two types of sliding, because for each type, the speed is controlled by the most unfavorable obstacle size. Thus, setting $S_A = S_B$ and eliminating L (assuming the ratio L'/L as given), Weertman obtained for the sliding velocity S

$$S = S_A = S_B = \left(\frac{2BCD}{3H\rho}\right)^{1/2} \left(\frac{\tau}{2}\right)^{(1+n)/2} \left(\frac{L'}{L}\right)^{1+n} \tag{7.3.16}$$

The main result of the above analysis is that sliding of glaciers can and will indeed take place. Weertman, using reasonable values (viz. $L'/L = 4$, H = 80 cal/g,

$D = 0.005\,\text{cal/deg/s/cm}$, $\tau = 1$ bar, $B = 0.017\,\text{bar}^{-4.2}\,\text{year}^{-1}$, $n = 4.2$) for the constants in (7.3.16), obtained a value of about 1 m/a. This is far less than what is observed in natural glaciers. The situation can be improved by slightly changing the roughness value (L′/L), because of the high value of n attached to the latter.

Modifications of the above theory have been attempted by Lliboutry (1959, 1967), who noted that the idealized glacier bed considered by Weertman (1957) represents too gross an oversimplification. Thus, Lliboutry suggested that, instead of a bed with square obstacles, one should consider a bed made up of parallel sine waves whose equation is

$$z = \frac{a}{2}\sin\left(\frac{2\pi x}{\lambda} - \varphi\right). \tag{7.3.17}$$

Here, z is the coordinate normal and x the coordinate parallel to the flow direction of the glacier.

In order to approach reality, one can allow different phase angles φ for different intervals on the y-axis. The quantity r

$$r = \frac{a}{\lambda} \tag{7.3.18}$$

may be called the *rugosity* of the bed.

The ice exerts a pressure upon the bed which will be greater in front of the waves than behind; thus Lliboutry assumes a harmonic variation for this pressure:

$$\sigma_x = \rho g h + \frac{\Delta\sigma}{2}\cos\left(\frac{2\pi x}{\lambda} - \varphi\right), \tag{7.3.19}$$

where $\rho g h$ is the mean pressure and $\Delta\sigma/2$ the maximum variation therefrom. Here, h is as usual the thickness of the glacier, ρ the density of the ice, and g the gravity acceleration. The quantity σ_x is a principal stress; the other two principal stresses are

$$\sigma_y = \sigma_z = \rho g h. \tag{7.3.20}$$

The mean frictional traction τ is (setting the sine equal to the tangent)

$$\tau = \frac{1}{\lambda}\int_0^\lambda \sigma\,\frac{dz}{dx}\,dx = \frac{1}{\lambda}\int_0^\lambda \frac{\Delta\sigma}{2}\,\frac{a\pi}{\lambda}\cos^2\left(\frac{2\pi x}{\lambda} - \varphi\right)dx \tag{7.3.21}$$

$$= \frac{\pi}{4}r\Delta\sigma$$

with r again being the rugosity as given by (7.3.18)

The above sliding theories have been extended and reviewed by Budd (1970), Kamb (1970), Morland (1976), Lliboutry (1979), and by Balise and Raymond (1985). These theories provide an explanation of the sliding of a glacier over its bed. The surface stress (τ above) leads to the production of glacial polish. It does not, however, account for the high sliding velocities observed in natural glaciers (cf. Sect. 1.7.2.3).

7.3.3.4 Subglacial Water; Surges

Since the above processes cannot possibly produce the sliding velocities of 30–3000 m per year (cf. Sect. 1.7.2.3), Lliboutry (1987) searched for a modification thereof which might produce higher velocities. He found such a modification in the possibility that the glacier might detach itself from its bed. The picture would then be that represented in Fig. 106. This means, in effect, that the glacier bed is rough in only one direction.

According to Formula (7.3.19), the minimum normal pressure of the glacier bed is

$$\sigma_{zmin} = \rho gh - \frac{\Delta\sigma}{2} = \rho gh - \frac{2\tau}{\pi r}. \tag{7.3.22}$$

If a detachment from the bed is to occur, this pressure must become zero, which yields

$$\tau = \frac{\pi}{2} r\rho gh. \tag{7.3.23}$$

However, if detachment occurs, it appears likely that the resulting cavity fills itself with *water* which will be subject to an equilibrium pressure p at a temperature blow 0 °C. Hence, Eq. (7.3.23) should in fact be written as follows

$$\tau = \frac{\pi}{2} r(g\rho h - p). \tag{7.3.24}$$

The quantity $g\rho h - p$ is an "effective" stress, representing the excess of the prevaling stress over the prevailing hydrostatic pressure.

The above considerations lead to a process of sliding which has entirely different characteristics from the processes considered earlier. Lliboutry (1959) envisaged it as occurring in such a fashion that the ice rests over the length x (see Fig. 106) on the protuberances ($x \ll \lambda$), producing a rise Z in the underside of the glacier, equal to

$$Z = \frac{x^2}{2R_s}, \tag{7.3.25}$$

where R_s is the radius of curvature at the top of the protuberance, which according to (7.3.17) is equal to

$$R_s = \frac{\lambda^2}{2\pi^2 a}. \tag{7.3.26}$$

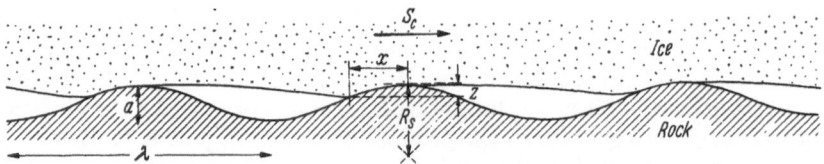

Fig. 106. Detachment of a glacier from its bed. (After Lliboutry 1959)

Hence

$$Z = \pi^2 a \frac{x^2}{\lambda^2}. \tag{7.3.27}$$

The contact force between the ice and the solid bed forms an angle $x/(2R_s)$ with the vertical whose tangent is the coefficient of friction. The average effective normal pressure is $\rho gh - p$, hence the frictional force is (since $x/2R_s$ is small):

$$\tau_c = \frac{(\rho gh - p)x}{2R_s} = (\rho gh - p)\frac{\pi^2 ax}{\lambda^2}. \tag{7.3.28}$$

Since the frictional stress τ_c reaches zero for $\rho gh = p$, it is evident that very high sliding velocities S_c are possible.

As representative values, Lliboutry claims that, if one assumes

$$\rho gh - p = 6.5\,\text{bar},$$

then the set

$$r = 0.1$$
$$\tau_c = 1.0\,\text{bar}$$
$$n = 3$$
$$a = 0.21\,\text{m}$$
$$S_c = 100\,\text{m/a} \tag{7.3.29}$$

forms a consistent set of numbers.

It is seen that one has at last a mechanism that provides for high sliding velocities. Extensions and refinements to the above simple theory have been made by Lliboutry (1987) himself and also by Iken (1981).

Sliding on subglacial water (Fowler 1987) may provide the explanation for the sudden surges of glaciers that have occasionally been observed (cf. Sect. 1.7.2.3). In fact, the problem of finding the cause of glacier surges has not yet been entirely solved. Thus, Clarke (1976) considered thermal and creep (Clarke et al. 1977) instabilities. General reviews of the problem have been published by Raymond (1987), Engelhardt (1987) and Sharp (1988a). Although the specific mechanism of surge initiation is still uncertain, the general consensus (with few exceptions) seems to be at the present time that surges are somehow connected with the basal hydraulic system of the glacier whose exact nature is not yet known.

7.3.3.5 Dynamics of Glacier Snouts

The problem of glacier snouts was studied at a very early date by Finsterwalder (1907) who, as noted in Sect. 7.3.3.2, based his analysis on a heuristic equation of the development of the ice surface. It is

$$[(n + 1)\kappa\Theta^n - a]\frac{\partial\Theta}{\partial x} + \frac{\partial\Theta}{\partial t} = -a. \tag{7.3.30}$$

Here Θ is the height of the surface above the base x (cf. Fig. 107) and a is the melting constant giving the amount of ice loss per unit time. The above picture is based on

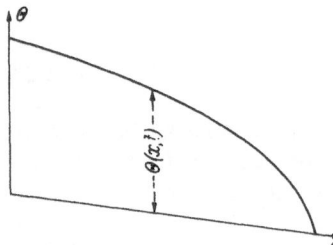

Fig. 107. Cross-section through the end of a glacier

the heuristic assumption that the ice velocity is given by

$$v = \kappa \Theta^n, \tag{7.3.31}$$

where κ depends on the bed slope of the glacier and n is a constant having a value between $\frac{1}{4}$ and $\frac{1}{2}$.

A numerical integration of the differential Eq. (7.3.30) has been given by Collatz (1955), choosing $n = \frac{1}{3}$, $\kappa = 0.075$, $a = \frac{1}{2}$. The result obtained is shown in Fig. 108.

The approach of Finsterwalder was put upon a somewhat more solid basis by Nye (1963a, 1963b). Accordingly, the advance and retreat of a glacier snout is but a special case of a general theory of the response of a glacier to climatic variations. The theory of such variations has to be primarily based on a mass balance equation. In addition, the central assumption is made that the rate of discharge in the glacier at a point is a definite function of the thickness of the glacier and of the slope of its surface at that point [this is a generalization of Eq. (7.3.31)]. The actual response of the glacier to changes in the rate of nourishment and wastage can then be studied by means of a linearized perturbation theory; it turns out that the behavior of the perturbation is given by a diffusivity equation with a source term and a mass transport term. This can then be applied to a hypothetical reasonable snout profile whose advance and retreat as a function of change in precipitation can be calculated.

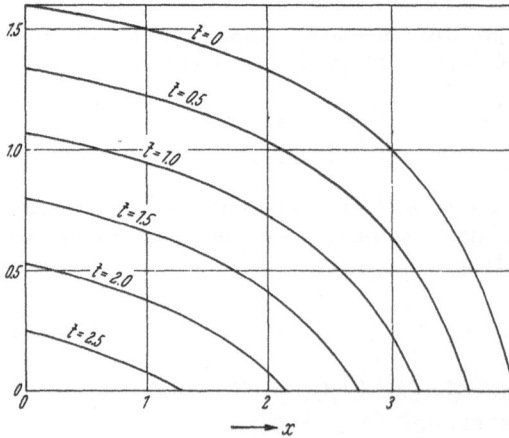

Fig. 108. Profile of glacier snout as a function of time. (After Collatz 1955)

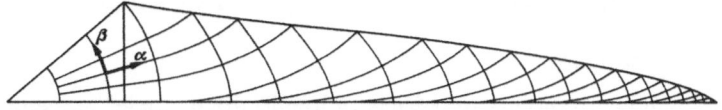

Fig. 109. Nye's plasticity solution for a glacier snout. (After Nye 1967)

Other studies of the mechanics of glacier snouts have been undertaken by Lliboutry (1956). These were based upon the assumption that the glacier behaves as a rheological material. It will be noted that the results obtained upon this basis, as presented in Sect. 7.3.3.3, do not allow for a stationary front to be possible. Lliboutry carried the solution to a further approximation, but the additional terms turned out to be quite negligible. It thus appears that, if the mechanics of ice is to be introduced [and not simply a semi-empirical equation like (7.3.30) is to be used], it is not sufficient to consider only the profile of the glacier snout, but that the effects of the valley sides have to be accounted for. This leads to a three-dimensional problem.

The problem can be solved more easily if, instead of a general flow law, the plasticity theory of ice flow is used. Upon this basis, Nye (1967) produced an equilibrium profile and slip line field of a glacier snout as shown in Fig. 109.

The stress state in the glacier tongue can be used for stability calculations; i.e., for an estimate and the prediction of the breakoff of an ice avalanche (Iken 1977).

7.3.3.6 Transverse Crevasses

We now come to the attempts that have been made at explaining the presence of *crevasses* in glaciers. In the longitudinal case such creavasses are transverse, reaching across the breadth of the glacier.

If we again revert to Nye's theory of plastic flow in glaciers, one may note that the stress solution given for active flow indicates [cf. Eq. (7.3.7)] that the upper layer of the glacier is under tensile stress which stays tensile to a depth d of (according to Nye 1951)

$$d = \frac{2h_0}{\sqrt{1 + 3\sin^2\alpha}}. \tag{7.3.32}$$

This might give rise to transverse fissures to this depth d, if the critical strain rate is exceeded (Holdsworth 1969).

The slip lines mentioned in Sect. 7.3.3.2 may also be associated with crevasses. If an actual displacement occurs along these lines, then it is obvious that this will result in crevasses in the case of active flow. These crevasses will again be transverse as we are dealing here with linear geometry.

Unfortunately, even "transverse" crevasses are often not straight, but crescent-shaped. This can still be handled by a semilinear theory, either by using curved coordinates or by perturbation theory. The procedure is generally one of using finite element calculations (Ott 1985).

7.3.3.7 Geomorphological Effects of Longitudinal Glacier Motion

1. Introduction

It remains to discuss the effects of longitudinal glacier movement on the terrain, with a view of possibly explaining the various features whose origin geomorphologists ascribe to the action of glaciers (cf. Sect. 1.7.3).

2. Valley glaciers

It stands to reason that the *longitudinal* profile of Alpine valleys can be explained by the *longitudinal* motion of glaciers. This concerns the step-shaped longitudinal profile of such valleys. With regard to the latter, we note that Nye's theory (Sect. 7.3.3.2) predicts that erosion should occur at places in the bed where the latter is concave. It thus appears that a glacier tends to deepen any existing hollows in its bed and thus to accentuate the existing relief. This is also the result of the instability principle of landscape evolution (cf. Sect. 5.2.3). The tendency to form stepped longitudinal profiles is thus inherent in valley glaciers (Röthlisberger 1967; Colman 1976). After the melting of the glacier, the hollows will be filled in by debris and the characteristic longitudinal profile of an Alpine valley results. Nye's theory may therefore be said to explain this profile.

Another effect of "longitudinal glacier erosion" is the scooping out (Holtedahl, 1967; Crary 1966) of fyords. As far as such fyords are drowned valleys, it simply is the glacial erosion as described above that might contribute to their deepening. However, the "glacier" may often, in effect, be a floating ice sheet, and then the abrasion will occur at the sides of the inlet. A theory for this process was supplied by Crary (1966).

3. Drumlins

Another feature that might possibly be explained by reference to the longitudinal motion of a glacier is the existence of drumlins (cf. Sect. 1.7.3.2). It has been noted that the form of drumlins is very close to that of a streamlined body (Flint 1957; Charlesworth 1957). Chorley (1959) has elaborated upon this supposition; he made comparisons of shapes of drumlins with that of aerofoils and found a good correspondence. Thus, drumlins are almost certainly streamlined subglacial landforms (Sharp 1984) in deformable beds. This also seems to be the consensus gleaned from a "drumlin symposium" in Manchester (Menzies and Rose 1987).

A further question conserns the reason why drumlins occur in fields. Whilst Smalley and Unwin (1968) regard the drumlin distribution simply as the result of random emplacements, Aario (1977) and Trenhaile (1975a) believe that there is a (negative) correlation with closeness to the ice edge: drumlin elongation and size seem to decline towards the margins of drumlin fields.

4. Roches moutonnées

The explanation given for drumlins has also been applied for roches moutonnées. However, this is almost certainly incorrect: such roches moutonnées consist of hard rock; their genesis must therefore be connected somehow with the *abrasive* action of the ice represented by the shearing stress of sliding ice [cf. Eq. (7.3.21)]. As noted, this stress can provide for "glacial polish" (Hallet 1979; Karlen 1981). The bottom stress of the ice can also cause parabolic tear fractures in the bed (Ficker et al. 1980; Wintges and Heuberger 1980; Wintges 1982). An even more effective rock-attacking force would be provided by the horizontal frontal stress caused in an (elastic) hump due to glacier flow (Morland and Morris 1977). This would be sufficient to cause destruction of the rock; the actual form of the "humps" (roches moutonnées) would again be a consequence of the principle of instability (Sect. 5.2.3).

5. Depositional features

Finally, we come to the depositional features caused by longitudinal glacier flow. These are mainly *moraines* which are simply debris left after the retreat of glaciers.

Special problems occur when advances and retreats of glaciers occur over such moraines. Thus, if a moraine is hit after the retreat of a glacier by its next advance, push-up features result in consequence of the frontal thrust of the ice (Grimmel 1976; Haeberli 1979). On retreat of a glacier from its frontal moraine, the latter may become unstable because of the missing support, and undergo ground-ice slumps (Lewkowicz 1987). An odd and special depositional feature is the formation of perched blocks on glaciers (Patterson 1984).

7.3.4 Three-Dimensional Movement of Ice

7.3.4.1 General Remarks

Thus far, we have dealt only with the longitudinal movement of glaciers. However, the three-dimensional motion of ice is also particular geomorphological significance. Thus, one would like to know what happens at the valley wall of a glacier valley, how a glacial cirque is excavated, and how ice caps spread at their edges. All these problems are three-dimensional.

7.3.4.2 Theories of Three-Dimensional Ice Movement

The mechanics of three-dimensional movements of masses with a complex rheology poses difficult mathematical problems. Reviews of the subject matter have been given, e.g., by Hutter (1982a, 1983).

In order to describe three-dimensional ice movement, one's first thoughts would be to try to generalize Nye's theory of plasticity to this case. However, it becomes obvious that one soon ends up in great mathematical difficulties. Unless the geometry is very simple (such as in circular ice caps where the problem, essentially, reduces itself again to a two-dimensional one), no solutions can be obtained on the basis of plasticity theory or a related theory (cf. Reeh 1982).

Therefore a different approach has been sought after by Matschinski (1958), who attempted to arrive at a differential equation of ice flow from purely logistic considerations. Thus, Matschinski postulated that any useful equation must be linear. Secondly, all functions of the coordinates must be invariants of coordinate transformations. Thirdly, the irreversibility of the ice flow process suggests that only odd-order time derivatives are permitted. Finally, no space derivatives of an order higher than the second should occur.

The above four conditions suffice to set up an equation of ice flow. Let us denote the vertical coordinate of the ice surface by Θ, the position coordinates by x and y. Then it is possible to show that the only possible combination of derivatives satisfying the four assumptions is:

$$\frac{\partial \Theta}{\partial t} + \varepsilon \frac{\partial \operatorname{lap} \Theta}{\partial t} = h \operatorname{lap} \Theta + F, \tag{7.3.33}$$

where h and ε are constants, and F is a function of t, x and y indicating external conditions.

Another "general" approach to the three-dimensional ice flow problem has been based on mass-balance considerations (Hutter 1982b). Otherwise, only numerical computations have met with any success (Yakowitz et al. 1985; Hutter et al. 1986). However, as usual, numerical approaches can deal only with individual cases, never with general features.

7.3.4.3 Ice Caps

A three-dimensional problem of ice movement which is amenable to theoretical treatment is the mechanics of ice caps. This problem, in fact, still reduces itself to a two-dimensional one if the ice cap is assumed to be wide. It has been investigated by Orowan (1951).

Thus, we consider a small element of a wide ice cap which is indicated by shading in Fig. 110. Assuming that the ice cap is spreading forward, and that the bed is perfectly rough, exerting a shear stress k (corresponding to the yield stress in plasticity theory) upon the ice, then the force F per unit width exerted by the bed on the ice is

$$F = k \, dx. \tag{7.3.34}$$

Another force arises from the normal pressures at the vertical faces of the element. This normal pressure decreases from approximately ρgh at the bottom to approximately zero at the surface; thus the average normal pressure is approximately $\frac{1}{2}\rho gh$. Because of the variation of h with x, this results in a radial force R (per unit width) which is

$$R = \tfrac{1}{2}\rho g(h + dh)^2 - \tfrac{1}{2}\rho g h^2 = \rho gh \, dh. \tag{7.3.35}$$

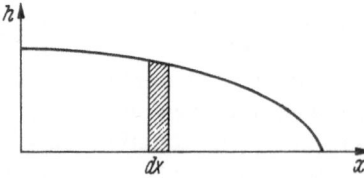

Fig. 110. Geometry of an ice cap

If we now write down the radial force balance equation, we obtain

$$F = k\,dx = \rho gh\,dh = R \tag{7.3.36}$$

or

$$\frac{dh}{dx} = \frac{k}{\rho g}\frac{1}{h} = \frac{h_0}{h} \tag{7.3.37}$$

if we set

$$h_0 = \frac{k}{\rho g}. \tag{7.3.38}$$

The differential equation (7.3.37) can be integrated to yield

$$h = \sqrt{2h_0(L - x)}, \tag{7.3.39}$$

where L is the length of the ice cap. Eq. (7.3.39) represents the equilibrium configuration of an ice cap. The surface cross-section of such an ice cap is represented by a parabola.

The above result is also valid for a circular ice cap (Nye 1952a) as long as the height h is small compared with the radius r where a measurement is made: the profile of a circular ice cap, at least at some distance from its center, is obtained by rotating a parabola around a line orthogonal to its axis. It can, in fact, be shown that both the above results are but special cases of a general theorem referring to an arbitrary ice cap resting upon an arbitrary bottom which is due to Nye (1952b) and which states that (1) in plan view flow takes place where the downward slope of the surface is greatest and that (2) the surface slope S is connected approximately with the ice thickness h at each point by the relation

$$h = h_0/S. \tag{7.3.40}$$

In the case of a wide ice cap, this leads to Eq. (7.3.39).

The above results are based upon an application of plasticity theory. However, it should be noted that it is, in fact, a *simplified* version of plasticity theory that has been used, the yield strength coming into play only at the very bottom of the ice cap. The stress state *within* the cap has not been investigated although it must be presumed that it represents an active Rankine state so that the force balance Eq. (7.3.36) is approximately applicable. Furthermore, the above theory does not give the time dependence of the surface. It must be assumed that conditions are such that somehow a steady state has been reached.

The above simple calculations can be improved upon if an appropriate "shallow ice approximation" (Hutter 1981) is introduced, implying that the vertical dimensions of the ice mass are small compared with the dimensions in the flow directions. The rheological law of ice flow may be that of plasticity or another one; the boundary conditions at the bed may include dry friction (Weertman 1966) or sliding on a basal water layer (Weertman 1961; Lliboutry 1969). Even under such assumptions, the mathematical problem to be solved is a complicated boundary value problem which is tackled by a perturbation solution. A further problem arises in that the ice cap motion may show features of instability (Weertman 1961; Birchfield 1977; Cary et al. 1979; Schubert and Yuen 1982). But even if one restricts oneself to stable solutions, it turns out that the evaluation of ice cap geometry from accumulation and bed form data is not possible. The argument is presented in the book of Hutter (1983) to which the reader is referred for further details.

7.3.4.4 Crevasses

Finally, it is of some interest to investigate the pattern of crevasses in multidimensional problems. We have already discussed crevasses in two-dimensional problems, where the former of necessity cannot be anything but *transverse* features. Here we shall try to see how the problems discussed earlier will be modified by the presence of an additional dimension.

The problem has been investigated long ago from a general standpoint by Hopkins (1862), but we shall follow here a treatment suggested by Nye (1952a) which is based upon his plasticity theory of glacial movement.

Nye's theory applies only to a long glacier, but it does take the effects of the valley wall into account. According to Nye's theory of glacier flow (cf. Sect. 7.3.3.2) the stress σ_x at the surface of the glacier may be compressive (passive flow) or tensile (active flow). In either case, the theory of plasticity predicts that any *normal* stress σ_s caused by the valley wall will be subject to the following inequality (cf. Hill 1950 p 129):

$$0 \lessgtr |\sigma_s| < \tfrac{1}{2}|\sigma_x|. \tag{7.3.41}$$

The only shear component at the surface of the glacier is τ_{sx} which is zero in the center and increases (in the absolute value) towards the sides. If all these stresses are taken together, one obtains for the surface lines across which the greatest tensile stress occurs, the lines shown in Fig. 111. The possible stress states at the margin of the glacier are shown in the small diagrams on top of each main drawing. Which stress state is realized, of course, depends on the interaction between the valley wall and the glacier.

7.3.4.5 Geomorphological Effects

1. General remarks

After having discussed the theory of three-dimensional glacier flow, it remains for us to examine the possibilities by which such flow can explain the glacial effects

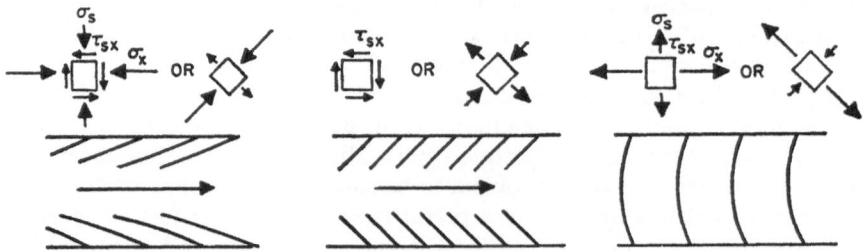

Fig. 111. Theoretical positions of crevasses in a glacier for three possible cases. (After Nye 1952a)

noted by geomorphologists (see Sects, 1.7.3.2 and 1.7.3). Part of these phenomena (viz. the longitudinal profile of Alpine valleys and the shape of drumlins) has already been explained by the theory of longitudinal glacier flow; here we shall investigate what can be said about the remaining features.

The features to be discussed are, on the one hand, *erosional* (cf. Sect. 1.7.3.2) and, on the other hand, *depositional* (cf. Sect. 1.7.3.3). We shall now discuss the individual cases.

2. Erosional effects

a) *The Transverse Profile of Glaciated Valleys.* Referring the reader to Sect. 1.7.3.2, we note that the task is to explain two phenomena: first the existence of a shoulder, and second the U-shape of the central trough.

Generally, it has been assumed that the U-form of a glacial valley is solely the effect of glacial erosion, possibly affecting and overdeepening tectonically predesigned synclinal fracture lines (cf. Schroeder-Lanz and Kinzl 1980; Hirano and Aniya 1988). Corresponding numerical calculations have been made by Harbor et al. (1988). However, Gerber (1945; see also Hantke 1978) maintains that the principle that all glacial valleys are U-shaped, all river valleys V-shaped, is wrong. Accordingly, the U-form is the result of debris accumulation and kame formation against the ice which form the "shoulder" on the "U", and not the result of erosion at all. Tectonic predesign, in fact, may play an *essential* role (Hantke, 1978).

b) *Glacial Cirques.* A further three-dimensional effect of ice motion is the formation of glacial cirques: a preexisting drainage basin or tectonically predesigned hollow represents a highly concave feature in which the erosive action of a glacier, according to Nye's (1965) theory has a maximum effect (cf. Sect. 7.3.3.2). Hence it should be expected that an existing basin should be subject to maximum erosion and should be thereby transformed into a cirque. Qualitatively, this theory has been substantiated by White (1970) and by Trenhaile (1975b); a corresponding mathematical treatment based on Weertman's (1957) sliding motion idea, extended to three-dimensional plastic rotating cirque glacier movement, has been devised by Körner (1983). The latter author gave pertinent stress conditions for the formation of critical spherical slide surfaces.

c) *Deep Erosion by Ice Sheets.* Ice sheets must also be assumed to cause three-dimensional erosional features on their beds. We have already mentioned roches moutonées in Sect. 7.3.3.7, where the latter were considered as two-dimensional features. However, such roches have also been studied as three-dimensional features: thus, Hooke and Iverson (1985) have made an experimental study of bump formation by ice. Whilst there were certain differences observed when a comparison was made with theoretical predictions, a fundamental similarity between experimental and theoretical shapes was confirmed.

Deep erosion by ice sheets has also been invoked as having substantially lowered the surface of shield areas during the Pleistocene, also creating the characteristic hummocky relief (White 1972). However, it is again probable (as with glacial valleys) that the erosion was slight and confined to the material that was already loose (Sugden 1976), and that *tectonic* predesign is paramount.

3. Depositional effects

a) *Till.* The material dumped by a glacier is quite generally called till (cf. Sect. 1.7.3.3). This material is unstratified and simply transported passively by the moving ice (sheet). The transport occurs on top and within the ice; it is simply left when the ice melts. The mechanics of till emplacement is therefore convection of the rock material by the moving ice. Till may also be present subglacially (Mills 1977; Hutter and Olunloyo 1981). It is then affected by Darcian water transport, sediment transport, consolidation, shear deformation, and comminution (Clarke 1987), and is finally dragged along by the glacier to form the characteristic unsorted deposits (Goldthwaite 1971; Van der Meer 1987).

b) *Crevasse-Filled Ridges.* Inasmuch as we have seen that crevasses form in longitudinal glaciers and in ice sheets, the former are liable to be filled by till materials which is again passively carried along and which preserves the crevasse pattern upon melting of the ice representing the aspect of small ridges.

c) *Lateral Moraines.* Like frontal moraines, three-dimensional moraines are also simply debris material left after the retreat of a glacier. The lateral moraines seem to represent possible stream lines of glacier flow. Interesting "lateral" moraine features arise at the confluence of two glaciers (Eyles 1977), leading to medial moraines (Eyles and Rogerson 1978) which trace the three-dimensional flow patterns of the erstwhile glaciers.

d) *Ice-Push Features.* The large-scale ice thrust features mentioned in Sect. 1.7.3.3 are not directly caused by the moving ice sheets, but represent the effect of the latter on preexisting extraglacial permafrost layers; they are thus geocryological in nature (cf. Sect. 7.5.5). However, smaller push features occur in frontal and lateral moraines leading to overthrusted soil profiles (Grimmel 1976).

7.3.5 Theory of Sea and Lake Ice

7.3.5.1 General Remarks

Bodies of water may be wholly or partly covered by ice. The ice may form in situ on the body of the water itself, or it may enter the latter from an expanding glacier or ice cap. In either case, geomorphological effects are caused by floating ice, primarily in nearshore areas, but also further out by the transport of boulders and other sediments to great distances.

A specific type of sediment deposition occurs in the form of varves; the latter are deposited off the edge of retreating sea or lake ice.

7.3.5.2 Nearshore Conditions

The most notable glacial effect in nearshore areas is the "calving" of ice masses into the water. The general problem of calving has been described by Meier and Post (1987), who distinguished between several cases: calving from surging glaciers, calving from "grounded" glaciers, etc. Calving has also beeen frequently observed in nature (e.g., Holdsworth 1971).

The calving of an ice mass into a body of water does not have a geomorphological effect by itself, and thus need not concern us here in detail (calving mechanisms have been treated analytically, e.g., in the monograph of Hutter 1983). Of interest are secondary effects caused by the calving ice masses: these are, on the one hand, surge waves in the body of water and, on the other hand, nearshore sedimentation (deposition of till and moraine material in the beach region).

The surge waves caused by the calving ice in bodies of water (Holdsworth 1973; Haeberli and Röthlisberger 1976) correspond to those caused by land slides and have been treated in that connection (cf. Sect. 6.2.2.5).

Specific sedimentation effects are more characteristic of the ice conditions. Contrary to the graded deposits laid down by turbidity currents, ice-laid deposits are unsorted and correspond to the till on land (Anderson et al. 1980a; Powell 1981).

7.3.5.3 Mechanics of Floating Ice

The mechanics of floating ice sheets is a special branch of glaciology and as such holds much interest in its own right. Reviews of the subject matter have been given by Hutter (1974, 1978; see also his monograph 1983). Applications are mainly important for the calculation of the strength and safety of ice covers on water bodies for the support of people and structures (Röthlisberger 1988).

Of some geomorphological interest are the formation of pressure ridges (Mock et al. 1972; Hibler et al. 1972) and crevasses (MacAyeal et al. 1986) on sea ice surfaces, although these do not refer to the surface proper of the solid earth. True geomorphological effects, however, have been caused by shelf ice: ice lobes extending into the sea have, on occasion, carried and deposited frontal moraines which were later exposed above the water by isostatic uplift (England et al. 1978).

7.3.5.4 Mechanics of Drifting Floes

Next, the motion of drifting floes can have a geomorphological effect.

The drift itself is initiated by wind stress (Smith 1972) and by ocean waves (Kristensen et al. 1982), the latter contributing to the deterioration and breakup of the ice.

The geomorphological action is caused by the transport of glacially eroded sediment (Anderson et al. 1980b) and of individual large boulders (Drake and McCann 1982). For the latter case, the ice-rafting competence can be calculated for possible floes and can be compared with actually observed (Dionne 1979) boulder assemblages.

7.3.5.5 Theory of Varve Formation

We have pointed out (Sect. 1.7.4.4) the existence of varves in periglacial seas. It is held that the thickness of a varve is simply representative of the sediment deposited in one year. Varves are thought to be deposited in glacial basins of the edge of the glacier. Since the glaciers have been receding since the conclusion of the last ice age, the distance from the sediment source (i.e., the edge of the glacier) is progressively increasing with time, creating thereby the thickness decrease of varves observed in varve sequences. A review of the above ideas is contained in a volume by Schlüchter (1978).

The above qualitative physical model has been put upon a quantitative basis by Scheidegger (1965). Accordingly, one assumes a basin of constant depth to be filled with water in a turbulent state. The basin, assumed as subject to a plane geometry, may extend on one side into infinity. On the opposite side is a sediment source of constant strength which slowly retreats into the opposite direction with a constant velocity. Within the basin itself there is a current with a constant offshore velocity component carrying water (and therewith sediment) away from the sediment source. This sediment-carrying water is turbulent. It is assumed that the turbulence is created at the sediment source and that is slowly decays as it is carried along with the current.

The decay laws of turbulence are well known [see Eq. (2.2.13)]. Then the sediment-carrying capacity (n particles per unit volume) of turbulent water with mean velocity fluctuation $\overline{u'^2}$ is given by [cf. Eq. (4.5.8)] with $\varphi \sim \overline{u'^2}$

$$n = C_1 e^{-C_2/\overline{u'^2}} \tag{7.3.42}$$

which yields, using (2.2.13) to eliminate u'

$$n = C_1 e^{-C_3 t^m}, \tag{7.3.43}$$

where C_1, C_2, ... etc. are constants. The rate of sedimentation is the negative derivative of this with regard to t

$$s = C_1 C_3 e^{-C_3 t} \qquad \text{for } m = 1 \text{ (initial range)}$$
$$s = \tfrac{5}{2} C_1 C_3 t^{3/2} e^{-C_3 t^{5/2}} \quad \text{for } m = \tfrac{5}{2} \text{ (terminal range).} \tag{7.3.44}$$

If we remember the assumption that the glacier retreats at a constant speed and that the speed of the offshore current is also constant, the above equations yield directly a quantity proportional to the varve thickness, as a function of time. The corresponding curves, plotted on semilog paper, yield almost straight lines which can be compared with the varve thickness decrease relations discussed in Sect. 1.7.4.4. There is evidently a good agreement between theory and observation. The only thing is that modifications may have to be introduced to account for stochastic variations (Agterberg and Bannerjie 1969), bottom-water currents (Gravenor and Coyle 1985), and possibly even for the action of bacteria (Dickman 1979).

7.4 Theory of Glaciohydrological Effects

7.4.1 Introduction

Glaciers (and ice sheets) interact with liquid (running) water at their surface, in their interior and on their bed, as well as in the "periglacial" region adjacent to the edge of the ice. A notable survey of glacial hydrology has been given by Röthlisberger

and Lang (1987). The glacial hydrological effects give rise to specific geomorphological features.

7.4.2 Supraglacial Flow

Supraglacial flow is mostly ephemeral and seasonal, caused by the warm-weather
melting in the summer. However, of greater interest are the perennial supraglacial
melt streams which typically exhibit meandering.

For such streams, channel incision depths can be explained by a simple model
taking ablation and melting into account (Ferguson 1973). Meander wavelength is
determined by various fluviohydrological variables, such as channel width, depth,
etc. (Parker 1975). Thus, conditions are much the same as in meandering plains
rivers, except that the interaction with the bed is through thermodynamic
processes on the ice rather than through the mechanical particle entrainment
known from rivers on land.

7.4.3 Intraglacial and Subglacial Drainage

7.4.3.1 General Theory

The drainage within and under a glacier is a subject of great complexity.

First, there is the seepage of the water through the glacier ice. Next, the
intraglacial drainage system forms an arborescent system of conduits leading to
the glacier bed; the conduits join together and form finally one or a few subglacial
tunnels which leave the glacier at a portal (Röthlisberger 1972; Röthlisberger and
Lang 1987).

The problem of water seepage through ice (and snow) has been studied by
Schommer (1978) and by Hantz and Lliboutry (1983). These authors achieved
reasonable correspondence with observations by the application of the theory of
flow through porous media with a free surface (cf. Sect. 2.3.2.3 and references given
there). In fact, Lliboutry (1983) objects to *any* other type of intraglacial flow, i.e., to
the arborescent conduit system of Röthlisberger mentioned above. According to
Lliboutry, channels can only form at the bottom of the ice.

Flow in ice conduits (where it is immaterial whether these are intra- or
subglacial) has been studied analytically by Röthlisberger (1972), Spring (1980),
and Spring and Hutter (1982), where the last authors produced a paper which is
mathematically most complete. Their treatment accounts for the balance of mass,
momentum, angular momentum, and energy, and ends up with a complicated set
of differential field equations that have to be solved for specific boundary
conditions. Compared with intraglacial conduits, the boundary conditions are
somewhat different for subglacial conduits (Röthlisberger and Lang 1987).
However, above a *permeable* bed such as till, the water in the ground and the
subglacial water may form a single aquifer whose pore pressure may equal the
overburden pressure of the ice (Shoemaker and Leung 1987). In that case, sheet
flow will occur along the bed.

7.4.3.2 Geomorphological Effects: Eskers

The subglacial tunneling system is commonly assumed to be the cause of the genesis of eskers. Indeed, a study of esker characteristics in terms of glacier physics has been made by Shreve (1985).

Accordingly, the formation of eskers occurs when the water pressure in subglacial tunnels approximates the weight of the ice overburden and when at the same time strong melting entrains debris into them. The trend of the eskers follows the sinuous course of the erstwhile subglacial channels. The sediment size distribution diagrams of esker sands have been noted to fall into the "fluvial" or "outwash" categories of sedimentation from water currents (Saunderson 1977a); the poor sorting of the sediments indicates fully developed tunnel flow (Saunderson 1977b). Such observations confirm the hypothesized tunnel origin of eskers.

7.4.4 Periglacial Runoff

7.4.4.1 General Features

The periglacial runoff is nothing but a fluvial process conditioned by glaciation (Church and Ryder 1972). Periglacial rivers are commonly braided (Ashworth and Ferguson 1986), entailing corresponding hydraulic relationships (cf. Sect. 4.7.6.3). As an additional feature, the discharge may be flood-like in consequence of a glacial (mini-) surge (Humphrey et al. 1986).

7.4.4.2 Periglacial Sedimentation

The periglacial runoff causes periglacial sedimentation.

The most obvious sedimentation effect is the formation of drift, which consists of layered strata in contrast to the unsorted till deposited by the ice. The drift sedimentation process corresponds entirely to that caused by rivers and sheet floods.

Some "drumlins" (Dardis 1985a; Dardis and McCabe 1987) and "moraines" (Dardis 1985b) may also have been affected by periglacial runoff and not only by glacial action: the stratified layers of sand, silt, and clay found in such features are characteristic of aqueous deposition. Similarly, kames are taken to have been caused by periglacial runoff: they are deposits laid by water against an ice margin, hence their asymmetrical (cf. Sect. 1.7.4.3) structure.

7.4.5 Glacial Lakes

Bodies of water can be dammed up by ice within and adjacent to glaciers. A particular feature of such bodies is that they are prone to periodic outbursts and floods because of the rheological instabilities occurring in the ice retaining them.

Because of the considerable danger represented by such floods, their occurrence has been studied. First of all, there is the possibility of monitoring the occurrence of floods statistically, like, e.g., that of severe weather events (Haeberli 1983). A more satisfactory procedure would be to attempt to set up dynamic conditions for the stability of the ice retaining the water, an approach, however, which owing to the complexity of the ice-water interaction does not seem to have been feasible to date. In fact, the initiation of glacier lake bursts is a subject of considerable controversy (cf. the discussion in Röthlisberger and Lang 1987).

The next step is that of calculating the evolution of the water burst through the ice channels if it *does* occur (Spring and Hutter 1981); for this, the ice conduit flow theory mentioned in Sect. 7.4.3.1 can be used. The final step is the calculation of the hydrograph equation for the flood event.

The geomorphological effects of such floods are similar to those of any type of flood. In particular, the role played by a glacial flood in the formation of Washington State Scablands has been explained in Sect. 3.4.4.3.

7.5 Theory of Geocryological Features

7.5.1 Introduction

Our final task in this chapter is to give an explanation of the genesis of features that are due to ground freezing effects. These include those features whose morphology has been described in Sect. 1.7.5, as well as the ice-pushed ridges described in Sect. 1.7.3.3.

A variety of theories have been proposed for an explanation of the occurrence of the phenomena under discussion. Unfortunately, as will be demonstrated below, these theories are only very qualitative and no numerical tests for their predictions seem ever to have been undertaken. The investigation of niveal effects has thus obviously not yet been brought to a close.

7.5.2 Periglacial Patterned Ground

7.5.2.1 General Remarks

We have mentioned in Sect. 1.7.5 that polygon patterns occur in the ground in periglacial areas. In fact, one has to distinguish between polygons formed by cracks, polygons formed by the sorting of stones, and polygons formed by ridging in soils. The genesis of these various features appears to be different.

7.5.2.2 Crack Polygons

The most common ground patterns in periglacial areas are crack polygons.

The genesis of crack polygons has been ascribed to contraction stress patterns by Leffingwell (1915). This hypothesis was examined from a mechanical standpoint by Lachenbruch (1962). Accordingly, small vertical fractures form in the frozen Arctic tundra in the winter owing to thermal contraction of the tundra surface.

Then, in the spring, water from the melting snow forms ice wedges in these cracks in the permafrost, giving rise to a cycle that repeats itself year after year.

The mechanical theory of Lachenbruch analyzes the mechanics of fracture, the stress before and after the crack formation, and also the multiple fracture patterns in a two-dimensional medium. Lachenbruch shows that polygons in patterned ground in permafrost can indeed be explained as contraction crack polygons formed in many media due to a decrease in volume. In permafrost regions, this contraction is due to cooling in the Arctic winter.

7.5.2.3 Sorting of Stones

Periglacial ground patterns are, on occasion, also formed by stones sorted into polygons and rings (e.g., Hallet and Prestrud 1986).

Explanations of such occurrences have been sought in the unstable growth of ice lenses (Kowalowski 1984) with attendant uneven upfreezing of clasts (Anderson 1988). Alternatively, convection cells driven by unstable density stratification in the aqueous phase and resulting in uneven melting of the underlying ice have been advocated for the sorting process (Ray et al. 1983). Most elegantly, Ahnert (1981) has ascribed the origination of patterns simply to a self-impeding random walk process (cf. the traffic-jam theory of river bed form formation presented in Sect. 4.6.2.4) during the upward migration of the stones during the freezing-thawing cycles.

7.5.2.4 Ridge Patterns in Soils

Finally, patterns also occur in periglacial soils (mainly in peat and muskeg) in form of ridges.

The development of such features is commonly ascribed to drainage impediment and resulting hydrological instabilities which lead to a ridge-and-pool system (Foster et al. 1983).

7.5.3 Slope Processes: Rock Glaciers

Slope processes in periglacial conditions (see review by Rapp 1986) are mainly initiated by the creep of seasonally and permanently frozen ground, i.e., by niveal solifluction. The theory of the latter has already been treated in Sect. 3.3.5.5 in connection with slopes in general.

A special type of feature on periglacial slopes are rock glaciers: these are large masses of supersaturated permafrost layers (Haeberli 1985). Their dynamics is still not entirely understood. Estimates of free water available, required pressures, and force balances have been made. For the actual motion, computer simulations have been made assuming a viscous or pseudoviscous rheology (Olyphant 1983); alternatively, the movement has been thought to be caused by hydrostatic overpressure (Giardino 1983). The problem has obviously not yet been solved.

7.5.4 Frost Heave Phenomena

7.5.4.1 General Remarks

As noted in Sect. 1.7.5.4, geomorphological frost-heave phenomena occur in the form of the larger-scale pingos and in form of smaller-scale palsas and similar features. These will have to be dealt with separately.

7.5.4.2 Pingos

Although the physiography of any one pingo appears superficially very much the same (cf. Sect. 1.7.5.4) as that of any other, there are indications that one must, in fact, discern between two genetically different types. This point has been particularly stressed by Müller (1959), who published a very extensive study of pingos.

According to Müller, the first pingo type which occurs mainly in Greenland (and hence has been called Greenland type) is an open ice-water system. Besides an ice lens, the interior of the pingo may contain unfrozen water. Isotopic analyses have yielded the result that the water is neither juvenile nor ancient, but identical to surface water. Therefore, Müller reasoned that pingos of the Greenland type are primarily caused by an artesian effect. The forces active in the genesis of a Greenland type pingo (as envisaged by Müller) are shown in Fig. 112.

The presence of springs in pingos certainly lends some credibility to Müller's artesian hypothesis. However, it is not certain whether the suggested mechanism is thermodynamically sound. Since pingos occur only above permanently frozen ground, and since the melting temperature of ice is lower in a porous system (the soil) than in a large container, an artesian tube contain water cannot exist side by side with frozen ground. Drillings do not seem to have disclosed any water

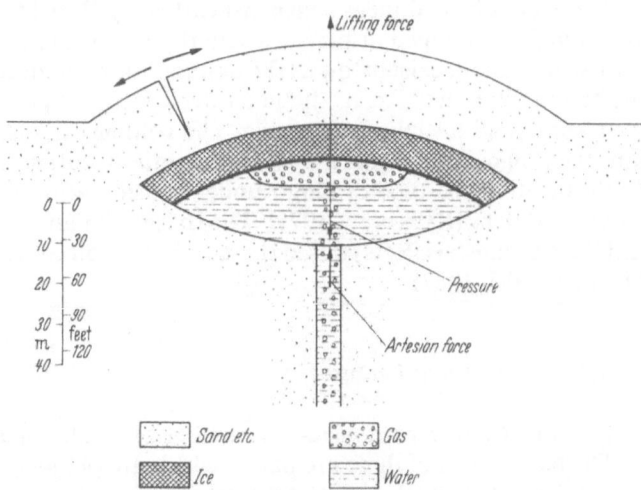

Fig. 112. Mechanism of a Greenland type pingo. (After Muller 1959)

chambers below the ice lens in a pingo, since the water appears to have been found somewhere near the top. The mechanism suggested by Müller, therefore, appears as physically questionable.

The same may be said for other attempts at an explanation of the genesis of Greenland-type pingos. Svetosarov (1934) thought that pingos are caused by the emergence juvenile water and Gussow (1954) proposed that they are remnants of Pleistocene ice masses. Both these hypotheses do not agree with the observed isotopic composition of water and ice collected from these pingos.

In view of the above remarks, the writer suggests that Greenland-type pingos are indeed thermodynamically open systems. However, the water oozing out does not necessarily have to come from below the ice lens. The phenomenon can be regarded as analogous to a giant frost boil caused by the mechanism of water freezing in a porous medium (cf. Sect. 2.3.4.3). The freezing temperature in a porous system (soil) is lower than in bulk masses of water. Hence, once an ice lens has been started, it will grow by the addition of water which is in thermal equilibrium with the ice in the frozen ground, but which is supercooled as soon as it leaves the porous medium. Since there is a general temperature gradient near the surface by which the temperature (in summer) increases upward, ice will melt at the top of the ice lens. Thus, water in the form of ice will move upward through the ice lens. If the process is equilibriated over the seasons, then the pingo will retain its size. Otherwise it will grow or shrink so as to achieve an energy balance. The melted water may escape immediately at the top of the pingo or collect in chambers inside the ice (but not below the ice lens), where it may be confined so as to be subject to relatively high pressures which, upon release, may give rise to a superficially artesian phenomenon.

The second type of pingos observed by Müller (1959) has been termed *Mackenzie type* by him. The structure of these pingos (with a central ice lens) is almost identical to that of the Greenland-type pingos, but there is evidence that the ice lens was formed at a definite time, up to 28 000 years ago. There is also evidence that the ice was at one time water which contained vegetation. This prompted Müller, who followed ideas suggested earlier by Porsild (1938), to postulate that these pingos developed at places where there was a lake at one time. In general, permafrost cannot exist below a lake of some 300 m or more in diameter. As silt and vegetation fill in the lake, its diameter decreases to a point where permafrost can exist below, and the lake freezes over. The volume expansion of the freezing water causes a pingo. Even if this model is not entirely correct in its details, there is no doubt but that the type of pingo under consideration is caused by an upward-growing ice lens (Ryckborst 1975). The thermodynamics of the growing ice mass and the mechanical effects caused thereby on the soil cover have been calculated by Mackay (1985, 1987).

7.5.4.3 Smaller-Scale Features

Under this heading we consider a number of small-scale periglacial features.

The best known of these are palsas which are pingo-type structures in bogs (cf. Sect. 1.7.5.4). Inasmuch as they contain, like pingos, a core of frozen ice, the theory of their genesis must be similar.

Further features are mud boils which are roundish (1–3 m in diameter) bare soil patches which form on perennially frozen till (Shilts 1978). They seem to be activated by cryostatic pressures induced above the frozen layers (Mackay and MacKay 1976); in a quiescent state they appear as frost hummocks (Mackay 1980).

Seasonal frost mounds are common in periglacial areas (Pollard and French 1984). Their dynamics is described by the theory outlined in Sect. 2.3.4.3.

Finally, cavities arise during the liquid stage of periodic or erstwhile ground frost. It is well known that Pleistocene "dead ice" has given rise to large present depressions and also to smaller "kettle holes" (Clark 1969). On a smaller scale, Pleistocene frost heaving has caused hollows that are filled with peat in areas that are now ice-free (Horn and Semmel 1985).

7.5.5 Ice-Thrust Features

In our discussion of physical geomorphology we have mentioned niveal pressure features (see Sect. 1.7.3.3). The generally accepted theory of the genesis of these features assumes that a thick permafrost layer existed ahead of the advance of the ice. Upon breaking, parts of this layer were pushed forward by the advancing ice over the permafrost below. This process produced the ridges (Rutten 1960; Van der Wateren 1985). Rutten noted that, for the establishment of the required permafrost layer, the drainage pattern has to be *toward* the advancing ice. Since the drainage in the plains of the United States (as opposed to the drainage in Canada) was southward, away from the ice, this explains the absence of ridges in that country.

On a small-scale basis, ice push also caused thrust features in moraines, as well as ephemeral ridges in the periglacial areas of today (Mackay and MacKay 1977; Oerlemans 1984). Particularly noteworthy features of this type are caused by *surging* glaciers (Sharp 1988b).

Furthermore, the thrust of the ice stripped the sediments off the Canadian and Fennoscandian Shields (Aber 1982).

References

Aario, R.: Geo J. 1(6), 65 (1977)
Aber, J.S.: Bull. Geol. Soc. Denmark 30, 79 (1982)
Ackroyd, P.: Z. Geomorph. 30(1), 1 (1986)
Agterbers, F.P. and I. Banerjee: Can. J. Earth Sci. 6(4), 625 (1969)
Ahneri, F.: Trans. Jpn. Geom. Un. 2(2), 301 (1981)
Anderson, J.B. and 4 others: Geol. 88, 399 (1980a)
Anderson, J.B., E.W. Dormack and D.D. Kurtz: J. Glaciol. 25, 387 (1980b)
Anderson, S.P.: Bull. Geol. Soc. Am. 100, 609 (1988)
Ashworth, P.J. and R.I. Ferguson: Geogr. Ann. 68A(4), 361 (1986)
Bakkehoi, S., U. Domaas and K. Lied: Ann. Glaciol. 4, 24 (1983)
Balise, M.J. and C.F. Raymond: J. Glaciol. 31 (1985)
Birchfield, G.E.: J. Geoph. Res. 82(31), 4909 (1977)
Birnie, R.V.: J. Glaciol. 18(78), 77 (1977)
Budd, W. F.: J. Glaciol. 9, 29 (1970)
Cary, P.W., G.K.C. Clarke and W.R. Peltier: Can. J. Earth Sci. 16(1), 182 (1979)

Charlesworth, J.K.: The Quaternary era with special reference to its glaciation. London: Arnold (1957)
Chorley, R.J.: J. Glaciol. 3, 339 (1959)
Church, M. and J.M. Ryder: Bull. Geol. Soc. Am. 83(10), 3059 (1972)
Clark, M.J., and M. Seppälä: Arctic and Alpine Res. 20(1), 97 (1988)
Clark, R.P.K.: J. Glaciol. 8, 485 (1969)
Clarke, G.K.: J. Glaciol. 16, 231 (1976)
Clarke, G.K.: J. Geoph. Res. 92(89), 9023 (1987)
Clarke, G.K., U. Nitsan and W.S.B. Paterson: Rev. Geoph. Space Phys. 15(2), 235 (1977)
Collatz, L.: Numerische Behandlung von Differentialgleichungen. 2nd ed. Berlin Göttingen Heidelberg: Springer (1955) See p. 288
Colman, S.M.: Z. Geomorph. 20(3), 297 (1976)
Conway, H. and J. Abrahamson: J. Glaciol. 34(117), 170 (1988)
Crary, A.P.: Bull. Geol. Soc. Am. 77, 911 (1966)
Daly, C.: Progr. Phys. Geogr. 8(2), 157 (1984)
Dardis, G.F.: Geogr. Ann. 67A, 13 (1985a)
Dardis, G.F.: Earth Surf. Proc. and Landf. 10, 483 (1985b)
Dardis, G.F. and A.M. McCabe: In: Drumlin symposium, eds. J. Menzies and J. Rose, p. 225. Rotterdam: Balkema (1987)
De Quervain, M.: Z. Gletscherk. Glazialgeol. 9(1–2), 189 (1973)
Dickman, M.D.: Quat, Res. 11, 113 (1979)
Dionne, J.C.: Maritime Sed. 15, 5 (1979)
Drake, J.J. and S.B. McCann: Can. J. Earth Sci. 19, 748 (1982)
Engelhardt, H.: Geowiss. uns. Zeit. 5(6), 212 (1987)
England, J., R.S. Bradley and G.H. Miller: J. Glaciol. 20, 393 (1978)
Eyles, N.: Can. J. Earth Sci. 14(12), 2807 (1977)
Eyles, N. and R.J. Rogerson: J. Glaciol. 20, 99 (1978)
Ferguson, R.I.: Bull. Geol. Soc. Am. 84, 251 (1973)
Fermi, E.: Thermodynamics. New York: Prentice-Hall (1937)
Ficker, E., G. Sonntag and E. Weber: Z. Gletscherk. Glazialgeol. 16(2), 25 (1980)
Finsterwalder, S.: Z. Gletscherkd. 2, 81 (1907)
Fitzharris, B.B. and I.F. Owens: J. Glaciol. 30(106), 308 (1984)
Flint, R.F.: Glacial and Pleistocene geology. New York: J. Wiley & Sons (1957)
Foster, D.R. and 3 others: Nature 306, 257 (1983)
Fowler, A.C.: J. Geoph. Res. 92(89), 9111 (1987)
Frontard, M.: C.R. Acad. Sci. (Paris) 174, 526 (1922)
Gardner, J.: Arctic and Alpine Res. 2(2), 135 (1970)
Gerber, E.: Mitt. Aarg. Natf. Ges. 22, 1 (1945)
Giardino, J.R.: Z. Geomorph. 27(3), 297 (1983)
Goldthwaite, R.P. (ed.): Till: a symposium. Columbus: Ohio State Univ. Press (1971)
Gravenor, C.P. and D.A. Coyle: Can. J. Earth Sci. 22, 291 (1985)
Grimmel, E.: Eiszeitalter u. Gegenw. 27, 69 (1976)
Gussow, W.C.: Bull. Am. Assoc. Petrol. Geol. 38, 2225 (1954)
Haeberli, W.: Z. Gletscherk. Glazialgeol. 11(2), 203 (1975)
Haeberli, W.: Geogr. Ann. 61A(1–2), 43 (1979)
Haeberli, W.: Ann. Glaciol. 4, 85 (1983)
Haeberli, W.: Mitt. Vers.-Anst. Wasserbau & c. ETH Zürich 77, 1(1985)
Haeberli, W. and H. Röthlisberger: Z. Gletscherk. u. Glazialgeol. 11(2), 221 (1976)
Haeberli, W., W. Schmid and D. Wagenbach: Z. Gletscherkunde 24(1), 1 (1988)
Haeberli, W. and 3 others: In: Glacier fluctuations and climatic change, ed. J. Oerlemans, p. 77, Amsterdam: Kluwer (1989)
Haefeli, R. and H.v. Sury : Publ. Int. Assoc. Sci. Hydrol. 114, 342 (1975)
Hallet, B.: J. Glaciol. 23, 39 (1979)
Hallet, B. and S. Prestrud: Quat. Res. 26, 81 (1986)
Hantke, R.: Eiszeitalter Bd. 1, p. 70. Thun: Ott (1978)
Hantz, D. and L. Lliboutry: J. Glaciol. 29, 227 (1983)
Harbor, J.M., B. Hallet and C.F. Raymond: Nature 333, 347 (1988)
Hestnes, E.: Ann. Glaciol. 6, 1 (1985)
Hibler, W.D., W.F. Weeks and S.J. Mock: J. Geoph. Res. 77(30), 5954 (1972)

Hill, R.: Mathematical theory of plasticity. Oxford: Clarendon Press (1950)
Hirano, M. and M. Aniya: Earth Surf. Proc. and Landf. 13(8), 707 (1988)
Holdsworth, G.: J. Glaciol. 8, 107 (1969)
Holdsworth, G.: Can. J. Earth Sci. 8(2), 299 (1971)
Holdsworth, G.: J. Glaciol. 12, 235 (1973)
Holtedahl, H.: Geogr. Ann. 49A, 188 (1967)
Holzhauser, H.: Geogr. Helv. 37(2), 115 (1982)
Hooke, R.L. and N.R. Iverson: Geogr. Ann. 67A(3–4), 187 (1985)
Hope, J.: Mitt. Forstl. Bundesvers.-Anst. Wien 159, 267 (1988)
Hopkins, W.: Philos. Trans. R. Soc. 152(2), 677 (1862)
Horn, M. and A. Semmel: Geol. Jb. Hessen 113, 83 (1985)
Huber, T.P.: Earth Surf. Proc. Landf. 7(2), 109 (1982)
Humphrey, N., C. Raymond and W. Harrison: J. Glaciol. 32, 195 (1986)
Hutter, K.: Mitt. Vers. Anst. Wasserb., Hydrol. Glaziol. ETH Zürich 11, 1 (1974)
Hutter, K.: Mitt. Vers. Anst. Wasserb. Hydrol. Glaziol. ETH Zürich 28, 1 (1978)
Hutter, K.: J. Glaciol. 27, 39 (1981)
Hutter, K.: Ann. Rev. Fluid Mech. 14, 87 (1982a)
Hutter, K.: Geoph. Astroph. Fluid Dyn. 21, 201 (1982b)
Hutter, K.: Theoretical glaciology. Dordrecht: Reidel (1983)
Hutter, K. and V.O.S. Olunloyo: Ann. Glaciol. 2, 29 (1981)
Hutter, K., S. Yakowitz and F. Szidarowsky: J. Glaciol. 32, 139 (1986)
Iken, A.: J. Glaciol. 19, 595 (1977)
Iken, A.: J. Glaciol. 27, 407 (1981)
Kamb, B.: Rev. Geoph. Space Phys. 8(4), 673 (1970)
Karlen, W.: J. Glaciol. 27, 190 (1981)
Kasser, P. and M. Allen: Houille Blanche 1976 (6/7), 467 (1976)
Komar, P.D.: J. Geol. 92, 133 (1984)
Körner, H.J.: Z. Gletscherk. Glazialgeol. 19(2), 103 (1983)
Kowalowski, A.: Quat. Stud. Poland 5, 75 (1984)
Kristensen, M., V.A. Squire and S.C. Moore: Nature 297, 669 (1982)
Laatsch, W., B. Zenke and J. Dankerl: Forstl. Forsch.-ber. München (Univ.) 47, 9 (1981)
Lachenbruch, A.H.: Spec. Pap. Geol. Soc. Am. 70, 1 (1962)
Lang, H. and G. Patzelt: Z. Gletscherk. Glazialgeol. 7(1–2), 39 (1971)
Leffingwell, E. De K.: J. Geol. 23, 635 (1915)
Lewkowicz, A.G.: Can. J. Earth Sci. 24(6), 1077 (1987)
Lied, K. and S. Bakkehoi: J. Glaciol. 26(94), 165 (1980)
Lliboutry, L.: Ann. Géoph. (Paris) 12(4), 245 (1956)
Lliboutry, L.: Ann. Géoph. (Paris) 15(2), 250 (1959)
Lliboutry, L.: Publ. Assoc. Int. Hydrol. Sci. 79, 33 (1967)
Lliboutry, L.A.: Sci. J. 5(4), 51 (1969)
Lliboutry, L.: J. Glaciol. 23, 67 (1979)
Lliboutry, L.: J. Glaciol. 29, 216 (1983)
Lliboutry, L.: J. Geoph. Res. 92(B9), 9101 (1987)
MacAyeal, D.R. and 3 others: J. Geoph. Res. 91(B8), 6177 (1986)
Mackay, J.R.: Geogr. Bull. 7(1), 1 (1965)
Mackay, J.R.: Can. J. Earth Sci. 17(8), 996 (1980)
Mackay, J.R.: Can. J. Earth Sci. 22(10), 1452 (1985)
Mackay, J.R.: Can. J. Earth Sci. 24, 1108 (1987)
Mackay, J.R. and D.K. MacKay: Can. J. Earth Sci. 13(7), 889 (1976)
Mackay, J.R. and D.K. MacKay: Can. J. Earth Sci. 14(10), 2213 (1977)
Martin, S.: Z. Gletscherk. Glazialgeol. 10, 89 (1974)
Martin, S.: Z. Gletscherk. Glazialgeol. 13(1–2), 127 (1978)
Matschinski, M.: Publ. Assoc. Hydrol. Sci. 47, 213 (1958)
McClung, D.M., J.O. Larsen and S.B. Hansen: Can. Geotech. J. 21, 250 (1984)
McClung, D.M. and K. Lied: Cold Reg. Sci. Technol. 13(2), 107 (1987)
Meier, M.F. and A. Post: J. Geoph. Res. 92(B9) 9051 (1987)
Menzies, J. and J. Rose (ed.): Drumlin symposium Manchester 1985. Rotterdam: Balkema (1987)
Mills, H.H.: Bull. Geol. Soc. Am. 88(6), 824 (1977)

Mock, S.J., A.D. Hartwell and W.D. Hibler: J. Geoph. Res. 77(30), 5945 (1972)
Morland, L.W.: J. Glaciol. 17, 447 (1976)
Morland, L.W. and E.M. Morris: J. Glaciol. 18, 67 (1977)
Müller, F.: Medd. Groenl. 153(3), 1 (1959)
Müller, P.: Mitt.-Vers.-anst. Wasserb., Hydrol. u. Glaziol. ETH Zürich 95, 3 (1988)
Nye, J.F.: Proc. Roy. Soc. A 207, 554 (1951)
Nye, J.F.: J. Glaciol. 2, 82 (1952a)
Nye, J.F.: Nature 169, 529 (1952b)
Nye, J.F.: J. Glaciol 2, 512 (1955)
Nye, J.F.: Proc. Roy. Soc. A 239, 113 (1957)
Nye, J.F.: Proc. Roy. Soc. A 275, 87 (1963a)
Nye, J.F.: Geoph. J. Roy. Astron. Soc. 7, 431 (1963b)
Nye, J.F.: J. Glaciol. 5, 661 (1965)
Nye, J.F.: J. Glaciol. 6, 695 (1967)
Oerlemans, J.: Z. Gletscherk. Glazialgeol. 20, 107 (1984)
Olyphant, G.A.: Bull. Geol. Soc. Am. 94, 499 (1983)
Onesti, L.J.: Ann. Glaciol. 6, 23 (1985)
Orowan, E.: J. Glaciol. 1(5), 231 (1951)
Ott, B.: Mitt. Vers.-Anst. Wasserbau & c., ETH Zürich 80, 10 (1985)
Parker, G.: Water Resour. Res. 11(4), 551 (1975)
Paterson, W.S.B.: The physics of glaciers. Oxford: Pergamon (1969)
Patterson, E.A.: J. Glaciol. 30, 296 (1984)
Pollard, W.H. and H.M. French: Can. J. Earth Sci. 21(10), 1073 (1984)
Porsild, A.E.: Geogr. Rev. 28(1), 46 (1938)
Powell, R.D.: Ann. Glaciol. 2, 129 (1981)
Prandtl, L.: Z. Angew. Math. Mech. 3, 401 (1923)
Rapp, A.: Progr. Phys. Geogr. 10(1), 53 (1986)
Ray, R.J. and 3 others: J. Glaciol. 29, 317 (1983)
Raymond, C.F.: J. Geoph. Res. 92(B9), 9121 (1987)
Reeh, N.: J. Glaciol. 28, 431 (1982)
Reynaud, L.: Z. Gletscherk. Glazialgeol. 13(1–2), 155 (1977)
Röthlisberger, H.: Proc. General. Ass. Bern, Int. Ass. Sci. Hydrol., Vol. Snow and ice p. 87 (1967)
Röthlisberger, H.: J. Glaciol. 11, 177 (1972)
Röthlisberger, H.: Tragverhalten von Eis. Basel: Schweiz. Lebensrettungsgesellschaft (1988)
Röthlisberger, H. and H. Lang: in Glacio-fluvial sediment transfer, eds. A.M. Gurnell and M.J Clark,
     p. 207. New York: Wiley (1987)
Rutten, M.G.: Am. J. Sci. 258, 293 (1960)
Ryckborst, H.: J. Hydrol. 26, 303 (1975)
Saunderson, H.C.: Z. Geomorph. 21(1), 44 (1977a)
Saunderson, H.C.: Sedimentology 24, 623 (1977b)
Scheidegger, A.E.: Bull. Int. Assoc. Sci. Hydrol. 10(1), 68 (1965)
Scheidegger, A.E.: Physical aspects of natural catastrophes. Amsterdam: Elsevier (1975)
Scheiwiller, T.: Mitt. Vers.-Anst. Wasserb., Hydrol., Glaziol. ETH Zürich 81, 1 (1986)
Scheiwiller, T., and K. Hutter: Mitt. Vers.-Anst. Wasserb., Hydrol., Glaziol. ETH Zürich 58, 1 (1982)
Scheiwiller, T. and K. Hutter: J. Glaciol. 29(102), 283 (1983)
Scheiwiller, T., K. Hutter and F. Hermann: Ann. Geoph. 5B(6), 569 (1987)
Schlüchter, C. (ed.): Varves and moraines. Rotterdam: Balkema (1978)
Schmidt, R.A.: Rev. Geoph. Space Phys. 20(1), 39 (1982)
Schommer, P.: Z. Gletscherk. Glazialgeol. 14(2), 173 (1978)
Schroeder-Lanz, H. and H. Kinzl: In: Colloquium Trier 1980.05. 15—17, p. 83, Rotterdam: Balkema
     (1980)
Schubert, G. and D.A. Yuen: Nature 296, 127 (1982)
Sharp, M.: Prog. Phys. Geogr. 8(2), 249 (1984)
Sharp, M.: Progr. Phys. Geogr. 12, 349 (1988a)
Sharp, M.: Progr. Phys. Geogr. 12, 533 (1988b)
Shilts, W.W.: Can. J. Earth Sci. 15(7), 1053 (1978)
Shoemaker, E.M. and H.K.N. Leung: J. Geoph. Res. 92(B6), 4935 (1987)
Shreve, R.L.: Bull. Geol. Soc. Am. 96, 639 (1985)

Smalley, I.J. and D.J. Unwin: J. Glaciol. 7, 377 (1968)
Smith, S.D.: J. Geoph. Res. 77(21), 3886 (1972)
Spring, U.: Mitt. Vers.-anst. Wasserb & c. ETH Zürich 48, 3 (1980)
Spring, U. and K. Hutter: Cold Reg. Sci and Technol. 4, 227 (1981)
Spring, U. and K. Hutter: Int. J. Eng. Sci. 20(2), 327 (1982)
Sugden, D.E.: Geology 4(10), 580 (1976)
Svetosarov, I.M.: Probl. Sovet. Geol. 4(10), 119 (1934)
Tollner, H.: Alpenver.-JB. 99, 101 (1974)
Trenhaile, A.S.: Ann. Assoc. Am. Geogr. 65(2), 297 (1975a)
Trenhaile, A.S.: Ann. Assoc. Am. Geogr. 65(4), 517 (1975b)
Vallon, L.: Z. Gletscherk. Glazialgeol. 13(1–2), 57 (1977)
Van der Meer, J.J.M. (ed.): Tills and glaciotectonics, Rotterdam: Balkema (1987)
Van der Wateren, D.F.M.: Bull. Geol. Soc. Denmark 34, 55 (1985)
Vischer, D. (ed.): Mitt.-Vers. Anst. Wasserb., Hydrol. Glaziol. ETH Zürich 94, 1 (1988)
Voellmy, A.: Schweiz, Bauztg. 73(12), 159 (1955)
Washburn, A.L.: Geocryology: a survey of periglacial processes and environments, 2nd edn. New York: Wiley (1980)
Weertman, J.: J. Glaciol. 3, 33 (1957)
Weertman, J.: J. Geoph. Res. 66, 3783 (1961)
Weertman, J.: J. Glaciol. 5, 287 (1964)
Weertman, J.: J. Glaciol. 6, 191 (1966)
White, W.A.: J. Geol. 78, 123 (1970)
White, W.A.: Bull. Geol. Soc. Am. 83, 1037 (1972)
Wintges, T.: Z. Geomorph. Suppl. 43, 161 (1982)
Wintges, T. and H. Heuberger: Z. Gletscherk. Glazialgeol. 16(2), 157 (1980)
Yakowitz, S., K. Hutter and F. Szidarowsky: Z. Gletscherk. Glaziolgeol. 21, 283 (1985)

8 Theory of Aeolian and Desert Features

8.1 Introduction

This, the last chapter of the book, will be concerned with geomorphological forms which are characteristic of the arid and semiarid regions of the world. In this, the layout of the chapter will follow closely and will be keyed to the exposition of the morphological aspects of such forms presented in Section 1.8.

8.2 Theory of Aeolian Features

8.2.1 The Significance of Wind Action

The present section (8.2) will deal with the theory of the geomorphological action of *wind*. This action is caused by the fact that wind is able to pick up and transport loose particles over very large distances.

There are in essence two modes of transportation of particles by the wind: some particles may be held in suspension for long periods by the *turbulence of the wind alone*, others have to *return to the ground* at short intervals. Particles which remain suspended in the air for long periods of time are commonly referred to as *dust* particles, the other as *sand* particles (Warren 1979; Willetts 1983).

The two modes of transportation by wind are fundamentally different and it will thus be necessary to treat them separately.

Wind action is particularly important in deserts where a variety of features are caused by it. After studying the physics of particle movement by wind, it will be our endeavor to present the theories of the origin of the various features in question, as they exist to date.

8.2.2 The Physics of Sand Movement

8.2.2.1 General Remarks

We shall first turn our attention to the motion of sand (as contrasted with dust). The classical investigations on this subject have been undertaken by Bagnold (1941), who described them in a monograph. Consequently, the present exposition can be held brief, the reader being referred to Bagnold's excellent book for most details.

8.2.2.2 Wind Velocity near the Ground

In order to study the motion of blown sand, it will first of all be necessary to obtain an idea about the wind velocity in those regions into which the moving sand is likely to penetrate.

It turns out (from field observations) that any wind velocity which is large enough to start sand grains moving is such that the wind will be in turbulent motion. Since air and water are simply two *examples* of *fluids*, one would expect that the same velocity distributions laws hold for streaming air as for streaming water. We have deduced earlier (cf. Sect. 4.2.3) the velocity distribution for water in turbulent motion near a wall: we have shown that in that case, Karman's logarithmic law of velocity distribution holds [cf. Eq. (4.2.30)]. It stands to reason that the same law applies when air is streaming by a wall in turbulent motion (Sutton 1943). This supposition has been checked in wind-tunnel experiments by Bagnold, who found a full confirmation thereof. He writes the law in the following fashion

$$u = 5.75 \, u_* \log_{10} \frac{z}{k}, \tag{8.2.1}$$

where z is the vertical coordinate, k a constant indicative of the surface roughness, and u_* the drag velocity

$$u_* = \sqrt{\frac{\sigma_m}{\rho}}. \tag{8.2.2}$$

Here, σ_m is the bottom drag force and ρ the density of the air. From experimental investigations, one was able to deduce that the constant k is approximately 1/30 of the diameter of the particles causing the surface roughness.

The theory of flow of turbulent air over the ground is therefore entirely analogous to the theory of flow of turbulent water in a channel.

If the blowing wind is sand-laden, it may be expected that formula (8.2.1) has to be modified since it stands to reason that the sand content will affect the flow behavior of the wind. Bagnold shows that, under these conditions, the altered velocity distribution u' can be described as follows

$$u' = 5.75 \, u'_* \log_{10} \frac{z}{k'} + u_t, \tag{8.2.3}$$

where k' is approximately equal to 0.3 cm for a fine uniform sand and equal to 1 cm for dune sand. Furthermore, u_t is the "impact threshold velocity" and u' a reaction velocity whose significance will be described more fully in the text section.

8.2.2.3 Grain Movement

Just like the bottom particles in a river, the sand grains acted upon by wind are subject to essentially two forces: first to the gravity force and second to the drag force caused by the wind. In order to start the sand grains moving, the drag exerted by the wind must achieve a critical value. Corresponding to Eq. (4.4.27) for water, the fundamental equation of the drag theory for the sand grains may be written as follows

$$v_{cr} = A \sqrt{\frac{\delta - \rho}{\rho} gd}, \qquad (8.2.4)$$

where δ is the density of the sand (2.65 g/cm^3 for quartz), ρ the density of air, g the gravity acceleration, and d the diameter of the grains. The quantity A is a constant approximately equal to 0.1. Furthermore, v_{cr} is the critical drag velocity which will just start the grains moving. According to Bagnold, the above [Eq. (8.2.4)] holds for sand grains in air exceeding 0.2 mm in diameter. The deduction of Eq. (8.2.4) can be performed in the same fashion as the deduction of the corresponding equation for the movement of bottom particles in rivers [cf. Eq. (4.4.27)]. The above equation can be made valid for sand grains with a diameter smaller than 0.2 mm provided the "constant" A is adjusted accordingly: the smaller the grains, the larger the constant.

Once grain movement is started by the wind, it can be maintained by a wind velocity blow the critical value which would be necessary to dislodge resting particles. This is due to the fact that sand grains move by *saltation* caused by impact. When grains strike the ground, their momentum may be sufficient to start other particles moving. Thus, in addition to a critical drag velocity of the wind, there is also an impact threshold velocity u_t which is sufficient to keep th above-mentioned "impact" mechanism operative indefinitely. Upon being struck, a grain may rise from the ground at any angle, usually almost vertically if it is to rise to any height at all. Its terminal forward velocity will be close to that of the wind v_w, its terminal downward velocity will be close to its settling velocity (cf. Sect. 4.4.2.2) v_s. Hence, the angle β by which it will strike the ground is

$$\tan \beta = \frac{v_s}{v_w}. \qquad (8.2.5)$$

Typical grain paths (after Bagnold) are shown in Fig. 113.

For the impact threshold velocity u_t, Bagnold gives the following equation

$$u_t = 5.75 \, A \sqrt{\frac{\delta - \rho}{\rho} gd} \, \log_{10} \frac{k'}{k}, \qquad (8.2.6)$$

where $A = 0.08$. This equation has been obtained from the drag equation (8.2.4); inserting the latter (v_{cr} denoting a shear velocity) into (8.2.1) and calculating the velocity for $z = k'$ yields an equation for the impact threshold velocity. In other words, k' is that height from which on downward the velocity no longer changes if

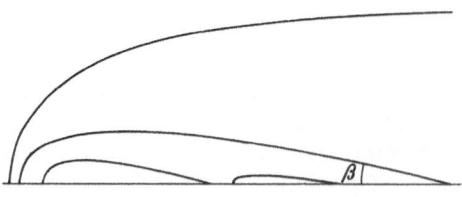

Fig. 113. Typical grain paths. (After Bagnold 1941)

the impact mechanism is fully developed. It should be noted, however, that k′ is, in fact, a parameter which has been adjusted a posteriori so that the above equations become self-consistent. The relations between v_{cr} and u_t are shown in Fig. 114 (after Bagnold), which also takes account of the change of the coefficient A for small grain sizes mentioned earlier.

In addition to saltation, another mechanism of sand movement may also occur. This has been called "*reptation*" or "*surface creep*". This mechanism refers to grains being pushed along the surface by the impact of the saltating grains; the grains in reptation never actually leave the surface. Bagnold notes that about one-quarter of the grains move in reptation, the rest in saltation.

It remains to relate the rate of sand movement to the wind velocity. Bagnold does this by assuming that each grain moving in saltation extracts all of its maximum forward momentum from the air. Thus, the loss of momentum from the air to move a sand mass q_s per unit time and width is $q_s v$, where v is the velocity attained by the grains. The loss of momentum is distributed over the length l where l is the length of a saltation jump so that the loss of momentum per unit time and area equals $q_s v/l$. However, this quantity must be equal to the boundary stress σ_m; hence

$$\sigma_m = q_s \frac{v}{l}. \tag{8.2.7}$$

Using (8.2.2), this yields

$$\rho u_*^2 = q_s \frac{v}{l}. \tag{8.2.8}$$

Bagnold now makes the arbitrary assumption that

$$l/v = Bu_*/g,$$

where B is a constant which depends on the surface. It then follows

$$q_s = \frac{B}{g}\rho u_*^2. \tag{8.2.9}$$

The formula can be modified by writing u'_* [corresponding to (8.2.3) embodying the reaction of the sand load onto the wind velocity] instead of u_*. The constant B

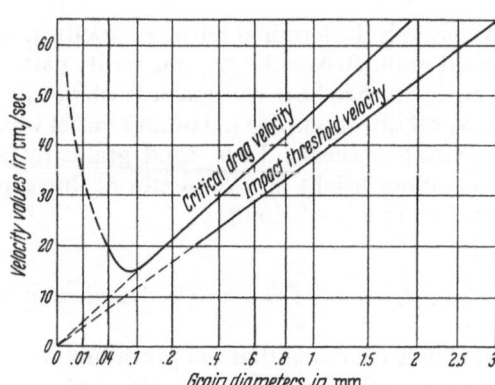

Fig. 114. Relation between critical drag velocity and impact threshold velocity. (After Bagnold 1941)

can be adjusted in such a manner as to express that only 3/4 of the sand are transported in saltation, the rest in reptation. Since reptation and saltation are proportional to each other, a formula of the type of (8.2.9) can be written down to express the *total* sand movement.

A slight, insignificant modification in Bagnold's formulas has been made by Kawamura (1951), and a general model of sediment transport by wind (encompassing all the various mechanisms discussed by Bagnold) has been set up by Anderson and Hallett (1986). Hsu (1973) confirmed Bagnold's calculations by computing aeolian sand transport from shear velocity measurements and comparing the former with actually observed transport rates.

8.2.3 Geomorphological Effects of Blown Sand

8.2.3.1 Outline of Sand Action

Our task is now to apply the basic physics of sand movement to an explanation of the various desert features described in Sect. 1.8.2.2.

In order to do this, we shall have to analyze first the distribution of sand concentration with height above the ground in a sand storm. This will yield a means to explain the grading of grain size distributions in some desert features. After these preliminaries have been dealt with, it will be possible to give an explanation for some of the small-scale (ripples) and large-scale (dunes) desert features.

8.2.3.2 Distribution of Sand Concentration in a Storm

In a sand storm, grains driven by the wind rarely ascend higher than 1 m above the ground. The average height is generally much less, of the order of 10 or 20 cm. The top of the sand cloud appears to have a rather sharp edge.

The occurence of the sharp edge of the sand cloud can be explained by the fact that sand grains traveling in a wind of velocity v will reach a terminal velocity equal to c

$$c = \sqrt{v^2 + w^2}, \tag{8.2.10}$$

where w is the terminal settling velocity in question. Thus, the velocity by which a sand grain strikes the ground on its path is c. If it bounces off the ground, the maximum height it can reach is obviously attained if it leaves the ground in a vertical direction. The maximum initial velocity will be equal to c if the impact is completely elastic. Thus, sand grains (of a given size) can only reach a finite maximum height in a sand storm; this explains the existence of a sharp upper surface of a sand cloud.

8.2.3.3 Grading of Grain Size Distribution on the Ground

It stands to reason that the prevailing wind strength will have an effect upon the grain size distribution on the surface of a sandy area. Unfortunately, it has not been possible to date to devise an exact analytical theory for determining the grain size distribution that is to be expected in a given area under the prevailing climatic conditions. It has only been possible to make more or less qualitative statements regarding the general processes that are at work.

Thus, according to Bagnold (1981), during the evolution of a sand storm, there is a definite sequence (or "cycle") of the events which are taking place. As the wind increases at the beginning of a storm, the fine particles are picked up. This leaves a certain surface roughness of the ground. Trikalinos (1928) referred to a surface in such a state as being in a "state of flocculation". In a wind of steady strength, the intensity of sand flow increases downward until equilibrium is reached. This means that grains can be picked up only from a limited area. At a given wind strenth, grains only up to a certain size can be picked up, as was shown in the discussion of Sect. 8.2.2.2. Thus, the area from which sand is removed moves continually downwind.

At the end of the cycle the sand picked up by the wind will have to be deposited somewhere. If this occurs due to the wind losing its strength, one speaks of *true sedimentation*: the falling grains have no longer the strength to keep up their saltating motion. There are, however, other possibilities by which deposition can take place. One of them is *accretion,* which is due to the surface becoming so irregular that the local wind strength in the lee of the irregularities is too small to carry the grains forward. This causes obstacles to grow larger by the accretion of sand thereto. During accretion, the grains may creep along the surface after impact until they find a quiet hollow in which to come to rest.

The final mode of deposition is by *encroachment*, which is a large-scale version of accretion. It is connected with large-scale discontinuities in the surface, such as steps. A step encroaches downwind because grains are sheltered in its lee from aerial entrainment.

The validity of the above general and qualitative statements (as noted above, a more accurate theory does not exist) has been checked experimentally by Bagnold (1941), who made observations of the grading changes (i.e., changes of grain size distribution) of various sands under a variety of conditions. It appears that the observed grading changes conform well with those expected. For the details of the experiments, the reader is referred to Bagnold's book.

8.2.3.4 Surface Ripples

As with the grain size distribution curves of desert sand, no analytical theory exists which could claim to explain the existence of *sand ripples* adequately. As was noted earlier (cf. Sect. 1.8.2), sand ripples are distinguished from larger features not only by their size, but also by the fact that in ripples the grading of the sand is such that the coarsest material is found at the crests, the finer material in the troughs. In larger desert features the reverse is true.

A good early review of the various attempts that have been made at achieving an explanation of the existence of desert sand ripples has been given by Trikalinos (1928). Bagnold (1941) in his book has, in fact, little to add to Trikalinos' article.

The earliest attempts at an explanation of ripples, due to Darwin (1884) and Cornish (1897), assumed that the mechanism is analogous to that responsible for the formation of ripples in a river bed. Accordingly, some mechanism involving resonant turbulence or a "traffic jam" (cf. Sect. 4.6.2.4) may be invoked. However, although experiments involving fluvial ripples and wind ripples yield the result that the two phenomena are superficially similar, it turns out that there is a fundamental difference between them: In wind ripples the grains are sorted with regard to size,in fluvial ripples this is not the case. Therefore, there seems to be a fundamental difference between the two kinds of ripples.

Other research workers (Baschin 1899; Solger 1910), attempting an explanation of sand ripples (due to wind), have assumed that there is a fundamental resemblance between the rippling of a water surface and the rippling of a sand surface by wind. However, it can again be shown that this resemblance is merely superficial and cannot be used be basis for an explanation of the phenomenon of sand ripples. If a very uniform sand is taken, no sand rippes will form, although, if the phenomenon were analogous to the creation of water waves, this should be the case. An attempt by Exner (1927) to treat the origin of ripples as an instability phenomenon in a layered fluid in which the density increases downward, suffers from the same deficiency as the attempts by Baschin and Solger.

It remains to try to explain the formation of desert ripples by assuming that they are a phenomenon which is truly due to the differential movability of grains of various sizes. This is an assumption which has been advanced by Trikalinos. It was later also maintained, independently of Trikalinos (as it appears) by Bagnold. Thus, ripples can form only if there is a spread in grain sizes in the sand. Ripples are ephemeral phenomena which change their shape, size, and orientation with the prevailing wind direction. There seems to be a causal connection between ripple spacing and a characteristic path length in saltation. On this basis, Anderson (1987) set up a conceptual model of the sand movement process which led to a more complete model of ripple formation.

8.2.3.5 Large-Scale Effects

We now turn our attention to the large-scale geomorphological effects caused by the motion of desert sand. In this connection, it is particularly the phenomenon of wandering *dunes* which captures one's imagination. However, in order to explain this phenomenon, it will first of all be necessary to have a close look at the mechanism of deposition of sand under conditions in which the direction and the strength of the wind vary. This will yield a possibility of explaining the two types of dunes (barchan dunes and seif dunes; cf. Sect. 1.8.2.2) which have been observed in nature. Finally, the slip face on a barchan dune will be discussed. Most of the pertinent investigations into the mechanism of formation of dunes have been undertaken by Bagnold (1941), who gives an extensive description thereof in his book.

Starting with the mechanism of sand deposition we note that the intensity q of sand flow was given earlier (cf. 8.2.6) as equal to

$$q = B\frac{\rho}{g}u_*^3, \qquad\qquad\qquad (8.2.11)$$

where B is a constant which depends on the surface, ρ is the density of the air, g is the gravity acceleration, and u_* is the drag velocity. If for some reason on a coarse, pebbly (with sand between the pebbles) surface a sand patch has been formed, the constant B will have a value B_1 on the coarse surface and B_2 on the sand patch with

$$B_1 > B_2. \tag{8.2.12}$$

This yields the result that deposition must take place over the sand patch, at a rate q_d equal to

$$q_d = q_1 - q_2 = (B_1 - B_2)\frac{\rho}{g}u_*^3. \tag{8.2.13}$$

This indicates that a sand patch, once it has been started, should grow by increasing its height, i.e., it will tend to form a dune.

In the formation of sand accumulations, one must distinguish between winds of various strengths. A "gentle" wind will be able to entrain particles from the fine-sand surface, but not from the coarse (pebbly) surface. A "strong" wind will be above the threshold velocity for both types of surfaces. It is the "strong" winds which tend to build up dunes, the "gentle" winds have the opposite effect.

It is convenient to separate the total sand flow into flow caused by gentle winds $Q^{(g)}$ and flow caused by strong winds $Q^{(s)}$ since the effects of the two wind types are somewhat different. It then appears as logical to hypothesize (Bagnold 1941) that longitudinal (or seif) dunes occur if the directions of $Q^{(s)}$ and $Q^{(g)}$ differ, and that barchan dunes occur if the directions of $Q^{(s)}$ and $Q^{(g)}$ nearly coincide. In this connection, however, the terms "gentle" and "strong" refer to a surface studded with dunes: a gentle wind is one whose strength is such that the corresponding intensity of sand movement is less than that required to give the surface an equivalent roughness equal to the actual roughness of the surrounding area. In a "strong" wind the reverse is true.

Turning first to the longitudinal profile of a dune, we note that its leeward side is generally steeper than its windward side. If a dune is progressing downwind with a velocity c without changing its shape, the horizontal component of the sand removal at any point must exactly correspond to the velocity c, which yields

$$\frac{dQ}{\rho_{BS}} = dV = c\,dt\,dh, \tag{8.2.14}$$

where dQ is the mass (and dV the volume) per unit width of sand removed during the time dt, ρ_{BS} is the bulk density of the sand and dh is the height increase of the original dune corresponding to a shift by the distance $dx = c\,dt$ at the bottom (cf. Fig. 115). Hence the rate of sand removal per unit time (and width) must be

$$dq = \frac{dQ}{dt} = c\rho_{BS}\,dh \tag{8.2.15}$$

or

$$\frac{dq}{dx} = c\rho_{BS}\frac{dh}{dx} = c\rho_{BS}\tan\alpha. \tag{8.2.16}$$

If the leeward side of a dune becomes too steep, it will collapse and form a *slip face* corresponding to the angle of repose (cf. Sect. 3.3.2.3) for sand. All the sand

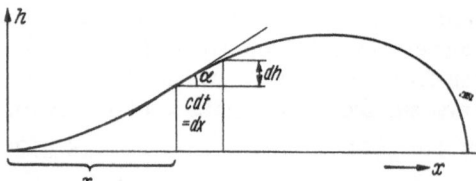

Fig. 115. Cross-section of a dune

passing over the rim of the slip face will contribute to its advance. If the rate of sand movement (per unit width) over the rim of the slip face be given by Q, and the slip face have the height h, we have from (8.2.15) (note dh = h)

$$c = \frac{Q}{\rho_{BS}h}.$$ (8.2.17)

If the above consideration be applied to a circular mound of sand which has a slip surface at the leeward side, it becomes obvious that, in a cross-wind section, the height of the slip face must decrease from a maximum at the middle to near zero at the extremities. If the sand flow Q is nearly the same across the dune, it follows immediately [from Eq. (8.2.17)] that the extremities must advance more rapidly than the center. Presumably, an equilibrium is reached if the extremities advance so far into a region sheltered by the rest of the dune so that the sand flow Q is reduced to such an extent that the whole dune proceeds at equal speed. This at once explains the crescentic shape of a barchan dune. The explanation holds if the wind blows predominantly always from the same direction. It may be noted that a somewhat more elaborate theory of barchan dune formation, based on corresponding mechanical assumptions, has been proposed by Ertel (1966); computer simulations of the various processes involved have been made by Howard et al. (1978) Wippermann and Gross (1986) and by Fisher and Galdies (1988).

If the wind is not unidirectional, it stands to reason that the barchan dune will develop into a seif dune. The procedure is illustrated in Fig. 116 (after Bagnold) where g denotes the direction of gentle winds and s the direction of strong winds. The argument is entirely qualitative: no exact mathematical treatment can be given. Nevertheless, Lancaster (1980, 1982) found supporting evidence in the field

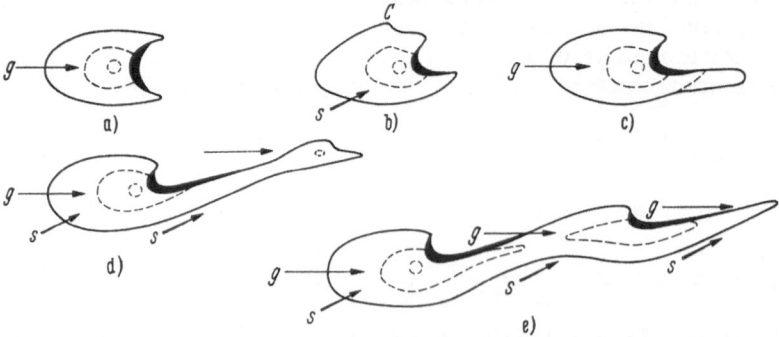

Fig. 116 a–e. Transition of a barchan dune into a seif dune. (After Bagnold 1941)

for Bagnold's theory. However, the latter has also been questioned: Verstappen (1968) assumed seif dunes to be due to different earlier climatic conditions, Tsoar (1984) maintained that the elongation of the barchan to seif dunes occurs in the lee direction of the *combined* gentle and strong wind vectors and Livingstone (1988) suggested that the origin of linear dunes lies in a modification of the near-ground air flow pattern by the dune itself. Such flow patterns could also explain the genesis of other rarer forms of dunes, such as star dunes.

Finally, we may mention a further few geomorphological effects caused by sand movement. Thus, blow holes are caused by wind erosion in sand areas; they have been simulated numerically by Jungerius (1984), who assumed that atmospheric gustiness is at the root of such features. Furthermore, the blown sand must be supposed to have a pronounced corrasive action. This explains the genesis of yardangs (cf. Sect. 1.8.2.2); it is also effective with regard to the grains themselves. Experimental investigations have been made by Kuenen (1960), who was able to show that aeolian abrasion of sand is far more effective than the aqueous abrasion of particles and may affect grains down to a diameter of 0.05 mm (cf. also Whalley et al. 1987).

8.2.3.6 System Theory

The geomorphological effects of sand action have also been treated by system theory.

Following the general principles outlined in Chapter 5, there have been, first of all, empirical studies of the variables that control the types of dunes that develop (Wasson and Hyde 1983; Mainguet 1983; Lancaster 1988); this corresponds, so to speak, to the regime theory of rivers.

The next level was to establish empirically the steady-state conditions for systems of sand features, such as sand seas (Fryberger and Ahlbrandt 1979) or of erg-type dune fields (Besler 1980; Mabbutt and Wooding 1983).

Finally, process-response diagrams have been constructed for the effects caused in the sand system by the change of one variable. This approach has been particularly successfully employed in studies by Besler (1983, 1985).

8.2.4 Physics of Dust Movement

8.2.4.1 Basic Principles

Dust particles, by definition, are so fine that their transport in the air occurs by *suspension*.

Natural dust is very similar to many man-made pollutants and therefore the movement of such particles has been studied fairly extensively by sanitary engineers (Magill et al. 1956; Frenkiel and Sheppard 1959; McCabe 1952).

The chief phenomenon influencing the movement of dust is that of atmospheric diffusion. We shall therefore first discuss the theory of atmospheric diffusion and then apply the latter to the problem of dust movement.

8.2.4.2 Theory of Atmospheric Diffusion

We shall first turn our attention to the general theory of atmospheric diffusion. A good review of this subject has, for instance, been given by Monin (1959). Accordingly, diffusion is a phenomenon which, in the lower layers of the Earth at least, can be attributed to *turbulence*. Simply using the empirical fact of the existence of atmospheric difusion, one can then, for purely heuristic reasons, write down the following diffusivity equation for the movement of a "dust" cloud:

$$\frac{\partial c}{\partial t} + u\frac{\partial c}{\partial x} = \frac{\partial}{\partial x}K_x\frac{\partial c}{\partial x} + \frac{\partial}{\partial y}K_y\frac{\partial c}{\partial y} + \frac{\partial}{\partial z}K_z\frac{\partial c}{\partial z}. \tag{8.2.18}$$

Here, x, y, z, are Cartesian co-ordinates (z is vertical), u is the wind velocity, assumed as parallel to the x-direction, K_x, K_y, and K_z are empirical diffusion coefficients, and c is the dust concentration. The diffusivity equation (8.2.18) can be justified theoretically by the existence of eddy diffusion in turbulent flow (cf. Sect. 2.2.2.2). It is assumed that the dust particles are so small that their motion is identical to that of the surrounding microscopic elemental volumes of air. The diffusivity equation as written above does not allow for the gravitational settling of the dust particles. If the latter is also to be taken into account, a term $-w(\partial c/\partial z)$ should be added to the left hand size, where w is the settling velocity.

The diffusion coefficients K_x, K_y, K_z are not constant. The term containing K_x is usually neglected in comparison with $u\,\partial c/\partial x$. The values of the other coefficients depend on the turbulence of the air. For further investigations, we shall use the Karman law for the turbulence in the air near the ground [cf. (8.2.1)] in the following form:

$$\frac{u(z)}{u_*} = \frac{1}{k}\text{lognat}\frac{z}{z_0}, \tag{8.2.19}$$

where u_* is the usual shear velocity given by (8.2.2) and z_0 is a constant. We now introduce the assumption that, near the Earth's surface, the shear stress is independent of height; we have then:

$$\sigma(z) = \sigma_m. \tag{8.2.20}$$

Based upon this assumption, one can deduce a value for the vertical eddy viscosity [cf. (2.2.9)]; for, one has (cf. Sect. 2.2.2.2) for a definition of symbols):

$$\sigma = \sigma_m = \rho\varepsilon\frac{\partial u(z)}{\partial z} = \rho\varepsilon\frac{u_*}{k}\frac{1}{z} \tag{8.2.21}$$

or (with 8.2.2)

$$\varepsilon = \frac{\sigma_m}{\rho u_*}kz = u_*kz. \tag{8.2.22}$$

If we assume, as stated above, that the dust particles contained in the air are transferred concurrently with the air, then the quantity ε must be equal to the vertical diffusion coefficient in the empirical equation (8.2.18), for, following a line of reasoning analogous to that in Sect. 4.5.2.2 we have for the vertical transfer of *momentum*:

$$\sigma = (-)\rho\varepsilon_m \frac{\partial u(z)}{\partial z} \qquad (8.2.23)$$

and for the vertical transfer of mass

$$F = -\varepsilon_s \frac{\partial c}{\partial z}, \qquad (8.2.24)$$

where F is the mass flux per unit time and unit area. However, by our assumption we have

$$\varepsilon_s = \varepsilon_m = \varepsilon. \qquad (8.2.25)$$

Thus, formulating the continuity condition yields indeed a diffusivity equation

$$\frac{\partial c}{\partial t} = -\frac{\partial F}{\partial z} = \frac{\partial}{\partial z}\varepsilon\frac{\partial c}{\partial z} \qquad (8.2.26)$$

with ε as diffusivity coefficient. Hence we find [with (8.2.22)]:

$$K_z = \text{const } u_* z. \qquad (8.2.27)$$

8.2.4.3 Light Particles

If we are faced with a concrete problem of dust movement in the air, it will be necessary to obtain the appropriate solutions of the diffusivity equation (8.2.18). For light particles, one can neglect the term originating from the settling velocity. Considering a steady-state motion in two dimensions, one has to solve the following diffusivity equation:

$$u(z)\frac{\partial c}{\partial z} = \frac{\partial}{\partial z}\left[K_z(z)\frac{\partial c}{\partial z} \right], \qquad (8.2.28)$$

where u(z) is given by (8.2.19) and $K_z(z)$ by (8.2.27). The boundary conditions for a line source are

$$\left.\begin{array}{ll} c = 0 & \text{for } x = \infty, z = 0 \\[2mm] K_z\dfrac{\partial c}{\partial z} = 0 & \text{for } z = 0, x > 0 \\[2mm] c = \infty & \text{for } x = z = 0 \\[2mm] \displaystyle\int_0^\infty uc(x, z)\, dz = Q & \text{for } x > 0. \end{array}\right\} \qquad (8.2.29)$$

The integration of Eq. (8.2.28) has to be achieved numerically as the result is not expressible in terms of simple functions. However, it is possible to give analytical solutions of the equation if u(z) and K_z are approximated by powers of z instead of being taken as given by Eqs. (8.2.19) and (8.2.27) respectively. Thus, let us set

$$u(z) = u_1 \left(\frac{z}{z_1} \right)^m, \tag{8.2.30}$$

$$K_z(z) = K_1 \left(\frac{z}{z_1} \right)^n, \tag{8.2.31}$$

then, the solution of (8.2.28) can be shown to be ($z_1 = $ unity; Sutton 1943)

$$c(x, z) = \frac{Q}{u_1 \Gamma(s)} \left[\frac{u_1}{(m - n + 2)^2 K_1 x} \right]^s \exp \left\{ - \frac{u_1 z^{m-n+2}}{(m - n + 2)^2 K_1 x} \right\} \tag{8.2.32}$$

with

$$s \equiv \frac{m + 1}{m - n + 2}. \tag{8.2.33}$$

It has been found empirically that for a smooth or short grass surface,

$$\begin{aligned} m &= \tfrac{1}{7} \\ n &= \tfrac{6}{7} \end{aligned} \tag{8.2.34}$$

and then we have

$$c(x, z) = \frac{Q}{(1\tfrac{2}{7})^{16/9} \Gamma(\tfrac{8}{9}) K_1^{8/9} u_1^{1/9} x^{8/9}} \exp \left\{ - \frac{u_1 z^{9/7}}{(1\tfrac{2}{7})^2 K_1 u} \right\}. \tag{8.2.35}$$

This gives the shape of a dust cloud caused by a line source of fine dust.

The results of the present section thus show how "light" particles (Goossens 1985) are diffused with the air. Since the diffusion goes on idenfinitely, it now becomes understandable that light dust will eventually reach the region where geostrophic flow takes place and that the dust can then be carried over very large distances.

8.2.4.4 Heavy Particles

If we turn our attention to the motion of heavy particles, we note that in this case, a gravity term has to be added in the basic diffusivity equation (8.2.18). The problem becomes then analogous to that of suspended sediment transportation in rivers and can be treated in the same manner (cf. Sect. 4.5.2).

The reader is therefore referred to Sect. 4.5.2 for further details.

8.2.4.5 Geomorphological Effects of Dust Movement

1. General Remarks

In the remarks on physical geomorphology (cf. Sect. 1.8.2.3), we have already stated in qualitative terms what the geomorphological effects of dust movement are: first the erosion of soil and desert dust in large quantities, and second the deposition of dust over large areas to form loess, the area of deposition being hundreds or thousands of miles removed from that of the origin of the dust.

2. Soil loss

It would be nice if it were possible to account for the *amount* of dust removed from or deposited in an area under given climatic conditions. Unfortunately, as far as the writer was able to ascertain, studies to determine this amount either theoretically or experimentally do not seem to have been undertaken as of yet.

3. Dust transport

This leaves one solely with the task of explaining theoretically the large distances over which dust movement has actually been observed to occur. In this regard, the atmospheric diffusion theory, applied to "light" particles (cf. Sect. 8.2.4.3) immediately furnishes the required explanation: a sufficient cross-wind with reasonably turbulent air can keep dust in suspension indefinitely. Thus the dust storms in the high latitude regions (Nickling 1978) as well as those off the Sahara (Harmattan winds: Morales 1986) are explained.

4. Loess deposition

The most spectacular geomorphological effect of dust movement is the deposition of large blankets of loess. Loess occurs at the margin of deserts (Tsoar and Pye 1987) as well as of (former) glaciated areas (Whalley et al. 1982). For loess to form, the dust deposition rate must exceed the erosion rate by wind and water and the rate of weathering by pedogenesis (Pye 1984). The mechanics of the deposition of loess has been discussed in detail by Goossens (1988), who stated the conditions that have to be satisfied as follows: (1) there must be enough material supplied so that a steady settling-down can take place, and (2) the downward forces (weight and turbulence) must exceed the upward-turbulent forces. Turbulence thus causes both upward and downward particle movement, but the higher the turbulence, the more particles become trapped on the ground during their downward fluctuations. On this basis, Goossens was able to calculate loess sedimentation rates in consequence of wind patterns caused by surface irregularities. He showed theoretically (and confirmed experimentally) that increased loess sedimentation occurs on a hill on the concave part of the windward slope, less sedimentation on the convex part of the windward slope, and no sedimentation at all on the leeward slope.

8.2.5 Wind Transport of Volcanic Materials

8.2.5.1 Ash Flows and Ash Falls

In Sect. 1.8.2.4 we noted that airborne transport on a large scale may also occur after volcanic eruptions. There are two types of such transport: ash flows (nuées ardentes) and ash falls.

8.2.5.2 Ash Flows

Turning first to ash flows, we note that nuées ardentes have much in common with *turbidity currents* in the sea (cf. Sect. 6.2.3). However, a mixture of air, gravel, and sand does not flow downhill on a moderate slope as does a turbidity current in water or a nuée ardente. An additional effect must therefore be operative to account for the mobility of a nuée ardente. The various theories that have been advanced have been reviewed, for instance, by McTaggart (1960).

Accordingly, the gas emission hypothesis, first stated by Anderson and Flett (1903), is the most widely accepted theory of the mobility of a nuée ardente. According to this theory, gas is emitted from the lava and rocks thrown out by the volcanic eruption while they are cooling, which is so plentiful that it keeps the individual particles in suspension. The gas emission hypothesis was accepted by very many authors; Reynolds (1954) added the specific statement that this gas emission produces *fluidization* in the cloud.

In order to keep a fluidized bed in suspension, the pressure drop Δp across it has to be at least as great as the (buoyant) weight of the bed (Leva et al. 1951)

$$\Delta p = L(1 - p)(\rho_{sed} - \rho_{fluid})g, \tag{8.2.36}$$

where L is the depth of the bed, P its porosity, ρ_{sed} the density of the particles, ρ_{fluid} the density of the fluid, and g the gravity acceleration.

In order to correlate the pressure drop Δp with the flow velocity, we take one of the available correlations, in particular, we take the correlation proposed by Kling (1940) for turbulent flow at low Reynolds numbers:

$$\lambda = 94/(Re)^{0.16}, \tag{8.2.37}$$

where λ is the "friction factor"

$$\lambda = \frac{2\delta\Delta p/L}{P^2v^2\rho_{fluid}} \tag{8.2.38}$$

(with d = particle diameter, v = pore velocity of fluid), and Re is the Reynolds number

$$Re = \frac{Pv\rho_{fluid}d}{\eta}, \tag{8.2.39}$$

where η is the viscosity of the fluid.

Solving for v, this yields

$$v^{1.84} = \left(\frac{P\rho_{fluid}}{\eta}\right)^{0.16} \frac{2(1 - P)(\rho_{sed} - \rho_{fluid})g}{94P^2\rho_{fluid}} d^{1.16}. \tag{8.2.40}$$

Setting $P = 0.2$, $\rho_{fluid} = \rho_{air} = 2.77 \times 10^{-4}\,g/cm^3$ (at 1.000 °C), η (of air) $= 491 \times 10^{-6}$ cgs (at 1.000 °C); $\rho_{sed} = 2.5\,g/cm^3$, $g = 980\,cm/s^2$, this yields

$$v = 3100 \cdot d^{0.63}\,cm/s. \tag{8.2.41}$$

In order to fluidize a nuée containing particles of 1 cm in diameter, a flow of air of a velocity of 31 m/s is required. This is a very high velocity and it is very doubtful indeed whether enough gas can be released from the lava to maintain it.

Therefore, modifications had to be suggested in the fluidization process. Thus, Shreve (1968) advanced the hypothesis that the pyroclastic flow develops a cushion of gas at the bottom upon which it slides. This process allows for a much greater mobility than mere flow in a fluidized state. The cushion of gas consists partly of air and partly of gas emitted from the debris. Furthermore, Fisher (1979) and also Walker and McBroome (1983) proposed that nuées ardentes are pyroclastic *surges* rather than flows: surges are low-concentration turbulent flows that can form from the collapse of an eruption column or directly from a crater. The low density of a surge would explain (at least qualitatively) the high mobility (Francis and Baker 1977) of pyroclastic flows.

8.2.5.3 Ash Falls

Turning now to the phenomenon of *ash falls*, we note that the characteristic downwind size decline of the deposits can be explained as a case of aeolian particle gradation. This hypothesis was advanced by Scheidegger and Potter (1968) and its mechanical implications were carefully studied. The physical model is that of a turbulent "slug" of air (containing suspended particles) being transported downwind. During the downwind motion, the turbulence decays according to standard laws. Thereby, the carrying capacity for sediment diminishes, and a sorting of the particles results: first the heavy particles, then the lighter and lighter ones fall out. The process is much like that considered in the theory of varve deposition (Sect. 7.3.5.5) except that, in addition to the *amount* of sediment deposited downcurrent, the particle *size* has to be considered. The theory yields an excellent correlation with observation.

8.3 Specific Desert Features

8.3.1 Introduction

Although the wind-caused morphological features are very common and impressive in desert regions, such features are not solely found in deserts: wind-generated dunes can be encountered in any sandy, wind-swept area.

Specific desert features are connected with the essential climatic conditions in deserts: their relative aridity. We shall now proceed to discuss the genesis of such features in the sequence in which their morphology was described in Sect. 1.8.3.

8.3.2 Small-Scale Features

8.3.2.1 Rock Varnish

Beginning with small-scale features, we focus our attention first on rock varnish.

In the section on morphology (1.8.3.2) we have noted that the varnish consists of manganese and iron compounds that are relatively resistant to aeolian abrasion.

Regarding the fixation of the iron and manganese compounds in the varnish there are essentially two opposing views: Dorn and Oberlander (1981, 1982) assume fixation by biological (bacterial) agents, whereas Smith and Whalley (1988, 1989) postulate an inorganic process, admitting, however, that some varnishes may be of biotic origin. Thus, the problem of rock varnish has obviously not yet been solved.

8.3.2.2 Laterite and Other Crusts

Iron deposits thicker than those in rock varnish are found in the form of laterite crusts at the surface of desert soils.

The phreatic processes indicated in Sect. 1.8.3.2 (leaching of iron compounds below the water table and precipitation in the zone between high and low water tables) seem to be fully confirmed (cf. Venkatesh and Roman 1981; McFarlane 1987). Questions remain, however, regarding the actual kinetics of the (ferric) cations in the silicate solids (Herbillon and Stone 1984). Thus, the detailed chemistry of the laterite formation process is not yet fully understood.

Other crusts form simply by precipitation of ever-saturated solutions. This explains the formation of "calcretes" found in porous sandstones (Bermudez 1981), as well as the deposition of evaporites.

8.3.2.3 Desiccation Cracks

The genesis of crack polygons found in arid areas is similar to that of polygons in glacial areas: they are due to intrinsic ground contraction caused by a volume decrease in the material (cf. Sect. 7.5.2.2). In arid areas, the contraction is brought about by desiccation (Mahaney and Boyer 1988) and particularly by periodic wetting and drying.

Under certain circumstances, the cracks become filled with wind-blown sand which leads to a fixation of their pattern (Blume 1987).

On clayey slopes, contraction cracks can initiate microrill channels leading to the characteristic badland drainage pattern (Haigh 1978).

8.3.3 Island Mounts

Island mounts (inselbergs) belong undoubtedly to some of the most impressive features found in deserts (cf. Sect. 1.8.3.3). Their genesis has been the subject of some speculation (cf. Kesel 1973; Thomas 1978; Birot 1978; Twidale 1981a; Stingl and Garleff 1987 for reviews). Thus, they have been regarded as results of sandblasting and sculpturing by desert winds, as representations of structural landforms, and as (possibly structurally predesigned) degradation stumps.

Sculpturing by sandblasting caused by desert winds has been advocated by Ollier and Tuttenham (1962) as the cause of inselbergs. Although erosion (corrasion) of stones or outcrops protruding well above the ground by wind is undoubtedly a factor, this does not go much beyond the creation of minor "ventifacts" (Anderson 1986). Wind abrasion does not seem to be effective much above 1 m above the ground. Thus, another mechanism for the genesis of inselbergs must be sought.

Therefore, inselbergs are presently commonly regarded as "structural landforms" (Twidale 1971, 1981b; Jese 1973; Selby 1977). This contention means that inselbergs are either the result of preexisting joint patterns (Moeyersons 1977; Twidale and Bourne 1978; Nicholas and Dixon 1986) or else the result of stress-relief sheeting (Selby 1977; Ollier 1978).

Naturally, the structural patterns act as a predesign for the attack of weathering and degrading agents (Twidale 1986a); the latter would then sculpture out the bizarre forms.

Following up the last idea, inselbergs could thus also be regarded as "degradation stumps" which generally correspond to a structural predesign. Theoretical estimates of the relation of rock strength to morphology of the resulting features have been made by Selby (1982), Cook (1983), and Pye (1986).

A more extreme view is that inselbergs may simply be erosion remnants of harder rocks (Mensching 1978); e.g., exposed surfaces of stocks of batholiths (Twidale 1986b). In fact, all of the mentioned possibilities may be variously realized.

8.4 Semidesert Features

8.4.1 Introduction

We have already mentioned (in Sect. 1.8.4) that the "semidesert" features are conditioned by the relatively rare but important occurrences of precipitation.

8.4.2 Intermittent Water Flows

Even in deserts, water is an important agent. The latter can be encountered in intermittent water courses (wadis) or as intermittent sheet floods.

Regarding intermittent streams, it can be stated that the latter act, during active phases, very much like ordinary rivers. Thus, statements can be made regarding discharge, stream power, sediment transport, and bed forms just as for rivers: a comprehensive study of such features has been made for channels in the extremely arid South Negev area (Lekach and Schick 1983), where it was found that the ordinary river conditions can be applied.

More interesting, therefore, are the larger-scale floods that occur in deserts at infrequent intervals; they activate the ephemeral flood channels (Baker et al. 1983) and also sculpture desert dunes (Besler 1976). Equally, they give rise to talluvial (talus-alluvial) processes (Frostick and Reid 1982).

In summary, it can be said that the intermittent water occurrences in deserts act very much like the nonintermittent flows elsewhere on land.

8.4.3 Badlands

8.4.3.1 General Remarks

Finally, we discuss briefly the phenomenon of badlands (cf. Sect. 1.8.4.3 regarding the morphology), which are characterized by a high drainage density, weird erosion pillars, and the absence of vegetation.

8.4.3.2 Drainage Density in Badlands

Badlands are characterized by a high drainage density. In fact, the reason for this high drainage density has been the subject of much speculation (Davies and Pearce 1981; Dardis 1989).

Regarding gully initiation in badlands, we have already mentioned a hypothesis (cf. Sect. 8.3.2.3): desiccation cracks formed during arid phases (particularly in clayey materials) act as triggers for rills and later gullies (Haigh 1978; Imeson 1986). Since the density of desiccation cracks is always high, this would explain the observed drainage densities in badlands.

Other, less specific speculations (DePloey 1974; Bowyer-Bower and Bryan 1986) ascribe the gullying in some fashion to the lithological properties of the badland materials (mainly clay): the friability of the latter would be conducive to the lack of vegetation and the consequent formation of many rills.

8.4.3.3 Pyramidal Badland Structures

Regular pyramidal structures occur in badland areas as earth pillars, mesas, and buttes, which are quite naturally explained as erosional features caused by rain. The considerable stability of such structures seems to be due to an intrinsic strength of the consolidated sediments of which they consist: these are stronger than the valley alluvium (Whalley 1976).

More interesting is the origin of hoodoos, structures with an overhanging "hat" (see Sect. 1.8.4.3 for their description).

If one tries to assume an origin of the hoodoos which would be analogous to that of the pyramidal structures referred to above, one is at once faced with the problem as to how the water gets around the "rim" of the "hat" of the hoodoos so as to wash out their "neck". One might think that the causes for the water turning the corner are surface forces. However, it is the writer's contention that the phenomenon is analogous to that encountered when tea being poured from a teapot runs down the underside of the spout rather than straight on into the cup. This phenomenon has been called the *teapot effect*; it is not due to surface forces, interfacial tensions, or such like, but is a consequence of the prevailing flow potentials (Scheidegger 1958).

In fact, one can demonstrate (Keller 1957) that flow around a plate is a possible potential flow and that the atmospheric pressure p_0 can support flow on the underside of such a plate if (cf. Fig. 117)

$$h < p_0/(\rho g), \tag{8.4.1}$$

where h is the thickness of the flow, ρ the density, and g the gravity acceleration.

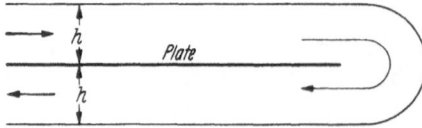

Fig. 117. Flow around a plate. (After Keller 1957)

However, although a horizontal flow on the underside of a plate can exist, such a flow is obviously an unstable flow: eventually it will detach itself and drop off downward. One must therefore make a corresponding stability analysis which can be done as follows.

Introducing a perturbation at the end of the plate, one can calculate the distance L at which it will have grown by the factor e. This distance depends on the interfacial tension T between air and water. Furthermore, it depends on the form of the original perturbation. Of interest is that distance L which is the smallest in all the modes of instability that can occur. The expression for this minimum distance cannot be written down in closed form, but in two limit cases this is possible (Keller 1957) with u = horizontal velocity:

$$\text{if} \quad \frac{\rho g h^2}{T} \ll 1, \quad \text{then} \quad L = \frac{2u}{g} \sqrt{\frac{T}{\rho h}}, \tag{8.4.2}$$

$$\text{if} \quad \frac{\rho g h^2}{T} \gg 1, \quad \text{then} \quad L = \frac{u}{\sqrt{2}} \left(\frac{27T}{\rho g^3} \right)^{1/4}.$$

In the case of hoodoos, the eroding agent is water. In the case of water, one has T = 80 dynes/cm, $\rho = 1$ g/cm^3, g = 980 cm/s^2; thus

$$\text{a) if} \quad 12h^2 \ll 1, \quad \text{then} \quad L = \frac{u}{h^{1/2}} \times 0.0183, \tag{8.4.3}$$

$$\text{b) if} \quad 12h^2 \gg 1, \quad \text{then} \quad L = u \times 0.0275, \tag{8.4.4}$$

where all units are in the c.g.s.-system.

The distance L, as has been explained above, is that distance in which the most significant disturbance grows by the factor e as stated above. In hydrodynamic stability theory, it is usually assumed that the instability will become predominant (i.e., the flow will detach itself) in a distance equal to ten times L. It turns out that the case (b) applies if h is greater than about 1/3 cm. Then

$$10\,L \cong u \times 0.28\,\text{cm} \tag{8.4.5}$$

irrespective of the thickness h of the flow. It is difficult to estimate the velocity u in the flow. In a good cloudburst it will probably reach about 1– 2 m/s at the edge of the overhang. This means that the flow can continue on the underside for about 28–56 cm before detaching itself. According to earlier remarks about the mechanism of erosion, this distance of 28–56 cm is the distance by which the "hat" of the hoodoos can overhang, for, in order to erode the soft material below, the water must obviously first reach it.

It thus appears that the values postulated above from a discussion of the teapot effect are in good agreement with those actually found in the hoodoos measured. This would serve to substantiate the theory proposed here.

References

Anderson, R.S.: Bull. Geol. Soc. Am. 97, 1270 (1986)
Anderson, R.S.: Sedimentology 34, 943 (1987)
Anderson, R.S. and B. Hallett: Bull. Geol. Soc. Am. 97, 523 (1986)

Anderson, T. and J.S. Flett: Philos. Trans. R. Soc. Lond. A-200, 353 (1903)
Bagnold, R.A.: The physics of blown sand and desert dunes. London: Methuen (1941)
Baker, V.R. and 3 others: Nature 301, 502 (1983)
Baschin, O.: Z. Ges. Erd. (Berl.), 34, 408 (1899)
Bermudez, F.L.: Estud. Geograf. 42, 89 (1981)
Besler, H.: Mitt. Basler Afrika Bibl. 15, 83 (1976)
Besler, H.: Stuttgarter Geogr. Stud. 96, 1 (1980)
Besler, H.: Z. Geomorph. Suppl. 45, 287 (1983)
Besler, H.: Stuttgarter Geogr. Stud. 105, 11 (1985)
Birot, P.: Z. Geomorph. Suppl. 31, 42 (1978)
Blume, H.P.: Z. Geomorph. 31(4), 443 (1987)
Bowyer-Bower, S. and R.B. Bryan: Z. Geomorph. Suppl. 60, 161 (1986)
Cooks, J.: Z. Geomorph. 27(4), 483 (1983)
Cornish, V.: Geogr. J. (Lond.) 9, 278 (1987)
Dardis, G.F.: Catena Suppl. 14, 1 (1989)
Darwin, G.H.: Proc. Roy. Soc. A34, 18 (1884)
Davies, R.H. and A.J. Pearce (eds.): Erosion and sediment transport in Pacific rim steeplands
    symposium. Christchurch (New Zeal.): I.A.H.S. Publ. 132 (1981)
DePloey, J.: Z. Geomorph. Suppl. 21, 177 (1974)
Dorn, R.I. and T.M. Oberlander: Z. Geomorph. 25(4), 420 (1981)
Dorn, R.I. and T.M. Oberlander: Progr. Phys. Geogr. 6(3), 317 (1982)
Ertel, H.: Monatsber. Dt. Akad. Wiss. 8, 713 (1966)
Exner, F.M.: Geogr. Ann. 9, 80 (1927)
Fisher, P.F. and P. Galdies: Comput. Geosci. 14(2), 229 (1988)
Fisher, R.V.: J. Volcanol. Geothermal Res. 6(3–4), 279 (1979)
Francis, P.W. and C.W. Baker: Nature 270, 164 (1977)
Frenkiel, F.N. and P.A. Sheppard (eds.): Atmospheric diffusion and air pollution. New York: Academic
    Press (1959)
Frostick, L.E. and I. Reid: Z. Geomorph. Suppl. 44, 53 (1982)
Fryberger, S.G. and T.S. Ahlbrandt: Z. Geomorph. 23(4), 440 (1979)
Goossens, D.: Catena 12, 373 (1985)
Goossens, D.: Catena 15, 179 (1988)
Haigh, M.J.: Z. Geomorph. 22(4), 457 (1978)
Herbillon, A.J. and W.E.E. Stone: Geo-Eco-Trop 8(1–4), 63 (1984)
Howard, A.D. and 3 others: Sedimentology 25(3), 307 (1978)
Hsu, S.A.: J. Geol. 81, 739 (1973)
Imeson, A.C.: Z. Geomorph. Suppl. 60, 115 (1986)
Jese, L.K.: Z. Geomorph. 17(2), 194 (1973)
Jungerius, P.D.: Earth Surf. Proc. Landf. 9, 509 (1984)
Kawamura, R.: Rept. Inst. Sci. Technol., Univ. Tokyo 5 (3/4) (1951)
Keller, J.B.: J. Appl. Phys. 28, 859 (1957)
Kesel, R.H.: Rev. Geomorph. Dyn. 22(3), 97 (1973)
Kling, G.: Z. Ver. Dtsch. Ing. 84, 85 (1940)
Kuenen, P.H.: J. Geol. 68, 427 (1960)
Lancaster, N.: Z. Geomorph. 24(2), 160 (1980)
Lancaster, N.: Progr. Phys. Geogr. 6(4), 475 (1982)
Lancaster, N.: Geology 16, 972 (1988)
Lekach, J. and A.P. Schick: Catena 10, 267 (1983)
Leva, M. et al.: Fluid flow through packed and fluidized systems. Washington: Bureau of Mines Bulletin
    No. 504 (1951)
Livingstone, I.: Geography 73(2), 105 (1988)
Mabbutt, J.A. and R.A. Wooding: Z. Geomorph. Suppl. 45, 51 (1983)
Magill, P.L. et al.: Air pollution handbook, New York: Mc Graw-Hill (1956)
Mainguet, M.: Z. Geomorph. Suppl. 45, 265 (1983)
Mahaney, W.C. and M.B. Boyer: Geogr. Phys. Quat. 42(1), 89 (1988)
McCabe, L. (ed.): Air pollution. New York: Mc Graw-Hill (1952)
McFarlane, M.J. (ed.): Laterites. Z. Geomorph. Suppl. 64 (1987)
McTaggart, K.C.: Am. J. Sci. 258, 369 (1960)

Mensching, H.: Z. Geomorph. 30, 1 (1978)
Moeyersons, J.: Z. Geomorph. 21(1), 14 (1977)
Monin, A.S.: Adv. Geoph. 6, 29 (1959)
Morales, C.: Climatic Change 9(1–2), 219 (1986)
Nicholas, R.M. and J.C. Dixon: Z. Geomorph. Suppl. 30(2), 167 (1986)
Nickling, W.G.: Can. J. Earth Sci. 15(7), 1069 (1978)
Ollier, C.D.: Z. Geomorph. 22(3), 249 (1978)
Ollier, C.D. and W.G. Tuttenham: Z. Geomorph. 5(4), 257 (1962)
Pye, K.: Loess Lett. 11, 5 (1984)
Pye, K.: Catena 13, 47 (1986)
Reynolds, D.L.: Am. J. Sci. 252, 577 (1954)
Scheidegger, A.E.: Geofis. Pura Appl. 41, 101 (1958)
Scheidegger, A.E. and P.E. Potter: Sedimentology 11, 163 (1968)
Selby, M.J.: Madoqua 10(3), 171 (1977)
Selby, M.J.: Earth Surf. Proc. Landf. 7, 489 (1982)
Shreve, R.L.: Bull. Geol. Soc. Am. 79, 653 (1968)
Smith, B.J. and W.B. Whalley: Earth Surf. Proc. Landf. 13, 251 (1988)
Smith, B.J. and W.B. Whalley: Earth Surf. Proc. Landf. 14, 171 (1989)
Solger, F.: Forsch. Z. Dtsch. Landes- u. Volksk. 19(1), (1910)
Stingl, H. and K. Garleff: Z. Geomorph. Suppl. 66, 65 (1987)
Sutton, O.G.: Micrometerology. New York: Mc Graw-Hill (1943)
Thomas, M.F.: Geomorph. Suppl. 31, 1 (1978)
Trikalinos, J.: Peterm. Geogr. Mitt. 74(9–10), 226 (1928)
Tsoar, H.: Z. Geomorph. 28(1), 99 (1984)
Tsoar, H. and K. Pye: Sedimentology 34(1), 139 (1987)
Twidale, C.R.: Structural landforms. Cambridge MA: M.I.T. Press (1971)
Twidale, C.R.: Rev. Geom. Dyn. 30(2), 49 (1981a)
Twidale, C.R.: Geogr. J. 147(1), 54 (1981b)
Twidale, C.R.: Geol. Rdsch. 75(3), 769 (1986a)
Twidale, C.R.: Geogr. Ann. 68A(4), 399 (1986b)
Twidale, C.R. and J.A. Bourne: Z. Geomorph. Suppl. 31, 111 (1978)
Venkatesh, S.V. and R.K. Roman (eds.): Laterisation processes, Proc. Internat. Seminar on Laterisation
    Processes Trivandrum. Rotterdam: Balkema (1981)
Verstappen, H.: Z. Geomorph. 12, 200 (1968)
Walker, G.P.L. and L.A. McBroome: Geology 11(10), 571 (1983)
Warren, A.: In: Process in geomorphology, ed. C. Embleton and J. Thornes, p. 325. London: Arnold
    (1979)
Wasson, R.J. and R. Hyde: Nature 304, 337 (1983)
Whalley, W.B.: Stud. Geomorph. Carpatho-Balc. 10, 49 (1976)
Whalley, W.B., J.B. Marshall and B.J. Smith: Nature 300, 433 (1982)
Whalley, W.B. and 3 others: In: Desert sediments, ed. L. Frostick and I. Reid. Geol. Soc. Spec. Publ. 35,
    129 (1987)
Willetts, B.: Sedimentology 30(5), 669 (1983)
Wippermann, F.K. and G. Gross: Boundary-Layer Meteorol. 36(4), 319 (1986)

Subject Index